Power Systems

Electrical power has been the technological foundation of industrial societies for many years. Although the systems designed to provide and apply electrical energy have reached a high degree of maturity, unforeseen problems are constantly encountered, necessitating the design of more efficient and reliable systems based on novel technologies. The book series Power Systems is aimed at providing detailed, accurate and sound technical information about these new developments in electrical power engineering. It includes topics on power generation, storage and transmission as well as electrical machines. The monographs and advanced textbooks in this series address researchers, lecturers, industrial engineers and senior students in electrical engineering.

Power Systems is indexed in Scopus

Hassan Haes Alhelou · Nikos Hatziargyriou ·
Zhao Yang Dong
Editors

Power Systems Cybersecurity

Methods, Concepts, and Best Practices

Springer

Editors
Hassan Haes Alhelou
Department of Electrical and Computer Systems Engineering
Monash University
Clayton, VIC, Australia

Nikos Hatziargyriou
Department of Electrical and Computer Engineering
National Technical University of Athens
Athens, Greece

Zhao Yang Dong
School of Electrical and Electronics Engineering
Nanyang Technological University
Singapore, Singapore

ISSN 1612-1287 ISSN 1860-4676 (electronic)
Power Systems
ISBN 978-3-031-20359-6 ISBN 978-3-031-20360-2 (eBook)
https://doi.org/10.1007/978-3-031-20360-2

This Springer imprint is published by the registered company Springer Nature Switzerland AG
The registered company address is: Gewerbestrasse 11, 6330 Cham, Switzerland

Preface

Due to environmental concerns and energy security risks, modern power systems are being through the transition toward the smart grid concept, which enables the high penetration level of renewable energies and motilities the high digitalization rate that requires new information and communication infrastructures. Renewable energy resources are mainly inverter-based sources that introduce new problems and technical issues to modern power systems that should be properly addressed. Such energy sources highly decrease the system's total inertia which puts the system's stability and security in danger and weakens the grid against different events including cyberattacks. The main aspect of the transition to smart grids is the replacement of traditional infrastructure that was based on SCADA systems with new ones that are being built based on wide-area measurement systems (WAMS), wide sensor networks (WSN), and internet of things (IoT) technologies. In one hand, the recent advances in the measurements and communications systems, i.e., phasor measurement units (PMUs) and wide-area monitoring systems (WAMSs), open a promising environment for developing new control and protection approaches that can help solve the renewable energy and modern power system operation technical issues. On the other hand, these new infrastructures are vulnerable to cyberattacks which put the system's cybersecurity under the danger, where, in some cases, a small false injected data as an attack could result in a serious blackout. A clear example of such events, the 2015 Ukraine blackout which has been proven that it happened due to cyberattack based on false-date injection scenario. This dangerous event and other recent cyberattack events against energy systems in different countries around the globe have highly motivated research activities on the power systems cybersecurity.

Cyberattackers are highly skilled professionals working as teams who are capable of generating very complex and coordinated attacks using sophisticated tools. Therefore, power system operators in WAMS centers need to be equipped with robust tools that can detect and classify the cyberattack signals instantly. The WAMS centers should be supplied with new operation, control, and protection approaches that can take cyberattack issues into account. For instance, it is obvious that all control systems in modern power systems should be replaced with new tolerant/resilient controllers against cyberattacks. The new wide-area monitoring systems provide the modern power systems with real-time measurements of electric quantities. Based on these quantities, the power system cybersecurity can be monitored and assessed online and in a real-time manner. Therefore, cybersecurity including the stability and security of modern energy systems can be guaranteed at a suitable level using new wide-area tolerant/resilient control and protection schemes. Developing such schemes can bring several advantages to both the system operators and consumers. Furthermore, this will help in completing the transition to smart grids and adopting new emerging concepts safely, such as demand response programs, virtual power plants, the integration of aggregated electric vehicles (EVs), and other smart appliances with the grid, with no delays due to their high vulnerability to cyberattacks. Thus, modern power systems require cyberattack detectors and tolerant/resilient operation and control schemes that can take cyberattacks into account, to be used online in the monitoring centers, the control rooms, the aggregators of EVs and other appliances, the virtual power plants, and protection schemes. There is a great research gap related to modern power systems cybersecurity in the presence of high penetration distributed generating units and renewable energy resources. Therefore, this book aims to do a complete state of the art and propose new control and protection schemes that can improve the overall stability and cybersecurity of the future power systems.

This book contains eighteen unique chapters that cover the different aspects of power system cybersecurity. First Part of the book, comprising 5 chapters, starts with chapters that comprehensively review and survey the cybersecurity of power systems and concludes with an in-depth analysis and overview of power system vulnerabilities and situational awareness in modern power systems. Second Part of the book, consisting of 3 chapters, presents a comprehensive understanding of cybersecurity of modern power system control with a focus on how to defend against cyberattacks on automatic generation control and frequency control in modern systems. Third Part of the book, comprising 2 special chapters, provides the theoretical and experimental required analysis of cybersecurity in power system protection. Fourth Part, which consists of 3 chapters, discusses the advances in cyberattack detection for modern power systems and smart grids. The book's Fifth Part, which contains 5 chapters, presents the special topics in cybersecurity of modern power systems and concludes with a study on cybersecurity standards for power systems. In more detail, the chapters of the book and their unique contributions are summarized below.

Chapter A Comprehensive Review on Cyber-Attack Detection and Control of Microgrid Systems provides a comprehensive review on cyberattack detection and control of microgrid systems. Due to the fast progress of Microgrid (MG) systems and the development of advanced computing technologies and communication networks—all of which enhance the efficiency and reliability of power networks—MGs are at the risk of various cyberattacks which can eventually lead to different glitches in the power distribution networks. There are many different kinds of cyberattacks, some of which are the False Data Injection Attack, Denial of Service, Stealth Attack, and Covert Attack. The common goals of these attacks are to cause power outages, economic loss, and even system instability. Cyberattacks could infiltrate MGs through the communication links, local controllers, or master control channels. In this chapter, a thorough review of the types of cyberattacks and the problems caused by them in MGs has been presented, and some methods of cyberattack detection, resilient control system design, and countermeasures against such attacks have been discussed. Numerous research works have already investigated the subject of cyberattacks on both the Direct-Current (DC) and Alternating-Current (AC) MG systems. These studies can be divided into two main categories: (a) detection and mitigation approaches, and (b) resilient control system designs. Several subclasses of each of these categories, along with their advantages and disadvantages have been thoroughly investigated in this chapter. In the first category, after detecting a compromised agent, an active or passive mitigation mechanism is activated to prevent the spread of the agent's destructive effects to the whole system. This may impose some strict limitations on the MGs. In the second category, by developing the distributed attack-resilient control protocols, the resilience of a MG system against potential attacks/faults/noises is enhanced to the point where no detection and mitigation action will be required.

Chapter Cyber Vulnerabilities of Modern Power Systems investigates cyber vulnerabilities of modern power systems. Modern power systems heavily rely on Internet of Things (IoT) and emerging wide-area sensor networks that expose them to cyber vulnerabilities such as network failures and cyberattacks. Some practical network failure examples include North America (2003) due to state estimator and alarm system failure, Austria (2013) due to network congestion caused by a software bug, and Switzerland (2005) due to information overload. Ukraine's power system went down in December 2015 leaving thousands of homes and facilities out of power due to a cyberattack caused by a malware, identified as BlackEnergy, in control center computers. Such failures and cyberattacks will leave majority of customers without power supply and may cause significant damage to highly sensitive and mission-critical equipment. In case of power electronics-intensive microgrids, the after-effects of the cyberattacks are even more detrimental due to comparatively weaker and fragile distribution grid, highly dynamic source and load profiles, and meager generational inertia. Cyber vulnerabilities are divided into two main categories, i.e., cyberattacks and network failures. An overview of such cyber vulnerabilities, practical limitations of modern power systems, relevant prevention measures, and a case study are presented in this chapter.

Chapter Cyber-Physical Security in Smart Power Systems from a Resilience Perspective: Concepts and Possible Solutions reviews the cyber-physical security in smart power systems from a resilience perspective: concepts and possible solutions. Vital energy infrastructures in today's evolving society are becoming more complex by distributed networks of cyber-physical systems (CPS). Although CPS provides a great deal of flexibility in the exploitation of critical infrastructure, they can also create other security-based threats which should be suitably addressed in the design and development stage. Considering this issue, power systems should be properly resistant to adverse events so that their operation is not easily impacted by severe conditions. At the same time, energy systems should be flexible enough to adapt to severe disruptions without losing or breaking down completely, and they should be able to recover in the least time after solving the problem. This critical feature of power systems is generally called resilience. This chapter provides an overview of various resilience methods from the perspective of CPS in smart power systems and microgrids. In order to provide a detailed study, in this chapter, cyber-physical structures, threats, and security issues in smart power systems are introduced. Vulnerabilities of control systems and important methods for detecting and countering cyberattacks with defenses mechanism in smart power systems are presented. Possible solutions to improve cyber-physical resilience of smart power systems and some important optimization techniques used in smart power systems and microgrids are provided, which can be utilized and developed by researchers in this field.

Chapter Cybersecurity Challenges in Microgrids: Inverter-Based Resources and Electric Vehicles surveys cybersecurity challenges in microgrids: inverter-based resources and electric vehicles. Power systems are shifting into a new paradigm of connectivity. This new paradigm facilitates massive integrations of different energy resources and loads, e.g., renewable energy and electric vehicle (EV) charging stations, throughout a microgrid with cyber interfaces. The integration of these energy resources and EVs can significantly enhance the microgrid efficiency, reliability, resiliency, and address environmental concerns. Several emerging technologies are required to prepare the cyber-physical infrastructure for this highly integrated grid. This new grid model is highly dependent on the cyber interfaces and communication infrastructure that makes the system vulnerable to cyber disruptions and threats. Considering the interconnected nature of these microgrids, cyber disorders may lead to disastrous consequences; therefore, it is essential to know the vulnerabilities of the grid and design tools to identify and mitigate these threats. This chapter specifically focuses on cybersecurity issues of the microgrid with inverter-based resources (IBRs) and EV charging stations. Due to complexity and unknown behavior of grids with IBRs and EV charging stations, a variety of cyber risks can impact the grid. Therefore, this chapter demonstrates the vulnerabilities of this new grid architecture, including IBRs and EV charging stations. Besides, it discusses the effective model-based, Artificial Intelligence (AI)-based techniques, and some technologies to detect and mitigate these cyber threats.

Chapter Improving Cybersecurity Situational Awareness in Smart Grid Environments looks at improving cybersecurity situational awareness in smart grid environments. Demand response (DR) and peer-to-peer (P2P) energy trading in smart grids use distributed architectures and multiple data sources to enable more consumer involvement. Given the reliance on and behind-the-meter data and the distributed and heterogeneous setups, these data and processes are prone to various cybersecurity attacks. Hence, the identification of security risks and continuous situational awareness are essential to establish system trust and resilience. In such a multi-layered, distributed system, data origin and the steps for processing, modifying, and aggregating data are highly significant. Data provenance denotes metadata describing data derivation throughout the different layers of the system. Tracking data provenance can provide valuable information on data history and lineage. However, while provenance generates metadata for data history, security-relevant information to estimate relevant risks is not addressed. This chapter emphasizes the need for security-aware data provenance in residential DR and P2P energy trading. Based on the existing Prov-IoT model for security-aware provenance in the Internet of Things applications, we present a refined model with entities and metadata specific to smart grids and microgrids. This instantiation named *Prov-IoT-MG*, demonstrates the importance and necessity of security-aware provenance graphs for continuously estimating risks against man-in-the-middle, false data injection, and load altering attacks. We illustrate how Prov-IoT-MG graphs can be generated and evaluated at run-time and are useful in providing up-to-date information on active security controls and other security-relevant information. Finally, we discuss how these graphs help to improve the resilience of grid processes with higher situational awareness.

Chapter Hybrid Physics-Based and Data-Driven Mitigation Strategy for Automatic Generation Control Under Cyber Attack proposes a novel technique for defending against cyberattacks on frequency control systems. The fusion of information, data, control, and electric power facilitates the electric cyber-physical system (ECPS). In the ECPS, though the high-level cyber-physical interaction and integration increase the flexibility and efficiency of power system operation, the accompanied cyber perils gradually endanger the system's security. In fact, several energy sector-targeted attacks, including the infamous Ukraine power grid hack, have shown the power system vulnerability in cybersecurity incidents. Known as the critical power-balancing operation in real-time control systems, automatic generation control (AGC) is a typical ECPS application. Measurements from remote sensors may be manipulated by attackers when telemetered to the AGC center, thereby disrupting the balance of power and frequency stability. In this situation, cyberattack-tolerant AGC plays an important role in the face of cybersecurity threats. To achieve the cyberattack-tolerant AGC, some strategies including the physics-based and data-driven attack mitigation schemes have been employed. In this chapter, inspired by the fault detection, diagnosis, and reconfiguration in fault-tolerant control theories, a hybrid physics-based and data-driven mitigation model is developed for AGC under false data injection attacks (FDIAs). A mathematical model is derived to reveal the causal relation between the FDIA signal and compromised AGC measurement data. Then, data-driven approaches are employed to establish the mapping between the compromised measurement data and power compensation. Finally, the compensation-based mitigation model is developed.

Chapter Data-Driven Cyber-Resilient Control of Wide Area Power Systems suggests a novel method for securing frequency control in wide-area power systems. With the uptake of advanced communication technologies, monitoring and operation of power systems have been experiencing a paradigm shift. Many centralized algorithms can now be implemented in a distributed scheme. For example, local decisions can make faster actions resulting in a resilient power grid. Despite these advantages, these communication networks have made power networks vulnerable to cyberattacks. Therefore, only attack-resilient algorithms can be reliably implemented using these technologies. In this chapter, a Data-Driven Cyber-Resilient Control (DDCRC) method is proposed for the frequency stability of wide-area power grids. In the proposed method, the automatic generation control signals are generated without using a predefined model such that the frequency stability of the power system in the presence of Deception Attacks (DA) is guaranteed. Simulation results on a three-area power grid show the efficiency and superiority of the proposed method.

Chapter Cyberattack-Resilient Control in Multi-area Power Generation investigates cyberattack-resilient control in multi-area power generation. While automatic generation control of multi-area power systems has better dynamic performance, it intrinsically shares the distributed control and communication frameworks and is thus inevitably vulnerable to cyberattacks. To better explore the merits of multi-area power systems, and at the same time, guarantee the security of power infrastructures and the safety of operators, countermeasures to cyberattacks must be taken. In this chapter, one new cyberattack-resilient control method suitable for automatic power generation is systematically revealed. Different from the existing techniques, this chapter presents a completely novel, unexplored resilient control framework for power generation systems under cyberattacks. This framework is equipped with two security layers. Specifically, the first security layer is about off-line tolerant control gains seeking, and the second security layer concerns online cyberattack detection. In the first security layer, different categories of cyberattack models can be presumed by specific concerns, with which the composite power system model is derived. Leveraging the Lyapunov method, the tolerant control gains with expected performance indexes can be sought. In the second security layer, the received signals from neighboring areas are carefully examined by the real-time cyberattack detection procedure, before executing control actions, to protect the communication network from cyberattacks. With the proposed cyberattack-resilient control method embedded in each control unit, the secure operation of multi-area power systems is well promoted.

Chapter Cyber-Security of Protection System in Power Grids—Part 1: Vulnerabilities and Counter-Measures is the first part of cybersecurity in power system protection. The recent trend to expand the use of Information Technology (IT) in power networks has made the electric grid potentially vulnerable to cyberattacks. Protection systems are among the most critical cyber-vulnerable components, as they directly affect the integrity and stability of power systems. On this basis, this chapter reviews and investigates the cyber vulnerabilities of protection systems in power networks. These vulnerabilities can be categorized into three main groups: (i) vulnerabilities of substations, (ii) vulnerabilities of Communication-Assisted Protection (CAP) schemes, and (iii) vulnerabilities of Wide-Area Protection (WAP) schemes. This chapter also discusses the techniques that can be used to detect cyberattacks against protection schemes and differentiate them from faults.

Chapter Cyber-Security of Protection System in Power Grids—Part 2: Case Studies on Securing Line Current Differential Relays is the second part of a two-part chapter that thoroughly discusses cyber vulnerabilities of protection systems in power networks. Part I comprehensively explained different cyberattacks and the taxonomy of attack detection for power system protection applications. In this Part, as an example, the working principle and cybersecurity of Line Current Differential Relays (LCDRs) are studied. Additionally, False Data Injection Attacks (FDIAs) and Time Synchronization Attacks (TSAs) against LCDRs are formulated in this chapter. The reasons for focusing on LCDRs are that (i) this type of relay is highly vulnerable to cyberattacks due to its dependence on communication media and the Global Positioning System (GPS) for collecting time-synchronized remote measurements from all terminals of the line in its protection zone, and (ii) LCDRs are widely used for protecting major transmission lines carrying giga-watts of power. Finally, this chapter elaborates on two types of methods, i.e., a Machine Learning (ML)-based and a model-based, to address the cybersecurity problem of this family of Communication-Assisted Protection (CAP) schemes.

Chapter Semi-supervised Deep Learning-Driven Anomaly Detection Schemes for Cyber-Attack Detection in Smart Grids is entitled as Semi-supervised deep learning-driven anomaly detection schemes for cyberattack detection in smart grids. Modern power systems are continuously exposed to malicious cyberattacks. Analyzing industrial control system (ICS) traffic data plays a central role in detecting and defending against cyberattacks. Detection approaches based on system modeling require effectively modeling the complex behavior of the critical infrastructures, which remains a challenge, especially for large-scale systems. Alternatively, data-driven approaches, which rely on data collected from the inspected system, have become appealing due to the availability of big data that supports machine learning methods to achieve outstanding performance. This chapter presents an enhanced cyberattack detection strategy using unlabelled data for ICS traffic monitoring and detecting suspicious data transmissions. Importantly, we designed two semi-supervised hybrids deep learning-based anomaly detection methods for intrusion detection in ICS traffic of smart grid. The first approach is a Gated recurrent unit (GRU)-based stacked autoencoder (AE-GRU), and the second is constructed using a generative adversarial network (GAN) model with a recurrent neural network (RNN) for both generator and discriminator that we called GAN-RNN. The employment of GRU and RNN in AE and GAN models is expected to improve the ability of these models to learn the temporal dependencies of multivariate data. These models are used for feature extraction and anomaly detection methods (Isolation forest, Local outlier factor, One-Class SVM, and Elliptical Envelope) for cyberattack in power systems. These approaches only employ normal events data for training without labeled attack types, making them more attractive for detecting cyberattack in practice. The detection performance of these approaches is demonstrated on IEC 60870-5-104 (aka IEC 104) control communication that is often utilized for substation control in smart grids. Results showed that GAN-GRU and AE-GRU-based LOF methods achieved enhanced detection with an averaged F1-score of 0.98, among others.

Chapter Vertical Approach Anomaly Detection Using Local Outlier Factor focuses on vertical approach anomaly detection using local outlier factor. Detection of anomalies based on smart meter data is crucial to identify potential risks and unusual events at an early stage. In addition, anomaly detection can be used as a tool to detect unwanted outliers, caused by operational failures and technical faults, for the pre-processing of data for machine learning, to detect concept drift as well as enhancing cybersecurity in smart electrical grid operations. It is known that anomalies are defined through their contextual appearance. Hence, anomalies are divided into point, conceptual, and contextual anomalies. In this work, the contextual anomaly detection is examined through a novel type of load forecasting known as vertical approach. This chapter explores the use of anomaly detection in the relevant learning systems for machine learning in smart electrical grid operation and management through data from New South Wales region in Australia. The presented vertical time approach uses seasonal data for training and inference, as opposed to the continuous time approach that utilizes all data in a continuum from the start of the dataset until the time used for inference. It is observed that the Local Outlier Factor identifies different local outliers given different vertical approaches. In addition, the local outlier factor score varies vertically. An anomaly is defined as a deviation from an established normal pattern. Spotting an anomaly depends on the ability to defy what is normal. Anomaly detection systems aim at finding these anomalies. Anomaly detection systems are in high demand, despite the fact that there is no clear validation approach. These systems rely on deep domain expertise.

Chapter A Modular Infrastructure for the Validation of Cyberattack Detection Systems studies a modular infrastructure for the validation of cyberattack detection systems. We propose a framework for the evaluation of cyberattack detection systems in which theoretical results can be tested in a realistic setup. We emulate a power control infrastructure, an attacker, and a monitoring system. In this controlled environment, through a modular approach, it is possible to evaluate a variety of detection models: we inject adversarial activity, collect logs from the systems, analyze such logs and produce evidence that are later processed by artificial intelligence models that can raise alerts, and give diagnostic or predictive information. In particular, we test our framework with detection models based on Dynamic Bayesian Networks, that consider the evolution of adversarial activities over time. The testbed allows us to effectively test the adequacy of the detection mechanisms for early warning of suspicious events; currently, it includes man-in-the-middle attacks and false data injection.

Chapter A Novel Self-learning Cybersecurity System for Smart Grids presents a novel self-learning cybersecurity system for smart grids. The dynamic nature of cyberattacks, as well as the incorrect predictions of Artificial Intelligence (AI)-based cybersecurity intrusion detection systems, are major impediments to the efficient protection of critical infrastructures' Information Technology (IT) and Operational Technology (OT) systems, including Electrical Power Energy Systems (EPES). This phenomenon is caused by the cyberattack detection models, which lose their effectiveness over time. The variability of cybersecurity threats makes it difficult to establish and implement a specific model, which detects all types of attacks accurately. The key to system protection is the integration of a cybersecurity framework, which simultaneously addresses new potential threats, predictions, and utilizes the best performing model, according to the most recent data. This work proposes a self-learning engine which is based on the SPARK data analytics framework and is integrated into a cybersecurity platform. The self-learning module provides the opportunity for annotating data to correct misclassifications or to add intelligence regarding previously unknown attacks. Initially, the domain experts submit annotated data through a Visual Analytics (VA) & monitoring system, to start the retraining process. Three (3) machine learning (ML) methods Random Forest (RF), Logistic Regression (LR), and K-nearest neighbors (KNN)—as well as one (1) Deep Learning (DL) method-SDAE—are dynamically compared in terms of F1-score and accuracy. After the completion of the retraining process, the best-performing model replaces the existing one and labels the incoming data. The dynamic nature of the self-learning module implies that it gets annotations from users anytime, compares the methods during the retraining process, and assigns data labeling to the most accurate model.

Chapter Cyber-Resilience Enhancement Framework in Smart Grids researches on cyber-resilience enhancement framework in smart grids. Reliability, resilience, and Quality of Service are essential features of modern electric power system operations that reflect the transition of electric energy infrastructure toward smart grid deployment. Leveraging the Software Defined Networking technology and other cybersecurity cutting-edge technologies and algorithms, this work aims to bring innovation in the modern Electrical Power and Energy System environment. To this end, we propose a cyber-resilience enhancement framework with the aim to modernize the traditional electrical grid and provide solutions in the domains of voltage and frequency restoration, cybersecurity, and network Quality of Service. Based on the results, the framework is able to detect accurately cyberattacks and perform network path re-allocation by maximizing the Quality of Service in a more accurate way than other state of art algorithms.

Chapter SOAR4DER: Security Orchestration, Automation, and Response for Distributed Energy Resources proposes a SOAR4DER: Security Orchestration, Automation, and Response for Distributed Energy Resources. Monitoring data and control functionality presented by interoperable photovoltaic (PV) inverters and other Distributed Energy Resources (DER) can be used to improve site maintenance, prognostics, and grid operations. Unfortunately, DER communications present attack vectors which could lead to power systems impacts. Since adversary capabilities continually improve, avoiding catastrophic consequences requires intelligent intrusion detection and remediation systems that consider both physical and cyber features. New Security Orchestration, Automation, and Response (SOAR) technologies are equipping cyber-defenders with new capabilities to autonomously respond to network and host-based system alerts, threat hunting results, and cyber intelligence data streams. In this Chapter, we present a novel SOAR approach for DER systems, called SOAR4DER, that ingests data from multiple Intrusion Detection Systems (IDSs) to quickly block attacks and revert DER systems to known good states. Our implementation used a collection of IDS technologies on a Bump-in-the-Wire (BITW) device which incorporated physical and cyber data to detect abnormal and potential malicious behaviors. Multiple SOAR playbooks then used the IDS data streams to automatically defend the system. Laboratory testing of the SOAR4DER system showed detection and response times under 30 s for all adversary reconnaissance, denial-of-service attacks, malicious Modbus commands, brute force logins, and machine-in-the-middle attacks.

Chapter A Study on Cyber-Physical System Architecture for Smart Grids and Its Cyber Vulnerability presents a study on cyber-physical system architecture for smart grids and its cyber vulnerability. Technological advances resulted in the overwhelming usage of electronic gadgets across the world which further resulted in increased load demands. The obvious challenges are blackouts, overloads, and voltage sags. One possible and trusted solution for these problems is the Smart grid. A power network supported by digital communication technology is termed as a Smart grid. But this solution is not as simple as it seems to be. There are a few challenges associated with this solution like cyber vulnerabilities and cyberattacks. Therefore, an effort is made here to review comprehensively the research work carried out to date, encompassing different heuristic detection and estimation techniques. It is a convenient factor that issues in Smart grid like relay protection, Power flow control, grid security and reliability can effectively be modeled and efficiently be analyzed. An effort is made to identify the dependencies among Cyber Physical System controls in this work. There is an immediate need to protect the communications and computations carried out by the digital communication equipment in Smart grids from cyberattack. An overview of current research efforts in expanding the Smart grid applications and their infrastructure security is presented here. The conclusions submitted here present the further scope of the research work.

Chapter A Study on Cybersecurity Standards for Power Systems concludes with a study on cybersecurity standards for power systems. In order to have explicit cybersecurity measures and best cybersecurity practices in power systems, a number of cybersecurity standards containing generic sets of rules/guidelines for maintaining proper cybersecurity hygiene are being designed and developed. However, appropriate study, proper understanding of scope, applicability, and implementation guidelines of these cybersecurity standards for power systems are not adequately being exercised and presented to date. In this chapter, a study is presented to understand the scope and applicability of cybersecurity standards and derive implementation guidelines in power systems. The proposed study discussed and analyzed relevant cybersecurity standards for understanding the scope, applicability, and cybersecurity process improvement for individual systems/devices deployed in the present-day power systems and also for a whole system setup such as substation/plant, communication setup, and local/central control center.

Clayton, Australia

Hassan Haes Alhelou
Hassan.haesalhelou@monash.edu

Athens, Greece

Nikos Hatziargyriou
nh@power.ece.ntua.gr

Singapore

Zhao Yang Dong
zydong@ieee.org

Contents

About the Editors

Dr. Hassan Haes Alhelou is a Senior Member of IEEE, and is currently with the Department of Electrical and Computer Systems Engineering, Monash University, Clayton, VIC 3800, Australia. Previously, He was with the School of Electrical and Electronic Engineering, University College Dublin (UCD), Dublin 4, Ireland between 2020–21, with Sultan Qaboos University (SQU) in Oman, and with Isfahan University of Technology (IUT), Iran. He is permanently a Professor and faculty member with Tishreen University, SY. Dr. Alhelou was included in the 2018 & 2019 Publons and Web of Science (WoS) list of the top 1% best reviewer and researchers in the field of engineering and cross-fields over the world. He was the recipient of the Outstanding Reviewer Award from many journals, e.g., Energy Conversion and Management (ECM), ISA Transactions, and Applied Energy. He was the recipient of the best young researcher in the Arab Student Forum Creative among 61 researchers from 16 countries at Alexandria University, Egypt, 2011. He also received the Excellent Paper Award 2021/2022 from IEEE CSEE Journal of Power and Energy Systems (SCI IF: 3.938; Q1). He has published more than 200 research papers in high-quality peer-reviewed journals and international conferences. His research papers received more than 3500 citations with H-index of 33 and i-index of 78. He authored/edited 15 books published in reputed publishers such as Springer, IET, Wiley, Elsevier, and Taylor & Francis. He serves as an editor in a number of prestigious journals such as IEEE Systems Journal, Computers and Electrical Engineering (CAEE-Elsevier), IET Journal of Engineering, and Smart Cities. He has also performed more than 800 reviews for highly prestigious journals, including IEEE Transactions on Power Systems, IEEE Transactions on Smart Grid, IEEE Transactions on Industrial Informatics, IEEE Transactions on Industrial Electronics, Energy Conversion and Management, Applied Energy, International Journal of Electrical Power & Energy Systems. Dr. Alhelou has participated in more than 15 international industrial projects over the globe and secured more than $1M funds. His major research interests are Power Systems, Power System Dynamics and Stability, Power System Cybersecurity, Power System Operation and Control, Dynamic State Estimation, Frequency Control, Smart Grids, and Practical Microgrids.

Prof. Nikos Hatziargyriou (Life Fellow, IEEE) is since 1984 with the Electrical and Computer Engineering School, National Technical University of Athens (NTUA), Professor in Power Systems, since 1995, and Professor Emeritus, since 2022. He has authored the book "Microgrids: Architectures and Control" and more than 300 journal publications and 600 conference proceedings papers. He is included in the 2016, 2017, and 2019, Thomson Reuters lists of the top 1% most cited researchers. He is 2020 Globe Energy Prize laureate and recipient of the 2017 IEEE/PES Prabha S. Kundur Power System Dynamics and Control Award and the 2023 IEEE Herman Halperin Electric Transmission and Distribution Award. From 2015 to 2019, he was the Chair and the CEO of the Hellenic Distribution Network Operator. From 2007 to 2012, he was Executive Vice Chair and Deputy CEO of the Public Power Corporation, responsible for the Transmission and Distribution and the Islands Divisions. He was the Chair of the Power System Dynamic Performance Committee of IEEE and Editor in Chief of the IEEE Transactions on Power Systems. He is currently EiC-at-Large for IEEE PES Transactions. He is an honorary member of CIGRE and was the Chair of CIGRE SC C6 "Distribution Systems and Distributed Generation." He was the Chair and Vice-Chair of the EU Technology and Innovation Platform on Smart Networks for Energy Transition (ETIP SNET). He has participated in more than 60 R&D projects funded by the EU Commission, electric utilities, and manufacturers.

Prof. Zhao Yang Dong is an endowed professor at Nanyang Technological University, Singapore. He is also the cluster director of Internet of Energy and Digital Grids with ERI@N. His previous roles include SHARP professor and Director of UNSW Digital Grid Futures Institute at the University of New South Wales, Director of ARC Research Hub for Integrated Energy Storage Solutions; Ausgrid Chair Professor and Director of the Ausgrid Centre for Intelligent Electricity Networks providing R&D support for the Smart Grid, Smart City national demonstration project of Australia. His research interests include power system planning and stability, smart grid and smart cities, renewable energy systems, and the electricity market. He is a Fellow of IEEE.

A Comprehensive Review on Cyber-Attack Detection and Control of Microgrid Systems

Hamidreza Shafei, Li Li, and Ricardo P. Aguilera

Abstract Due to the fast progress of Microgrid (MG) systems and the development of advanced computing technologies and communication networks—all of which enhance the efficiency and reliability of power networks—MGs are at the risk of various cyber-attacks which can eventually lead to different glitches in the power distribution networks. There are many different kinds of cyber-attacks, some of which are the False Data Injection Attack, Denial of Service, Stealth Attack, and Covert Attack. The common goals of these attacks are to cause power outage, economic loss, and even system instability. Cyber-attacks could infiltrate MGs through the communication links, local controllers, or master control channels. In this chapter, a thorough review of the types of cyber-attacks and the problems caused by them in MGs has been presented, and some methods of cyber-attack detection, resilient control system design, and countermeasures against such attacks have been discussed. Numerous research works have already investigated the subject of cyber-attacks on both the Direct-Current (DC) and Alternating-Current (AC) MG systems. These studies can be divided into two main categories: (a) detection and mitigation approaches, and (b) resilient control system designs. Several subclasses of each of these categories, along with their advantages and disadvantages has been thoroughly investigated in this chapter. In the first category, after detecting a compromised agent, an active or passive mitigation mechanism is activated to prevent the spread of the agent's destructive effects to the whole system. This may impose some strict limitations on the MGs. In the second category, by developing the distributed attack-resilient control protocols, the resilience of a MG system against potential attacks/faults/noises is enhanced to the point where no detection and mitigation action will be required.

Keywords Microgrid · Cyber-attack · Cyber-security · Attack detection · Resilient control

H. Shafei (✉) · L. Li (✉) · R. P. Aguilera
School of Electrical and Data Engineering, University of Technology Sydney, Sydney, Australia
e-mail: Hamidreza.shafei@student.uts.edu.au

L. Li
e-mail: Li.Li@uts.edu.au

H. Haes Alhelou et al. (eds.), *Power Systems Cybersecurity*, Power Systems,
https://doi.org/10.1007/978-3-031-20360-2_1

Abbreviations

AC	Alternating Current
ANN	Artificial Neural Network
BLIA	Bias Load Injection Attack
BP	Back Propagation
CKF	Cubature Kalman Filter
CNN	Conventional Neural Network
CPL	Constant Power Load
CPS	Cyber-Physical Systems
DC	Direct Current
DER	Distributed Energy Resource
DNN	Deep Neural Network
DoS	Denial of Service
DDoS	Distributed Denial of Service
DTL	Deep Transfer Learning
EKF	Extended Kalman Filter
FDIA	False Data Injection Attack
GA	Genetic Algorithm
KL	Kullback-Liebler
LMI	Linear Matrix Inequality
LSTM	Long Short-Term Memory
MAS	Multi-Agent Systems
MG	Microgrid
MITM	Man-in-the-Middle
MPC	Model Predictive Control
NN	Neural Network
PI	Proportional Integral
RESs	Renewable Energy Resources
RNN	Recurrent Neural Network
SCADA	Supervisory Control and Data Acquisition
SMC	Sliding Mode Control
SMO	Sliding Mode Observer
SVM	Support-Vector Machine
UKF	Unscented Kalman Filter
UIO	Unknown Input Observers

1 Introduction

Due to the use of Renewable Energy Resources (RESs), Microgrids (MGs) are not only reliable and economical, but they also provide environmental benefits [1]. They offer more efficient and reliable operation than conventional power grids. For better integration of RESs, battery energy technology is introduced [2], which is very important to recuperate the supply's reliability and improve the power plant's efficiency. According to [3], the worldwide power demand increases by 2.5% each year and creates a significant deficiency that conventional power generation technology can hardly fulfil. In addition, the huge and undesirable power losses that occur in transmission lines due to their resistance emphasize the importance of MG development. The unique features of MGs, i.e., their at-demand proximity, high efficiency, and quick installation, make them a practical solution for today's growing power demands.

Due to their importance and ability to provide the needed power, MGs have been the subject of various studies. The recent technological advances in control, computing, and communications will allow the MGs to be more distributed and computer-networked. As a result, they will become more prone to cyber-attacks. Any type of attack on the exchanged data signals will have serious consequences, such as economic loss, power outage, and system instability.

In order to prevent the system performance degradation resulting from different types of cyber-attacks, proactive and novel security schemes have to be developed and implemented. The main focus of this chapter is to review the techniques that have been previously developed to deal with the cyber-attacks on MGs. According to the literature, there are mainly two approaches in this respect. In the first scheme, after detecting the compromised agents, a mitigation mechanism is activated to prevent the spread of the destructive effects of cyber-attacks to the whole system [4]. To this end, Liu et al. [5] proposed an attack-resilient cooperative scheme by developing a properly designed observation network. In this approach, by monitoring the behaviour of all the neighbours of a distributed generation (DG), the misbehaving DGs will be isolated from the network. One of the most serious drawbacks of this strategy is the strict restrictions that have to be imposed. Fawzi et al. [6] showed that it is impossible to restore the states of an attacked system when more than half of its agents are attacked. A common strategy to alleviate the effects of cyber-attacks is to isolate the attacked agents from the whole system. However, by doing so, the network's connectivity and, as a result, the whole system's performance can be degraded. In view of the abovementioned drawbacks of the first approach, a second method for dealing with cyber-attacks has been developed. In this scheme, without having to detect and correct the compromised agents, the considered MGs can be made resilient against attacks, faults, and noises by just using a distributed attack-resilient control system [7]. By doing so, the resiliency of MGs will be achieved through distributed attack-resilient control protocols against potential noises/faults/attacks, without the need to detect, identify and correct/remove misbehaving agents. Unlike the noise and fault signals, which can be assumed to be bounded, the attack signals are not bounded, as they are intentionally designed to maximize their damage [6]. In this regard, Zuo et al. [8] developed a distributed resilient containment-based control system to regulate the voltage magnitudes of the MGs against unbounded attacks on different channels. Table 1 illustrates a comparison between this review chapter and the previously published review papers.

Table 1 A comparison between this review chapter and previously published review papers

References	Main points
Mohan et al. [9]	Investigating the effects of cyber-attacks on just load frequency control systems
Nejabatkhah et al. [10]	Focusing on detection and protection schemes against FDIAs, but not on MGs
Xu [11]	Investigating the risk of FDIAs in power generation and distributed systems
Bhamare et al. [12]	Studying the cybersecurity problem with a machine learning-based approach
Sánchez et al. [13]	Just focusing on all control schemes that have been used for attack mitigation
Mahmoud et al. [14]	Surveying on methods for attack detection, especially DoS, deception, and replay attacks
Current work	Studying two different schemes to deal with cyberattacks in MGs: attack detection & resilient control system design

As can be seen, there is still a lack of a comprehensive literature review about the general approaches that deal with various types of cyber-attacks and the strategies employed to mitigate the impacts of such attacks on MGs. Therefore, a wide-ranging survey about the abovementioned issues has been provided in this chapter.

The rest of this chapter has been structured as follows: different types of cyber-attacks, their relevant mathematical expressions, and the importance of studying cyber-security are discussed in Sect. 2. In Sect. 3, the first approach for dealing with cyber-attacks (i.e., attack detection design) is investigated, and the subclasses of this approach are also reviewed. Then, in Sect. 4, the design of a resilient control system for the cyber-security of MGs is discussed. In each section, the advantages and disadvantages of the presented approaches are also outlined. Finally, in Sect. 5, the future directions of this research area are discussed, and this work is concluded in Sect. 6. For a better understanding of this review chapter, a flowchart of its structure is depicted in Fig. 1.

2 Various Kinds of Cyber-Attacks and the Importance of Studying Cyber Security

Thanks to the tremendous progress in the field of communication technology, a fundamental revolution has occurred in Cyber-Physical Systems (CPSs) [15]. Since becoming more distributed and network-integrated, the CPSs have become more prone to cyber threats and different types of network anomalies [16]. Cyber-attacks mostly occur on two fronts: software, and hardware. The software attacks generally lead to communication interference, manipulation, communication latency, etc. The

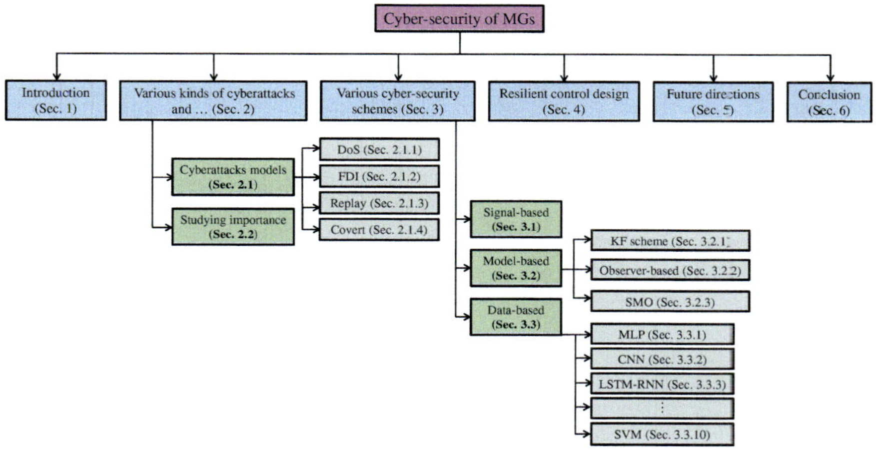

Fig. 1 Organization of the chapter

hardware attacks usually cause physical damage to sensors or modify the sensor information. Of these two types of attacks, the hardware attacks are the most dangerous ones [17].

The first step in dealing with cyber-attacks is to identify and model them. There are various kinds of cyber-attacks, each with its own particular goals and consequences. For example, some attacks may affect the actuators and controllers, some may focus on sensors, some may try to block the transmission data channels, and some others may attempt to falsify the information signals. As is depicted in Fig. 2, the cyber-attacks could target different channels of a CPS and thus degrade the performance of a MG system by adversely affecting one of its essential features below (i.e., availability, integrity, timeliness, and confidentiality). The dotted lines denote the affected signals, while the solid lines represent the unaffected ones.

In addition, Fig. 3 depicts the framework of a MG in the presence of various kinds of cyberattacks through different channels. To get familiar with MGs, see the more recent review papers [18–20], as well as the references cited therein.

Availability: Availability concerns the accessibility and availability of data at all times [21], which may be lost under DoS attacks.

Fig. 2 Cyber-attacks on different channels

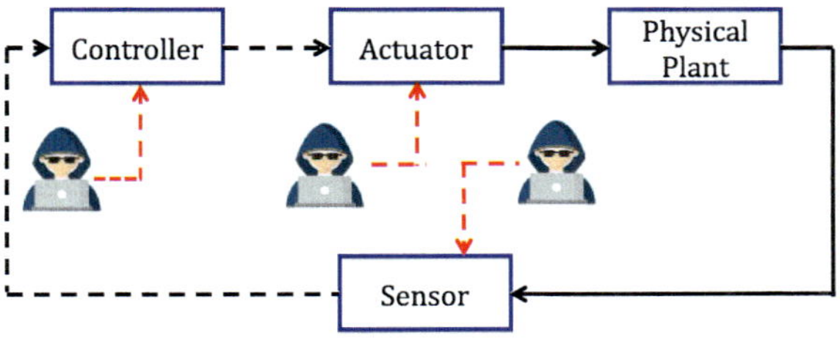

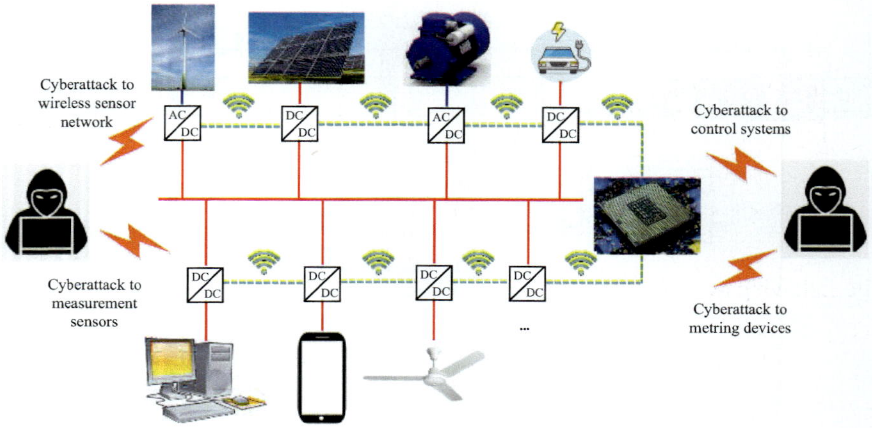

Fig. 3 The framework of a MG in the presence of various cyberattacks

Integrity: This feature demands that the data be clean and trustworthy. The process of generating, transmitting, and storing data must be genuine and free of any corruption [22]. The Replay and FDIAs may compromise the essential feature of data integrity.

Timeliness: This crucial feature concerns the real-time processing, generation, and availability of data [23], which may be lost with replay attacks.

Confidentiality: Confidentiality indicates the availability of data to authorized users only [24], and it could be adversely affected by the Man-in-the-Middle (MITM) type of attacks.

Brief descriptions of different types of cyber-attacks, their mathematical models, and their impacts on the MG systems have been provided in the following sections. Due to the diversity of cyber-attacks, it is vital to know the characteristics of each one and to find practical means to mitigate their effects.

2.1 Cyber Attack Models

2.1.1 DoS Attack

To disrupt the availability of data, potential intruders try to make the communication networks inaccessible. DoS attacks can be performed in two ways, flooding services or crashing services. In flood attacks, aiming to slow down and eventually stop the services, attackers send too much traffic for the server to buffer. In addition, it can occur by jamming signals to block the communication channels. To this end, the data from a sensor or actuator can be prevented from reaching their intended destination. Thus, a DoS attack on a sensor or actuator can be mathematically formulated with

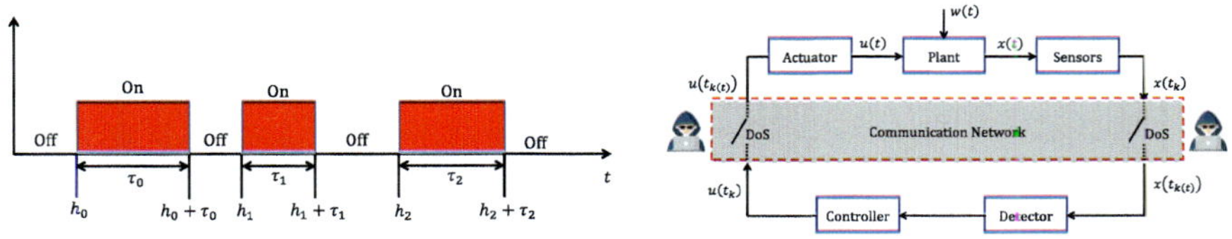

Fig. 4 DoS attack representation [9]

the following availability function [25]:

$$p(t) = \begin{cases} 1 \; for \; t \in \Pi_N \\ 0 \; for \; t \in \Pi_D \end{cases} \quad (1)$$

where $\sigma_D = \{h_0, h_1, \cdots, h_k, \cdots\}$ is the time sequence in a DoS attack, and $\sigma_I = \{\tau_0, \tau_1, \cdots, \tau_k, \cdots\}$ is the time sequence in the DoS attack duration, where $h_{k+1} - h_k > \tau_k > 0$ means that there is a normal communication duration between two consecutive DoS attacks [26]. In the above equation, $\Pi_D(h, t)$ denotes all the time intervals in which the DoS attacks occur over $[h, t)$, where $t > h$. On the contrary, $\Pi_N(h, t)$ represents all the time intervals in which there is normal communication over $[h, t)$. It is clear that $\Pi_N(h, t) \cap \Pi_D(h, t) = \varnothing$ and $\Pi_N(h, t) \cup \Pi_D(h, t) = [h, t)$. . In (1), $p(t) = 0$ means the presence of a DoS attack and $p(t) = 1$ denotes the absence of a DoS attack. Figure 4 depicts the structure of a DoS attack, where $x(t) \in R^n$, $u(t) \in R^p$, $y(t) \in R^m$, and $w(t) \in R^n$ are the state variables, control input, output signal, and the process noise, respectively.

2.1.2 FDI Attack

To destroy the data integrity, attackers try to alter the states of a system by injecting false data into sensors [27, 28]. FDIA is launched through sensor spoofing, intrusion of communication links, magnetic field injection attacks, GPS spoofing, etc. In the systems with unstable modes, the dynamic FDIAs attempt to modify the system measurements so that some of the unstable system modes become unobservable. Similar to DoS attacks, FDIAs can also be applied to both the actuator and the sensor channels, and they are formulated as follows [9]:

$$\begin{aligned} \hat{u}_j(t) &= u_j(t) + u_{a_j}(t) for \; j = 1, 2, \ldots, p \\ \hat{y}_i(t) &= y_i(t) + y_{a_i}(t) for \; i = 1, 2, \ldots, m \end{aligned} \quad (2)$$

where $\hat{u}_j(t)$ and $\hat{y}_i(t)$ are the disrupted control input and output signals, respectively. $u_j(t)$ and $y_i(t)$ are the desired control input and measurement. In addition, $u_{a_j}(t)$ and $y_{a_i}(t)$ are the FDIA signals to the actuators and sensors which are defined as follows:

$$u_{a_j}(t) = \begin{cases} 0 \; for \; t \notin \tau_{a_j} \\ \lambda_j \mathcal{F}_j(.) \; for \; t \in \tau_{a_j} \end{cases}$$

$$y_{a_i}(t) = \begin{cases} 0 \; for \; t \notin \tau_{a_i} \\ \lambda_i \mathcal{F}_{\text{i}}(.) \; for \; t \in \tau_{a_i} \end{cases} \tag{3}$$

In the above equation, τ_a is the attack time period, and λ is an attack parameter. $\mathcal{F}(\cdot)$ can either be a time/state-dependent function or completely independent.

2.1.3 Replay Attack

A replay cyber-attack targets both the data integrity and the timeliness features of a system. This attack can occur practically through monitoring and recording sensor measurements during the invasion. Then, attackers use them instead of the actual measurements. In this type of attack, the attackers record a data sequence for a certain amount of time and then replay these readings in the system to deceive the operators. In the replay attacks, the attackers do not need any information about the system being targeted. To model this type of attack in a sensor channel, one can write

$$a_y(t) = -Cx(t) + y(t - \tau) \tag{4}$$

In the above equation, $0 < \tau < t$ and $a_y(t)$ is the replay attack in the sensor channel. Also, $x(t) \in R^n$ and $y(t - \tau) \in R^m$ are the state of the system and the sensor data gathered through monitoring, respectively. Two different stages should be considered in this procedure: (1) Monitoring phase, and (2) Replay phase. In the monitoring phase ($0 \leq t < t_0$), the collected measurements are stored in $J(t)$, and thus [9]

$$\begin{aligned} y_a(t) &= 0 \\ J(t) &= \Gamma^y.y(t) \end{aligned} \tag{5}$$

In this phase, cyber-attack is not applied yet (i.e. $y_a(t) = 0$). While in the replay phase, the gathered data from the monitoring stage are sent to the controller. So, we have

$$\begin{aligned} y_a(t) &= J(t - t_0) \\ J(t) &= J(t - 1) \end{aligned} \tag{6}$$

The same procedure can be applied to model the replay attacks in an actuator channel. The structure of this type of attack is depicted in Fig. 5, where u_a and y_a are the actuator and sensor attack signals, respectively. In addition, d is the external disturbance acting on the plant, and u is the infected control signal.

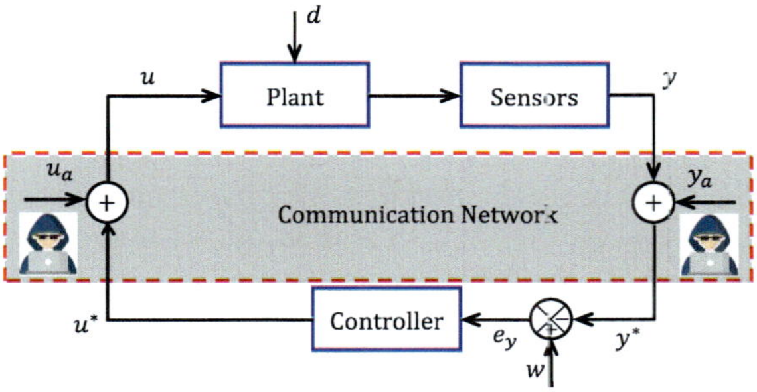

Fig. 5 Replay attack representation

2.1.4 Covert Attack

The covert attack is another hard-to-detect cyber-attack in which the attackers modify the measurements of a system in such a way that it cancels out all the effects of the attack on the system dynamics. To practically apply this kind of attack, attackers cancel the effect of attacks by calculating the system's output response and subtracting it from the measurement readings. In this type of attack, the attackers apparently have access to both the actuator and sensors. Therefore, a complete knowledge of the system is required. To mathematically model the covert attack, the following two steps should be taken [29].

1. The actuator of the system is manipulated by the attackers as follows:

$$\tilde{u}_t^i = u_t^i + a_t \tag{7}$$

In the above equation, u_t^i is the original control law, a_t is the corrupted signal injected by attackers, and $\tilde{u}_t^i$ is the corrupted control signal which will alter the system states at time $t + 1$, as follows:

$$\tilde{x}_{t+1}^i = x_t^i + B_i a_t \tag{8}$$

As a result, the output signal changes to the following:

$$\tilde{y}_{t+1}^i = y_{t+1}^i + C_i B_i a_t \tag{9}$$

2. In this step, the sensor measurements are modified so that the attack cannot be detected. Thus,

$$\check{y}_{t+1}^i = \tilde{y}_{t+1}^i - \Upsilon \tag{10}$$

In the above equation, $\Upsilon = C_i B_i a_t$. By doing so, the altered measurements (i.e., $\check{y}_{t+1}^i$) get to be exactly equal to the original measurements in the absence of an attack (i.e., y_{t+1}^i). Therefore, this kind of attack is disguised.

2.2 *Why is Studying Cybersecurity Important?*

A MG system is a typical CPS in which the physical units and the cyber environments work together [30]. In such systems, a supervisory control and data acquisition (SCADA) system is in charge of management and control [31]. In the early days of such systems, the SCADA was secure when they had no communication with the outside [32]. With the advancement of technology, the next generation of SCADA systems (known as the distributed SCADA) emerged, which could be integrated with local area networks. Eventually, in the most recent generation of SCADA systems (the networked SCADA), the wide area networks can also be accessed. Compared to the previous SCADA generations, the networked SCADA systems enjoy lower cost, more straightforward installation, and easier maintenance, to name but a few. In spite of these positive features, these networked systems are highly susceptible to cyberattacks [33].

Considering the key role of the SCADA in controlling and managing sensitive industries like smart grid systems, its security against cyber-attacks has attracted a great deal of attention [34]. In this respect, the NERC CIP[1] plan was established in 2006 with the aim of protecting the power control systems [35], the IEC TC57[2] was established to undertake the development of various security standards [36], the NIST[3] proposed the necessary procedures for securing the SCADA systems [37], and the ISA[4] presented its security protocols for industrial manufacturing [38].

A large number of cyber-attacks on CPSs have occurred in recent years. For example, in January 2003, intruders attacked the Dacis-Besse power plant in Oak Harbour (Ohio State, USA) with a slammer worm and shut down its safety display system [38]. In 2010, a malicious computer worm, known as the Stuxnet virus, targeted the SCADA system of the Natanz nuclear site in Iran and caused serious damage to numerous nuclear centrifuges [39]. In 2011, the SCADA system at an Illinois water plant was breached, and as a result, the supply pumps were disabled [40]. In another attack in December 2015, the power grids in Ukraine were totally disrupted for hours by several FDI attacks [41]. A complete review of the previously occurring cyber-attacks is presented in [42]. Table 2 summarizes some of the major cyber-physical attacks that have occurred in the energy sector. These examples out of a long list of cyber-attacks clearly highlight the great importance of studying and researching the cyber-security matters related to MGs. A complete list of cyberattack events can be found in [43].

---

[1] North American Electric Reliability Corporation Critical Infrastructure Protection.

[2] International Electrotechnical Commission Technical Committee.

[3] National Institute of Standards and Technology.

[4] International Society of Automation.

Table 2 Some important events resulting from Cyber-attacks

Year	Location	Attacked targets	Attack type	Impact
2022	Finland	Minister of defence and foreign affairs	Distributed DoS attack	All affairs were suspended for 4 h
2022	Israeli	Telecommunication provider	DDOS attack	Multiple Israeli government websites were taken offline
2021	Australia	Telecommunications systems	A malicious code	Recorded all communication data
2021	Norway	Energy Technology Company Volue	Ransomware attack	Shutdown of water & water treatment facilities in 200 municipalities
2019	California and Utah, USA	Grid operators	DDOS attack	Disrupted their operation but not any outages
2015	Kiev, Ukraine	The breaker settings were attacked by the "BlackEnergy" malware	FDI Attack	More than 225,000 customers were affected for hours by the blackout
2013	New York, USA	The Bowman Dam was attacked when it was under maintenance	FDI Attack	Gaining remote access to the information on water levels and flow rates

3 Various Cyber-Security Schemes

As was mentioned earlier, there are two main approaches for dealing with cyber-attacks, attack detection and mitigation, as well as the implementation of a resilient control system. The signal-based attack detection method implemented in micro-grids is achieved by monitoring the signals in the communication links in real-time. However, in model-based detection schemes, cyberattacks can be detected by utilizing the mathematical model of the system. In addition, data-based detection methods typically rely on machine learning or statistical mechanisms to infer a model of the system from historical data and measurement signals [44]. This section thoroughly discusses these two schemes and their positive and negative features.

3.1 Signal-Based Methods

In a signal-based method, the potential attacks can be detected by monitoring the state signals in real-time. For example, any discrepancy between the frequency of a system and its desired values could signal an attack [45]. In this paper, the detection and isolation of the attacked communication links are realized on time. Without prior knowledge of system dynamics, Beg et al. [46] detected the FDI and DoS attacks on a complex DC MG system by using a temporal signal logic formula and comparing the system voltages and currents with their pre-set values. By monitoring the secondary sub-layer outputs and tracking the attacks' changes, a cooperative vulnerability factor was proposed to identify the FDIAs on the voltage measurements [28]. To identify the attacked nodes under a FDIA, a consensus algorithm based on the discordant elements was developed in [47]. By examining the discrepancy between the received and forecasted measurements, Dong et al. [48] proposed a short-term state forecasting scheme for detecting FDIAs, which is integrated with the ∞-norm and L_2-norm residual measurement analysis. In another work, Madichetty et al. [17] proposed a low-computational-burden approach based on the adaptive state observer for estimating the system currents and voltages in the presence of FDIAs. In this paper, by analysing the error between the estimated and actual signals, the FDI attacks are detected in real-time.

As a sub-branch of the signal-based methods, the weighted least-square scheme is a reliable and efficient technique for attack detection in smart grids [49]. Using the graph signal processing technique, Drayer and Routtenberg [50] devised an attack detection scheme for the AC MGs. By comparing the maximum norm of the graph Fourier transform of an estimated grid state with a predefined threshold, they could detect the presence of FDIAs. In this approach, the high-frequency components related to the large eigenvalues of the Laplacian matrix are filtered out. The same approach is also used to detect the data integrity and data availability attacks on power generation systems [51]. In another study, Hu et al. [52] introduced a residual skewness coefficient technique for detecting the stealth attacks via signal analysis. This method suffers from its dependence on the manual adjustment of parameters.

One of the major shortcomings of the signal-based methods is that the relationship between the measured data and the control signal is not investigated, which is necessary for reliable detection [44]. Also, the limitations of such schemes in detecting the stealth attacks are mentioned in [53].

3.2 Model-Based Methods

Another general approach developed for attack detection in the MGs is the use of mathematical model-based methods. The Kalman filter [54], state estimation [55], observer-based detection [56], and the sliding mode observer (SMO) [57] schemes are some of the more common techniques in this category. In the face of model uncertainties, a lack of reliable measurements is the main limitation of robust cyber-attack detection methods. Using both the graph and system-based theoretical approaches, the fundamental limitations in the monitoring of the CPSs under attack are investigated in [58]. Each of these schemes is discussed along with its positive and negative features in this section.

3.2.1 Kalman Filter-Based Techniques

The Kalman filter (KF) is a common scheme used for state estimation. In this method, accurate state estimation can be achieved by applying an optimal recursive algorithm and using the previously-estimated states and the current measurements. Since previous measurements are not needed in this approach, it is considered as an efficient state estimation strategy. The KF method can be conducted in two steps: (a) the prediction step, and (b) the correction step [54].

Several filtering techniques have already been proposed for the estimation of dynamic states, such as the Extended Kalman Filter (EKF) [59], Particle Filter [60], Unscented Kalman Filter (UKF) [61], and the Cubature Kalman Filter (CKF) [62]. For example, Manandhar et al. [4] used the KF to estimate the state variables of a system. Then, the estimated states as well as the system readings were utilized by a χ^2-detector to detect the arising system faults and various kinds of attacks. In another paper, the KF method was employed to estimate the expected measurements, which were then used to obtain the deviations between the actual and the estimated values [63]. In this process, a Chi-square detector and the cosine similarity matching technique were applied to detect the cyber-attacks. For the real-time attack detection and state estimation in the presence of FDIA and measurement noise, Miao et al. [64] respectively employed a Chi-square test-based adaptive secure algorithm and an adaptive UKF scheme. Sargolzaei et al. [65] developed a KF-based observer for estimating the states of a system. In order to accurately deal with FDI attacks, they proposed a Neural Network (NN) architecture, whose weights were updated via an EKF. In another work, Soltani et al. [54] presented four different types of KFs (i.e., original KF, Adaptive KF, Fuzzy KF, and Fuzzy adaptive KF) for state estimation. These KF methods were integrated into a framework through an ordered weighted averaging operator, whose weighting factors are updated by a real covariance matrix and a theoretical covariance matrix. Abbaspour et al. [66] devised a resilient detection and mitigation-based controller for load frequency control against FDIAs. They proposed a Luenberger observer along with an Artificial Neural Network (ANN) enhanced by an EKF for online anomaly detection. The same approach was used for the detection of time-varying unknown FDIAs on sensors [67]. Li et al. [62] employed a robust CKF to dynamically estimate the states of a system under both the FDI and DoS attacks in real-time. Their simulation results showed the superiority of the robust CKF over the CKF. A comparison between CKF and nonlinear observers was conducted in [68] to show the performance of each of these schemes in the dynamic estimation of system states in the presence of cyber-attacks and model uncertainties.

In another study, Adeli et al. [69] used a trust-based UKF and a modified secure node strategy to estimate the states of nonlinear systems under cyber-attacks. To prevent the spreading of the compromised data, they developed a cluster-based fusion approach. The shortcomings of the traditional KF approach in detecting the stealth attacks were investigated in [53], where a novel attack detection technique based on KF was also proposed. Using the upper bound of filtering error covariance minimization, a modified UKF was proposed in [70] for a specific class of nonlinear systems under sensor saturation and randomly occurring FDIAs. This paper also provides the sufficient conditions for the exponentially bounded filtering error. By employing a KF decomposition framework, the secure estimation of system states for linear time-invariant Gaussian systems under sparse integrity attacks was achieved in [71], and the sufficient conditions under which the proposed estimator is still stable were established.

In general, apart from the positive features of KF schemes in providing reasonable state estimation when the precise mathematical model of a system is available, the performance of the KF-based methods deteriorates in the absence of a good model or in the presence of external disturbances. In addition, the initial conditions of the considered states and covariance matrices can affect the estimation quality.

3.2.2 Observer-Based Methods

The observer-based schemes for state estimation and attack detection usually use a system's deterministic state–space mathematical model. A few examples of the wide-ranging observer-based methods include the high-gain observers [72], Kullback-Liebler (KL) observer [73], and the unknown input observer (UIO) [74]. These observers have been widely explored and used in various studies and applications.

The UIO is an observer-based scheme that detects potential cyber-attacks on the CPSs by estimating the system states. For example, an UIO was developed in [56] to detect the compromised agents in a distributed manner. In this study, the attacked signal was considered as an external input with no prior information. Then, the sufficient and necessary conditions for observing the system misbehaviours were established. Mustafa et al. [75] developed an attack detection scheme based on the KL divergence criterion for AC MGs. The KL divergence factors were then used to mitigate the effects of cyber-attacks by determining the trustworthiness of the information received from the other agents. Also, to inform the other agents of the reliability of the transmitted data, each agent generates a self-belief factor and communicates it with its neighbors. By comparing the KL criteria for both the healthy and the attacked systems, the FDIAs on the power grids can be detected [76]. The same approach has been employed to detect the DC MGs' attacks [77].

To secure the CPSs against the periodic DoS attacks in both the measurement and control channels, an $\mathcal{H}_\infty$ observer-based output feedback controller was developed in [78]. In this paper, by considering the DoS attacks with a cyclic dwell-time switching characteristic, an augmented system is achieved as a discrete-time cyclic dwell-time switched system with an unstable and a stable subsystem. For the resulting augmented system, the proposed observer-based controller is developed in a piecewise linear form. Attack detection and attack mitigation by a finite-time convergent observer-based output feedback controller were also investigated in [79], where two different cases are considered: (a) when all the sensors are targeted, and (b) when some of the sensors are attacked. By using the proposed observer with the adaptive gains, the applied sensor attacks are reconstructed. In this procedure, the corrupted measured outputs are cleared of the malicious content, and the output feedback controller uses these cleaned signals to mitigate the effects of the cyber-attacks.

In another article, Wang et al. [80] employed an UIO to detect the FDIAs on MGs under system uncertainties. To deal with the computational limitation of the predefined threshold, an interval residual-based detection criterion was proposed in this work. Moreover, for the undetectable FDIAs, an algorithm based on the local logic judgment matrix was applied to both detect and isolate the mentioned attacks. By solving a robust $\mathcal{H}_\infty$ multi-objective optimization problem and using an independent and cooperative detector, the existing anomalies in the multi-agent systems (MASs) subjected to both the FDIAs on their communication links and the physical faults in their facilities were detected [81]. By doing so, a compromise was made between robustness to disturbance and sensitivity to attack.

To detect the covert attacks on large-scale interconnected systems, Barboni et al. [82] proposed a distributed model-based observer. By augmenting an original system with a switched auxiliary one, the covert attacks were detected via a switched Luenberger observer [83]. In another study, Das and Madichetty [84] designed a Luenberger observer to estimate the DC-DC converter current based on the input voltage. This scheme has some positive features, such as the reduction of sensor requirement and computational burden as well as performance improvement. Gao et al. [85] developed a novel observer with a set of K-filters to detect the deceptive cyber-attacks in the sensor and actuator channels of the high-order nonlinear CPSs.

With the purpose of detecting the cyber-attacks in the presence of disturbance and measurement noise, Yan et al. [86] developed a new FDIA detection scheme by decomposing a large-scale MG into several interconnected subsystems and designing a set of dynamic reduced-order observers to generate residual signals, which were then compared with the adaptive detection thresholds using a prescribed performance benchmark. To deal with the biased load attacks, Wang et al. [87] proposed an unknown input interval-based observer that could mitigate the influences of the regionally-interconnected information and disturbances. They introduced a novel detection criterion in order to eliminate the prior threshold limitations.

Achieving cyber-security via the dynamic state estimation schemes was also reported in [88]. In this work, first, the incoming cyber-attacks are detected by the dynamic observers and by eliminating the effects of unknown inputs. Then a new attack-tolerant control scheme is used to eliminate the destructive effects of cyber-attacks on the operation and performance of power systems. The bias load injection attack (BLIA) is another kind of FDIA, which targets the vulnerable generator loads. Due to the destructive impact of the BLIA on MG systems, it is very important to develop a secure scheme to detect and mitigate this type of attack. In this regard, Wang et al. [89] proposed a three-stage approach to detecting the BLIAs. In this method, by applying a subregion division technique, first, the complexity of attack detection in a large-scale grid system is reduced. Then, a robust adaptive detection method is employed to accurately estimate the physical dynamics of the system. Finally, a logical judgment matrix is used to cope with the undetectability of sensor attacks under system vulnerability.

To estimate the states of continuous time-invariant linear systems under sparse sensor attacks, An et al. [90] developed a supervisory state observer by utilizing a bank of candidate nonlinear sub-observers and applying a switching logic. In this scheme, a monitoring function is used to select the active sub-observers at every moment. In another work, Yang et al. [91] studied the estimation of distributed states in large-scale power systems under both the FDI and DoS attacks through the neighborhood coordination scheme. In this approach, the DoS attacks are compensated via a measurement predictor, where the FDIAs are treated as the measurement uncertainties.

Besides the positive features of the observer-based approaches in detecting cyber-attacks, it is challenging to be applied to more complex systems because they utilize the exact mathematical model of the system.

3.2.3 Sliding Mode Observer

A common method of state estimation in MGs is the sliding mode observer (SMO) approach, which has intrinsic robustness against uncertainties, noise, disturbances, and cyber-attacks [92–99]. In this approach, the system states are estimated in a finite time by forcing them toward a stable manifold in the presence of uncertainties and disturbances. Despite the positive features of the SMO approach, including its simple implementation, disturbance rejection ability, and robustness, it suffers from chattering due to its high-frequency switching around the sliding surface [100, 101]. In the following, we will review some research works related to attack detection by SMO.

Saha et al. [92] proposed a SMO-based fault diagnosis technique for estimating the sensor measurement errors arising from a possible cyber-attack or sensor faults in the DC MGs. To ensure a resilient system operation, the estimated errors are used to initiate a mitigation action. Most of the previously developed SMO-based methods assume a linear MG system, which is a restrictive assumption. In fact, in the presence of constant power loads (CPLs), the assumption of system linearity will not be valid anymore. In this regard, Cecilia et al. [102] proposed a distributed nonlinear observer that can robustly detect the FDIAs in both the cyber links and current sensors. In this work, the system equations are transformed into a form suitable for observer design purposes. For estimating the states of a system in the presence of time delay, a sliding mode estimation-based control system was presented in [93], which estimates the states of a MG system and also predicts the time delays. Furthermore, to ensure the system stability and the accuracy of the estimated time delays, the estimation error was considered as a disturbance to the Sliding Mode Control (SMC) system and used to update it adaptively. For estimating the system states and detecting the sensor attacks, Ao et al. [94] proposed two adaptive SMOs, whose parameters are updated online. They also established the sufficient conditions for attack detection.

In another study, Rinaldi et al. [95] proposed a distributed adaptive dual-layer super-twisting SMO to isolate, reconstruct, and mitigate the adverse effects of both the disturbances and communication attacks. In this paper, similar to [94], the observer gains are adjusted with an adaptive law. After reconstructing the attacks, a standard Proportional Integral (PI) controller is employed to mitigate their destructive effects. The same procedures have also been used with the other types of SMO schemes, such as the integral SMO [103–105], terminal SMO [106], and the high-order SMO [107, 108]. To deal with the cyber-attacks on sensors, a SMO was proposed in [99], which is sensitive to the sensor attacks. This observer alarms the operators when at least one of the MG subsystems is under the sensor attack. Furthermore, to isolate the exact subsystems under attack, another SMO is proposed which is sensitive to the attack on the i th sensor but robust against external disturbances and sensor attacks other than those on the i th agent.

Yang et al. [109] presented a SMO-based state and attack reconstruction scheme with the system augmentation and Linear Matrix Inequality (LMI) approach for detecting the FDI attacks in both the sensor and actuator channels. Li et al. [110] investigated an augmented state observer that uses a terminal integral adaptive SMC to estimate a system's states under sensor and actuator FDI attacks. Su et al. [97] modelled the MGs under the dynamic load-altering attacks and developed a robust SMO that generates a residual signal. The cyber-attacks are then detected by comparing this signal with a threshold value. A linear parameter-varying SMO with adaptive parameters that uses an equivalent output error injection algorithm to estimate the system states and reconstruct the sensor faults in the presence of erroneous scheduling parameter information was explored in [111]. By considering the unknown inputs and the sensor attacks, a linear discrete-time state-space and a sparse vector were devised to model the system and the cyber-attacks, respectively [98]. To estimate the states of the system under unknown inputs, a new model in the descriptor form that uses an iterative approach was developed. In this model, a novel sparse SMO that employs a projection operator is proposed for estimating the system states affected by the cyber-attacks.

The weakness of the SMO-based methods in detecting the intelligent attacks with changing distributions can be considered as their drawback. Furthermore, in the presence of unknown system disturbances (e.g., voltage oscillations, load variations, and neighboring voltage variations), these methods may not be able to achieve a reliable state estimation [112]. Table 3 summarizes some of the recently published works in the field of cybersecurity with model-based schemes.

As mentioned, the model-based detection approaches are designed to compare the predicted and actual sensor measurements. In this scheme, detecting intelligent attacks in high-dimension systems can be considered its primary challenge. The idea of developing data-based schemes for attack detection has been proposed to deal with this issue, which will be considered in the next section.

3.3 Data-Based Methods

The data-based methods are another approach for dealing with cyber-attacks in MGs. These methods use historical data and measurement signals to learn the features of

Table 3 A brief review of the attack detection and mitigation studies using the model-based approach

No	Study	Year	Attack type	Method	Advantages	Disadvantages
1	Saha et al. [92]	2018	FDI and Replay attack	Using a SMO to detect attacks	A complete and general model of MGs is considered. The effectiveness of the method against different undesired situations are examined	This scheme can only deal with bounded cyberattacks; Required the exact model of the system
2	Abianeh et al. [113]	2021	FDIAs	With localized communication link quality observer & multi-objective SMC	This method can deal with unbounded and extreme cyberattacks The vulnerability of the method to various FDIs and time delay is very low	Besides being strongly model-dependent, this scheme is applicable to just FDIAs
3	Shi et al. [114] and Li et al. [115]	2021	FDIAs	Adaptive distributed observer	The vulnerability of this method against FDIs is low. High reliability and scalability with robustness against cyber-attacks	This scheme is highly vulnerable to time delays and other kinds of cyber-attack
4	Sahoo et al. [116]	2021	FDIAs	Using localized observer	By this method, other kinds of cyber-attacks can also be detected	They just considered frequency FDIAs This method is vulnerable to time delay
5	Li et al. [62]	2019	FDI & DoS attacks	Robust Cubature KF	Dynamic estimation of states in case that FDA & DoS attacks are applied	This method is validated with a simple system

(continued)

Table 3 (continued)

No	Study	Year	Attack type	Method	Advantages	Disadvantages
6	Afshari et al. [117]	2020	Time-varying and unbounded Deception attack	Using distributed state observer (DSO) to estimate the states and attack signals	The effects of battery are considered The proposed method can deal with unbounded cyberattack	This method is just for AC MGs, and deception attacks
7	Barboni et al. [82]	2020	Covert attacks	Robust model-based observer	Using 2 robust decentralized & distributed observers based on the H_∞ solved by LMI. Besides the covert attacks, disturbances are also considered. Applicable to large-scale systems	The uncertainties and attacks should be bounded. The model accuracy affects the detection results
8	Poudel et al. [77]	2020	FDIAs	Distributed Kullback-Liebler observer	Using an interior-belief factor to determine the reliability and an exterior-belief value to check the trustworthiness of the information	Exact model of the system is required
9	Cecilia et al. [57]	2021	FDIAs and Uncertainties	Extended Astolfi/Marconi plus linear Luenberger observers	Using 2 adaptive observers for observable and unobservable state estimation under FDIAs and CPL and other undesired conditions	Not applicable for other kinds of attacks

(continued)

Table 3 (continued)

No	Study	Year	Attack type	Method	Advantages	Disadvantages
10	Cecilia et al. [102]	2021	FDIAs and Uncertainties	High-order SMO along with a nonlinear open loop observer	Using a distributed HOSMO for the observable states and an open-loop observer for the unobservable states under FDIAs and CPL condition	The estimation of the power line currents relies on an open-loop integration which is not tuneable
11	Sahoo et al. [28]	2019	FDIs and Stealth attack	Using cooperative vulnerability factor framework	The capability of proposed method in detecting stealth attacks Proposing a CVF factor to alarm the attacked agents	The method should be developed to be applicable to AC MGs
12	Yan et al. [93]	2017	Time delay	Estimation error used as a disturbance in SMO	Controlling a system with several kinds of delays with a SMO-based SMC	Requires the exact model of a system
13	Manandhar et al. [4]	2014	DoS	KF with X^2 detector	Resilient to spoof attacks; with low computational complexity	Dependent on model accuracy; vulnerable to disturbances; cannot be used for nonlinear systems

a MG system. Some of the more popular intelligent cyber-security schemes, such as the deep reinforcement learning, deep transfer learning, the generative adversarial network, and so on, will be discussed in this section. These techniques can be generally classified into the unsupervised, semi-supervised, supervised, and the reinforcement learning schemes.

For detecting the FDIAs with a machine learning-based approach, Ozay et al. [118] analysed the performances of the supervised, semi-supervised learning and the online learning algorithms in the detection of FDIAs on MGs. In another research, Habibi et al. [119] developed an intelligent-based method for detecting the FDI attacks and identifying the attacked agents in DC MGs. They proposed a time-series analysis and a nonlinear auto-regressive NN-based exogenous model to estimate the DC currents and voltages. These researchers proposed a Model Predictive Control (MPC)/ANN scheme for FDI attack detection in another article [120], where a MPC plays the role of attack mitigator, while an ANN detects the FDIAs on the DC MGs. In [121], an artificial intelligent method with a nonlinear autoregressive exogenous model based on NN is proposed for detecting the FDIAs on the voltage and current sensors. This method suffers from a large computational burden in real-time implementations.

As was mentioned previously, a drawback with the model-based detection methods is that they require an exact mathematical model of a system. To deal with this problem, researchers have proposed the idea of using the data-based techniques. Compared to the traditional machine learning techniques, the chief merit of the deep learning schemes is that they can better learn from large security datasets. A comprehensive review of the above-mentioned intelligent attack detection methods has been provided in the following sections.

3.3.1 Multi-layer Perceptron (MLP)

With a supervised learning algorithm, the MLP is a feedforward ANN that can be considered as the base architecture of the Deep Neural Networks (DNNs). The structure of this network is illustrated in Fig. 6. Back propagation (BP) is a widely-used supervised learning scheme for training the feedforward NNs. This approach has been used frequently in the field of cyber-attack detection. For example, Felipe et al. [122] applied this method as a real-time attack detection approach with low computation cost. This technique has also been employed for the detection of malicious botnet traffic attacks [123].

Due to the sensitivity of the MLP approach to the feature scales, this method requires sufficient numbers of neurons, hidden layers, and iterations, which increases the computation cost. However, this scheme is capable of dealing with nonlinear systems even in the online learning tasks and real-time operations [124]. By integrating a multilayer perceptron with a deep random NN, Huma et al. [125] developed a hybrid deep random NN method for cyber-attack detection. This scheme is 98–99% effective in classifying the different types of cyber-attacks.

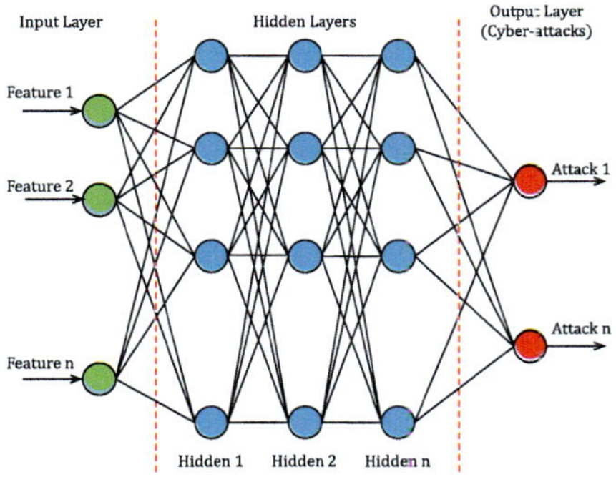

Fig. 6 MLP network structure for attack detection

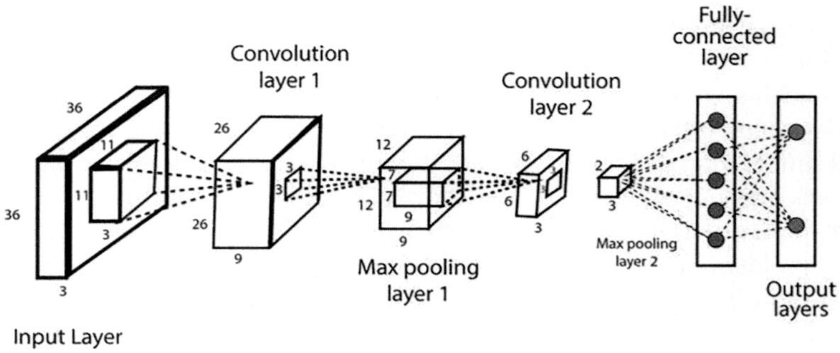

Fig. 7 CNN structure with multiple convolution and pooling layers

3.3.2 Convolutional NN (CNN)

The CNN is another deep learning scheme that can learn without the need for manual feature extraction. The structure of this network, which comprises multiple layers, is depicted in Fig. 7.

The optimized parameters in each layer of the CNN help reduce the network complexity. In addition, the "dropout" layer is used to deal with the problem of over-fitting. This network has been extensively employed in various cyber-security applications, such as DoS attack detection [126] and malware detection [127]. For example, Yanmiao et al. [128] devised a highly-accurate deep learning approach of low complexity based on multi-CNN fusion for the detection of intruders. With the aim of detecting the injected data measurements, He et al. [129] proposed a deep learning framework based on the CNN and Long Short-Term Memory (LSTM) network, which can monitor the data measurements and the network-level features. To detect the possible cyber-attacks, a multi-CNN with the NSL-KDD evaluation dataset was presented in [128]. The good performance, high accuracy and low complexity of this model were verified by experimental results. Compared to the ANN, the CNN method is more expensive computationally, but it can be implemented automatically and without human intervention.

3.3.3 Long Short-Term Memory Recurrent Neural Network (RNN)

As another type of ANN, The RNN can process an input sequence and retain its states while processing the next input sequence. The feedback loops in the recurrent layer of RNN help sustain the data in the memory. The LSTM network is a specific type of RNN with a memory cell for long-time data storage. An example of this network is depicted in Fig. 8 [130]. The LSTM scheme is also used for attack detection [131] and time-based botnet detection [132] purposes. By employing this technique, researchers have been able to detect the FDI attacks in the measurement channel [133] and classify the attacks by means of a hybrid deep learning method [134]. To

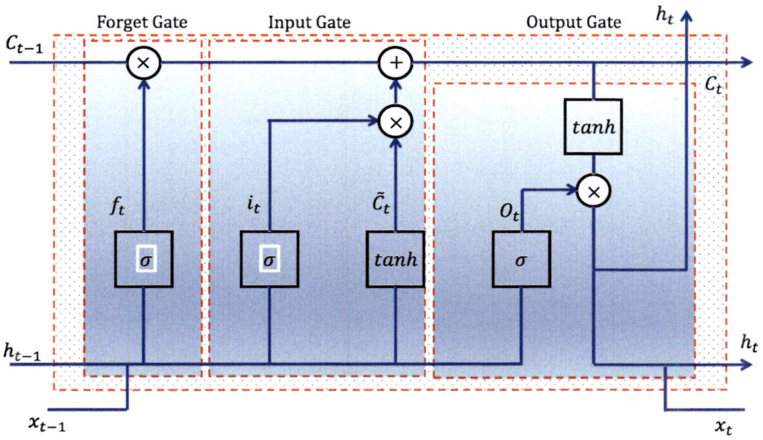

Fig. 8 The LSTM network structure

optimize the RNNs and CNNs, Vasan et al. [135] proposed a robust malware detection model based on advanced ensemble learning, which has very high precision and low computational burden. By integrating the LSTM modules into an ensemble detector, an advanced deep learning model was proposed in [136] for detecting the new kinds of cyber-attacks with high accuracy. A LSTM-RNN model requires a lot of time and resources to get fully trained; so it is able to improve the detection process and precision.

3.3.4 Auto Encoder (AE)

The AE is another type of ANN which can be trained in an unsupervised fashion [137]. However, to reduce the dimension sizes, it ignores the noise effects. The AE network has three different components: encoder, code, and decoder (Fig. 9). An advantage of this method in the propagation stage is its ability to extract the useful information and filter out the useless features [138].

As far as cyber-security is concerned, an effective security model can be built with a deep AE network, since it enjoys the minimum number of security features. For example, Yousefi-Azar et al. [139] devised a learning technique based on the AE network for the classification and detection of network attacks and anomalies. In another study, Binghao and Guodong [140] proposed a stacked sparse auto-encoder to improve the attack detection. Therefore, in view of its ability to capture the main data features, the AE approach can be useful in the field of cyber-security.

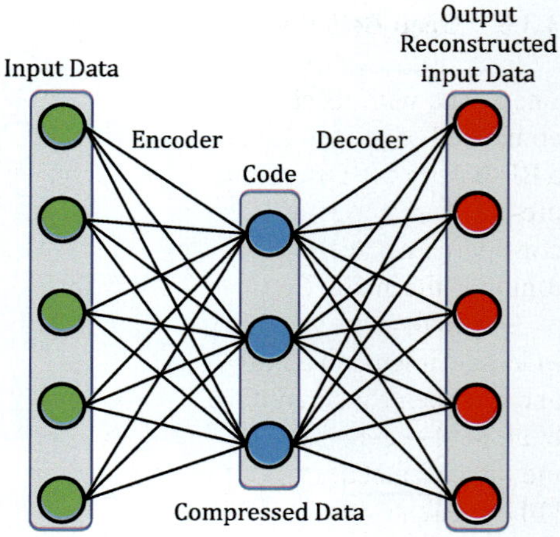

Fig. 9 AE network structure with its components

3.3.5 Restricted Boltzmann Machine (RBM)

The Boltzmann Machine (BM) is an unsupervised deep learning model with two types of nodes: visible and hidden. Among the various kinds of BMs, the RBM is a model with limited connections between the layers. By reducing the number of connections, the training algorithm becomes more efficient than the other types of BM networks [141] (Fig. 10).

The RBM method has already been used in cyber-security matters. For instance, Fiore et al. [142] applied this technique for the detection of cyber-attacks, where the effectiveness of the proposed model was evaluated by combining the positive features of the generative models with the ability of the RBM to deduce new information from incomplete training data. In another study, the accuracy of DoS attack detection was improved by means of the RBM scheme [143]. By eliminating the noises from the input data and extracting new information from them, Seo et al. [144] could enhance the performance of this model in the attack detection tasks.

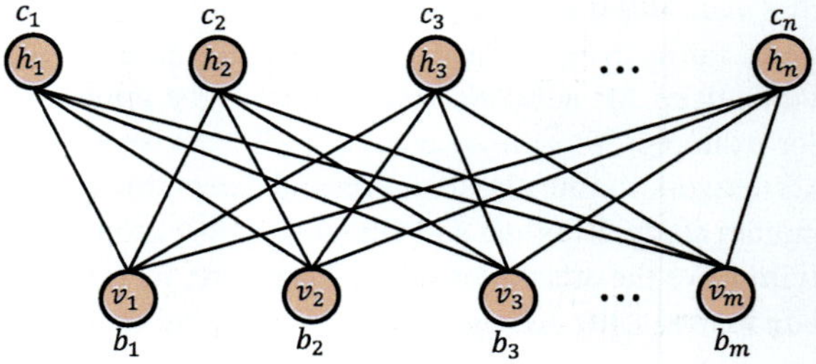

Fig. 10 The RBM structure with n hidden and m visible nodes

3.3.6 Deep Belief Network (DBN)

The DBN, with its structure depicted in Fig. 11, is a probabilistic generative model composed of several RBMs. As is shown, this model consists of multiple RBMs and a BP-NN. In the DBN scheme, a two-stage training process is implemented: first, the pre-training of the model by an unsupervised layer-wise learning technique, which is considered as a contrastive divergence based training process; and second, the fine-tuning of the model by means of the BP-NN and a supervised learning approach [145].

Due to the unique classification and feature extraction abilities of the DBN model in large-dimension applications, it plays a prominent role in cyber-security domains and has been considered in numerous relevant research studies. For example, Salama et al. [146] developed the DBN model as a feature reduction scheme and used it in the attack detection tasks. A DBN-based attack detection strategy was reported in [147], and its results were compared with those of the Support-Vector Machine (SVM) scheme, which revealed a higher accuracy and speed of attack detection by the DBN method. To achieve the same objectives, Peng et al. [145] proposed an optimization approach to DBN. For dealing with newer types of cyber-attacks, self-adaptive attack detection techniques have been developed. In this regard, Zhang et al. [148] presented a cyber-attack detection scheme based on the deep belief network and genetic algorithm (GA). In this work, the optimal numbers of the hidden layers and neurons are adaptively determined by the GA in order to boost the attack detection precision.

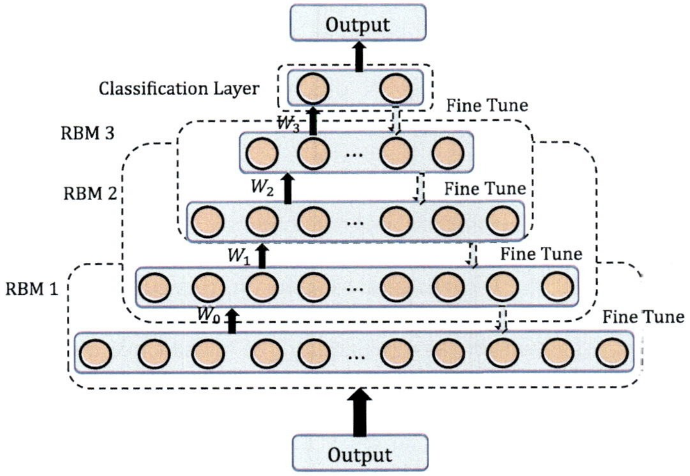

Fig. 11 The DBN network structure

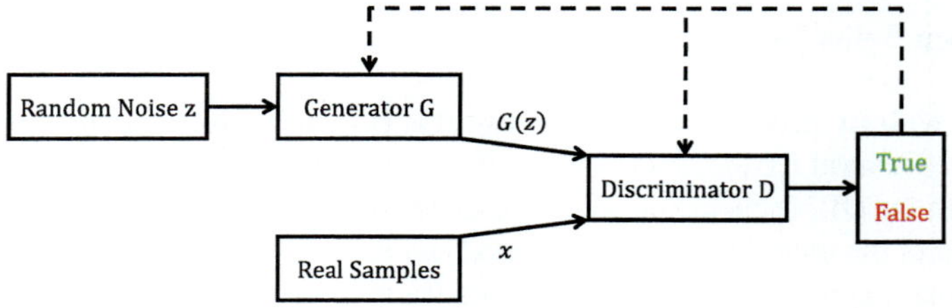

Fig. 12 The GAN network structure

3.3.7 Generative Adversarial Network (GAN)

In the GAN model, two NNs, a discriminator (D) and a generator (G), are trained to compete with each other [149]. The structure of this learning scheme is depicted in Fig. 12. The GAN framework has been widely used in the domain of cyber-security to make the learning models more robust against intruders that try to manipulate the data systems. For example, Jin-Young et al. [150] devised a novel learning framework based on the deep AE-based transmitted GANs and achieved a classification accuracy of over 95%. In this scheme, the fake malware could be distinguished from the real ones. By balancing the real datasets with additionally generated data, Merino et al. [151] proposed a technique which improves the detection of the cyber-attacks and is suitable for all kinds of learning schemes, including the reinforcement learning as well as the fully-supervised, semi-supervised, and unsupervised approaches. In this method, to balance the previously unbalanced datasets, the model generates additional data that mimic those distributed by different kinds of attacks.

3.3.8 Deep Transfer Learning (DTL)

The DTL is a powerful and common method for solving the problems that lack adequate training data. In this method, it is not necessary to train the artificial intelligent models, since it is able to train the NNs with just a small amount of data [152]. As is shown in Fig. 13, the DTL scheme uses the pre-trained models for the tasks in the target domain. Considering the different cases, the DTL can be classified into three sub-classes: (a) inductive transfer learning, (b) transductive transfer learning, and (c) unsupervised transfer learning [153].

Thanks to the positive features of the DTL scheme, including its shorter training, time, and higher output accuracy with fewer training data, it has attracted a lot of attention in the cyber-security field of study. For example, network attack detection was achieved in [154] with a ConvNet model based on the DTL scheme. Xianwei et al. [155] presented a novel semi-supervised DTL network for attack detection. In this paper, to avoid the disassembling of the files, a byte classifier is proposed that extracts the needed byte features from the programs and classifies them with an RNN. To improve the byte classifier accuracy from 94.72% to 96.90%, a new ASM

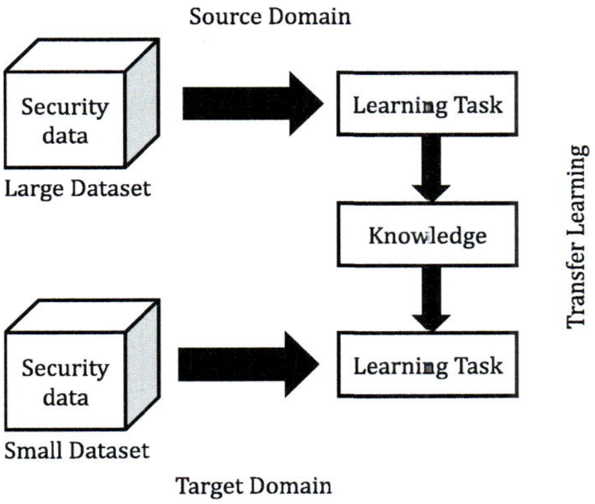

Fig. 13 The DTL structure

classifier with a combination of ASM features is designed in this paper. In another work, Vu et al. [156] presented a DTL scheme for detecting the cyber-attacks in the IoT devices with a high level of precision. Their proposed method can learn from the data collected from various labeled and unlabeled sources. Since this learning scheme can speed up the training process with high efficiency and precision even with small datasets, it has received a great deal of attention from the researchers in the cyber-security field.

3.3.9 Deep Reinforcement Learning (DRL)

The DRL is another kind of machine learning approach in which the model can learn from its previous actions. This learning scheme, which is depicted in Fig. 14, integrates the reinforcement and the deep learning methods together [157].

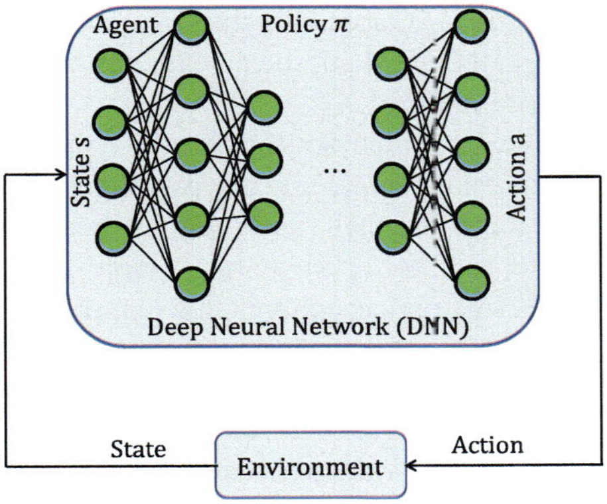

Fig. 14 The DRL structure

The DRL scheme has been previously studied in the cyber-security related research works. For instance, using a labeled dataset, several supervised DRL algorithms for the detection of cyber-attacks were presented in [158]. By conceptually modifying the conventional DRL algorithms by the rewarding of the detection errors in the training phase, the cyber-attack detection process could be made faster and more precise than the traditional machine learning schemes. To deal with the inability of the conventional methods in maintaining a balance between high accuracy and low false positive rate against the new and unknown attacks, a DRL-based adaptive attack detection algorithm was proposed in [159]. By combining a deep learning model with an intrusion detection strategy, Li et al. [160] proposed a deep migration learning-based scheme for cyber-attack detection. Hughes et al. [161] proposed a model-free DRL method for intrusion detection and evaluated its performance by stopping two distinct multi-stage attack scenarios on a virtualized test-bed.

3.3.10 Support-Vector Machine Learning

By using the SVM learning datasets to build a series of hyperplanes, the SVM approach can be employed for cyber-attack detection in large-dimensional spaces. In this regard, two machine-learning-based methods (i.e., the supervised anomaly detection and the unsupervised distributed SVM) have been proposed to detect the stealth attacks in the AC MGs [162]. Lee et al. [163] developed a 98%-accurate lightweight machine learning scheme for cyber-attack detection, in which an SVM algorithm is used for feature extraction and a deep auto encoder is applied for attack classification. Raman et al. [164] presented an adaptive and robust attack detection method in which the Hypergraph-based GA is used for parameter selection and the SVM technique is applied for feature extraction. In another study, Tang et al. [165] developed an algorithm based on the SVM that uses a nonlinear input data scaling scheme to detect the cyber-attacks. By comparing the results of the SVM method and the ANN model, they demonstrated the superiority of their proposed method in detecting the applied cyber-attacks. Also, a lightweight attack detection scheme based on the SVM approach, which is suitable for the low-resource IoT devices, was proposed in [166].

Table 4 summarizes the various types of the above-mentioned learning schemes with their advantages and disadvantages. Despite the many advantages of these intelligent methods, they suffer from a heavy computational load, which makes them inappropriate for applications to large-scale distributed MGs.

This section has conducted a comprehensive overview of cybersecurity from the artificial NN and deep learning point of view. Reviewing recent studies in this domain has discussed how different kinds of NNs and deep learning methods can be used for cybersecurity solutions. The proposed security approach should possess the relevant deep learning modeling based on the data characteristics.

Table 4 A brief review of the cyber-security studies using the data-based approach

No	Study	Year	Attack type	Method	Advantages	Disadvantages
1	Habibi et al. [167]	2021	FDIAs	Using artificial NN to detect cyberattacks	This method can also be used when all units are under attacks Deal with coordinated high domains & unfairly FDIAs	It can detect coordinated FDI attacks on current channels
2	Habibi et al. [119]	2020	FDIAs	Using NARX NN method for attack detection	It can detect FDI attacks in both current and voltage channels	This method needs strong data base for training. It can only deal with constant FDIAs
3	He et al. [129]	2017	FDIAs	Deep learning-based Conditional Deep Belief Network	Resilient to predicted FDIAs and GPS spoofing; suitable for large-dimensional data applications	High computational complexity; vulnerable to unforeseen attacks; low accuracy to abrupt attacks
4	Hughes et al. [161]	2022	DoS	Deep Reinforcement Learning	Do not need a system's model; suitable for solving complex problems; tuned by a small number of training episodes	High computation cost, and requires faster technologies for training purposes
5	Kavousi Fard et al. [168]	2021	FDIAs	Using "Prediction interval" for the lower and upper estimation model	Highly accurate FDIA detection	High computational cost
6	Dehghani et al. [169]	2021	Constant FDIAs	Deep Machine Learning & Wavelet Singular Values Approach	Using a novel optimization scheme based on Gray Wolf to detect FDIAs FDIA detection is achieved precisely and robustly	Real-time implementation of this method to MGs is a challenge

(continued)

Table 4 (continued)

No	Study	Year	Attack type	Method	Advantages	Disadvantages
7	Gao et al. [155]	2020	Malware threats	Semi-supervised transfer learning with a Recurrent NN	Increased training speed; high efficiency and precision with small datasets; Not need to extract sensitive data	This algorithm needs to be optimized for classification. For attack detection, it should be improved

4 Resilient Control Design Approach

In spite of the many advantages of distributed control systems, such as greater flexibility, scalability, low communication burden, and faster system response, their vulnerability against cyber-attacks is still a challenging issue that has to be fully addressed. That's why the cyber-security field has become more important than ever [170]. Considering the devastating consequences of the recent cyber-attacks, the subject of cyber-security, similar to the topic of attack detection and mitigation, has attracted the attention of many researchers in the power industry. Given the various sources of uncertainties (e.g., the plug-and-play functionality of DGs, changes in microgrid topology, and uncertain and unknown dynamics), resilient control systems are very important means of controlling the MGs. In the presence of various uncertainties and cyber-attacks, the resilient controllers can provide some relief in the form of bounded control signals, reference tracking with zero steady-state error, and an acceptable transient response [171].

The resilient control of MGs has been achieved by many techniques such as SMC [172], adaptive control [173], robust H_∞ [174], fuzzy controller [175], learning-based approach [176], etc. For example, Duan et al. [177] proposed a four-step attack-resilient distributed controller based on a detection-mitigation algorithm to immunize the DC MGs against the data integrity attacks. Unlike the grid-connected MGs, whose primary objective is to achieve proper power-sharing among all the DGs, in the islanded MGs, there should be fair power-sharing as well as accurate frequency and voltage regulation [178]. In this regard, a cyber-attack-resilient distributed controller is proposed for dealing with both the time-varying and successive attack signals in the communication links, local controllers, and master controller. The proposed method uses the periodically intermittent communication to detect and isolate the adversaries. The consensus-based secondary control systems have been employed to detect the FDIAs in both the DC [47] and AC MGs [116]. In these works, the cyber-attacks are detected through an asynchrony index, a detection scheme which is highly dependent on the chosen threshold. In another study, Abhinav et al. [179] developed a resilient distributed secondary cooperative controller for the islanded AC MGs in the presence of cyber-attacks. The performance of the proposed controller was evaluated under the sensor and actuator attacks as well as the attacks on the communication links. In this study, a distributed observer-based controller was developed for dealing with the sensor/actuator attacks, while a trust/confidence-based control system was designed for mitigating the effects of cyber-attacks on communication links. Also, an attack detection scheme for partial primal–dual-based distributed frequency control was proposed in [180].

Although the terms "robustness" and "resilience" seem to have similar meanings, they actually refer to two different features in the context of power systems. In fact, resilience refers to a system's active reactions in real-time to unexpected and extreme events. While robustness refers to the preservation of a controller's desired performance in the presence of a given range of disturbances in a passive manner. The works that have adopted a detection and isolation scheme for boosting the resiliency of control systems have the following drawbacks: (a) they make a compromise between the timeliness of the real attack mitigation when tuning the isolation criterion and the controller's robustness to normal disturbances, and (b) the number of cyber-attacks that can be mitigated is limited [177]. Some works have focused on improving the robustness of the distributed control system. For instance, Ding et al. [181] proposed a distributed resilient secondary control system for a battery energy storage system under the DoS attacks. To ensure the controller's robustness against the DoS attacks, an acknowledgment attack detection scheme and a communication recovery mechanism were applied to repair the connectivity-paralyzed communication topology. By using the weighted mean subsequence reduced algorithm for each DER, the corrupted information received from other neighbouring agents could be discarded and, as a result, the security of a proposed distributed secondary controller against the FDI attacks could be ensured [182]. A distributed resilient control system based on an adaptive technique was employed in [183] to restore the frequency and voltage of an AC MG system, achieve a fair real power-sharing, and balance the state-of-charge with multiple ESSs under abnormal conditions. To ensure the stability of an AC microgrid system against unbounded attacks, Zou et al. [8] developed a fully distributed resilient controller for the secondary frequency regulation and voltage containment by introducing a virtual resilient layer with hidden networks. However, they assumed that the hidden network is secure, which may not be true in the real applications. To address the threat of cyber-attacks in the AC MGs, a resilient distributed optimal frequency control system that uses a leader–follower consensus protocol was proposed in [170], where an auxiliary networked system is connected with the original control system.

The resilience of MGs against infrastructure failures has been enhanced by a novel master–slave control system [184]. A resilient SMC system that guarantees the exponential stability of MGs against the DoS attacks was proposed in [185]. Baghaee et al. [186] developed a decentralized robust mixed H_2/H_∞ voltage controller for disturbance rejection and set point tracking purposes. The LMI scheme was employed for solving a multi-objective optimization problem, where the fuzzy logic approach was also applied to improve the performance of the controller against the nonlinear loads and small/large disturbances.

Li et al. [187] proposed a resilient control system based on an adaptive control scheme for the islanded AC MGs whose communication links are under cyber-attacks. The resiliency of the proposed method in voltage and frequency restoration and its power sharing capability were confirmed through computer simulations. To model the time-varying uncertainties of the power outputs and loads in multiple RESs, Lin et al. [188] proposed a novel offline probabilistic model and an online updating scheme. In this paper, a resilient stochastic optimization model that incorporates the concept of dynamic MGs is proposed to restore the voltage and frequency values. To achieve the sufficient conditions for a secure consensus among the neighboring agents in the absence of continuous communication due to the DoS attacks, Xu [189] proposed a novel resilient leader-following event-triggered control protocol. Most of the research studies on resilient control systems consider the sensor and actuator attacks separately with complex nonlinear systems, which increases the computational burden and makes it difficult to analyze and control the attacked MGs. To deal with this problem, Zhao et al. [190] designed a resilient control system based on feedback linearization for the MGs under both the sensor and actuator attacks. In this work, after simplifying the mathematical model of a targeted MG system, they used a hybrid resilient control scheme based on an adaptive terminal SMC.

A resilient cooperative output-regulated controller for the heterogeneous linear MASs with unknown switching exosystem dynamics under the DoS attack was proposed in [191]. In this paper, they designed a distributed resilient observer for estimating the auxiliary states and distributed controllers for individual agents. Song et al. [192] developed an adaptive resilient controller based on the fractional-order command filtered backstepping technique for the nonlinear systems affected by time-delays and unknown state-dependent deception attacks. To make the MASs resilient against both the actuator and sensor attacks, a resilient consensus control system based on a secure dynamic event-triggered scheme was proposed in [193]. In this work, a distributed adaptive compensator was also developed to predict the unavailable system states. To guarantee the stable operation of the wind turbines in redistributing the generated power among the converters under the system faults and cyber-attacks, a resilient DC voltage controller was developed [194]. Zuo et al. [195] investigated a fully distributed resilient controller for a multi-group network system under the unknown and unbounded FDIAs on both the cyber-physical and the virtual layers. In another study, Sadabadi et al. [196] proposed a resilient controller that doesn't need any information about the nature or location of an attack. In this work, a Lyapunov-based approach and the graph theory were applied to ensure average voltage regulation and proportional load sharing in the DC MGs affected by the FDIAs.

To rectify the effects of sensor faults in the islanded MGs, Saha et al. [197] proposed a SMO integrated with an H_∞ output feedback control law. The purpose of designing the SMO was to estimate the sensor faults, which were then used to develop the proposed robust fault-tolerant controller. A fully-distributed resilient control strategy based on an output formation containment scheme was proposed to control a high-order heterogeneous multigroup system under unknown and unbounded attacks [198]. In this paper, the authors assumed that the unknown unbounded attacks can affect the actuators, the local state feedback, and the communication channels of each agent. A novel switching event-triggered resilient control scheme based on the droop control and MPC was proposed at the primary control level to achieve the frequency/voltage restoration as well as power sharing in the islanded MGs [199]. Moreover, a distributed consensus pinning-based control scheme was provided at the secondary level to manage the transient deviations and the response speed. Finally, to improve the immunity of the proposed system against severe accidents, a resilient control technique was designed by integrating the primary and the secondary control levels.

Ma et al. [200] demonstrated that with a resilient neuroadaptive dynamic control technique based on the Gaussian radial basis function NN, the semi-global uniformity and the ultimate boundedness of all the signals under the FDI and DoS attacks, unknown control gains, system uncertainties, and output constraints can be guaranteed. A cascade 2 degree of freedom controller integrating a PI and an internal model controller was proposed in [201] as a resilient control scheme for managing the sustained operation of the MGs under the parametric uncertainties and unbalanced loads. A brief summary of the resilient control schemes in the field of cyber-security along with their merits and drawbacks has been provided in Table 5.

5 Future Direction

The cybersecurity of MGs is still at its early stage of progress, which makes it become a hot topic. Proposing and developing new and practical methods to mitigate the effects of cyber-attacks on MGs is an inevitable challenge that needs to be more considered. Some potential research topics regarding this problem have been proposed in the following.

- Developing a secure filter based on attacked measurement outputs is suggested as future work to achieve acceptable security performance. The existing KF methods can obtain the minimal variance of the filtering errors by using exact knowledge of noise statistics. However, since the statistical characteristics of signals transmitted by the attacker cannot be obtained, this assumption is usually not applicable to MGs [202].
- For practical applications, it is highly recommended to consider MGs under various kinds of attacks simultaneously.
- Developing a **general** resilient control system that can ensure the stability of MGs against multiple cyberattack models has paramount importance, which has not been considered yet. Most current methods are developed specifically for one specific attack and under particular conditions.
- To achieve accurate results from the data-based approach in cyber-attack detection, developing pre-processing methods or proposing new data preparation techniques is required to clean data from missing and ambiguous values, meaningless data, and outliers.

Table 5 A brief review of the resilient control system design approaches

No	Study	Year	Attack type	Method	Advantages	Disadvantages
1	Sahoo et al. [203]	2020	MITM	Multilayer event-driven detection based on diverging factor	Using a diverging factor to locate the compromised link(s); Good robustness against MITM attacks and disturbances	It can be developed for dealing with other attacks; the performance is prone to model uncertainties
2	Chen et al. [204]	2022	DoS attack	Resilient distributed control based on PI & average consensus schemes	Providing average voltage regulation & load power sharing under line impedance and CPL, where the max allowed CPL is increased	Resiliency drops against large time delays; it is effective only on the DoS attacks
3	Zhou et al. [45]	2020	FDIA on three channels	Signal-based approach by observing the switching frequency signal	Dealing with the cyberattacks on communication links; local and master controllers; resilient against time-varying attacks	Medium computation load; its performance is dependent on cyber graph connectivity
4	Liu et al. [26]	2021	FDI and DoS attacks	Controller based on a combined error with a pinning gain	Not needing any information about the cyber-attacks by proposing an adaptive gain; considering both the FDI and DoS attacks	Do not consider the control signals constraints, No robustness against time-delays and disturbances
5	Bidram et al. [182]	2019	Bounded FDIAs	Weighted Mean Subsequence Reduced algorithm	Time-varying communication graphs are used to improve the resiliency; has low processing load	Medium computation burden; depends largely on the number of attacked cyber links/nodes

(continued)

Table 5 (continued)

No	Study	Year	Attack type	Method	Advantages	Disadvantages
6	Sahoo et al. [205]	2020	Bounded FDIAs	Adaptive discord element & quick mitigation via an event-driven algorithm	Simple design; Heterogeneous to attack detection/mitigation; Resilient to single point of failure; PnP; has reduced communication cost	–
7	Zuo et al. [8]	2020	Unbounded FDIAs	Fully distributed controller consisting of a virtual MAS	Unbounded attacks are considered; attacks on various channels are considered; the controller does not require any global information	Has high computation load; its resiliency is maintained when less than $N/2$ nodes are under attack
8	Li et al. [115]	2021	FDIA	A novel adaptive resilient secondary control scheme	The proposed scheme is robust against FDIAs; The bounds of faults and attack do not need to be known;	Controller is highly vulnerable to other cyber-attacks and time-delays; considering one channel fault only
9	Zhao et al. [190]	2021	FDIA to actuators/sensors	Composite resilient controller based on terminal SMO with adaptive gains	Considering simultaneous attacks on actuators and sensors; easy to analyse; low computation cost; simple implementation	An exact model of the system is required
10	Song et al. [192]	2022	DoS attack	Adaptive resilient control based on fractional-order backstepping scheme	Mitigating the effects of input delays and unknown sensor deception attacks; all the signals in the closed-loop systems are semi-globally UUB	High sensitivity to other kinds of attacks

- As was mentioned, each approach has its merits and demerits. Integrating various methods from different schemes can be considered a novel idea to benefit from the positive features of each technique. For instance, by developing a hybrid method based on the integration of model-based and data-based techniques, one can enjoy the low computation burden of the model-based approach; at the same time, the exact model of the system is not required.
- Most resilient control systems studied for controlling MGs against cyber-attacks, especially FDIAs, do not consider the communication channel time delays, which can quickly destroy the system's stability. Thus, this critical factor should be considered in the resilient control system for future work.
- Most works mainly focus on the cybersecurity of either DC or AC MGs. However, attack detection and mitigation of hybrid MGs have not been addressed well. Therefore, another future work that can be considered is attack detection and resilient control system design for hybrid MGs.
- Due to the low inertia of power electronics-based DERs, the frequency stability in hybrid MGs is one of the main concerns that must be considered. This situation can even be worse in the presence of cyber-attacks. Thus, proposing a proper and robust controller for attack detection and mitigation on the frequency control of hybrid MGs can be an important future topic.
- Like frequency control, any voltage variations on each side of a hybrid MG can transfer to the other side. As a result, it can deteriorate the stability of the system. Thus, voltage regulation of hybrid MGs under cyber-attacks is a challenging issue that has to be considered.

6 Conclusion

The importance of improving the security of CPSs against various types of cyber-attacks has increased more and more in recent years. A comprehensive review of the recently published works in the cybersecurity domain has been conducted in this chapter, which provides an overall picture of the historical, current, and future developments in this area. For this purpose, a new division regarding the existing approaches for addressing this problem in MGs has been proposed. Therefore, after introducing some of the most common cyberattack models, such as FDIA, DoS, covert, and replay attacks, we have reviewed the existing schemes that have been developed so far for cyber-attack detection, including signal-based, model-based, and data-based schemes. By conducting an extensive survey for each of these methods, their advantages, disadvantages, and challenges in practice have been illuminated. Then, we have focused on the second approach of dealing with this problem, i.e., resilient control system design, by which the robustness and resiliency of the MGs against cyberattacks can be guaranteed. Finally, the future direction in this field of study has been expressed.

References

1. H.B. Eluri, M.G. Naik, Challenges of res with integration of power grids, control strategies, optimization techniques of microgrids: a review. Int. J. Renew. Energy Res. (IJRER) **11**(1), 1–19 (2021)
2. M. Upasani, S. Patil, Grid connected solar photovoltaic system with battery storage for energy management, in *2018 2nd International Conference on Inventive Systems and Control (ICISC)*, (IEEE, 2018)
3. S. Grid, *IEEE Vision for Smart Grid Controls: 2030 and Beyond*
4. K. Manandhar et al., Detection of faults and attacks including false data injection attack in smart grid using Kalman filter. IEEE Trans. Control Netw. Syst. **1**(4), 370–379 (2014)
5. Y. Liu et al., An attack-resilient cooperative control strategy of multiple distributed generators in distribution networks. IEEE Trans. Smart Grid **7**(6), 2923–2932 (2016)
6. H. Fawzi, P. Tabuada, S. Diggavi, Secure estimation and control for cyber-physical systems under adversarial attacks. IEEE Trans. Autom. Control **59**(6), 1454–1467 (2014)
7. N.M. Dehkordi et al., Distributed noise-resilient secondary voltage and frequency control for islanded microgrids. IEEE Trans. Smart Grid **10**(4), 3780–3790 (2018)
8. S. Zuo et al., Resilient networked AC microgrids under unbounded cyber attacks. IEEE Trans. Smart Grid **11**(5), 3785–3794 (2020)
9. A.M. Mohan, N. Meskin, H. Mehrjerdi, A comprehensive review of the cyber-attacks and cyber-security on load frequency control of power systems. Energies **13**(15), 3860 (2020)
10. F. Nejabatkhah et al., Cyber-security of smart microgrids: a survey. Energies **14**(1), 27 (2020)
11. Y. Xu, A review of cyber security risks of power systems: from static to dynamic false data attacks. Prot. Control Modern Power Syst. **5**(1), 1–12 (2020)
12. D. Bhamare, et al., Cybersecurity for industrial control systems: A survey. Comput. Secur. **89**, 101677 (2020)
13. H.S. Sánchez et al., Bibliographical review on cyber attacks from a control oriented perspective. Annu. Rev. Control. **48**, 103–128 (2019)
14. M.S. Mahmoud, M.M. Hamdan, U.A. Baroudi, Modeling and control of cyber-physical systems subject to cyber attacks: a survey of recent advances and challenges. Neurocomputing **338**, 101–115 (2019)
15. A. Villalonga, et al., Cloud-based industrial cyber–physical system for data-driven reasoning: A review and use case on an industry 4.0 pilot line. IEEE Trans. Ind. Inf. **16**(9), 5975–5984 (2020)
16. S. Tan et al., New challenges in the design of microgrid systems: Communication networks, cyberattacks, and resilience. IEEE Electrif. Mag. **8**(4), 98–106 (2020)
17. S. Madichetty, S. Mishra, Cyber attack detection and correction mechanisms in a distributed DC microgrid. IEEE Trans. Power Electron. **37**(2), 1476–1485 (2021)
18. L. Ortiz et al., A review on control and fault-tolerant control systems of AC/DC microgrids. Heliyon **6**(8), e04799 (2020)
19. S. Sen, V. Kumar, Microgrid control: a comprehensive survey. Annu. Rev. Control. **45**, 118–151 (2018)
20. F. Martin-Martínez, A. Sánchez-Miralles, M. Rivier, A literature review of microgrids: a functional layer based classification. Renew. Sustain. Energy Rev. **62**, 1133–1153 (2016)
21. R. Shirey, *Internet security glossary*. 2000, FYI 36, RFC 2828, May
22. F. Pasqualetti, F. Dorfler, F. Bullo, Control-theoretic methods for cyberphysical security: geometric principles for optimal cross-layer resilient control systems. IEEE Control Syst. Mag. **35**(1), 110–127 (2015)
23. P.B. Hansen, *Operating System Principles* (Prentice-Hall, Inc., 1973)
24. B. Zhu, A. Joseph, S. Sastry, A taxonomy of cyber attacks on SCADA systems, in *2011 International Conference on Internet of Things and 4th International Conference on Cyber, Physical and Social Computing*. (IEEE, 2011)
25. D. Liu, D. Ye, Cluster synchronization of complex networks under denial-of-service attacks with distributed adaptive strategies. IEEE Trans. Control of Netw. Syst. (2021)

26. X.-K. Liu et al., Resilient control and analysis for dc microgrid system under DoS and impulsive FDI attacks. IEEE Trans. Smart Grid **12**(5), 3742–3754 (2021)
27. Y. Liu, P. Ning, M.K. Reiter, False data injection attacks against state estimation in electric power grids. ACM Trans. Inf. Syst. Secur. (TISSEC) **14**(1), 1–33 (2011)
28. S. Sahoo et al., A stealth cyber-attack detection strategy for DC microgrids. IEEE Trans. Power Electron. **34**(8), 8162–8174 (2018)
29. D. Li, et al., Deep learning based covert attack identification for industrial control systems, in *2020 19th IEEE International Conference on Machine Learning and Applications (ICMLA)* (IEEE, 2020)
30. C. Wang et al., Impacts of cyber system on microgrid operational reliability. IEEE Trans. Smart Grid **10**(1), 105–115 (2017)
31. M. Keshk et al., An integrated framework for privacy-preserving based anomaly detection for cyber-physical systems. IEEE Trans. Sustain. Comput. **6**(1), 66–79 (2019)
32. L. Yang, X. Cao, J. Li, A new cyber security risk evaluation method for oil and gas SCADA based on factor state space. Chaos, Solitons Fractals **89**, 203–209 (2016)
33. S. Liu et al., Extended Kalman filtering for stochastic nonlinear systems with randomly occurring cyber attacks. Neurocomputing **207**, 708–716 (2016)
34. W. Granzer, F. Praus, W. Kastner, Security in building automation systems. IEEE Trans. Ind. Electron. **57**(11), 3622–3630 (2009)
35. K. Hemme, Critical infrastructure protection: maintenance is national security. J. Strat. Secur. **8**(3), 25–39 (2015)
36. F. Cleveland, IEC TC57 security standards for the power system's information infrastructure–beyond simple encryption, in *Transmission and Distribution Conference and Exhibition* (2005)
37. K. Stouffer, J. Falco, K. Scarfone, *Guide to Industrial Control Systems (ICS) Security*. (National Institute of Standards and Technology, 2008)
38. A. Cardenas, et al., Challenges for securing cyber physical systems, in *Workshop on Future Directions in Cyber-Physical Systems Security* (Citeseer, 2009)
39. J.P. Farwell, R. Rohozinski, Stuxnet and the future of cyber war. Survival **53**(1), 23–40 (2011)
40. X. Cao, et al., The geological disasters defense expert system of the massive pipeline network SCADA system based on FNN, in *Asia-Pacific Web Conference* (Springer, 2012)
41. D.U. Case, *Analysis of the Cyber Attack on the Ukrainian Power Grid*. Electricity Information Sharing and Analysis Center (E-ISAC),vol. 388 (2016), pp. 1–29
42. K.E. Hemsley, E. Fisher, *History of Industrial Control System Cyber Incidents*. Idaho National Lab.(INL), Idaho Falls, ID (United States, 2018)
43. C.F. Strategic, I. Studies, *Significant Cyber Incidents Since 2006*. Center for Strategic and International Studies, (2021)
44. S. Tan et al., False data injection cyber-attacks detection for multiple DC microgrid clusters. Appl. Energy **310**, 118425 (2022)
45. Q. Zhou et al., A cyber-attack resilient distributed control strategy in islanded microgrids. IEEE Trans. Smart Grid **11**(5), 3690–3701 (2020)
46. O.A. Beg et al., Signal temporal logic-based attack detection in DC microgrids. IEEE Trans. Smart Grid **10**(4), 3585–3595 (2018)
47. S. Sahoo et al., On detection of false data in cooperative DC microgrids—a discordant element approach. IEEE Trans. Ind. Electron. **67**(8), 6562–6571 (2019)
48. J. Zhao et al., Short-term state forecasting-aided method for detection of smart grid general false data injection attacks. IEEE Trans. Smart Grid **8**(4), 1580–1590 (2015)
49. C.-H. Lo, N. Ansari, CONSUMER: a novel hybrid intrusion detection system for distribution networks in smart grid. IEEE Trans. Emerg. Top. Comput. **1**(1), 33–44 (2013)
50. E. Drayer, T. Routtenberg, Detection of false data injection attacks in smart grids based on graph signal processing. IEEE Syst. J. **14**(2), 1886–1896 (2019)
51. H. Tu et al., A hybrid cyber attack model for cyber-physical power systems. IEEE Access **8**, 114876–114883 (2020)

52. Y. Hu et al., Detecting stealthy attacks against industrial control systems based on residual skewness analysis. EURASIP J. Wirel. Commun. Netw. **2019**(1), 1–14 (2019)
53. Y. Liu, L. Cheng, Relentless false data injection attacks against Kalman filter based detection in smart grid. IEEE Trans. Control Netw. Syst. (2022)
54. S. Soltani et al., Improved estimation for well-logging problems based on fusion of four types of Kalman filters. IEEE Trans. Geosci. Remote Sens. **56**(2), 647–654 (2017)
55. B. Li et al., Detecting false data injection attacks against power system state estimation with fast go-decomposition approach. IEEE Trans. Ind. Inf. **15**(5), 2892–2904 (2018)
56. J. Yan, F. Guo, C. Wen, Attack detection and isolation for distributed load shedding algorithm in microgrid systems. IEEE J. Emerg. Select. Top. Ind. Electron. **1**(1), 102–110 (2020)
57. A. Cecilia et al., On addressing the security and stability issues due to false data injection attacks in DC microgrids—an adaptive observer approach. IEEE Trans. Power Electron. **37**(3), 2801–2814 (2021)
58. F. Pasqualetti, F. Dörfler, F. Bullo, Attack detection and identification in cyber-physical systems. IEEE Trans. Autom. Control **58**(11), 2715–2729 (2013)
59. A. Paul, I. Kamwa, G. Joos, Centralized dynamic state estimation using a federation of extended Kalman filters with intermittent PMU data from generator terminals. IEEE Trans. Power Syst. **33**(6), 6109–6119 (2018)
60. J.M. Pak et al., Improving reliability of particle filter-based localization in wireless sensor networks via hybrid particle/FIR filtering. IEEE Trans. Ind. Inf. **11**(5), 1089–1098 (2015)
61. G. Anagnostou, B.C. Pal, Derivative-free Kalman filtering based approaches to dynamic state estimation for power systems with unknown inputs. IEEE Trans. Power Syst. **33**(1), 116–130 (2017)
62. Y. Li, Z. Li, L. Chen, Dynamic state estimation of generators under cyber attacks. IEEE Access **7**, 125253–125267 (2019)
63. D.B. Rawat, C. Bajracharya, Detection of false data injection attacks in smart grid communication systems. IEEE Signal Process. Lett. **22**(10), 1652–1656 (2015)
64. K. Miao, W.A. Zhang, X. Qiu, An adaptive unscented Kalman filter approach to secure state estimation for wireless sensor networks. Asian J. Control (2022)
65. A. Sargolzaei et al., Detection and mitigation of false data injection attacks in networked control systems. IEEE Trans. Ind. Inf. **16**(6), 4281–4292 (2019)
66. A. Abbaspour et al., Resilient control design for load frequency control system under false data injection attacks. IEEE Trans. Ind. Electron. **67**(9), 7951–7962 (2019)
67. A. Chattopadhyay, U. Mitra, Security against false data-injection attack in cyber-physical systems. IEEE Trans. Control Netw. Syst. **7**(2), 1015–1027 (2019)
68. J. Qi, A.F. Taha, J. Wang, Comparing Kalman filters and observers for power system dynamic state estimation with model uncertainty and malicious cyber attacks. IEEE Access **6**, 77155–77168 (2018)
69. M. Adeli, et al., Distributed trust-based unscented Kalman filter for non-linear state estimation under cyber-attacks: The application of manoeuvring target tracking over wireless sensor networks (2021)
70. J. Lu et al., Unscented Kalman filtering for nonlinear systems with sensor saturation and randomly occurring false data injection attacks. Asian J. Control **23**(2), 871–881 (2021)
71. X. Liu, Y. Mo, E. Garone, Local decomposition of Kalman filters and its application for secure state estimation. IEEE Trans. Autom. Control **66**(10), 5037–5044 (2020)
72. J. Kim et al., Detection of sensor attack and resilient state estimation for uniformly observable nonlinear systems having redundant sensors. IEEE Trans. Autom. Control **64**(3), 1162–1169 (2018)
73. Y. Mao et al., Novel stealthy attack and defense strategies for networked control systems. IEEE Trans. Autom. Control **65**(9), 3847–3862 (2020)
74. Y. Yamamoto, N. Kuze, T. Ushio, Attack detection and defense system using an unknown input observer for cooperative adaptive cruise control systems. IEEE Access **9**, 148810–148820 (2021)

75. A. Mustafa et al., Detection and mitigation of data manipulation attacks in AC microgrids. IEEE Trans. Smart Grid **11**(3), 2588–2603 (2019)
76. G. Chaojun, P. Jirutitijaroen, M. Motani, Detecting false data injection attacks in AC state estimation. IEEE Trans. Smart Grid **6**(5), 2476–2483 (2015)
77. B.P. Poudel et al., Detection and mitigation of cyber-threats in the DC microgrid distributed control system. Int. J. Electr. Power Energy Syst. **120**, 105968 (2020)
78. Y. Zhu, W.X. Zheng, Observer-based control for cyber-physical systems with periodic DoS attacks via a cyclic switching strategy. IEEE Trans. Autom. Control **65**(8), 3714–3721 (2019)
79. S. Nateghi, Y. Shtessel, C. Edwards, Cyber-attacks and faults reconstruction using finite time convergent observation algorithms: electric power network application. J. Franklin Inst. **357**(1), 179–205 (2020)
80. X. Wang et al., Detection and isolation of false data injection attacks in smart grid via unknown input interval observer. IEEE Internet Things J. **7**(4), 3214–3229 (2020)
81. Y. Li, H. Fang, J. Chen, Anomaly detection and identification for multiagent systems subjected to physical faults and cyberattacks. IEEE Trans. Ind. Electron. **67**(11), 9724–9733 (2019)
82. A. Barboni et al., Detection of covert cyber-attacks in interconnected systems: A distributed model-based approach. IEEE Trans. Autom. Control **65**(9), 3728–3741 (2020)
83. C. Schellenberger, P. Zhang, Detection of covert attacks on cyber-physical systems by extending the system dynamics with an auxiliary system, in *2017 IEEE 56th Annual Conference on Decision and Control (CDC)* (IEEE, 2017)
84. D. Das et al., Luenberger observer based current estimated boost converter for PV maximum power extraction—A current sensorless approach. IEEE J. Photovolt. **9**(1), 278–286 (2018)
85. R. Gao et al., K-filter-based adaptive output feedback control for high-order nonlinear systems subject to actuator and sensor attacks. Int. J. Robust Nonlinear Control **32**(6), 3469–3484 (2022)
86. J. Yan, G.-H. Yang, Y. Wang, Dynamic reduced-order observer-based detection of false data injection attacks with application to smart grid systems. IEEE Trans. Ind. Inf. (2022)
87. X. Wang et al., Detection and localization of biased load attacks in smart grids via interval observer. Inf. Sci. **552**, 291–309 (2021)
88. H.A.H. Alhelou, P. Cuffe, A dynamic state estimator based tolerance control method against cyberattack and erroneous measured data for power systems. IEEE Trans. Ind. Inf. (2021)
89. X. Wang et al., Detection and location of bias load injection attack in smart grid via robust adaptive observer. IEEE Syst. J. **14**(3), 4454–4465 (2020)
90. L. An, G.-H. Yang, Supervisory nonlinear state observers for adversarial sparse attacks. IEEE Trans. Cybern. (2020)
91. J. Yang, W.-A. Zhang, F. Guo, Adaptive distributed Kalman-like filter for power system with cyber attacks. Automatica **137**, 110091 (2022)
92. S. Saha et al., Sensor fault and cyber attack resilient operation of DC microgrids. Int. J. Electr. Power Energy Syst. **99**, 540–554 (2018)
93. H. Yan et al., A novel sliding mode estimation for microgrid control with communication time delays. IEEE Trans. Smart Grid **10**(2), 1509–1520 (2017)
94. W. Ao, Y. Song, C. Wen, Adaptive cyber-physical system attack detection and reconstruction with application to power systems. IET Control Theory Appl. **10**(12), 1458–1468 (2016)
95. G. Rinaldi et al., Adaptive dual-layer super-twisting sliding mode observers to reconstruct and mitigate disturbances and communication attacks in power networks. Automatica **129**, 109656 (2021)
96. Z. Zhang, Y. Niu, J. Song, Input-to-state stabilization of interval type-2 fuzzy systems subject to cyberattacks: an observer-based adaptive sliding mode approach. IEEE Trans. Fuzzy Syst. **28**(1), 190–203 (2019)
97. Q. Su et al., Observer-based detection and reconstruction of dynamic load altering attack in smart grid. J. Franklin Inst. **358**(7), 4013–4027 (2021)
98. C. Wu et al., Secure estimation for cyber-physical systems via sliding mode. IEEE Trans. Cybern. **48**(12), 3420–3431 (2018)

99. L. Ye, F. Zhu, J. Zhang, Sensor attack detection and isolation based on sliding mode observer for cyber-physical systems. Int. J. Adapt. Control Signal Process. **34**(4), 469–483 (2020)
100. V. Utkin, J. Guldner, J. Shi, *Sliding Mode Control in Electro-Mechanical Systems* (CRC Press, 2017)
101. G. Chen, Y. Song, Y. Guan, Terminal sliding mode-based consensus tracking control for networked uncertain mechanical systems on digraphs. IEEE Trans. Neural Netw. Learn. Syst. **29**(3), 749–756 (2016)
102. A. Cecilia et al., Detection and mitigation of false data in cooperative dc microgrids with unknown constant power loads. IEEE Trans. Power Electron. **36**(8), 9565–9577 (2021)
103. M. Lv, et al., An integral sliding mode observer for CPS cyber security attack detection. Chaos Interdiscip. J. Nonlinear Sci. **29**(4), 043120 (2019)
104. M. Ahmadizadeh, A. Shafei, M. Fooladi, A recursive algorithm for dynamics of multiple frictionless impact-contacts in open-loop robotic mechanisms. Mech. Mach. Theory **146**, 103745 (2020)
105. A. Shafei, H. Shafei, Considering link flexibility in the dynamic synthesis of closed-loop mechanisms: a general approach. J. Vib. Acoust. **142**(2) (2020)
106. L. Xu et al., A continuous terminal sliding-mode observer-based anomaly detection approach for industrial communication networks. Symmetry **14**(1), 124 (2022)
107. H. Wang, et al., High-order observer-based sliding mode control for the isolated microgrid with cyber attacks and physical uncertainties. Complexity **2020** (2020)
108. H.R. Shafei, M. Bahrami, H.A. Talebi, Disturbance observer-based two-Layer control strategy design to deal with both matched and mismatched uncertainties. Int. J. Robust Nonlinear Control **31**(5), 1640–1656 (2021)
109. H. Yang, et al., Sparse actuator and sensor attacks reconstruction for linear cyber-physical systems with sliding mode observer. IEEE Trans. Ind. Inf. (2021)
110. M. Li et al., Adaptive sliding-mode tracking control of networked control systems with false data injection attacks. Inf. Sci. **585**, 194–208 (2022)
111. L. Chen, C. Edwards, H. Alwi, Sensor fault estimation using LPV sliding mode observers with erroneous scheduling parameters. Automatica **101**, 66–77 (2019)
112. S. Cui et al., Coordinated data-injection attack and detection in the smart grid: A detailed look at enriching detection solutions. IEEE Signal Process. Mag. **29**(5), 106–115 (2012)
113. A.J. Abianeh et al., Cyber-resilient sliding-mode consensus secondary control scheme for Islanded AC microgrids. IEEE Trans. Power Electron. **37**(5), 6074–6089 (2021)
114. M. Shi et al., Observer-based resilient integrated distributed control against cyberattacks on sensors and actuators in islanded AC microgrids. IEEE Trans. Smart Grid **12**(3), 1953–1963 (2021)
115. X. Li, Q. Xu, F. Blaabjerg, Adaptive resilient secondary control for islanded AC microgrids with sensor faults. IEEE J. Emerg. Select. Top. Power Electron. **9**(5), 5239–5248 (2020)
116. S. Sahoo, Y. Yang, F. Blaabjerg, Resilient synchronization strategy for AC microgrids under cyber attacks. IEEE Trans. Power Electron. **36**(1), 73–77 (2020)
117. A. Afshari et al., Resilient synchronization of voltage/frequency in AC microgrids under deception attacks. IEEE Syst. J. **15**(2), 2125–2136 (2020)
118. M. Ozay et al., Machine learning methods for attack detection in the smart grid. IEEE Trans. Neural Netw. Learn. Syst. **27**(8), 1773–1786 (2015)
119. M.R. Habibi et al., Detection of false data injection cyber-attacks in DC microgrids based on recurrent neural networks. IEEE J. Emerg. Select. Top. Power Electron. **9**(5), 5294–5310 (2020)
120. M.R. Habibi, et al., Secure MPC/ANN-based false data injection cyber-attack detection and mitigation in DC microgrids. IEEE Syst. J. (2021)
121. S. Zhao, F. Blaabjerg, H. Wang, An overview of artificial intelligence applications for power electronics. IEEE Trans. Power Electron. **36**(4), 4633–4658 (2020)
122. F. de Almeida Florencio, et al., Intrusion detection via MLP neural network using an arduino embedded system, in *2018 VIII Brazilian Symposium on Computing Systems Engineering (SBESC)* (IEEE, 2018)

123. Y. Javed, N. Rajabi. Multi-layer perceptron artificial neural network based IoT botnet traffic classification, in *Proceedings of the Future Technologies Conference* (Springer, 2019)
124. F. Pedregosa, et al., Scikit-learn: machine learning in Python. J. Mach. Learn. Res. **12**, 2825–2830 (2011)
125. Z.E. Huma et al., A hybrid deep random neural network for cyberattack detection in the industrial internet of things. IEEE Access **9**, 55595–55605 (2021)
126. B. Susilo, R.F. Sari, Intrusion detection in IoT networks using deep learning algorithm. Information **11**(5), 279 (2020)
127. J. Yan, Y. Qi, Q. Rao, Detecting malware with an ensemble method based on deep neural network. Secur. Commun. Netw. **2018** (2018)
128. Y. Li et al., Robust detection for network intrusion of industrial IoT based on multi-CNN fusion. Measurement **154**, 107450 (2020)
129. Y. He, G.J. Mendis, J. Wei, Real-time detection of false data injection attacks in smart grid: a deep learning-based intelligent mechanism. IEEE Trans. Smart Grid **8**(5), 2505–2516 (2017)
130. C. Jiang et al., A mixed deep recurrent neural network for MEMS gyroscope noise suppressing. Electronics **8**(2), 181 (2019)
131. J. Kim, et al., Long short term memory recurrent neural network classifier for intrusion detection, in *2016 International Conference on Platform Technology and Service (PlatCon)* (IEEE, 2016)
132. D. Tran et al., A LSTM based framework for handling multiclass imbalance in DGA botnet detection. Neurocomputing **275**, 2401–2413 (2018)
133. X. Niu, et al., Dynamic detection of false data injection attack in smart grid using deep learning, in *2019 IEEE Power & Energy Society Innovative Smart Grid Technologies Conference (ISGT)* (IEEE, 2019)
134. M.M. Hassan et al., A hybrid deep learning model for efficient intrusion detection in big data environment. Inf. Sci. **513**, 386–396 (2020)
135. D. Vasan et al., MTHAEL: Cross-architecture IoT malware detection based on neural network advanced ensemble learning. IEEE Trans. Comput. **69**(11), 1654–1667 (2020)
136. M. Saharkhizan et al., An ensemble of deep recurrent neural networks for detecting IoT cyber attacks using network traffic. IEEE Internet Things J. **7**(9), 8852–8859 (2020)
137. I. Goodfellow, Y. Bengio, A. Courville, *Deep Learning* (MIT Press, 2016)
138. W. Liu et al., A survey of deep neural network architectures and their applications. Neurocomputing **234**, 11–26 (2017)
139. M. Yousefi-Azar, et al., Autoencoder-based feature learning for cyber security applications, in *2017 International Joint Conference on Neural Networks (IJCNN)* (IEEE, 2017)
140. B. Yan, G. Han, Effective feature extraction via stacked sparse autoencoder to improve intrusion detection system. IEEE Access **6**, 41238–41248 (2018)
141. R. Memisevic, G.E. Hinton, Learning to represent spatial transformations with factored higher-order Boltzmann machines. Neural Comput. **22**(6), 1473–1492 (2010)
142. U. Fiore et al., Network anomaly detection with the restricted Boltzmann machine. Neurocomputing **122**, 13–23 (2013)
143. Y. Imamverdiyev, F. Abdullayeva, Deep learning method for denial of service attack detection based on restricted boltzmann machine. Big data **6**(2), 159–169 (2018)
144. S. Seo, S. Park, J. Kim. Improvement of network intrusion detection accuracy by using restricted Boltzmann machine, in *2016 8th International Conference on Computational Intelligence and Communication Networks (CICN)*. (IEEE, 2016)
145. P. Wei et al., An optimization method for intrusion detection classification model based on deep belief network. IEEE Access **7**, 87593–87605 (2019)
146. M.A. Salama et al., Hybrid intelligent intrusion detection scheme, in *Soft computing in industrial applications*. (Springer, 2011), pp.293–303
147. F. Qu, et al., An intrusion detection model based on deep belief network, in *Proceedings of the 2017 VI International Conference on Network, Communication and Computing* (2017)
148. Y. Zhang, P. Li, X. Wang, Intrusion detection for IoT based on improved genetic algorithm and deep belief network. IEEE Access **7**, 31711–31722 (2019)

149. I. Goodfellow, et al., *Generative adversarial nets*. Advances in neural information processing systems (2014), p. **27**
150. J.-Y. Kim, S.-J. Bu, S.-B. Cho, Zero-day malware detection using transferred generative adversarial networks based on deep autoencoders. Inf. Sci. **460**, 83–102 (2018)
151. T. Merino, et al., Expansion of cyber attack data from unbalanced datasets using generative adversarial networks, in *International Conference on Software Engineering Research, Management and Applications* (Springer, 2019)
152. K. Weiss, T.M. Khoshgoftaar, D. Wang, A survey of transfer learning. J. Big data **3**(1), 1–40 (2016)
153. S.J. Pan, Q. Yang, A survey on transfer learning. IEEE Trans. Knowl. Data Eng. **22**(10), 1345–1359 (2009)
154. P. Wu, H. Guo, R. Buckland, A transfer learning approach for network intrusion detection, in *2019 IEEE 4th International Conference on Big Data Analytics (ICBDA)*. IEEE (2019)
155. X. Gao et al., Malware classification for the cloud via semi-supervised transfer learning. J. Inf. Secur. Appl. **55**, 102661 (2020)
156. L. Vu et al., Deep transfer learning for IoT attack detection. IEEE Access **8**, 107335–107344 (2020)
157. K. Arulkumaran et al., Deep reinforcement learning: a brief survey. IEEE Signal Process. Mag. **34**(6), 26–38 (2017)
158. M. Lopez-Martin, B. Carro, A. Sanchez-Esguevillas, Application of deep reinforcement learning to intrusion detection for supervised problems. Expert Syst. Appl. **141**, 112963 (2020)
159. K. Sethi, et al., Deep reinforcement learning based intrusion detection system for cloud infrastructure, in *2020 International Conference on COMmunication Systems & NETworkS (COMSNETS)* (IEEE, 2020)
160. D. Li et al., IoT data feature extraction and intrusion detection system for smart cities based on deep migration learning. Int. J. Inf. Manag. **49**, 533–545 (2019)
161. K. Hughes, K. McLaughlin, S. Sezer, A model-free approach to intrusion response systems. J. Inf. Secur. Appl. **66**, 103150 (2022)
162. M. Esmalifalak et al., Detecting stealthy false data injection using machine learning in smart grid. IEEE Syst. J. **11**(3), 1644–1652 (2014)
163. S.J. Lee et al., IMPACT: impersonation attack detection via edge computing using deep autoencoder and feature abstraction. IEEE Access **8**, 65520–65529 (2020)
164. M.G. Raman et al., An efficient intrusion detection system based on hypergraph-Genetic algorithm for parameter optimization and feature selection in support vector machine. Knowl. Based Syst. **134**, 1–12 (2017)
165. X. Tang, S.X.-D. Tan, H.-B. Chen, SVM based intrusion detection using nonlinear scaling scheme, in *2018 14th IEEE International Conference on Solid-State and Integrated Circuit Technology (ICSICT)* (IEEE, 2018)
166. V. Shakhov, et al., On Lightweight method for intrusions detection in the Internet of Things, in *2019 IEEE International Black Sea Conference on Communications and Networking (BlackSeaCom)* (IEEE, 2019)
167. M.R. Habibi et al., Decentralized coordinated cyberattack detection and mitigation strategy in DC microgrids based on artificial neural networks. IEEE J. Emerg. Select. Top. Power Electron. **9**(4), 4629–4638 (2021)
168. A. Kavousi-Fard, W. Su, T. Jin, A machine-learning-based cyber attack detection model for wireless sensor networks in microgrids. IEEE Trans. Ind. Inf. **17**(1), 650–658 (2020)
169. M. Dehghani et al., Cyber-attack detection in dc microgrids based on deep machine learning and wavelet singular values approach. Electronics **10**(16), 1914 (2021)
170. Y. Liu et al., Robust and resilient distributed optimal frequency control for microgrids against cyber attacks. IEEE Trans. Ind. Inf. **18**(1), 375–386 (2021)
171. M. Shafiee-Rad et al., Robust decentralized voltage control for uncertain DC microgrids. Int. J. Electr. Power Energy Syst. **125**, 106468 (2021)
172. L. Xu, Y. Chen, M. Li, Sliding mode resilient control for TOD-based servo system under DoS attack. Int. J. Control Autom. Syst. **20**(2), 526–535 (2022)

173. R. Sakthivel et al., Resilient sampled-data control for Markovian jump systems with an adaptive fault-tolerant mechanism. IEEE Trans. Circuits Syst. II Express Briefs **64**(11), 1312–1316 (2017)
174. Y. Xu, et al., Resilient and robust control for event-triggered uncertain semi-Markov jump systems against stochastic cyber attacks. Int. J. Robust Nonlinear Control
175. X. Xie, J. Lu, D. Yue, Resilient stabilization of discrete-time Takagi-Sugeno fuzzy systems: dynamic trade-off between conservatism and complexity. Inf. Sci. **582**, 181–197 (2022)
176. Y. Seyedi, H. Karimi, J. Mahseredjian, A data-driven method for prediction of post-fault voltage stability in hybrid AC/DC microgrids. IEEE Trans. Power Syst. (2022)
177. J. Duan, W. Zeng, M.-Y. Chow, Resilient distributed DC optimal power flow against data integrity attack. IEEE Trans. Smart Grid **9**(4), 3543–3552 (2016)
178. H. Zhang et al., Distributed load sharing under false data injection attack in an inverter-based microgrid. IEEE Trans. Ind. Electron. **66**(2), 1543–1551 (2018)
179. S. Abhinav et al., Synchrony in networked microgrids under attacks. IEEE Trans. Smart Grid **9**(6), 6731–6741 (2017)
180. L.-Y. Lu et al., Intrusion detection in distributed frequency control of isolated microgrids. IEEE Trans. Smart Grid **10**(6), 6502–6515 (2019)
181. L. Ding et al., Distributed resilient finite-time secondary control for heterogeneous battery energy storage systems under denial-of-service attacks. IEEE Trans. Ind. Inf. **16**(7), 4909–4919 (2019)
182. A. Bidram et al., Resilient and cybersecure distributed control of inverter-based islanded microgrids. IEEE Trans. Ind. Inf. **16**(6), 3881–3894 (2019)
183. C. Deng et al., Distributed resilient control for energy storage systems in cyber–physical microgrids. IEEE Trans. Ind. Inf. **17**(2), 1331–1341 (2020)
184. T. Ding et al., A resilient microgrid formation strategy for load restoration considering master-slave distributed generators and topology reconfiguration. Appl. Energy **199**, 205–216 (2017)
185. C. Wu et al., Active defense-based resilient sliding mode control under denial-of-service attacks. IEEE Trans. Inf. Forensics Secur. **15**, 237–249 (2019)
186. H.R. Baghaee et al., A decentralized robust mixed $h_ {{2}}/h_ {{{\infty}}}$ voltage control scheme to improve small/large-signal stability and frt capability of islanded multi-der microgrid considering load disturbances. IEEE Syst. J. **12**(3), 2610–2621 (2017)
187. X. Li, et al. Adaptive resilient secondary control for microgrids with communication faults. IEEE Trans. Cybern. (2021)
188. C. Lin et al., Dynamic MGs-based load restoration for resilient urban power distribution systems considering intermittent RESs and droop control. Int. J. Electr. Power Energy Syst. **140**, 107975 (2022)
189. Y. Xu, Resilient secure control of networked systems over unreliable communication networks. IEEE Trans. Ind. Inf. (2021)
190. Y. Zhao, et al., Adaptive resilient control of cyber-physical systems under actuator and sensor attacks. IEEE Trans. Ind. Inf. (2021)
191. C. Deng, D. Zhang, G. Feng, Resilient practical cooperative output regulation for MASs with unknown switching exosystem dynamics under DoS attacks. Automatica **139**, 110172 (2022)
192. S. Song et al., Adaptive resilient control design for nonlinear time-delay systems against unknown state-dependent deception attacks. Int. J. Robust Nonlinear Control **32**(4), 2159–2182 (2022)
193. Y. Yang, Y. Qian, W. Yue, A secure dynamic event-triggered mechanism for resilient control of multi-agent systems under sensor and actuator attacks. IEEE Trans. Circuits Syst. I Regular Papers (2021)
194. P. Meng, et al., Resilient DC voltage control for islanded wind farms integration using cascaded hybrid HVDC system. IEEE Trans. Power Syst. (2021)
195. S. Zuo, D. Yue, Resilient containment of multigroup systems against unknown unbounded FDI attacks. IEEE Trans. Ind. Electron. **69**(3), 2864–2873 (2021)
196. M.S. Sadabadi, S. Sahoo, F. Blaabjerg, Stability-oriented design of cyberattack-resilient controllers for cooperative DC microgrids. IEEE Trans. Power Electron. **37**(2), 1310–1321 (2021)

197. S. Saha, S. Gholami, M.K.K. Prince, Sensor fault-resilient control of electronically coupled distributed energy resources in Islanded microgrids. IEEE Trans. Ind. Appl. **58**(1), 914–929 (2021)
198. S. Zuo, D. Yue, Resilient output formation containment of heterogeneous multigroup systems against unbounded attacks. IEEE Trans. Cybern. (2020)
199. Y. Shan, A. Pan, H. Liu, *A switching event-triggered resilient control scheme for primary and secondary levels in AC microgrids.* ISA Trans. (2022)
200. X. Ma, et al., Consensus tracking control for uncertain non-strict feedback multi-agent system under cyber attack via resilient neuroadaptive approach. Int. J. Robust Nonlinear Control
201. T. Vigneysh, N. Pachauri, V. Suresh, Internal model-based cascaded control approach for multi-functional grid interactive converters. IEEE Access **10**, 19862–19874 (2022)
202. Z. Wang et al., Centralized security-guaranteed filtering in multirate-sensor fusion under deception attacks. J. Franklin Inst. **355**(1), 406–420 (2018)
203. S. Sahoo, T. Dragičević, F. Blaabjerg, Multilayer resilience paradigm against cyber attacks in DC microgrids. IEEE Trans. Power Electron. **36**(3), 2522–2532 (2020)
204. X. Chen et al., Distributed resilient control against denial of service attacks in DC microgrids with constant power load. Renew. Sustain. Energy Rev. **153**, 111792 (2022)
205. S. Sahoo, T. Dragičević, F. Blaabjerg, Resilient operation of heterogeneous sources in cooperative dc microgrids. IEEE Trans. Power Electron. **35**(12), 12601–12605 (2020)

Cyber Vulnerabilities of Modern Power Systems

Asad Ali Khan and Omar A. Beg

Abstract Modern power systems heavily rely on Internet-of-Things (IoT) and emerging wide-area sensor networks that expose them to cyber vulnerabilities such as network failures and cyber-attacks. Some practical network failure examples include North America (2003) due to state estimator and alarm system failure, Austria (2013) due to network congestion caused by a software bug, and Switzerland (2005) due to information overload. Ukraine's power system went down in December 2015 leaving thousands of homes and facilities out of power due to a cyber-attack caused by a malware, identified as BlackEnergy, in control center computers. Such failures and cyber-attacks will leave majority of customers without power supply and may cause significant damage to highly sensitive and mission critical equipment. In case of power electronics-intensive microgrids, the after-effects of the cyber-attacks are even more detrimental due to comparatively weaker and fragile distribution grid, highly dynamic source and load profiles, and meager generational inertia. Cyber vulnerabilities are divided into two main categories, i.e., cyber-attacks and network failures. An overview of such cyber vulnerabilities, practical limitations of modern power systems, relevant prevention measures, and a case study is presented in this chapter.

1 Introduction

In recent years, research has focused more on the impact of cyber-attacks on power systems. Modern power grids with highly integrated communication and computation capabilities rely more heavily on data transfers compared with their conventional counterparts. Advancements in information and communication technology,

A. A. Khan
Department of Electrical Engineering, The University of Texas at San Antonio,
San Antonio, United States
e-mail: asad.khan@my.utsa.edu

O. A. Beg (✉)
The University of Texas Permian Basin, 4901 E University Blvd, Odessa, Texas, TX 79762,
United States
e-mail: beg_o@utpb.edu

H. Haes Alhelou et al. (eds.), *Power Systems Cybersecurity*, Power Systems,
https://doi.org/10.1007/978-3-031-20360-2_2

and development of internet-of-things (IoT)-based sensors, have improved situational awareness, decision support, and control of the electrical grid [1]. A smart grid facilitates the integration of renewable energy resources such as wind turbines and photovoltaics to enhance the reliability of the system by minimizing environmental impact from carbon emissions and decreasing energy losses in the power grids. In this chapter, modern power systems refer to microgrids and smart grids that rely heavily on modern communication and computation technology. A microgrid is a self-contained small-scale distribution system that includes a variety of distributed generators and energy storage devices. It can operate in either a grid-connected or islanded mode. The smart grid's key benefits are intended to be increased system resilience and improved environmental impact. Resilience refers to an entity's ability to withstand and recover rapidly from unexpected events [2].

Electric power systems have evolved from a unidirectional generation and transmission model to a more distributed architecture that supports both traditional generation and distributed energy resources (DER). The development of information and communication technologies, automated control systems, IoT-based sensing, and embedded controllers have principally facilitated the transformation of conventional power networks to cyber-physical systems (CPS). Smart grids are typical CPS that include digital, analog, and physical components. The smart grid is an energy generation, delivery, and consumption system based on communication and information technologies. Better planning, more flexible control, cyber-secure operations, system-wide optimization, transactive energy systems, improvements in power quality, system reliability advancements, resiliency, interoperability, and cleaner energy generation are typical characteristics of modern power systems [3].

Such power systems function as a result of a significant interaction between information and communication technology, and physical devices. The modern power systems incorporate extensive communication networks and intelligent electronic devices (IEDs) that act as an interface between cyber and physical layer of the system [4]. Power systems are vulnerable to cyber threats due to their widespread use of information and communication technology as shown in Fig. 1. For example, in 2015, the Ukraine power grid was attacked and substation breakers were opened by malicious attackers. The attacker was able to get illegal access to the control centers due to malware inserted through the communication lines. Following that, the data was gathered to establish the key lines in the regional grid and malicious commands were transmitted to trip these lines, resulting in a massive power outage. The restoration effort was further slowed down by subsequent hacking of the control system [5, 6].

Malicious attackers can disrupt the information exchange process, causing power outages, economic loss, and system instability. Power grids are facing severe cybersecurity issues as the smart grids evolve and communication networks become more interconnected [7, 8]. Modern power systems are more exposed and vulnerable to attacks due to their interconnectivity and complexity. Vulnerability is a measure of the system's weakness involving a series of cascading events that may include line or generator outages, malfunctions or unwanted actions of protective relays, information or communication system failures. Integration of information network

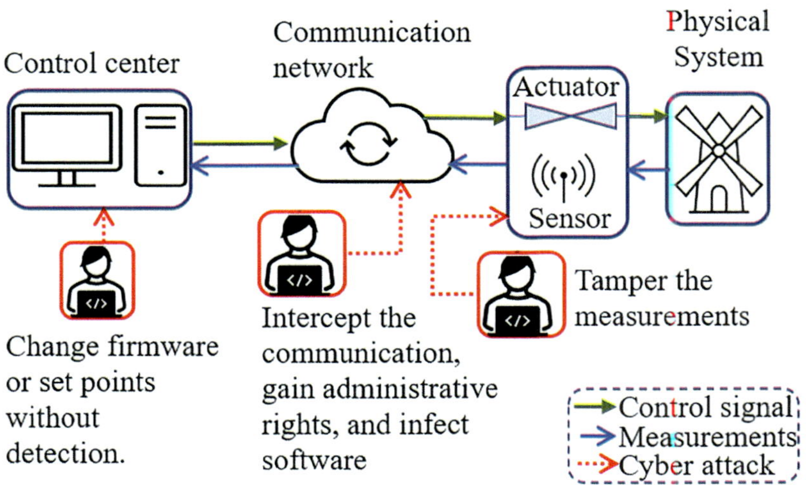

Fig. 1 Illustrative diagram of cyber vulnerabilities of modern power systems is shown

broadens the attack surface in modern power systems. The ability of a system to foresee, plan for, and adapt to changing situations, as well as to endure, respond to, and recover quickly from disruptions, is known as resilience. System vulnerability and resilience must be assessed and measured before proactive steps can be taken to protect power system operations [9].

Smart grids include a wide range of intelligent equipment and communication infrastructures, all of which are vulnerable to cyber-attacks. Malicious attacks on critical mechanisms such as unit commitment, demand response, and distributed control architecture will have potentially disastrous consequences [10, 11]. Integration of DER on a large scale, such as energy storage, distributed generation, and electric vehicles, will have a significant impact on the stability that can be vulnerable to cyber-attacks. The costs of an attack on the electrical system often include equipment damage, power disruptions, and economic losses. The diversity and complexity of modern power system's architecture have demonstrated the potential threats to their security and resilience [12]. It is evident that the existing power networks are becoming more smart by incorporating advanced communication and computing tools. This has significantly improved the productivity of smart grids. On the other hand, it has created serious cyber security concerns since a large number of intelligent devices are connected through communication networks. Cyber-attacks on such devices have a direct impact on the reliable and secure operation of modern power systems. Some of the major type of cyber-attacks targeting the smart grids include man-in-the-middle attacks, distributed denial-of-service-attacks, jamming attacks, and false data injection attacks [13–15].

This chapter provides a comprehensive overview of the cyber vulnerabilities and the mitigation strategies from the architectural viewpoint of the modern power sys-

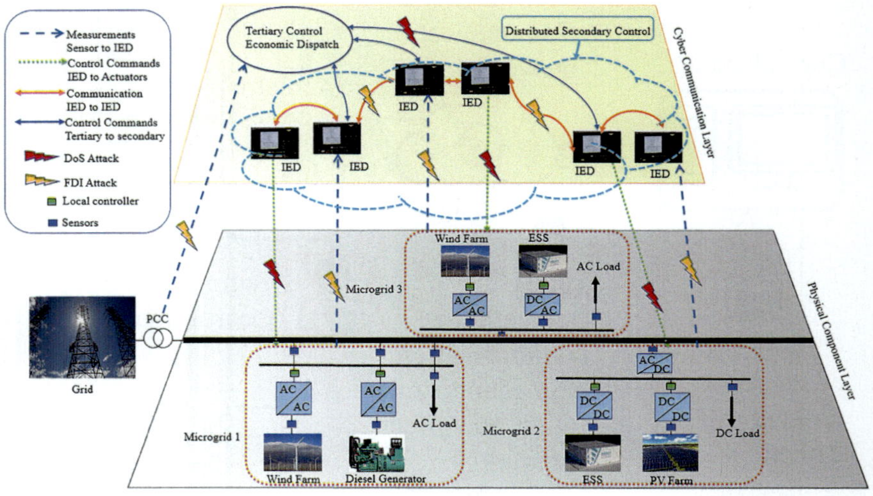

Fig. 2 Architecture of networked microgrids with cyber vulnerabilities is shown. The networked microgrids are connected to the main grid at point of common coupling (PCC). The primary level control is implemented locally in the physical layers to control the converters. The distributed secondary level control is implemented using IEDs that communicate with neighboring agents to cooperatively achieve the control objective that is received from the tertiary level controller. The sensor measurements and communication among IEDs are vulnerable to FDI attacks, whereas the control commands being sent to the actuators are potential target of DoS attacks

tems. The rest of the chapter is organized as follows: Sect. 2, discusses the architecture of the modern power systems including distributed control-based smart grids. Classification of various cyber-attacks targeting the modern power systems along with some prevention methods are presented in Sect. 3. Finally, Sect. 4 concludes this chapter.

2 Architectures of Smart Grids

The growing integration of numerous distributed and renewable energy resources, along with enhanced metering, control, and communication technologies, fosters a transformation from traditional power systems to modern cyber-physical smart grids. In smart grids, new advanced communication technologies, such as wireless networks, are important for achieving dependable and secure information interconnections across distributed microgrid clusters. Networked microgrids can contain a cluster of microgrids within a power distribution system and such networked microgrids are generally categorized as AC, DC, or hybrid microgrids [15].

Among these networked microgrids, AC microgrids are most popular so far, since they take advantage of topology of the existing power system. DC microgrids consists of DERs, energy storage system, and loads that are connected to a DC bus through power electronics-based converters. DC microgrids may offer lower operation cost than their AC counterpart due to negligible reactive power and eddy current losses. Furthermore, technical benefits of both AC and DC microgrids can be combined in a hybrid microgrid offering high flexibility [16]. Typical architecture of a networked microgrid containing both AC and DC microgrids is shown in Fig. 2. A constituent microgrid in a network contains, the power-electronics-based converters and microgrid controllers that are used to connect, energy storage system (ESS), wind turbine, and photo voltaic (PV) generator to a common bus. Each microgrid may operate in islanded or grid connected mode to exchange power with the main grid and with other microgrids within the network. For sustained and reliable operation of such modern power systems, advanced control architectures are being deployed to achieve stable power quality, minimize the operational cost, and to manage information sharing among interconnected power networks. A microgrid energy management system helps to obtain secure and economic operation by providing cooperative control of microgrids. The main functions of such an energy management system include information exchange, control and scheduling, security, and auxiliary services [17]. Typical networked microgrid architectures can be classified into centralized, decentralized, hybrid and nested structures.

1. **Centralized control**: All of the controllable generators and loads are controlled by a central controller in the centralized structure. The central controller is responsible for managing power interactions between microgrids and DER scheduling, as well as dispatch commands, control, and managing microgrid operation modes. Such a centralized architecture lacks the adaptability and is prone to single point of failure due to high computation power requirement [18].
2. **Decentralized system**: In a decentralized system, each microgrid is a self-contained unit with its own local controller for profit maximization. The microgrid's operation status is monitored by the local controller, which then independently determines the state of the DER and controllable loads. The decentralized structure may foster microgrid competition, reducing overall system performance. The exchange of information between microgrids would be used to create global microgrid operational control [19].
3. **Hybrid architecture**: This architecture with a central controller at system level and a local controller at microgrid level is designed to address the limitations of both centralized and decentralized architectures. Each microgrid's local controller is responsible for local power exchange and optimization, and only shares the net amount of power supply to the central controller. The central controller is responsible for resolving control inconsistencies and economic dispatch across microgrids, ensuring to maintain the global optimal operation of power network. The hybrid architecture offers a two level control, where local controller connects the customers to the central control that provides single level privacy protection for customers. Although such a structure offers flexibility and minimal operating costs, it is heavily reliant on a dependable communication network to function [20].

4. **Nested architecture**: This is a hierarchical structure that contains multiple control levels. Local agents are in charge of power sharing between generators, voltage management, and frequency regulation in the primary control layer. A control agent is expected to be installed on all active equipment, such as solar panels, energy storage systems, and loads. Droop controllers, which function independently on each unit, are frequently deployed to achieve primary level control objectives. After the primary control has completed, the secondary control layer is responsible for dispatching assets to restore frequency and voltage to nominal values. The tertiary frequency control layer dispatches units and may modify unit commitment to restore reserves depleted during primary and secondary frequency control operations. This type of three-layer architecture distributes control to a large number of agents throughout the system. In most circumstances, the failure of a single agent or a breakdown in communication between agents will not result in the system collapse [21].

2.1 Communication Requirement for Modern Power System Architectures

Smart grid is the future concept for modern power system to deliver electric power optimally by incorporating DERs. The core components of smart grid are embedded controllers, communication networks, power electronics-based devices, advanced control architectures, and sensing and measurement devices [22]. Traditional power systems are also being transformed into CPS with real-time sensing, dynamic control, and bidirectional information exchange between sensors and controllers. The integration of DERs into the modern power system poses numerous challenges in the communications networks. In order to incorporate more renewable and alternative energy sources, the communication infrastructure must be able to handle a rising volume of data traffic or service requests while also monitoring and controlling all of these nodes in real time. Modern power systems with DERs are utilizing four communication networks, namely peer-to-peer (P2P), mesh, aggregated, and nested architectures, as shown in Fig. 3 [15].

1. **P2P network**: As shown in Fig. 3a, in a P2P network each node directly communication with a central controller without having a local microgrid controller, where node represent DERs with communication capabilities. Central controller is responsible for information collection and sharing with energy management system. This structure offers the advantages of fast control, easy integration, and low operation cost. On the other hand it makes the system vulnerable to single point of failure to collapse the whole system in case central controller suffers any issues [23].
2. **Aggregated structure**: In this structure, local microgrid controllers are employed that connect DERs with the central controller, as shown in Fig. 3b. This structure allows to share the communication burden of central controller with the help

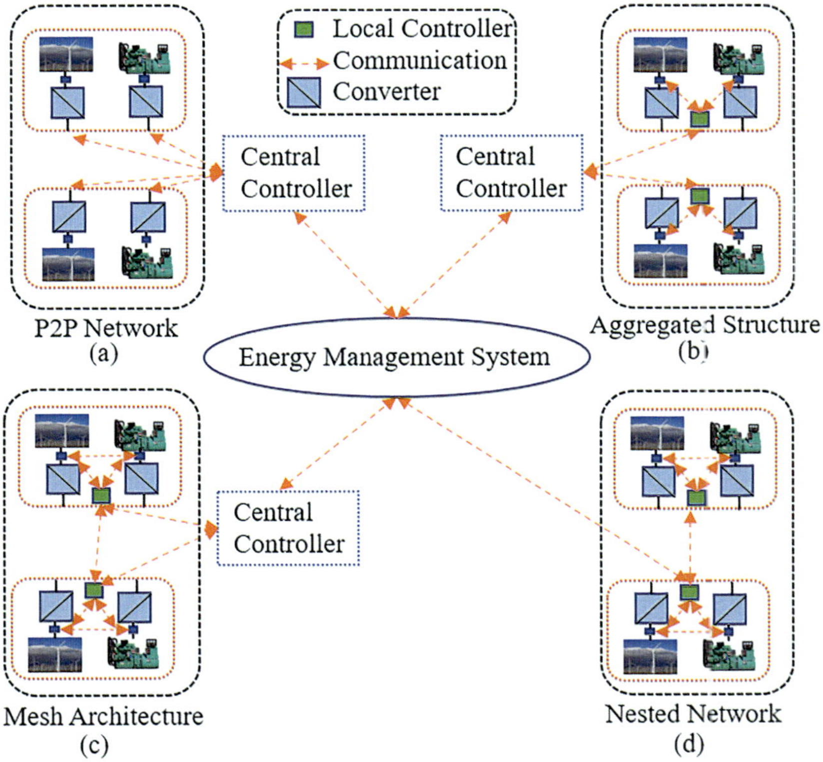

Fig. 3 Various types of communication architectures for microgrids are shown. **a** P2P network. **b** Aggregated structure. **c** Mesh structure. **d** Nested network

of local controllers and is suitable for large scale microgrids; although it is still vulnerable to single point of failure [24].

3. **Mesh structure**: As shown in Fig. 3c, in this structure, various DERs and local microgrid controllers are able to communicate with each other for information sharing. This offers redundancy in the communication that makes the power system more reliable and flexible. These redundant communication links help to tackle the single point of failure problem but increase the operation cost of the power system [25].
4. **Nested network**: In this network, the communication occurs only among adjacent local controllers, as shown in Fig. 3d. This structure efficiently solves the issue of single point of failure with decreased operation cost of the system [15].

2.2 *Communication Protocols*

The concepts and standards that the entities in a communication network must follow to achieve a secure and trustworthy information flow are referred to as communication protocols. Since microgrid operations need timely control actions, the IEC 61850 and IEEE 1646 standards give specifications for these requirements [26, 27]. There are two main types of communication technologies being used in the modern power systems;

1. **Wired communications**: This type of communications have traditionally been used to send data through the electrical grid because they perform better than wireless technologies in terms of robustness, reliability, security, and bandwidth. Wired technology, on the other hand, has a greater deployment cost and relevant network expansion strategies get more complicated. Serial connection RS-232/422/485 for supervisory control and data acquisition (SCADA) systems, Ethernet (IEEE 802.3 technology), bus-based technologies (e.g., ModBus, ProfiBus), and Power-Line Communication are all prominent wired technologies used in microgrids. With the exception of Ethernet, all technologies are often unable to deliver robust decentralized communications.
2. **Wireless communications**: Wireless technologies are improving their security capabilities, and because of the lower installation costs, they could be an appealing choice for distributed microgrid communication networks. Different wireless network technologies can be identified based on network coverage: WPAN (Wireless Personal Area Network), WLAN (Wireless Local Area Network), WMAN (Wireless Metropolitan Area Networks), and WWAN (Wireless Wide Area Network) [23].

Modern power systems are typical CPS that depend on communication networks and information sharing for stable and economic operation. On the other hand, these communication networks are vulnerable to cyber-attacks that may degrade the performance of the power system. The general impact of such attacks may result in leakage of privacy data, destabilization of power system, imbalance in power sharing, frequency and voltage deregulation, and compromised information sharing. The next section will cover the details about various cyber-physical attacks targeting the modern power system.

3 Cyber Vulnerabilities of Modern Power System

Modern power systems and associated communication networks are more vulnerable to cyber threats as a result of their extensive interconnected topology. A complete security architecture for modern power systems, covering all critical aspects required to assess and identify cyber-attacks, needs a large investigation that may incur huge financial burden. Various advanced technologies are being integrated into power systems, and the lack of comprehensive security standards could create a vulnerability for attackers to exploit system weaknesses [28]. Smart grid technologies have the potential to improve the scalability of complex power systems, which has a direct impact on the size of data exchange in terms of greater communication and computational requirements. This increased complexity in the architecture of modern power systems necessitates effective data and control management, and if not addressed properly, it may render the system more vulnerable to cyber-attacks. Interoperability between various modules of modern power systems mandates the use of communication structures and multiple information exchange protocols, which might leave the system vulnerable to cyber threats owing to missing data [13].

An overview of several cyber-attacks against modern power systems and relevant preventive techniques is presented as under.

3.1 Man-in-the-Middle Attack

In man-in-the-middle (MITM) attack, the adversary replays and possibly modifies communication between the sender and the receiver. In essence, the attacker creates an unauthorized communication link between two existing nodes and begins transmitting malicious data as shown in Fig. 4. While remaining undetected, the attacker can impersonate a legitimate participant, intercept and change message packets sent between two communication nodes, and insert new message packets [29]. The use of phasor measuring units (PMUs) and their measured synchrophasors in the dynamic monitoring of transient processes in power systems is extremely beneficial. The most prevalent protocol for transmitting data from PMUs is the User Datagram Protocol, which might raise the risk of MITM cyber-attacks if a cyber secured framework such as the Secure Sockets Layer is not employed. Wide Area Networks are also utilized to communicate between substations and control centers. If public communication networks are used for this reason, the network becomes exposed to cyber-attacks, and private communication networks might be accessed by adversaries through hardware access. Man-in-the-middle attacks frequently distort critical information sharing among nodes, such as control signals, sensor readings, and optimal dispatch rules, among several other factors. Such attacks may also target critical components of the power system to collect and disrupt control center personnel communications, which might be used to launch subsequent attacks [30]. The rising number of cyber-attacks on the power system needs the development of novel technologies to assure the system's resilience. The modern power system's two-way power flow controllability and transactive energy capabilities are largely reliant on a huge number of bidirectional power electronic converters, which must be flexible, rapid, and robust in order to sustain the grid during both normal and extreme operations. The implementation of improved security protocols in smart grids communication networks can significantly reduce the risk of MITM attacks [31]. This includes role-based access control in communication network, application of firewalls, and segmenting operational technology (OT) and information technology (IT) networks. The regular updating of software and firmware, securing the network with the application

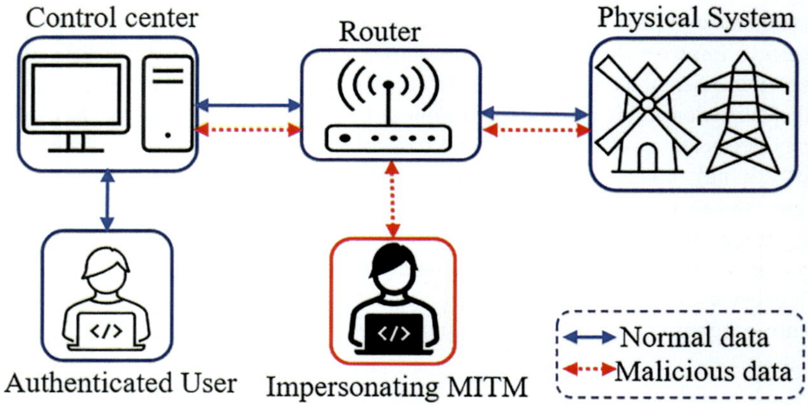

Fig. 4 Illustration of MITM attack is shown

of signature-based-intrusion detection, usage of strong passwords, and blocking any unused ports may help to eliminate the risk of unauthorized access to communication networks.

3.2 Denial-of-Service-Attack

The purpose of denial-of-service (DoS) attack is to block the access to a network service to its authorized user as shown in Fig. 5. This is achieved by obstructing communication between control centers and peripheral devices like sensors and actuators by jamming communication connections with spurious data and preventing normal data transfer. It is not necessary to have knowledge of the power system configuration or the capacity to modify control messages and measurements to launch such attacks. Operators were unable to deliver control commands to the actuators during the DoS attack on the Ukrainian power grid [32]. Malware can be installed in routers at substations to gain unauthorized access to PMU communications in wide area networks, making them vulnerable to DoS attacks. A successful attack renders a critical resource inaccessible to authorized users, compromising the power system's reliability, as all communication channels must be available for timely dispatch of control commands in the power system. An attacker may communicate with a compromised meter in the network after installing malware, posing a cyber threat to advance metering infrastructures (AMI). In large networks, IP spoofing can be used to disguise infected meters, finding it challenging to locate the source of the attack. It is possible to manipulate a large number of agents to transmit an excessive amount of communication packets to the system user. As a result of the flooded traffic, the authorized user will drop some of the authorized packets [13]. In a distributed control-based smart grid having n DERs, the DoS attack can be modeled as:

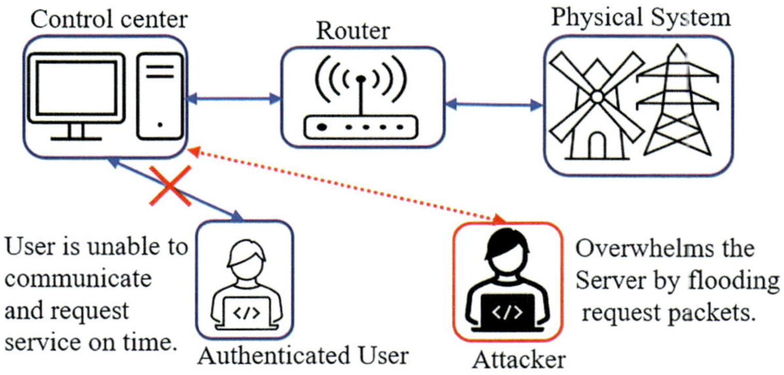

Fig. 5 Illustration of DoS attack is shown

$$[\eta_i]_{n\times 1} = \gamma_i(\tau)[\chi_i]_{n\times 1} + \gamma_j(\tau)[\chi_j]_{n\times 1}, \quad (1)$$

where η_i represents one of the compromised target nodes and χ_i, χ_j represent the neighboring nodes of the compromised node η_i. $\gamma_i(\tau)$, $\gamma_j(\tau)$ represent gain factor, and $(t_1 < \tau < t_2)$ is the time interval for the occurrence of a malicious cyber-attack, and $ij \in \mathbb{R}$ representing one of the DERs in the smart grid. The value of the gain factor is (1) if the communication link is not jammed otherwise its value is (0).

With the inclusion of extensive communication infrastructures, the DoS vulnerabilities are becoming a serious concern making the modern power systems a prime target for cyber-attacks. Many security vulnerabilities, including DoS attacks, have their roots in the system software's inherently risky behavior, especially at the operating system kernel level. Implementation of data filtering techniques, intrusion detection and prevention system, and cryptography authentications have huge potential in timely detection and mitigation of DoS in smart grids [33]. Modern networked control systems, which heavily rely on IoT based communication infrastructure, are linked from the downstream microgrid level to the upstream distribution system level. IoT devices enable the DER's controllers to coordinate and integrate in a more flexible manner, as well as improve overall network management. Data transmissions in such control networks are often secured in two ways: first, using IT data encryption and certificate authentication, and second, through the use of resilient control systems [34–36]. Smart grids communication network should follow the standard security protocols such as Transport Layer Security implementations, message authentication code, and multi-factor authentication to terminate any attempt for unauthorized connection. To improve robustness against cyber threats, an IoT-based digital twin for cyber physical networked microgrids is presented in [37]. The cloud-based digital twin platform is used to provide centralized management for the networked microgrids. The suggested digital twin includes a digital replica for both the physical and cyber layers, as well as their hybrid interactions, to ensure the proper and secure operation of the networked microgrid, and to detect DoS attacks on the control system.

3.3 False Data Injection Attack

Modern power systems have evolved into complex cyber-physical systems as sensor, computer, and communication networks have advanced rapidly. An attacker can either obtain economic profit or disrupt the power system by initiating a proper false data injection (FDI) attack. In a successful FDI attack, the attacker tampers with power system sensor measurements and injects false data into estimation of state vector such as voltage angles and magnitudes as shown in Fig. 6. As a result, critical power system activities such as economic dispatch and contingency analysis, which rely on state estimations, could be jeopardized [38, 39]. By hacking a subset of meters and sending modified measurements without changing the value completely, an attacker can compromise the state estimator's integrity. By gathering online data, market data, and power flow measurements, FDI attacks can be performed even if the attacker has no prior knowledge of the power system configuration [40]. The feedback signal of a distributed smart grid controller under FDI attack can be modeled as:

$$h(u_i(t)) = u_i(t) + \phi_i(t), \tag{2}$$

where $h(u_i(t))$ is the feedback signal after false data $\phi_i(t)$ is injected by attacker in the ith normal feedback signal of the controller $u_i(t)$. Several cases of FDI attacks can be modeled by updating $\phi_i(t)$, such as periodic, non-periodic, and simultaneous FDI attacks [41].

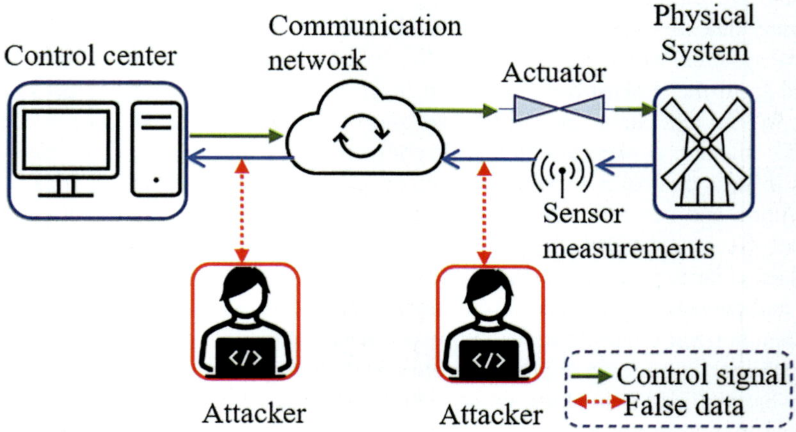

Fig. 6 Illustration of FDI attack is shown

Such FDI attacks also have the potential to destabilize or even completely disrupt the distributed control systems that support the modern power networks. Spoofing communication lines and changing signal levels over a distributed communication network are used to launch FDI attacks that tend to alter the output of AC microgrids, resulting in equipment damage. The range of malicious data injected is selected in the range of nominal system operating state to keep the FDI attack undetected. Various types of FDI attacks can be launched against distributed controller feedback signals by injecting a variety of false data into the communication network [42]. In a multi-agent system, cooperative control ensures that the agents in the network share information in order to achieve a shared goal. Furthermore, due to cooperative control protocols, the distorting effects of cyber-physical attacks in a single DER may spread to multiple DERs. Since timely identification of these attacks can prevent this undesirable microgrid operation, signal temporal logic (STL)-based detection has been presented in [43]. To detect and quantify the impact of cyber-attacks on DC microgrids, STL monitors output voltages and currents against defined operational limits. Model-based attack detection and mitigation strategies require complete mathematical modeling of the system, which can add complexity and computing burden; thus, data-driven intelligent schemes are being employed in microgrids. As a result, an intelligent mitigation strategy based on artificial neural networks is presented as a resilient secondary voltage control layer in [44]. This intelligent secondary voltage control is distributed in mechanism and implemented for AC microgrids. Under various cyber-attack scenarios targeting the controller's feedback signal, the designed control layer was able to maintain normal system operation.

3.4 Network Failures

A smart grid is a typical cyber-physical system and comprises a cyber layer and a physical layer. Physical layer is composed of physical entities to include DERs, power electronics devices, embedded controllers, and physical interconnections to form a power network. Cyber layer includes the communication devices and relevant communication protocols to form a communication network. The power network and the communication networks collectively form a smart grid. Network failures occur due to faults in any of the following in the power network or the communication network:

1. Physical devices (such as wind turbines, photovoltaics, power electronics devices)
2. Physical interconnections (such as transmission lines)
3. Software bugs in embedded controllers
4. Software bugs in communication devices or communication protocols.

A failure in the communication network will inherently induce failures in the power network, and such failures will propagate throughout the power network in a cascade. Cascade failures can result in the collapse of one (or both) of the networks [45]. When the power grid is under stress, failures in the communication network can have a significant impact on the total blackout probability. In smart grids, communication networks play a critical role in electricity delivery reliability. When the electrical grid is stressed due to network failures, further interconnected events may propagate within and across these networks, resulting in large-scale cascading failures and blackouts [46].

An example of the power network failure is the Texas power crisis in 2021 due to freezing cold during the winter. More than 4.5 million homes and businesses were left without electric power due to this severe weather. The state of Texas is dependent on various electricity sources, namely, natural gas, wind, coal, nuclear, photovoltaics, in the order of their share in the electricity produced. The freezing cold conditions in the region prevented all those electricity sources to produce enough electricity to meet the demand. As per Electric Reliability Council of Texas (ERCOT), partial grid breakdowns were initiated to prevent the complete blackout of the power infrastructure in Texas.

As experienced in the Texas power crisis, the power systems are becoming increasingly interconnected, and this trend is projected to continue in the smart grid of the future. Independent operators manage interconnected power systems such that each operator utilizes state estimation to estimate the state of the corresponding interconnected system. The Western Interconnect in the United States and the European Network of Transmission System Operators for Electricity in Europe are the two examples of interconnected power systems. Some cascade failures (e.g., the US North-East blackout in 2003 and Texas power crisis in 2021) have revealed that the safety of an interconnected electrical system is dependent on the safety of its constituent parts.

The destabilizing effects of FDI attacks on a microgrid having multiple DERs are presented in the next section.

3.5 Case Study

The architecture of inverter-based microgrid under study is shown in Fig. 7a. This test microgrid has a physical layer consisting of DERs and loads. The DERs are interconnected through RL lines and each DER is modeled by a voltage source inverter that provides three phase AC output through the LC filter. DERs are also connected to an AC bus through circuit breakers (CB). The primary controller is implemented locally at DERs through conventional droop control technique. The distributed cooperative control is implemented at the secondary level to mitigate the microgrid voltage and frequency deviations from their nominal values which are caused by the primary control. The cyber communication layer is the medium for information sharing between the components of the physical layer through the sparse communication network. The parameters of the test microgrid system are given in Table 1. Further details related to design of distributed control-based microgrids can be found in [47]. After setting up the model on real time digital simulator OPAL-RT, the FDI attacks are introduced in the test microgrid system. FDI attacks are

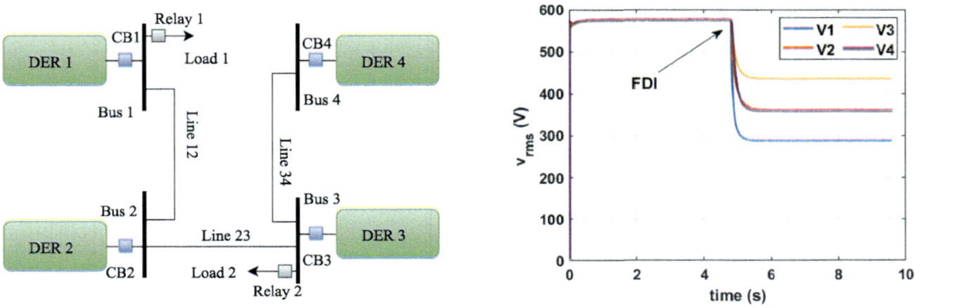

(a) 4 DERs-based microgrid under study

(b) RMS output voltage of all 4 DERs

Fig. 7 The architecture of test case study microgrid and destabilizing effect of FDI attack are shown

Table 1 Parameters of test case study microgrid are given

Microgrid			
Name	Parameter	Name	Parameter
L_{12}	(0.23+j318μ) Ω	V_{ref}	300 V
L_{23}	(0.35+j1847μ) Ω	L_{filter}	1.35 mh
L_{34}	(0.23+j318μ) Ω	C_{filter}	50 μF

initiated by spoofing the communication links and altering the signal values over the distributed communication network. Based on (2), the voltage feedback information is targeted for FDI and to keep the FDI attack undetectable, the range of malicious data injected is selected in the range of nominal system operating state. The destabilizing effect of FDI attack in the voltage sensor of DER in terms of output rms voltage are shown in Fig. 7b, where it can be seen that under such an attack, the output voltage of the compromised DERs deviates from the reference value. When system is experiencing an attack then it will not converge to a normal operating limit. The false data is injected at time $(t + \delta t)$, until time t the output of all the DER units is following the expected reference value. But right after the injection of false data, the output deviates from normal operating point. To further investigate the impact of FDI attacks, two different case studies are performed by considering various $\phi_i(t)$ as follows:

1. **Non-periodic attack:** A non-periodic attack is initiated by adding a constant multiple α of the desired signal $u_i(t)$ in $h(u_i(t))$ at certain time τ during the operation of the system, given as:

$$h(u_i(t)) = \begin{cases} u_i(t), & if \ \ t < \tau, \\ u_i(t) + \alpha * u_i(t), & if \ \ t > \tau. \end{cases} \tag{3}$$

Fig. 8 The destabilizing impact of FDI attacks in terms of the output voltage at *Load* 1 of the test microgrid is shown

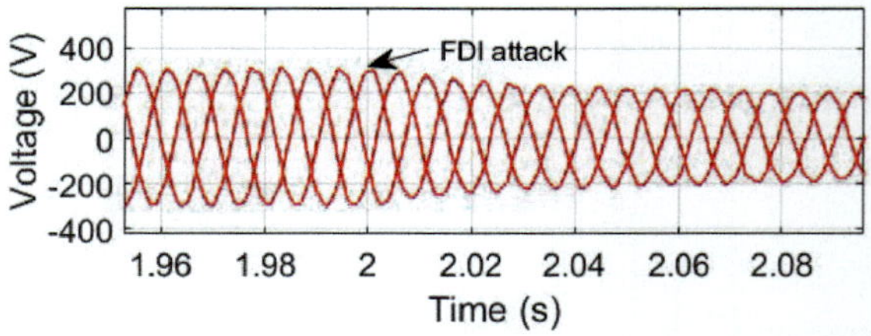

(a) Non-periodic FDI attack

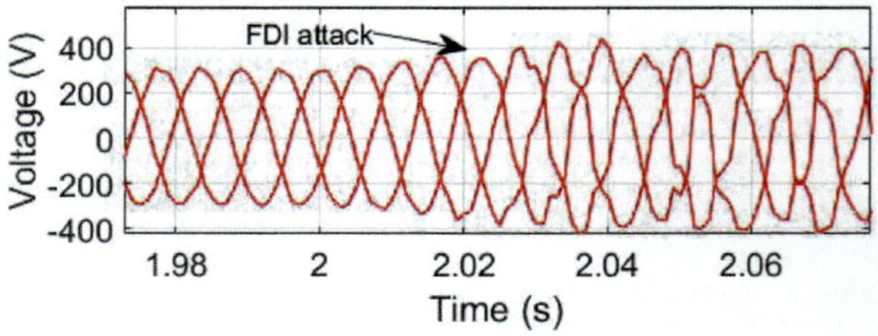

(b) Periodic FDI attack

2. **Periodic attack:** Periodic FDI attack is time varying in nature where a sinusoidal signal having time period (ωt) and amplitude β is injected to normal signal u_i, given by:

$$\phi_i(t) = \begin{cases} 0, & if \ \ t < \tau, \\ \beta sin(\omega t) * u_i(t), & if \ \ t > \tau. \end{cases} \tag{4}$$

The secondary voltage control of DER1 is selected, and a non-periodic FDI attack is initiated at the communication link sharing the voltage measurements of DER1. This cyber-attack is based on (3), where the attacker starts injecting the false data into the voltage measurements of DG1 with $\alpha = 0.5$, starting at $t = 2$ s. The microgrid operates in normal mode for $t < 2$ s, and after that, false data is injected to compromise the performance of secondary voltage control. The impact of non-periodic FDI attack is shown in Fig. 8a and it is evident that after the FDI attack, the output voltage at *Load* 1 deviates significantly from the normal operating voltage prior to the injection of false data. After that, the periodic FDI attack based on (4) is introduced on the voltage communication link of DER1. A periodic sinusoidal signal having amplitude $\beta = 0.5$ and frequency $\omega = 60$ Hz is added to the voltage signal coming from DER1 at $t = 2$ s. As shown in Fig. 8b, following the FDI attack, the output voltage at *Load* 1 gets destabilize with significant deviations from the desired normal voltage range. This may destabilize the system and DERs may run out of synchronized operation to meet the cooperative control objectives resulting in damages to the equipment and interruption in power supply to the loads. Further details regarding the implementation of FDI attacks and mathematical formulation can be found in [41, 44].

4 Conclusion

Modern power systems face cyber vulnerabilities due to interconnected architecture, extensive cyber communication layers, and distributed control systems. Cooperatively controlled smart microgrids rely on communication networks, IoT, and wide-area sensor networks making them vulnerable to cyber-physical attacks. Furthermore, due to cooperative control protocols, the distorting effects of cyber-physical attacks in a single DER may spread to multiple DERs. Since cyber-attacks can pose a serious threat to the modern power system's economic and physical security, it is critical for the system operator to devise effective defensive measures to safeguard the network. It is also challenging to completely account for the new dynamics and complexity of smart power systems in the domain of cyber-physical security. The majority of modern attack detection and mitigation solutions rely on information exchange between the system's cyber and physical layers. This dependency also creates a potential risk for successful cyber-attacks on power system communication networks. In real-world events, such as the 2003 US Northeast blackout, coordinated attacks resulting in sequential power loss led to blackouts. Both cyber and physical layers must work together in a modernized power system. The omnipresent threat to the smart grid's wide security vulnerabilities necessitates a detailed study and classification of attacks. This chapter provides architecture of a power system's operation as well as the vulnerabilities that an attacker could exploit. The existing attack strategies against various components are classified using the cyber-physical paradigm. The cyber vulnerabilities of modern power system are presented in this chapter such as MITM, DoS, network failures, and FDI attacks. An overview of current state-of-the-art techniques for detecting and mitigating cyber-attacks is offered, ranging from model-based methods to intelligent data-driven solutions. More physical devices are being incorporated into traditional power networks as smart grid technologies evolve, requiring the establishment of a vast communication network, posing a variety of challenges and opportunity for potential cyber-attacks. A case study depicting the destabilizing effects of cyber-attacks on a multi DER-based AC microgrid is also presented. The application of improved security protocols in smart grids communication networks, implementing multi-factor authentication, planning and practicing for potential breaches, and deploying resilient control strategies can significantly reduce the risk of cyber-attacks in modern power systems.

References

1. R. Liu, C. Vellaithurai, S.S. Biswas, T.T. Gamage, A.K. Srivastava, Analyzing the cyber-physical impact of cyber events on the power grid. IEEE Trans. Smart Grid **6**(5), 2444–2453 (2015). https://doi.org/10.1109/TSG.2015.2432013
2. Z. El Mrabet, N. Kaabouch, H. El Ghazi, H. El Ghazi, Cyber-security in smart grid: survey and challenges. Comput. Electr. Eng. **67**, 469–482 (2018)
3. I. Zografopoulos, J. Ospina, X. Liu, C. Konstantinou, Cyber-physical energy systems security: threat modeling, risk assessment, resources, metrics, and case studies. IEEE Access **9**, 29 775–29 818 (2021). https://doi.org/10.1109/ACCESS.2021.3058403

4. S. McLaughlin, C. Konstantinou, X. Wang et al., The cybersecurity landscape in industrial control systems. Proc. IEEE **104**(5), 1039–1057 (2016). https://doi.org/10.1109/JPROC.2015.2512235
5. Y. Xu, A review of cyber security risks of power systems: from static to dynamic false data attacks. Protect. Control Modern Power Syst. **5**(1), 1–12 (2020)
6. Y. Xiang, L. Wang, N. Liu, Coordinated attacks on electric power systems in a cyber-physical environment. Electric Power Syst. Res. **149**, 156–168 (2017)
7. F. Liberati, E. Garone, A. Di Giorgio, Review of cyber-physical attacks in smart grids: a system-theoretic perspective. Electronics **10**(10), 1153 (2021)
8. S. Mehrdad, S. Mousavian, G. Madraki, Y. Dvorkin, Cyber-physical resilience of electrical power systems against malicious attacks: a review. Current Sustain./Renew. Energy Rep. **5**(1), 14–22 (2018)
9. S. Paul, F. Ding, K. Utkarsh, W. Liu, M.J. O'Malley, J. Barnett, On vulnerability and resilience of cyber-physical power systems: a review. IEEE Syst. J. 1–12 (2021). https://doi.org/10.1109/JSYST.2021.3123904
10. H. Haes Alhelou, M.E. Hamedani-Golshan, T.C. Njenda, P. Siano, A survey on power system blackout and cascading events: research motivations and challenges. Energies **12**(4), 682 (2019)
11. Z. Mohammadi, S.J. Pinto, G. Panda, S. Thokchom, A survey of cyber security in smart microgrid, in *Sustainable Energy and Technological Advancements* (Springer, 2022), pp. 687–698
12. H. He, J. Yan, Cyber-physical attacks and defences in the smart grid: a survey. IET Cyber-Phys. Syst. Theory Appl. **1**(1), 13–27 (2016)
13. M.M. Pour, A. Anzalchi, A. Sarwat, A review on cyber security issues and mitigation methods in smart grid systems. SoutheastCon **2017**, 1–4 (2017). https://doi.org/10.1109/SECON.2017.7925278
14. S. Mavale, J. Katade, N. Dunbray, S. Nimje, B. Patil, Review of cyber-attacks on smart grid system, in *Proceedings of Third International Conference on Communication, Computing and Electronics Systems* (Springer, 2022), pp. 639–653
15. B. Zhou, J. Zou, C.Y. Chung et al., Multi-microgrid energy management systems: architecture, communication, and scheduling strategies. J. Modern Power Syst. Clean Energy **9**(3), 463–476 (2021)
16. Z. Li, M. Shahidehpour, F. Aminifar, A. Alabdulwahab, Y. Al-Turki, Networked microgrids for enhancing the power system resilience. Proc. IEEE **105**(7), 1289–1310 (2017)
17. Y. Han, K. Zhang, H. Li, E.A.A. Coelho, J.M. Guerrero, Mas-based distributed coordinated control and optimization in microgrid and microgrid clusters: a comprehensive overview. IEEE Trans. Power Electron. **33**(8), 6488–6508 (2017)
18. A. Ouammi, H. Dagdougui, L. Dessaint, R. Sacile, Coordinated model predictive-based power flows control in a cooperative network of smart microgrids. IEEE Trans. Smart Grid **6**(5), 2233–2244 (2015)
19. H. Gao, J. Liu, L. Wang, Z. Wei, Decentralized energy management for networked microgrids in future distribution systems. IEEE Trans. Power Syst. **33**(4), 3599–3610 (2017)
20. B. Zhao, X. Wang, D. Lin et al., Energy management of multiple microgrids based on a system of systems architecture. IEEE Trans. Power Syst. **33**(6), 6410–6421 (2018)
21. K.P. Schneider, S. Laval, J. Hansen et al., A distributed power system control architecture for improved distribution system resiliency. IEEE Access **7**, 9957–9970 (2019)
22. M. Yazdanian, A. Mehrizi-Sani, Distributed control techniques in microgrids. IEEE Trans. Smart Grid **5**(6), 2901–2909 (2014)
23. S. Marzal, R. Salas, R. González-Medina, G. Garcerá, E. Figueres, Current challenges and future trends in the field of communication architectures for microgrids. Renew. Sustain. Energy Rev. **82**, 3610–3622 (2018)
24. M. Saleh, Y. Esa, A.A. Mohamed, Communication-based control for dc microgrids. IEEE Trans. Smart Grid **10**(2), 2180–2195 (2018)

25. J. Yu, M. Ni, Y. Jiao, X. Wang, Plug-in and plug-out dispatch optimization in microgrid clusters based on flexible communication. J. Modern Power Syst. Clean Energy **5**(4), 663–670 (2017)
26. IEEE, IEEE standard communication delivery time performance requirements for electric power substation automation. IEEE Std **1646 2004**, 1–24 (2005)
27. D. Baigent, M. Adamiak, R. Mackiewicz, G. Sisco, IEC 61850 communication networks and systems in substations: an overview for users, SISCO Systems (2004)
28. N.K. Singh, V. Mahajan, Analysis and evaluation of cyber-attack impact on critical power system infrastructure. Smart Sci. **9**(1), 1–13 (2021)
29. P. Wlazlo, A. Sahu, Z. Mao et al., Man-in-the-middle attacks and defence in a power system cyber-physical testbed. IET Cyber-Phys. Syst.: Theory Appl. **6**(3), 164–177 (2021)
30. F. Li, X. Yan, Y. Xie, Z. Sang, X. Yuan, A review of cyber-attack methods in cyber-physical power system, in *IEEE 8th International Conference on Advanced Power System Automation and Protection (APAP)*, vol. 2019 (IEEE, 2019), pp. 1335–1339
31. A.K. Das, S. Zeadally, Data security in the smart grid environment, in *Pathways to a Smarter Power System* (Elsevier, 2019), pp. 371–395
32. H. Zhang, B. Liu, H. Wu, Smart grid cyber-physical attack and defense: a review. IEEE Access **9**, 29 641–29 659 (2021)
33. A. Huseinović, S. Mrdović, K. Bicakci, S. Uludag, A survey of denial-of-service attacks and solutions in the smart grid. IEEE Access **8**, 177 447–177 470 (2020). https://doi.org/10.1109/ACCESS.2020.3026923
34. S.N. Islam, Z. Baig, S. Zeadally, Physical layer security for the smart grid: vulnerabilities, threats, and countermeasures. IEEE Trans. Industr. Inf. **15**(12), 6522–6530 (2019). https://doi.org/10.1109/TII.2019.2931436
35. M.N. Kurt, Y. Ylmaz, X. Wang, Real-time detection of hybrid and stealthy cyber-attacks in smart grid. IEEE Trans. Inf. Forens. Secur. **14**(2), 498–513 (2019). https://doi.org/10.1109/TIFS.2018.2854745
36. M.Z. Gunduz, R. Das, Cyber-security on smart grid: threats and potential solutions. Comput. Netw. **169**, 107 094 (2020)
37. A. Saad, S. Faddel, T. Youssef, O.A. Mohammed, On the implementation of iot-based digital twin for networked microgrids resiliency against cyber attacks. IEEE Trans. Smart Grid **11**(6), 5138–5150 (2020)
38. H.T. Reda, A. Anwar, A. Mahmood, Comprehensive survey and taxonomies of false data injection attacks in smart grids: attack models, targets, and impacts. Renew. Sustain. Energy Rev. **163**, 112 423 (2022)
39. T.O. Olowu, S. Dharmasena, A. Hernandez, A. Sarwat, Impact analysis of cyber attacks on smart grid: a review and case study, in *New Research Directions in Solar Energy Technologies* (2021), pp. 31–51
40. G. Liang, J. Zhao, F. Luo, S.R. Weller, Z.Y. Dong, A review of false data injection attacks against modern power systems. IEEE Trans. Smart Grid **8**(4), 1630–1638 (2016)
41. A.A. Khan, O.A. Beg, M. Alamaniotis, S. Ahmed, Intelligent anomaly identification in cyber-physical inverter-based systems. Electric Power Syst. Res. **193**, 107 024 (2021)
42. C. Peng, H. Sun, M. Yang, Y.-L. Wang, A survey on security communication and control for smart grids under malicious cyber attacks. IEEE Trans. Syst. Man Cybern. Syst. **49**(8), 1554–1569 (2019). https://doi.org/10.1109/TSMC.2018.2884952
43. O.A. Beg, L.V. Nguyen, T.T. Johnson, A. Davoudi, Signal temporal logic-based attack detection in dc microgrids. IEEE Trans. Smart Grid **10**(4), 3585–3595 (2018)
44. A.A. Khan, S. Ahmed, O.A. Beg, Intelligent anomaly mitigation in cyber-physical inverter-based systems, in *IEEE Energy Conversion Congress and Exposition (ECCE)*, vol. 2021 (IEEE, 2021), pp. 1301–1306
45. S. Ruj, A. Pal, Analyzing cascading failures in smart grids under random and targeted attacks, in *2014 IEEE 28th International Conference on Advanced Information Networking and Applications, 2014* (2014), pp. 226–233. https://doi.org/10.1109/AINA.2014.32

46. R.A. Shuvro, Z. Wang, P. Das, M.R. Naeini, M.M. Hayat, Modeling impact of communication network failures on power grid reliability, in *North American Power Symposium (NAPS)*, vol. 2017 (2017), pp. 1–6. https://doi.org/10.1109/NAPS.2017.8107224
47. A. Bidram, F.L. Lewis, A. Davoudi, Distributed control systems for small-scale power networks: using multiagent cooperative control theory. IEEE Control Syst. Mag. **34**(6), 56–77 (2014). https://doi.org/10.1109/MCS.2014.2350571

Cyber-Physical Security in Smart Power Systems from a Resilience Perspective: Concepts and Possible Solutions

Mohammad Ghiasi, Zhanle Wang, Taher Niknam, Moslem Dehghani, and Hamid Reza Ansari

Abstract Vital energy infrastructures in today's evolving society are becoming more complex by distributed networks of cyber-physical systems (CPS). Although CPS provides a great deal of flexibility in the exploitation of critical infrastructure, they can also create other security-based threats which should be suitably addressed in the design and development stage. Considering this issue, power systems should be properly resistant against adverse events so that their operation is not easily impacted by severe conditions. At the same time, energy systems should be flexible enough to adapt to severe disruptions without losing or breaking down completely, and they should be able to recover in the least time after solving the problem. This critical feature of power systems is generally called resilience. This chapter provides an overview of various resilience methods from the perspective of CPS in smart power systems and microgrids. In order to provide a detailed study, in this chapter, cyber-physical structures, threats and security issues in smart power systems are introduced. Vulnerabilities of control systems and important methods for detecting and countering cyber-attacks with defenses mechanism in smart power systems are presented. Possible solutions to improve cyber-physical resilience of smart power systems and some important optimization techniques used in smart power systems and microgrids are provided, which can be utilized and developed by researchers in this field.

Keywords Smart power systems · Microgrid · Resilience · Cyber-physical systems · Security

---

M. Ghiasi (✉) · T. Niknam · M. Dehghani · H. R. Ansari
Department of Electrical and Electronic Engineering, Shiraz University of Technology, Shiraz, Iran
e-mail: m.ghiasi1@gmail.com

M. Ghiasi · Z. Wang
Electronic Systems Engineering, University of Regina, Regina, SK, Canada

H. Haes Alhelou et al. (eds.), *Power Systems Cybersecurity*, Power Systems,
https://doi.org/10.1007/978-3-031-20360-2_3

1 Introduction and Background

The first power grid was built in the late nineteenth century, and the electrical system had not changed much in a long time; but today, it has entered a new era. Over time, the number of power grids has increased to the point where today, there are thousands of power grids around the world, and these grids provide millions of megawatts of power needed by consumers. Since the power grid has evolved over time, the use of digital and smart technologies to improve the efficiency of the power grid has increased worldwide. This new digital technology enables two-way communication and ensures a direct connection between production and consumption. The same link between digital technologies and the power grid has led to a new concept called the smart grid. In this section, we will discuss the smart grid concept and its various aspects [1, 2].

1.1 The Necessity of Smart Power System

A complete power grid consists of power generation units, transmission lines, distribution lines, transformers, substations, consumers, etc. In order to have optimal use of equipment and access to a network that can be managed online, as well as having accurate forecasts of generation and consumption in the future, we need to change and smarten the network and equipment. These changes will, of course, complicate the network cyber and physically. In addition, we should consider one of the main drivers of this change to the smart grid, which has been the reduction of greenhouse gases. Therefore, replacing fossil fuel-based production sources with new, low-carbon or carbon-free technologies plays a key role in smart grids [3, 4].

In the past, renewable energy (RE) generations, especially from wind turbines (WT) and solar and photovoltaic (PV) panels were not as easy to control as traditional generators, because these generators used uncontrollable wind and solar sources. However, as a result of advancement in power electronics and controllable switching applications such as inverters/convertors, renewable generations have entered a new era of technology. Distributed generators (DGs) are small-scale of generators which include renewable generators (WT and PV), and diesel generator units. They are usually connected to low voltage sides of distribution networks and may have less controllability. Since power consumption in the network may increase unpredictably, or in the event of a power outage in the main grid, DGs are used to keep the network stable [5]. This electricity can be generated even by consumers during peak network hours. Therefore, in order to control the supply and demand, we need a system to share the information of the network online to the power control center. This platform can be provided by the Internet and communication protocols. Such network should also use intelligent algorithms to optimally control supply and demand. Therefore, we need a smart power system.

1.2 Smart Grid Definition

Smart grid is a network that uses bidirectional communication system with digital information and online control framework to improve the reliability, security and efficiency of the power grid, as well as dynamic optimization of network operations and resources. This network should have a robust cyber-physical security which benefits from the deployment and integration of distributed resources and generation including renewables, development and integration of demand response, leading demand sources and energy efficiency sources. In smart grids, the power companies automatically respond to the area affected by the failure and take necessary actions, because the components of the smart grid have enough data which are provided to the network [6, 7].

The main difference between the current network and the smart network of the future is that production and demand are transcendent. The current power grid consists of very large substations connected to a high voltage transmission system. The output of most of these substations can be changed according to the needs of the whole system. The global network is constantly issuing guidelines to ensure a balance between production and consumption. Since the amount of production is known and the amount of demand can be predicted, the amount of demand determines which substations work and with what amount of output [8]. To better understand the issue, we need to know smart grid architecture model (SGAM) frameworks which presented in several studies. In general, the SGAM framework is given to be able to design intelligent network applications in a specific architecture and to present the processes and technology used in the system from a cyber-physical perspective. It can also enable the validation of smart grid applications and their support by standards. Reference [9] presented SGAM framework which includes interoperability layers (Business Layer, Function Layer, Information Layer, Communication Layer, and Component Layer), SGAM Domains (Bulk Generation, Transmission, Distribution, DER, and Customer Premises), and SGAM Zones (Process, Field, Station, Operation, Enterprise, and Market). Figure 1 displays a typical structure of SGAM framework (Fig. 1).

1.3 Benefits and Challenges of Smart Grids

Based on the references [2, 10–12], some important benefits of a smart grid are as follows.

- Optimal transmission and distribution of electrical power
- Quick power recovery after system failure and sudden problems in lines and feeders
- Low cost of operation and maintenance
- Lower electricity tariffs due to reduced courier demand
- Facilitating the integration of renewables into the network

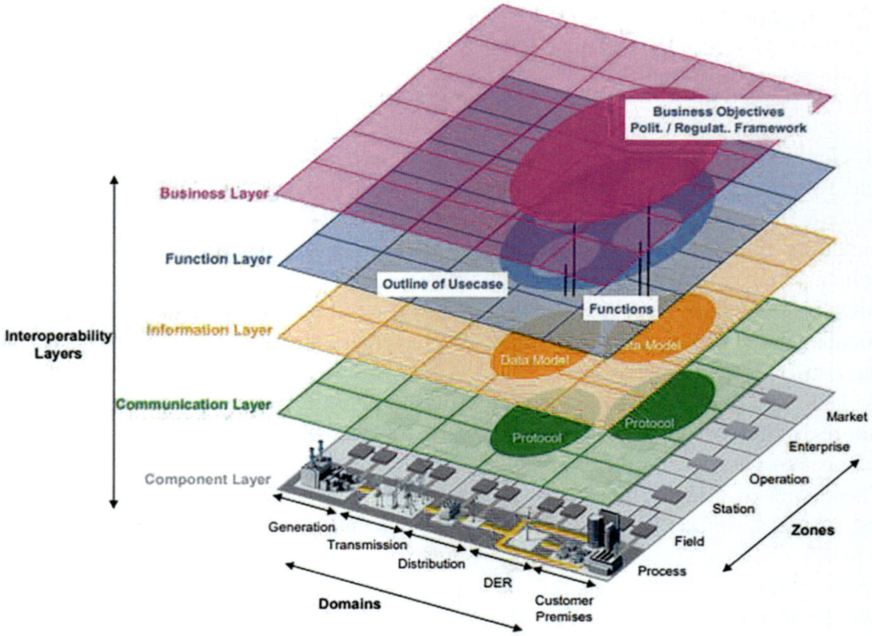

Fig. 1 A typical structure of SGAM framework [9]

- Increasing the versatility of transmission lines
- Network protection management in case of emergency
- Better demand response
- Improving power quality (PQ)
- Reducing greenhouse gas emission

The development of digital technology in smart grids ensures the reliability, efficiency, and access of consumers to production, which in turn leads to economic stability. However, some challenges should be considered. The important challenges of smart grids include the following:

- Privacy issues: The biggest concern in a smart grid is security issues. As we know, smart grids use smart meters that dynamically establish the connection between production and consumption. Some smart meters are easily hacked and may control the power source of a building or neighbors.
- Network fluctuations: Smart grid can intelligently manage generation and consumption because it uses smart control system, meters and sensors. But in the transmission section, power networks usually use automatic rely protection systems to manage switching operations which is not smart enough.
- Cyber-physical security issues: Having a developed and smart control system in the power network can considerably improve the stability of the whole system

against possible fluctuations. So far, in smart grids, attention has been paid to the sectors of power generation and energy consumption by consumers. If a large number of nodes are added to the network before the software intelligence for network control is developed, it leads to an oscillating smart grid. Intelligent control systems are required for a safe and secure power system operation. Due to the technological development based on microprocessors and computer-based methods in DC and AC microgrids, intelligent monitoring, control, and operation strategies, and their restructuring, smart power systems can be threatened by several cyber and physical attacks [13, 14].

Considering the importance of the microgrid concept in smart power systems, we also evaluate cyber and physical protection systems in microgrids. In general, a smart microgrid system can include batteries, super-capacitors, power electronics, fuel cells, PV systems, WT, loads, plug-in hybrid electric vehicles (PHEVs), smart sensors, and communication networks such as cable, fiber optic or wireless to send and receive data [2, 4, 15].

1.4 Cyber-Physical Security in Smart Power Systems

The goal of the new generation of smart grids is to provide high capacity, efficiency and reliability by combining physical-cyber technologies to existing electrical networks. However, due to the combination of information technology-based components and open port systems, physical and cyber risks in the smart grid will increase. Management and control of the electricity network and related equipment using the supervisory, control, and data acquisition (SCADA) network based on communication protocols and the use of fiber optics, wireless networks, or Internet platforms have entered a new phase [16, 17]. Therefore, the flexibility and resilience of the network in the design and operation phase should be given special attention to make the power system resilient. Methods for rapid detection of accidents or attacks and appropriate strategies to ensure the continuity of infrastructure services must be developed [18, 19].

With the increasing development of intelligent equipment and the widespread use of online devices connected to the Internet, the data exchanged on this platform has increased indescribably. These data, especially classified and private data, must be protected from cyber-attacks [20, 21]. Accordingly, various methods have recently been proposed by researchers for physical security of the power grids such as hardening equipment and creating multiple protective layers. Also, since any defect in the cyber system can cause damage to the physical system of the power network, experts have proposed solutions to identify and counter cyber-attacks to ensure the optimal and flexible design and management of energy infrastructure.

Unexpected events with low probability and high impact (HILP) that fundamentally affect the network have raised concerns about routine reliability and flexibility. As a critical infrastructure, the power systems are expected to be much more resilient to HILP accidents [22]. The ability of the power grid to withstand and be flexible against HILP events is called the resilience of power systems. This capability should provide sufficient energy to vital loads and speed up network recovery. In order to have a resilience power system, different parts of the network need to be changed and smartened to improve response to HILP-sensitive loads [23, 24].

Industrial control systems (ICS) are widely used in industrial and energy networks including electricity, water, natural gas and petrochemicals. ICS can effectively process complex data ranging from complex space station maneuvers to ground-level water monitoring. The main components of ICS include monitoring the SCADA, data collection and processing by remote terminal units (RTU) and programmable logic units (PLC). Using this system, professionals and users can change the status of ICS and their parameters. This model of system is increasingly susceptible to complex and targeted cyberattacks, which are usually exploited by exploiting vulnerabilities in devices and industrial control protocols. Many of the proposed network intrusion detection systems (IDS) focus only on identifying and describing cyber-attacks against the SCADA system, however, they cannot completely describe their true effect on energy systems. An IDS is a software or application or device that monitors a network traffic of a system for malicious activity or policy violations and responds to suspicious activity. Therefore, smart power systems with communication protocols and network-based equipment can be a good target for this type of cyber-attacks [25]. One of the parts of the power system that has received a lot of attention in the last decade for smartening is microgrids. So, the necessary strategies should be anticipated and considered in order to prevent threats and make this network resilient. Microgrids are basically classified into three different categories including DC, AC, and hybrid microgrids [26]. Microgrids also consist of various components, including interconnected loads and distributed energy sources (DERs), which have distribute coordinated controllers that develop the stability, reliability, and efficiency of these energy systems [27]. The emergence of the idea of hybrid DC-AC power grids is due to the increasing access to renewable energy sources [3, 28, 29].

But, what are the cyber security and cyber-attack? Given that the intelligent power grid uses information technology to intelligently deliver energy using a two-way communication, the inherent weakness of communication technology has exposed the system to numerous security threats. In the power network, hackers can change and manipulate values by accessing communication protocols, wireless or wired networks, computers and network control software, sensors, activators, meters, or by creating viruses, worms, and Trojans. Or information corruption that may destabilize the network or cause blackouts and result in severe financial damage. Figure 2 shows the main reasons and percentages to cause the recent recorded 140 interruptions globally [30, 31]. These physical and cyber reasons can destabilize the power system and bring it from normal to critical point. Timely performance before, during and after a crisis can effectively reduce interruption. These values indicate that the network should be resilient against cyber-physical attacks.

Also, as shown in Fig. 3, a smart power system can be decomposed into four main components: generation, transmission, distribution and consumption. Customers use electrical devices such as smart appliances and electric vehicles (EVs), and their

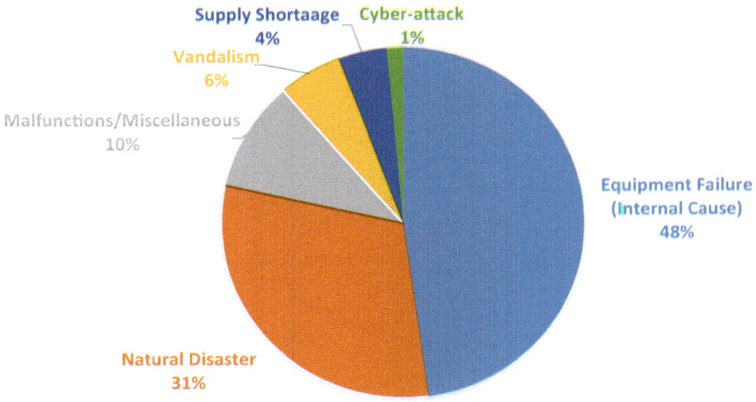

Fig. 2 Reasons for recent 140 interruptions and their frequency

power consumption is scaled with an advanced measuring device such as smart meters [32]. The smart meter is a key component of the advanced measurement infrastructure (AMI). Smart meters can be connected to a home area network (HAN) gateway through which it interacts with the network. We have shown a typical connection network in the Fig. 2 as the HAN gate. A neighborhood area network (NAN) can also be located in a substation, where several HANs are hosted. Power utility companies might utilize wide area networks (WAN) for connecting distributed NANs. Distributed and heterogeneous applications in the smart grid require different communication protocols. The smart network architecture shows the protocols used in each network. For instance, in HANs, home appliances use the ZigBee and Z-wave protocols. In the NAN, devices typically connect to the system via IEEE 802.11, IEEE 802.15.4, or IEEE 802.16 standards [33].

Connecting microgrids and power sources and transferring data between them can improve the performance of the central and distributed controllers, converters, and coordination; however, it can create malicious cyber-attack points [34, 35]. It is vital to identify and detect cyber malicious factors that can cause defects, unrealistic and real errors in measuring devices, monitors, and control devices. Therefore, such mechanisms can prevent real crisis and instability in the network such as blackout in the system and make the network resistant to malicious cyber-attacks. Methods such as signal processing, neural networks, artificial intelligence, deep learning and blockchain are effective to detect, prevent, and defend against cyber-attacks [14, 17, 36].

The rapid growth of technology in recent years in areas such as computer-based applications, power electronics, information technology and materials for smart power systems has been very promising [37, 38]. Using different types of new technologies in power systems has created emerging system security threats. The threats posed by human attacks on smart energy systems have become more serious

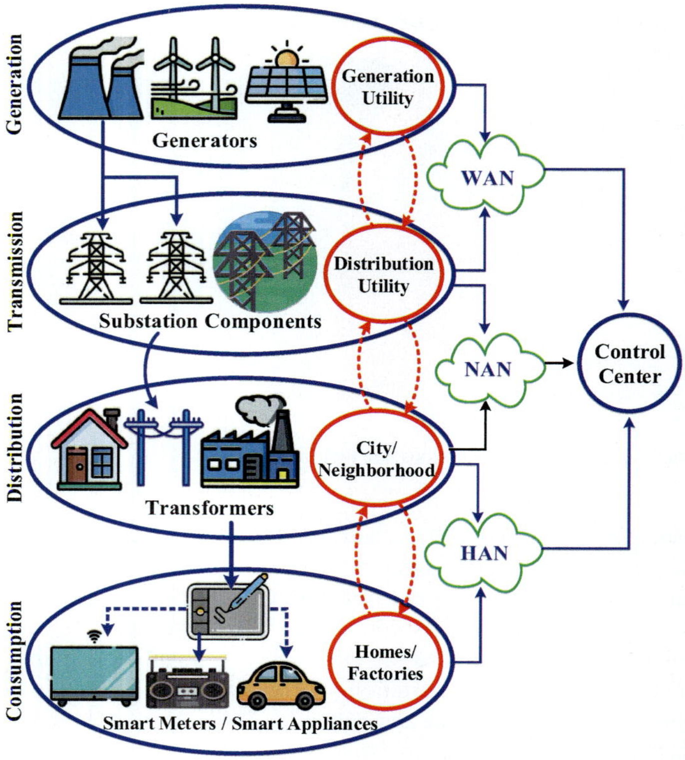

Fig. 3 An overview of cyber security in a smart power system

in recent years [39]. For example, the use of smart-grid-based communication equipment to further control them has increased the likelihood of cyberattacks, and the widespread use of renewable energy sources (RES) as power supplies in power networks has changed the system and posed various security challenges [34, 40, 41]. A resilient power system should be able to overcome failures with HILP. Considering the results of natural disasters, the main features of contingencies that affect the physical performance of the power systems are as follows [30, 42]

- The time and duration of such disturbances can be highly uncertain and might last for several days.
- They are dynamic in nature such that the events and sequence are very hard to forecast.
- Their occurrence might lead to the failure of several equipment in a short time. For example, transmission lines, towers and substations might be damaged.

Power grid optimization can effectively help make the grid more resilient. So, knowing effective methods for optimization as well as different network structures that can be resisted against natural disasters, sabotage or cyber-attacks will improve the overall performance of the system. Different optimization strategies have been used to face cyber-physical security threats in power systems. For example, optimal methods using meta-heuristic algorithms are proposed in [43]. Similarly, the authors of the papers [44, 45] examined various methods such as genetic algorithm (GA), in which the parameters were set with the concepts of fuzzy logic. There is a fundamental need to classify and evaluate current and emerging system security threats. Table 1 shows some of the important studies focused on cyber-physical security and threats and resiliency mechanisms related to the power systems in the last few years.

2 Concepts of Cyber-Attacks and Physical Threats in Power Systems

Power systems can be vulnerable to cyber-physical threats. This vulnerability can expose smart grid applications to many distinct cyber-physical threats, which can negatively affect network security. Proper identification of security threats and vulnerabilities makes it possible to determine appropriate countermeasures. Cyber-attacks can affect communication protocols, equipment, and power grid software and hardware. For example, by changing the values of current, voltage or frequency, they can cause the malfunction of protection relays which leads to blackouts in the network. Physical threats can also be a major cause of blackout, and these threats may be due to unforeseen factors such as floods, earthquakes, and hurricanes, or to deliberate sabotage such as bombings or disruption of transmission and distribution lines [53]. In recent blackouts, many similarities can be seen in various sources, for instance, the November 4, 2006 blackout in the western part of the European Union, which uses an interconnected network, and the lack of proper coordination and management of the transmission network leads to severe frequency deviations. As a result, the entire network was divided into three regions and the needs of more than 15 million European subscribers were cut off [58]. Another good example was the widespread blackouts in many parts of the United States in 2003. One of the catastrophic results was a sharp drop in water pressure due to a lack of backup power due to a lack of electricity for electric pumps. Also, all trains connected to New York were stopped, mobile telecommunications equipment has been disrupted, and the cable TV system was shut down [59].

There are several motives for threatening or attacking power systems such as economic reasons, entertainment, and terrorism. For example, manipulating the power meter to reduce the price of electricity bills, access to the central energy control systems or other vital resources and infrastructures are of these types of attacks. Expansion of smart grids, smart cities, smart homes that also use topics such as the Internet of Things (IoT) may also provide communication infrastructure for enemies. Physical attacks can also negatively affect cyber systems [60]. For example, the integrity of a cyber-physical system could be compromised by injecting false data over the network into a sensor or detector or by placing a faulty sensor next

Table 1 Some of the studies focused on cyber-physical security, threats and resiliency mechanisms in the last few years

Paper	Field of study	Main focus of the research	Sub-focus of research	Results
[46]	Power systems analysis against unexpected events	Reliability and resiliency	Resiliency values analysis	A review of conceptual strategies of power system resilience
[47]	Power systems resiliency	Disaster modeling	Natural disaster	New method against unexpected natural disasters
[48]	Power systems resiliency	Resilience improvement efforts with different regions	Analysis in different countries	Analytical comparisons of power systems resilience
[49]	Methods used in microgrids for improving resiliency	multi-energy microgrid and energy hubs	Advanced operation strategies	New method for physical resilience of multi energy system
[50]	Methods used in microgrids for improving resiliency	Resiliency improvement for AC/DC and hybrid microgrids	Centralized, decentralized, and hierarchical approaches	Improving physical resilience in AC/DC microgrids
[51]	Communication, event-based, and component resiliency	Communication resiliency	Networks analysis	Improving ICT structures
[52]	Communication, event-based, and component resiliency	Event-based resilience	Resiliency analysis of power components	New strategy for enhancing resilience of communication structure
[53]	CPSs	Renewable energies in microgrids	Reliability, resiliency and securing data	New method for improving CPS resilience
[54]	CPSs	Cyber-attacks in microgrids	DoS attacks and control for energy storage systems	New strategy for cyber-attack detection
[55, 56]	CPSs	Energy management system	Cyber-physical DC microgrids	Securing data in CPS
[14, 34, 35, 39]	False data injected Attacks	Signal processing	Blockchain, Hilbert Huang transform, fast Fourier transform	Different strategies for FDIA detection with signal processing

(continued)

Table 1 (continued)

Paper	Field of study	Main focus of the research	Sub-focus of research	Results
[57]	False data injected Attacks	Machine learning and deep learning	False data detection	Using wavelet transform and singular value decomposition based on ML in FDIA detection

to a static sensor. It is also possible to perform a combination of cyber-attacks on cyber-physical infrastructure in a hybrid manner. Therefore, since physical protection of all assets of large-scale physical systems such as smart grids and smart city is economically and practically impossible, there is a need to develop methods and algorithms to detect and combat hybrid attacks. Considering cyber-physical security in smart power systems, we have categorized examples of attacks on CPS into seven different categories, which are displayed in Table 2.

Although the performance and cybersecurity of smart power systems have been somewhat successful in defending against pure cyber or physical attacks, none of them alone can guarantee smart grid security due to combined attacks. Recent research has therefore focused more on the security challenges than ever before. For example, in articles [61, 62], the authors have proposed various strategies for adopting appropriate methods such as modelling of natural disaster, equipment vulnerability and network resistance analysis, and improving network resilience by considering the RES effects. Paper [63] presented the benefits of having microgrids for managing DERs to create power grid flexibility to HILPs. Cyber-attackers in smart power systems might be able to change values of network devices, sensors, or actuators,

Table 2 Classification of cyber-security and consequences in CPSs

Classification	Type of action and performance range
Confidentiality	Protecting the user's personal information in CPS; Preventing the attacker from trying to learn the normal condition of the network and affect the state of the physical system with eavesdropping to the communication channels
Integrity	Changing data, information or resources without permission
Availability	Existence of defects in computer technology, management, communications and equipment
Reliability/ Consistency	Confirmation of both parties involved, who are really the ones who pretend
Authenticity	Proving the identity of a subject or source
Non-repudiation	Proving the occurrence and application of events so that they cannot be denied later
Accountability	The ability to track the actions of an entity uniquely in the entity

compromising the metering devices or communicating infrastructure. As a result, compromised system conditions may disrupt the operation of the power grid, leading to physical or economic damage to power systems. For instance, in references [41, 64], studies were based on the invasion of these types of cyber-attacks. The threats of false data injection attacks (FDIAs) can be considered as leading challenges in estimating the system status amid common forms cyber-physical attacks in intelligent systems.

Without a progressive diagnostic plan, FDIA can secretly start multiple times and pose a significant threat to the network. Several research have been devoted to find possible strategies to building FDIA in smart microgrids. Much of the research on FDIA fabrication has focused on cyber-physical attacks on power grids through DC mode estimation based on regular systematic models [65]. For example, one known attack scenario is that when an attacker gains access to some of the electrical infrastructure information, which can change some of the measured values. In these approaches, the FDIA can successfully circumvent or modify conventional detection methods, and inject or otherwise incorrectly inject information into sensors or subsystem devices.

The control of smart energy systems is the responsibility of the power control centers (PCCs). In general, smart grids consist of a set of components including communication networks and protocols, sensors, actuators, meters, data processing units such as PLCs, and central computers. There are various general models to illustrate the structure of an intelligent control. We show a conceptual model that identifies vulnerable areas of the control system in smart grids in Fig. 4.

The vulnerabilities of the smart grids control system are divided into six general categories. These vulnerabilities can mostly have adverse effects on: 1- sensors and operators, 2- communication network equipment, 3- local controller like PLC or RTU, 4- control network communication protocols, 5- Local area network (LAN), 6- financial-commercial networks. The layered structure of this figure shows that the types of vulnerabilities of each layer are different from the microgrid control system, so each layer needs different and specific security measures.

Since control systems are commonly developed to detect, process, and reduce systematic and unforeseen errors, it is not surprising that this has been the subject of much research to predict, detect, and protect against attacks. Microgrids can be more prone to similar attacks found in the power grid, because their control and communication equipment may be more available to attackers. Denial of service (DoS) attacks, for example, raise a number of issues, but at the same time, these types of attacks can be detected by the system operator [67]. Therefore, operators will be able to quickly detect that the network has been attacked. Similarly, the severity of FDI attacks is largely attributed to the complexity and variability of the detection method on the control structure. FDI, which manages to infiltrate the network by maintaining authority and without changing system visibility disturbance alerts, known as stealth attacks, can create unpredictable stability issues, and even worse, it can be detected. It is practically impossible.

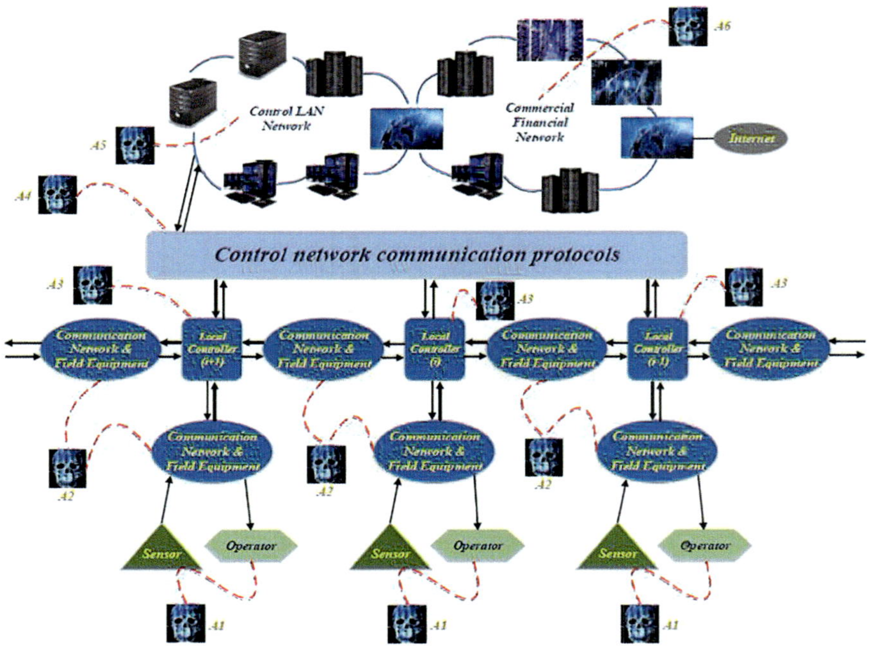

Fig. 4 Conceptual model of control system vulnerabilities in microgrids [66]

Several studies have identified and differentiated faults in control loops based on a detection strategy or working on preventive measures based on threat modeling, security analysis and protection strategy. One of the most common techniques utilized for detecting bad measurement data in transmission systems is the static state estimator (SSE). It is generally based on minimum weighted least squared (WLS) and is not immune to attacks. Figure 5 shows a schematic of FDI attack on a power grid control system [68].

3 Measuring the Resilience of Power Systems

The power systems enter the stage of durability from the decline of system performance to when the situation is established in new conditions. Network vulnerability is the maximum decrease in the network performance index compared to its value under normal conditions. So, it can be said that the degree of network vulnerability depends on the ratio of system compatibility to the perturbation. Figure 6 shows a concept diagram of response in conventional and resilient power systems. As can be seen, possible modes of resilience based on safety margins include prepare mode, withstand (alarm) mode, adopt mode, recover mode, and normal mode. The transition

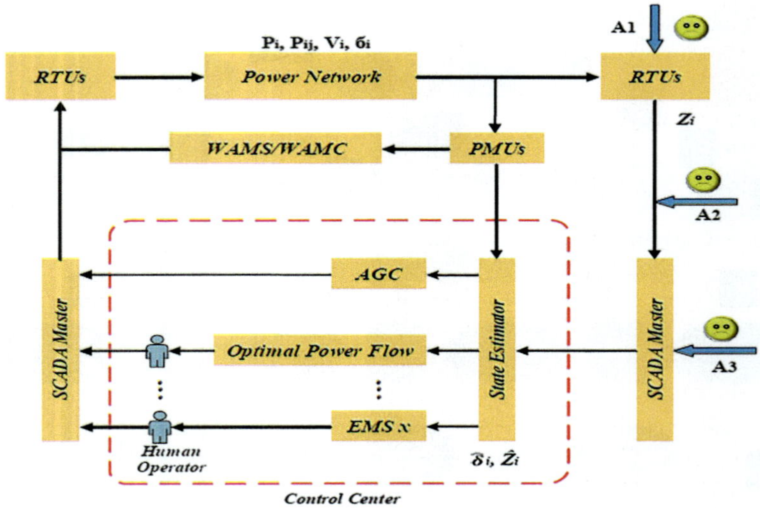

Fig. 5 Schematic of FDI attack on a power grid control system [66]

between these states is highly dependent on electrical events and the effectiveness of preventive and corrective measures. Different degrees of resilience are associated with each of these power system modes. The performance of the system does not decrease rapidly after the onset of perturbation in time, and due to the degree of endurance it has, it takes time for the performance of the system to decline. This stage of system behavior can be called the prevention stage, which depends on the level of awareness of the operator and the strength of the network [13].

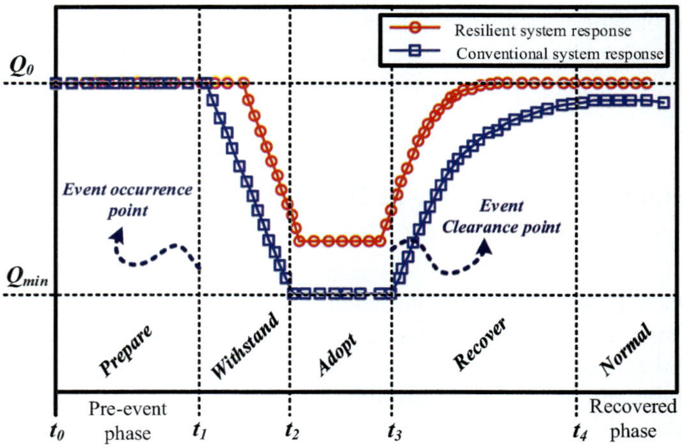

Fig. 6 Conceptual diagram of resilience and how it relates to other factors in power systems [31]

With more knowledge and more effective ways to deal with the system threats, the power grid system will be more flexible and resilient. Also, protective cyber-physical tools such as firewalls and effective detection methods can highly increase the system resilience.

One way to improve the resilience of the power system during the design and installation phase is to hardening the equipment against unexpected events. Also, installing DGs or small power plants in the distribution network is an effective way to minimize the number of blackouts. One of the advantages of installing DG is that in cases where the system is interrupted due to natural disasters and cannot supply energy to consumers from the air network, DGs enter the network and provide electricity to sensitive loads. This ensures that sensitive loads are fed, especially in islanding or microgrids states [13, 69, 70]. This process in turn leads to improved flexibility and physical resilience in the energy systems. The use of renewable energy sources including solar and wind, diesel generators and energy storage systems with an optimal management of energy production at suitable times and the use of techniques based on optimization algorithms, will be necessary. From a physical point of view, experience results from natural disasters illustrate that conventional power grids are fragile systems owing to the geographical extent of contingencies. In such systems, the failure of their equipment can lead to large blackout. So larger areas of the power system are expected to experience a higher probability of outages. Therefore, the power network might not be available on critical lines and in adjacent areas for several days. Clearly, this may not be the case for all of disasters, but there is evidence of such disasters around the world such as Hurricane Katrina, the earthquake and the tsunami in Japan, and the Sichuan earthquake in China [30, 63].

It should be noted that the complexity of power systems can lead to changes in the resilience of the electrical system and equipment. This reinforces the critical thinking about identifying the risk that a device will fail when challenged by a particular threat. Although this facilitates risk assessment, it makes it difficult to create a comprehensive picture of resilience for this sector or to prioritize risks. Identifying a more systematic framework of resilience, as well as specific criteria which can be utilized to control, monitor and evaluate resilience in the electrical system, is a rapidly evolving field of research.

4 Possible Solutions to Improve the Cyber-Physical Resilience of Power Systems

In this section, we explore approaches to improving cybersecurity in a smart grid. One of the most powerful ways to detect cyber-attacks with the ability to extract high-level features as well as against new attacks or small mutations is to use Fast Fourier Transform (FFT). In addition, the use of a combination of machine learning (ML) and FFT techniques can lead to better performance in detecting and counteracting cyber-attacks. Recent articles [57, 71, 72] have addressed the issue of false signal injected attack using FFT to detect the attacks. Noise in FDI attacks is different from noise in other forms of cyber-attacks. Overall, successful FDI attacks can change normal, corrupt, invalid, inaccurate, and bad data [73].

According to the mentioned features, in order to have a resilient power grid, the system must include the following features:

- Due to the high level of uncertainty, the diversity of supplier resources should be increased as much as possible, also the dependence on a limited set of power supply equipment should be avoided.
- Sufficient flexibility in the power system must be considered to respond quickly to unexpected events, and accordingly, operating methods should be adjusted, even over short periods of time.
- Due to the lack of available resources, priorities in order to supply various loads must be identified and specified.

Long-term planning and measure are needed to improve the system's resilience to climate change-related events. The following are some of these measures:

- Management and risk assessment in order to evaluate and prepare for such events
- Accurate estimation of the probable location and its severity of the event
- Upgrading emergency plans and preparedness
- Pruning trees and managing plant life to clear roadside transmission lines
- Undergrounding distribution and transmission power lines
- Upgrading and hardening equipment with stronger materials
- Expanding the redundancy of transmission lines with the construction of new facilities
- Redirecting transmission lines to regions that are less affected by severe weather
- Removing substations and switching facilities to areas that are less prone to flooding
- Proper storage and maintenance of masts and backup materials in a suitable place
- Increasing monitoring of environmental awareness via advanced controlling and forecasting tools

Considering the above mentioned items, we can understand the importance and impact of HILP events on the power system and the resulting economic and social risks. Hence, this important issue has created a fundamental need to address the issue of resiliency of CPS in power systems. Estimating unexpected events such as heavy snow or rain, floods, earthquakes and hurricanes and providing optimal solutions to deal with them can make the power grid resilient to such physical hazards. Using information to predict severe weather changes such as meteorological news and the use of radar can protect the network against unwanted disasters. Also, methods of damage assessment and network repair in the fastest possible time can be effective in reducing blackouts. Figure 7 summarizes the performance process of the power system resilience against disasters followed by rapid recovery to normal operation.

The power system must also be flexible enough to adapt to severe disruptions without losing power generation in power plants due to voltage or frequency or power instabilities, and must be designed to recover immediately after the problem

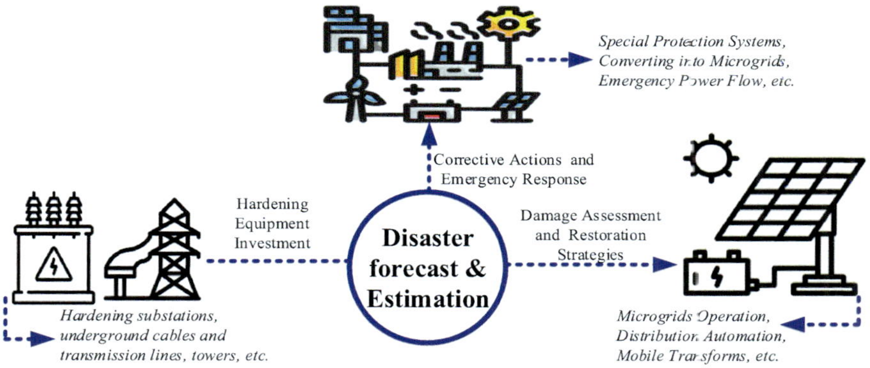

Fig. 7 The performance process of the power system against natural disasters

is resolved. Optimal network design with fast restoration capability and optimal control and management of renewable DG resources such as wind generators, PV or grid-connected diesel generators can effectively increase the stability, flexibility and resiliency of power systems [30, 74]. Also, the combination of wind and solar energy and the use of larger energy storage (ES) systems can effectively make the grid more resilient to physical threats, even helping to reduce pollution and greenhouse gas emissions [75, 76]. The optimal placement of DGs and the ES with effective controlling strategies can be effective ways to have a resilient power system [77, 78]. As a result, designing a system that can withstand such disasters or attacks and ensure the stability of the network is one of the main responsibilities of electricity power system operators. Therefore, finding and applying optimal strategies to face these conditions and recover power systems in shortest time are necessary. According to the newly published papers [79–81], the effective strategies include:

- Providing optimal methods in order to respond to the demand
- Optimal management of generation, transmission and distribution of electrical energy
- Optimization in energy storage methods
- Providing accurate methods for predicting unexpected events through machine learning, deep learning and neural networks
- Providing methods to detect cyber-attacks in different layers of physical cyber
- Using firewalls and increasing network security
- Improving and optimizing defense methods against hackers
- Providing different methods of fast recovery of power system
- Optimization in control networks

Converting conventional power grid into microgrids can be an effective way to recover power systems after attacks. In the event of temporary or permanent smart grid instability, an effective solution could be converting smart distribution network into several microgrids where additional switches, DGs and other physical devices are used to increase the resilience of the system. In order to convert a typical distribution network to microgrids during disasters, optimal location of switches is required. As shown in Fig. 8, the analysis of distribution networks into independent micro-networks has been proposed in the research [31]. As can be seen, the size of the distributed generators is the desired size and they are placed in optimal positions to decrease power loss, while maximizing the stability of the system through locally DGs. In the event of major disruptions, not all network loads may be used. Therefore, load priorities are set to ensure minimum service stability for critical loads.

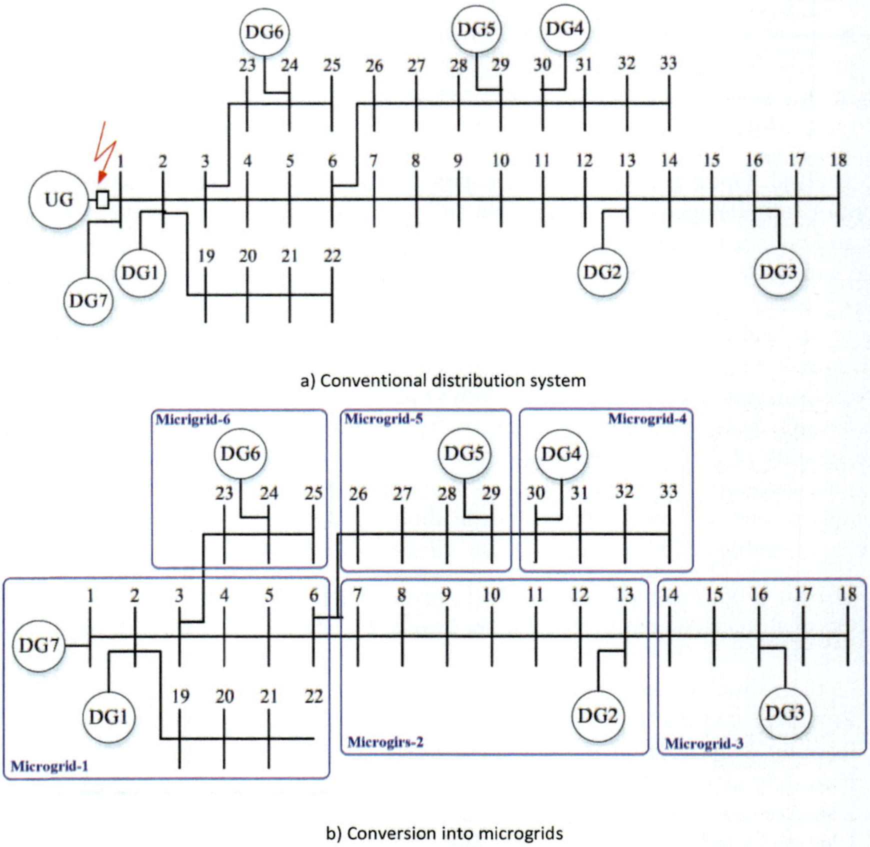

Fig. 8 Conventional distribution system conversion into microgrids

5 Conclusion

With the development of power systems technology, the use of computer and Internet-based programs has grown significantly in the last decade. This development has created concepts such as smart power systems, smart and microgrids, smart cities and smart homes. Considering this issue, management and control of the power systems and related equipment using the SCADA network based on communication protocols and the use of fiber optics, wireless network or Internet-based platform can be threatened by attackers. Therefore, the flexibility and resilience of the network in the design and operation phase should be given special attention. To achieve this goal, new approaches and solutions must be developed that in the first place guarantee the rapid detection of an unexpected event or attack, as well as the adoption of appropriate strategies to ensure the continuity of infrastructure services.

Accordingly, different methods such as equipment reinforcement, hardening and the creation of multiple protective layers for cyber-physical protection and strategies for detecting and countering cyber-attacks are provided by researchers. These methods can ensure the optimal and flexible design and management of critical system-based energy infrastructure of CPSs. As the number of devices of smart energy networks connected to the Internet increases every day, these devices and networks can become suitable targets for cyber-attacks that can remotely threaten the security of the power system, companies and citizens. So, it is essential to take appropriate measures for providing maximum resilience to unexpected cyber and physical events in smart power systems. In particular, smart grids should be able to provide such infrastructure to individuals if properly designed, controlled, and managed. In this chapter book, concepts of cyber-physical security in smart power systems from a resilience perspective with possible solutions have been reviewed in details.

References

1. H. Zareipour, Digital power grid: lessons from history to imagine the future [in my view]. IEEE Power Energ. Mag. **20**, 90–92 (2022)
2. M. Ghiasi, Detailed study, multi-objective optimization, and design of an AC-DC smart microgrid with hybrid renewable energy resources. Energy **169**, 496–507 (2019)
3. M. Ghiasi, M. Dehghani, T. Niknam, P. Siano, H. Haes Alhelou, A Detailed Analysis of the Barriers of Using Renewable Energies and Their Roles in Sustainable Development in Iran," in *Handbook of Smart Energy Systems*, ed. by M. Fathi, E. Zio, P.M. Pardalos (Springer International Publishing, Cham, 2022), pp. 1–24
4. M. Ghiasi, T. Niknam, M. Dehghani, P. Siano, H. Haes Alhelou, A. Al-Hinai, Optimal multi-operation energy management in smart microgrids in the presence of RESs based on multi-objective improved DE algorithm: cost-emission based optimization. Appl. Sci. **11**, 3661 (2021)
5. M. Ghiasi, S. Esmaeilnamazi, R. Ghiasi, M. Fathi, Role of renewable energy sources in evaluating technical and economic efficiency of power quality. Technol. Econ. Smart Grids Sustain. Energy **5**, 1 (2020)

6. M.H. Amini, H. Arasteh, P. Siano, Sustainable smart cities through the lens of complex interdependent infrastructures: panorama and state-of-the-art, in *Sustainable Interdependent Networks II: From Smart Power Grids to Intelligent Transportation Networks*, ed. by M.H. Amini, K.G. Boroojeni, S.S. Iyengar, P.M. Pardalos, F. Blaabjerg, A.M. Madni (Springer International Publishing, Cham, 2019), pp. 45–68
7. B. Arbab-Zavar, E.J. Palacios-Garcia, J.C. Vasquez, J.M. Guerrero, Smart inverters for microgrid applications: a review. Energies **12**, 840 (2019)
8. N. Bayati, A. Hajizadeh, M. Soltani, Protection in DC microgrids: a comparative review. IET Smart Grid **1**, 66–75 (2018)
9. J. Bruinenberg, L. Colton, E. Darmois, J. Dorn, J. Doyle, O. Elloumi, et al., CEN-CENELEC-ETSI smart grid co-ordination group smart grid reference architecture, in *CEN, CENELEC, ETSI, Technical Report*, vol. 23 (2012)
10. L. Sgrafetto, P. Emmerich, M. Klingel,Smart grids and data–exploring emerging data-driven solutions for distribution system operators, in *PESS 2020; IEEE Power and Energy Student Summit* (2020), pp. 1–6
11. H. Manoharan, Y. Teekaraman, I. Kirpichnikova, R. Kuppusamy, S. Nikolovski, H.R. Baghaee, Smart grid monitoring by wireless sensors using binary logistic regression. Energies **13**, 3974 (2020)
12. P. Kumar, S. Nikolovski, I. Ali, Introduction to the special section on innovative technologies for microgrid and smart-grid systems. Comput. Electr. Eng. **77**, 414–415 (2019)
13. M. Ghiasi, M. Dehghani, T. Niknam, H.R. Baghaee, S. Padmanaban, G.B. Gharehpetian et al., Resiliency/cost-based optimal design of distribution network to maintain power system stability against physical attacks: a practical study case. IEEE Access **9**, 43862–43875 (2021)
14. M. Ghiasi, M. Dehghani, T. Niknam, A. Kavousi-Fard, P. Siano, H.H. Alhelou, Cyber-attack detection and cyber-security enhancement in smart DC-microgrid based on blockchain technology and Hilbert Huang transform. IEEE Access **9**, 29429–29440 (2021)
15. M. Mohammadi, A. Kavousi-Fard, M. Dehghani, M. Karimi, V. Loia, H. H. Alhelou, et al., Reinforcing data integrity in renewable hybrid AC-DC microgrids from social-economic perspectives. ACM Trans. Sens. Netw. (2022)
16. M. Dehghani, M. Ghiasi, T. Niknam, K. Rouzbehi, Z. Wang, P. Siano et al., Control of LPV modeled AC-microgrid based on mixed H2/H∞ time-varying linear state feedback and robust predictive algorithm. IEEE Access **10**, 3738–3755 (2022)
17. M. Dehghani, M. Ghiasi, T. Niknam, A. Kavousi-Fard, M. Shasadeghi, N. Ghadimi et al., Blockchain-based securing of data exchange in a power transmission system considering congestion management and social welfare. Sustainability **13**, 90 (2021)
18. H.H. Alhelou, M.E.H. Golshan, N.D. Hatziargyriou, A decentralized functional observer based optimal LFC considering unknown inputs, uncertainties, and cyber-attacks. IEEE Trans. Power Syst. **34**, 4408–4417 (2019)
19. H.H. Alhelou, M.E.H. Golshan, N.D. Hatziargyriou, Deterministic dynamic state estimation-based optimal lfc for interconnected power systems using unknown input observer. IEEE Trans. Smart Grid **11**, 1582–1592 (2019)
20. M. Ghiasi, N. Ghadimi, E. Ahmadinia, An analytical methodology for reliability assessment and failure analysis in distributed power system. SN Appl. Sci. **1**, 44 (2018)
21. O.P. Mahela, B. Khan, H.H. Alhelou, P. Siano, Power quality assessment and event detection in distribution network with wind energy penetration using stockwell transform and fuzzy clustering. IEEE Trans. Industr. Inf. **16**, 6922–6932 (2020)
22. M.H. Amirioun, F. Aminifar, H. Lesani, M. Shahidehpour, Metrics and quantitative framework for assessing microgrid resilience against windstorms. Int. J. Electr. Power Energy Syst. vol. 104, pp. 716–723, 2019/01/01/ 2019.
23. H. Shirazi, M. Ghiasi, M. Dehghani, T. Niknam, M.G. Garpachi, A. Ramezani, Cost-emission control based physical-resilience oriented strategy for optimal allocation of distributed generation in smart microgrid, in *2021 7th International Conference on Control, Instrumentation and Automation (ICCIA)* (2021), pp. 1–6

24. M.A. Gilani, A. Kazemi, M. Ghasemi, Distribution system resilience enhancement by microgrid formation considering distributed energy resources. Energy **191**, 116442 (2020)
25. C. Sheng, Y. Yao, Q. Fu, W. Yang, A cyber-physical model for SCADA system and its intrusion detection. Comput. Netw. **185**, 107677 (2021)
26. M. Fathi, M. Ghiasi, Optimal DG placement to find optimal voltage profile considering minimum DG investment cost in smart neighborhood. Smart Cities **2**, 328–344 (2019)
27. M. Khodadadi Arpanahi, M. Kordi, R. Torkzadeh, H. Haes Alhelou, P. Siano, An augmented prony method for power system oscillation analysis using synchrophasor data. Energies **12**, 1267 (2019)
28. Z. Wang, U. Munawar, R. Paranjape, Stochastic optimization for residential demand response with unit commitment and time of use. IEEE Trans. Ind. Appl. **57**, 1767–1778 (2021)
29. K. Ginigeme, Z. Wang, Distributed optimal vehicle-to-grid approaches with consideration of battery degradation cost under real-time pricing. IEEE Access **8**, 5225–5235 (2020)
30. H. Qiu, W. Gu, Z. Wu, S. Zhou, G. Pan, X. Yang et al., Resilience-directional robust power dispatching of microgrids under meteorological disasters. IET Renew. Power Gener. **13**, 2084–2093 (2019)
31. A. Hussain, V.-H. Bui, H.-M. Kim, Microgrids as a resilience resource and strategies used by microgrids for enhancing resilience. Appl. Energy **240**, 56–72 (2019)
32. H. Shahinzadeh, J. Moradi, G.B. Gharehpetian, H. Nafisi, M. Abedi, IoT architecture for smart grids, in. International Conference on Protection and Automation of Power System (IPAPS) **2019**, 22–30 (2019)
33. M. Kuzlu, M. Pipattanasompom, S. Rahman, A comprehensive review of smart grid related standards and protocols, in *2017 5th International Istanbul Smart Grid and Cities Congress and Fair (ICSG)* (2017), pp. 12–16
34. M.R. Habibi, H.R. Baghaee, T. Dragiˇcevi, F. Blaabjerg, Detection of false data injection cyber-attacks in DC microgrids based on recurrent neural networks. IEEE J. Emerg. Select. Top. Power Electron. (2020)
35. M. Dehghani, T. Niknam, M. Ghiasi, P. Siano, H. Haes Alhelou, A. Al-Hinai, Fourier singular values-based false data injection attack detection in AC smart-grids. Appl. Sci. **11**, 5706 (2021)
36. W. Xu, J. Li, M. Dehghani, M. GhasemiGarpachi, Blockchain-based secure energy policy and management of renewable-based smart microgrids. Sustain. Cities Soc. **72**, 103010 (2021)
37. S. Mehrdad, S. Mousavian, G. Madraki, Y. Dvorkin, Cyber-physical resilience of electrical power systems against malicious attacks: a review. Current Sustain./Renew. Energy Rep. **5**, 14–22 (2018)
38. H. Davarikia, M. Barati, A tri-level programming model for attack-resilient control of power grids. J. Modern Power Syst. Clean Energy **6**, 918–929 (2018)
39. M. Dehghani, M. Ghiasi, T. Niknam, A. Kavousi-Fard, E. Tajik, S. Padmanaban et al., Cyber attack detection based on wavelet singular entropy in AC smart islands: false data injection attack. IEEE Access **9**, 1–1 (2021)
40. M.R. Habibi, H.R. Baghaee, T. Dragičević, F. Blaabjerg, False data injection cyber-attacks mitigation in parallel DC/DC converters based on artificial neural networks. IEEE Trans. Circuits Syst. II: Express Briefs (2020)
41. M. Dehghani, M. Ghiasi, T. Niknam, A. Kavousi-Fard, S. Padmanaban, False data injection attack detection based on Hilbert-Huang transform in AC smart islands. IEEE Access **8**, 179002–179017 (2020)
42. W. Yuan, J. Wang, F. Qiu, C. Chen, C. Kang, B. Zeng, Robust optimization-based resilient distribution network planning against natural disasters. IEEE Trans. Smart Grid **7**, 2817–2826 (2016)
43. A. Sadegheih, Optimal design methodologies under the carbon emission trading program using MIP, GA, SA, and TS. Renew. Sustain. Energy Rev. **15**, 504–513 (2011)
44. B.-Y. Qu, Y. Zhu, Y. Jiao, M. Wu, P.N. Suganthan, J. Liang, A survey on multi-objective evolutionary algorithms for the solution of the environmental/economic dispatch problems. Swarm Evol. Comput. **38**, 1–11 (2018)

45. M. Ghiasi, J. Olamaei, Optimal capacitor placement to minimizing cost and power loss in Tehran metro power distribution system using ETAP (A case study). Complexity **21**, 483–493 (2016)
46. Y. Wang, C. Chen, J. Wang, R. Baldick, Research on resilience of power systems under natural disasters—a review. IEEE Trans. Power Syst. **31**, 1604–1613 (2016)
47. G. Liu, T. Jiang, T.B. Ollis, X. Li, F. Li, K. Tomsovic, Resilient distribution system leveraging distributed generation and microgrids: a review, in *IET Energy Systems Integration* (2020)
48. E.W. Prehoda, C. Schelly, J.M. Pearce, US strategic solar photovoltaic-powered microgrid deployment for enhanced national security. Renew. Sustain. Energy Rev. **78**, 167–175 (2017)
49. A. Khodaei, Resiliency-oriented microgrid optimal scheduling. IEEE Trans. Smart Grid **5**, 1584–1591 (2014)
50. M.H. Amirioun, F. Aminifar, H. Lesani, Towards proactive scheduling of microgrids against extreme floods. IEEE Trans. Smart Grid **9**, 3900–3902 (2017)
51. C. Chen, J. Wang, F. Qiu, D. Zhao, Resilient distribution system by microgrids formation after natural disasters. IEEE Trans. Smart Grid **7**, 958–966 (2016)
52. A. Hussain, A.O. Rousis, I. Konstantelos, G. Strbac, J. Jeon, H.-M. Kim, Impact of uncertainties on resilient operation of microgrids: a data-driven approach. IEEE Access **7**, 14924–14937 (2019)
53. V. Venkataramanan, A. Hahn, A. Srivastava, CP-SAM: cyber-physical security assessment metric for monitoring microgrid resiliency. IEEE Trans. Smart Grid **11**, 1055–1065 (2020)
54. A. Afshari, M. Karrari, H.R. Baghaee, G.B. Gharehpetian, Resilient synchronization of voltage/frequency in AC microgrids under deception attacks. IEEE Syst. J. 1–12 (2020)
55. Z. Cheng, M.Y. Chow, Resilient collaborative distributed energy management system framework for cyber-physical DC microgrids. IEEE Trans Smart Grid **11**, 4637–4649 (2020)
56. M. Mohammadi, A. Kavousi-Fard, M. Dabbaghjamanesh, A. Farughian, A. Khosravi, Effective management of energy internet in renewable hybrid microgrids: a secured data driven resilient architecture. IEEE Trans. Industr. Inf. **18**, 1896–1904 (2021)
57. M. Dehghani, T. Niknam, M. Ghiasi, N. Bayati, M. Savaghebi, Cyber-attack detection in DC microgrids based on deep machine learning and wavelet singular values approach. Electronics **10**, 1914 (2021)
58. G. Maas, M. Bial, J. Fijalkowski, "Final report-system disturbance on 4 november 2006, in *Union for the Coordination of Transmission of Electricity in Europe, Technical Report* (2007)
59. U.-C. Force, S. Abraham, Us-canada power system outage task force, in *Final Report on the August,* vol. 14 (2004), p. 2003
60. I. Cvitić, D. Peraković, M. Periša, B. Gupta, Ensemble machine learning approach for classification of IoT devices in smart home. Int. J. Mach. Learn. Cybern. 1–24 (2021)
61. A.N. Tari, M.S. Sepasian, M.T. Kenari, Resilience assessment and improvement of distribution networks against extreme weather events. Int. J. Electr. Power Energy Syst. **125**, 106414
62. X. Liu, K. Hou, H. Jia, J. Zhao, L. Mili, Y. Mu et al., A resilience assessment approach for power system from perspectives of system and component levels. Int. J. Electr. Power Energy Syst. **118**, 105837 (2020)
63. E. Galvan, P. Mandal, Y. Sang, Networked microgrids with roof-top solar PV and battery energy storage to improve distribution grids resilience to natural disasters. Int. J. Electr. Power Energy Syst. **123**, 106239 (2020)
64. Q. Wang, W. Tai, Y. Tang, M. Ni, Review of the false data injection attack against the cyber-physical power system. IET Cyber-Phys. Syst.: Theory Appl. **4**, 101–107 (2019)
65. G. Liang, J. Zhao, F. Luo, S.R. Weller, Z.Y. Dong, A review of false data injection attacks against modern power systems. IEEE Trans. Smart Grid **8**, 1630–1638 (2016)
66. M. Ghiasi, M. Dehghani, T. Niknam, A. Kavousi-Fard, Investigating overall structure of cyber-attacks on smart-grid control systems to improve cyber resilience in power system. IEEE Smart Grid Newsl. (2020)
67. G.A. Jaafar, S.M. Abdullah, S. Ismail, Review of recent detection methods for HTTP DDoS attack. J. Comput. Netw. Commun **2019** (2019)

68. A. Kavousi-Fard, W. Su, T. Jin, A Machine-learning-based cyber attack detection model for wireless sensor networks in microgrids. IEEE Trans. Industr. Inf. **17**, 650–658 (2021)
69. H. Karimi, T. Niknam, J. Aghaei, M. GhasemiGarpachi, M. Dehghani, Switches optimal placement of automated distribution networks with probability customer interruption cost model: A case study. Int. J. Electr. Power Energy Syst. **127**, 106708 (2021)
70. W. Hughes, W. Zhang, A.C. Bagtzoglou, D. Wanik, O. Pensado, H. Yuan et al., Damage modeling framework for resilience hardening strategy for overhead power distribution systems. Reliab. Eng. Syst. Saf. **207**, 107367 (2021)
71. S. Liu, S. You, H. Yin, Z. Lin, Y. Liu, W. Yao et al., Model-free data authentication for cyber security in power systems. IEEE Trans. Smart Grid **11**, 4565–5456 (2020)
72. I. Giechaskiel, K. Rasmussen, Taxonomy and challenges of out-of-band signal injection attacks and defenses. IEEE Commun. Surv. Tutor. **22**, 645–670 (2019)
73. R. Deng, G. Xiao, R. Lu, Defending against false data injection attacks on power system state estimation. IEEE Trans. Industr. Inf. **13**, 198–207 (2015)
74. A. Arjomandi-Nezhad, M. Fotuhi-Firuzabad, M. Moeini-Aghtaie, A. Safdarian, P. Dehghanian, F. Wang, Modeling and optimizing recovery strategies for power distribution system resilience. IEEE Syst. J. 1–10 (2020)
75. P. Ge, F. Teng, C. Konstantinou, S. Hu, A resilience-oriented centralised-to-decentralised framework for networked microgrids management. Appl. Energy **308**, 118234 (2022)
76. J. Tobajas, F. Garcia-Torres, P. Roncero-Sánchez, J. Vázquez, L. Bellatreche, E. Nieto, Resilience-oriented schedule of microgrids with hybrid energy storage system using model predictive control. Appl. Energy **306**, 118092 (2022)
77. B. Li, M. Ghiasi, A new strategy for economic virtual power plant utilization in electricity market considering energy storage effects and ancillary services. J. Electr. Eng. Technol. (2021)
78. U.B. Tayab, F. Yang, M. El-Hendawi, J. Lu, Energy management system for a grid-connected microgrid with photovoltaic and battery energy storage system, in. Australian & New Zealand Control Conference (ANZCC) **2018**, 141–144 (2018)
79. M.Z. Oskouei, H. Mehrjerdi, D. Babazadeh, P.T. Baboli, C. Becker, P. Palensky, Resilience-oriented operation of power systems: Hierarchical partitioning-based approach. Appl. Energy **312**, 118721 (2022)
80. H. Wang, Y.-P. Fang, E. Zio, Resilience-oriented optimal post-disruption reconfiguration for coupled traffic-power systems. Reliab. Eng. Syst. Saf. **222**, 108408 (2022)
81. M. Abdelmalak, M. Benidris, Proactive generation redispatch to enhance power system resilience during hurricanes considering unavailability of renewable energy sources. IEEE Trans. Ind. Appl. (2022)

Cybersecurity Challenges in Microgrids: Inverter-Based Resources and Electric Vehicles

Tambiara Tabassum and Mohammad Reza Khalghani

Abstract Power systems are shifting into a new paradigm of connectivity. This new paradigm facilitates massive integrations of different energy resources and loads, e.g., renewable energy and electric vehicle (EV) charging stations, throughout a microgrid with cyber interfaces. The integration of these energy resources and EVs can significantly enhance the microgrid efficiency, reliability, resiliency, and address environmental concerns. Several emerging technologies are required to prepare the cyber-physical infrastructure for this highly integrated grid. This new grid model is highly dependent on the cyber interfaces and communication infrastructure that makes the system vulnerable to cyber disruptions and threats. Considering the interconnected nature of these microgrids, cyber disorders may lead to disastrous consequences; therefore, it is essential to know the vulnerabilities of the grid and design tools to identify and mitigate these threats. This chapter specifically focuses on cybersecurity issues of the microgrid with inverter-based resources (IBRs) and EV charging stations. Due to complexity and unknown behavior of grids with IBRs and EV charging stations, a variety of cyber risks can impact the grid. Therefore, this chapter demonstrates the vulnerabilities of this new grid architecture, including IBRs and EV charging stations. Besides, it discusses the effective model-based, Artificial Intelligence (AI)-based techniques, and some technologies to detect and mitigate these cyber threats.

Keywords Microgrid cybersecurity · Inverter-based resource · EV charging station · Vulnerability analysis · Emerging technologies · Detection and mitigation schemes

1 Introduction

The current trend of energy consumption reveals that enormous energy sources are required to meet societies' needs in the near future. According to North America Reliability Corporation (NERC) [1], the penetration of Distributed Energy Resources

T. Tabassum · M. R. Khalghani (✉)
4700 Research Way, Lakeland, FL 33805, USA
e-mail: Khalghani@ieee.org

H. Haes Alhelou et al. (eds.), *Power Systems Cybersecurity*, Power Systems,
https://doi.org/10.1007/978-3-031-20360-2_4

(DERs) is projected to exceed 60 GW total capacity through 2030, which is double the present integrated amount. The advancement in semiconductor devices eases the integration of DERs based on renewable energy resources to grids through power electronic converters, known as inverter-based resources (IBRs). In addition, the US Department of Energy [2] projected 1, 8. and 26 TWh of incremental energy needed to supply electric vehicles (EVs) for the 2030 low, medium, and high EV growth scenarios. Despite the myriad advantages of these two technologies, they need an interconnected infrastructure using cyber interfaces for real-time operation. This smart infrastructure can control and monitor both IBRs, and EVs distributed throughout the grid.

Protecting IBRs, such as photovoltaic (PV) systems, wind turbines, and EVs against cyber-physical threats and guaranteeing their security is becoming increasingly important as the transition to smart grids and microgrids continues. Cybersecurity threats can be particularly the primary concern in microgrids with a high penetration of power electronics converters that do not have inherent inertia. Such islanded microgrids can be exposed to cyber-attacks that, in turn, cause disastrous consequences on their transient and steady-state stability [3]. Cyber disruptions can be destructive on hybrid AC-DC microgrids too. Any cyber-attacks on one side of a hybrid AC-DC microgrid will sabotage the other side. For example, if a cyber-attack disrupts the frequency stability of the AC subgrid, it will disrupt DC voltage stability on the DC side via interconnected power converters. There is a significant gap in existing smart grid cybersecurity research to address these issues. Ongoing research activities primarily focus on cyber concerns affecting the grid reliability and availability rather than the performance and behavior of power electronics subsystems [4]. This trend is driven by the growing adoption of Internet-of-Things (IoT) applications. In smart grids, RES differs from EVs in terms of power output, penetration, and tight integration with many grid interfaces via grid-tied inverters, sensors, and communication devices. Moreover, since the usage of traditional energy sources is the root cause of the world's current issues, which include climate change, the depletion of fossil fuels, air pollution, and constantly rising energy prices, greenhouse gas emissions that contribute to global warming must be drastically reduced. Besides, the transport sector plays a crucial role in this regard; it accounts for roughly 23% of worldwide greenhouse gas emissions [5]. Due to their improved energy efficiency, lower carbon intensity for energy carriers, and support for various renewable energy sources, electric vehicles (EVs) and associated infrastructure may reduce these emission rates effectively. Because renewable energy sources are inherently intermittent; determining normal versus compromised behavior presents unique issues; as a result, cybersecurity algorithms must be more carefully built and adapted to detect attacks [6] because it is easier for the attacker to hide in this more unpredictable grid environment. For example, EV cybersecurity is best addressed by separating systems like infotainment from vehicle operations, but IBR cybersecurity relies on communications to determine operational settings and control. EV cybersecurity focuses on microgrid controllers and digital grid control and sensing, whereas IBR cybersecurity will concentrate on device-level to grid-level interactions, such as communications, grid controls, and power conversion [7]. As the performance of IBRs components deteriorates and networked power electronics systems continue to be subjected to botnet hacking by malicious software/processes, anomalous microgrid operating conditions such as reactive power output imbalance, irregular power flow, and grid instabilities, such as sub-synchronous resonance, may occur. These issues may cause more significant consequences in the main grid, e.g., failing some load supply or even blackout. If these threats are not discovered early enough in many safety-critical applications, they can result in catastrophic failures and significant financial losses. Therefore, these cyber anomaly types, impacts, and their detection and mitigation need to be well studied in power grids equipped with IBRs.

Various cyber-attacks have caused widespread power disruptions. The recent blackout in Utah was a wake-up call for everyone. A Utah-based Solar and Wind facility named sPower fell victim to Denial-of-Service (DoS) attack. The attack disconnected the power generation source from the central power grid and made the corporation the first renewable energy provider in the United States to be the target of a cyber-attack. According to an emergency report submitted to the Department of Energy by sPower, threat actors exploited a known issue in Cisco firewalls to disrupt communications channels for about 12 hours. [8]. According to the analysis, a blackout affecting 15 states in the United States would affect 93 million people, and cost between 243 billion and 1 trillion dollars [9, 10]. Several projects and programs have lately been launched in response to the disruptive consequences of cyber-attacks and the vulnerability of smart grids. The federal governments of the United States and Canada have begun working together on National Electric Grid Security and Resilience action plan to safeguard the developing electricity infrastructure against cyber-attacks. Furthermore, the US Department of Energy (DoE) has launched many programs addressing cyber-security concerns [11].

To the authors' best knowledge, very few studies exist that combine all the security aspects and challenges in a grid equipped with inverter-based resources and electric vehicle charging stations. This chapter provides a comprehensive review of microgrid cybersecurity. It reviews the state-of-the-art in microgrid cybersecurity, discusses different layers of cyber-physical paradigm, assesses cyber vulnerabilities and potential threats, and suggests improvements to the system by providing different attack detection and mitigation techniques Fig. 1.

A. Literature Review

The cyber-physical systems of smart grids and their security have been studied in this literature [12–16]. The necessity of cyber-security in operation and control of microgrids is highlighted in general in [14] and cyber vulnerabilities in microgrids, as well as the possible risks of cyber-attacks, are discussed. In [15], the cyber-physical electrical energy systems are comprehensively explored, and their essential scientific challenges, such as co-simulation, energy-information network interaction, communication system failure, and cyber-physical system security, are discussed. Security problems and vulnerabilities in controlling grid-tied voltage source converters

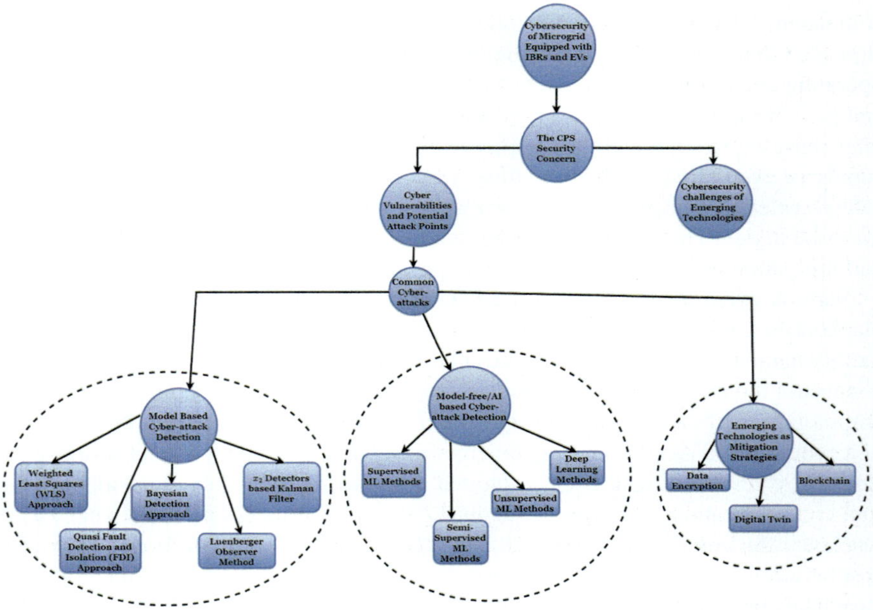

Fig. 1 Workflow in the literature

(VSCs) were explored in a recent study [7]. Cyber-attacks on VSCs in microgrids, high-voltage DC (HVDC), static synchronous compensators (STATCOM), and other systems are documented in [7]. Reference [16] investigates False-Data Injection (FDI) attacks in power systems. J. Zhang et al. suggested a framework for assessing cyber-attacks in a microgrid in [17]. The proposed method analyzes the effects of attack scenarios such as FDIs on the stability and performance of smart grids. Zhang et al. [18] investigated the vulnerabilities in a PV farm and presented machine learning and deep learning approaches for detecting cyber-attacks. The authors in [19–21] describe cyber-attack detection and diagnosis methods. Reference [22] offers a methodology for detecting cyber-attacks in a DC microgrid using invariants characterizing microgrid features. Reference [7] investigates stealthy cyberattack strategies in DC microgrids and introduces a cooperative vulnerability factor in secondary controllers based on the dynamic consensus algorithm to detect cyber occurrences. A statistical leverage score and binary matrix factorization are utilized in [19] to detect cyberattacks and defects in electric power grids using a revolutionary high-dimensional data-driven strategy. Reference [21] presents a strategy for detecting active deception attacks in microgrids. A resilient control strategy against DoS components of the Time Delay Switch (TDS) attack has been proposed in [23]. A fully distributed resilient control framework for secondary voltage and frequency restoration has been presented in [24]. Reference [25] proposes a resilient control strategy involving detection using adaptive discord elements and mitigation through an event-driven approach.

Machine learning and pre-defined-based methods are two categories of anomaly detection. Following the development of distributed voltage control architecture based on consensus algorithm for an isolated DC microgrid, [26] presented an analytical consistency-based anomaly detection mechanism. An Intrusion Detection System (IDS) based on the combination of network and grid data is proposed in [27]. Reference [28] proposed an artificial neural network based contextual anomaly detection method for low voltage microgrid. Reference [29] offers an anomaly detection technique to identify attacks on PV systems, such as PV disconnect, power curtailment, Volt-var attack, and reverse power flow in a section of the distribution grid with a specific amount of DER penetration. This method makes use of semi-supervised machine learning methods such as isolation forest, random forest with synthetic corruption, and principal component analysis (PCA), inverse-PCA with convex hulls. Reference [30] suggests a method for detecting cyberattacks on a PV system based on a fully connected neural network autoencoder. Reference [31] uses a deep learning method for fault identification in wind turbines that consists of long and short-term memory-stacked autoencoders, convolutional neural networks, and a softmax activation layer.

Since this modern energy grid is highly tied with cyber and communications interfaces to be efficiently operative, the system control design is more challenging. This control system is called Networked Control Systems (NCS), where the control procedure requires considerable cyber interfaces, including sensors, actuators, etc [32]. This chapter elaborates on the NCS development for IBRs and EV systems. The NCS must be constructed to detect the cyber anomalies and mitigate them. Section 2 discusses cyber-physical microgrid structure, cyber vulnerabilities, and potential cyber-attack points, different types of cyber-attacks are discussed in Sect. 3, then different types of detection and mitigation techniques are discussed in Sect. 4 and finally, the chapter is concluded in Sect. 5.

2 Cyber-Physical Microgrid Structure

Power electronics converters are used to interface dispersed generation, energy storage, and loads in smart microgrids. Information and communication technologies are strongly connected with the physical and electrical components, and their operations are tightly integrated with cyber system functions. In general, the cyber-physical paradigm of a microgrid has four layers (Fig. 2):

- *Physical Layer:* The physical layer of a microgrid consists of power generation, power lines, operating and protection equipment, e.g., circuit breakers, transformers, power electronic converters, and consumption modules, e.g., AC and DC loads.

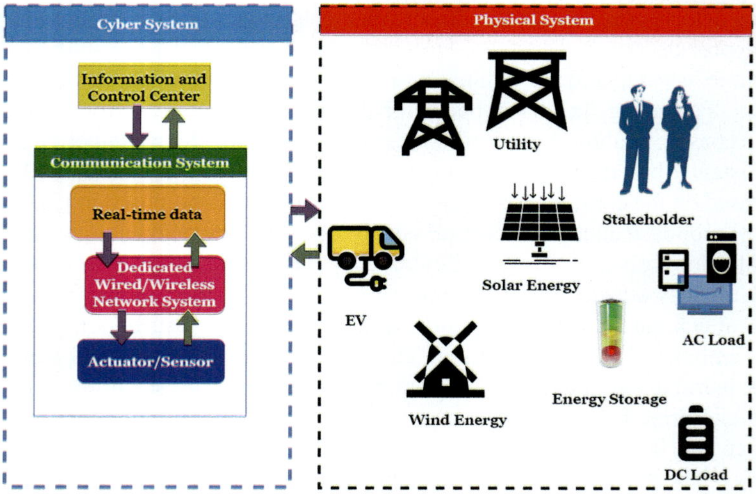

Fig. 2 Cyber-physical system of smart microgrid [11]

- *Sensor and Actuator Layer:* Sensors and measurement devices, and computing and processing resources make up the sensor and actuator layer. Sensors and measurement devices are utilized to monitor and control frequency, voltage, current, circuit breaker condition, etc. Generator controllers, distributed generation controllers, and circuit-breaker relays have actuators and control devices.
- *Communication Layer:* The communication layer, which includes routers, switches, and the communication medium, is in charge of information exchange between the various tiers. Depending on the system needs, smart microgrid communication systems can be wired or wireless.
- *Control and Management Layer:* The management layer is a centralized control system that oversees the microgrid's operation under various circumstances. This layer takes data from the measuring layer via the communication layer and generates control signals to ensure the optimal operation of smart microgrids. Control signals are once again supplied to actuators via the communication layer.

The modern cyber-physical microgrid model calls for a multi-layer control framework with different control time responses. As a result, the hierarchical architecture is implemented for microgrid control and management because of its intrinsic cascade structure, which allows numerous variables to be regulated almost independently [33, 34]. Hierarchical control structure comprises three control layers- Primary Control, Secondary Control, and Tertiary Control and their dependency on the communication network makes the system subject to cyber disruptions as shown in Fig. 3.

Primary control, also known as Droop control, is responsible for properly sharing power between DERs and addressing some of the power quality issues [3]. Since DERs are physically dispersed and management mostly relies on local measurements, communication in this control layer is often sparse or non-existent. Primary control

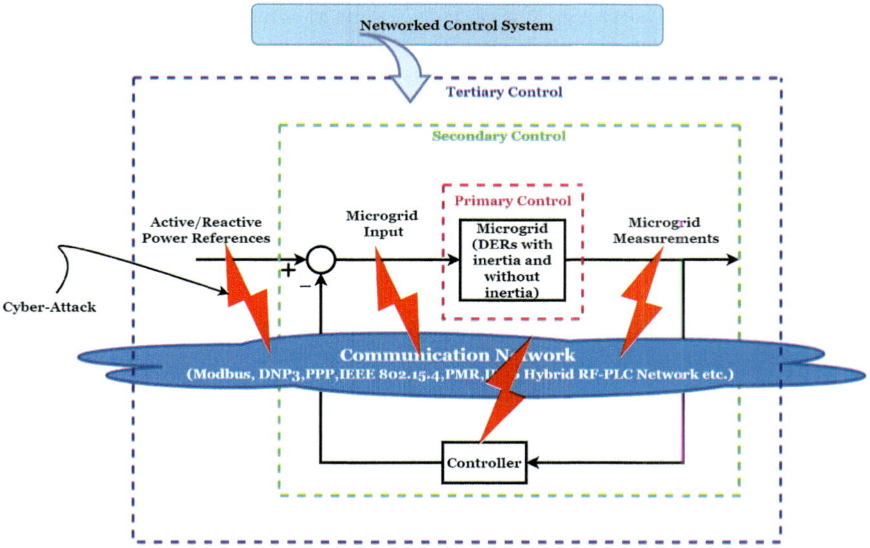

Fig. 3 Closed-loop control structure with cyber vulnerabilities

reacts to grid voltage and power mismatch with IBRs and spinning-based energy resources faster than secondary control. For DERs, the primary control addresses the issues mentioned earlier by interacting directly with power electronic converters, and this layer involves two control loops. The inner control loop maintains the regulation of inverter voltage and current output, and the outer control loop ensures safe and proper power-sharing between the DERs. As droop control is primarily independent of communication links, it is less susceptible to cyber intrusions.

The secondary control is in charge of correcting any voltage or frequency irregularities that the primary control failed to resolve. This control loop is more important in grids with higher penetration of Inverter-Based Resources (IBRs) and EV stations since they do not have enough inertia as spinning-based resources inherently have [35]. Because of time scale variations, the power references achieved by tertiary control are not necessarily consistent with the real system instantaneous power values [36]. Therefore, a mid-level control layer is needed to rectify any power mismatch. The secondary control modifies the power references obtained by tertiary control and sends them to the corresponding DERs to maintain grid stability and economic efficiency. Due to their high dependency on the communication channel, secondary control is highly susceptible to cyber-attacks.

Tertiary control manages energy similarly to the secondary control layer but in slower time samples on its constraints and objective functions. It is the highest control level in the hierarchical control structure which defines the optimal active and reactive power references of each Distributed Generator (DG), as well as set the price and energy level trade-offs [37], by taking economic and meteorological prediction data into account, with the electricity market to fulfill the balance between power generation and load consumption. Although the secondary control layer focuses on power quality and voltage and frequency regulation, the tertiary control layer focuses on microgrid economic factors, such as electricity market participation, spinning reserve management, and ancillary service provision [38].

3 Cyber Vulnerabilities and Potential Attack Points

A. Cyber Infrastructure Components and Vulnerable Spots

A microgrid network's vulnerability is the point where an attacker can access the network and attack the system. The microgrid is vulnerable to various threats since it connects to several domains via different protocols [39]. The factors that could make the grid more vulnerable to cyber-attacks are as follows:

- Increased installation of intelligent electronic devices (IEDs): The number of attack locations for attackers grows in lockstep with the number of devices in the network. The entire network system is affected even if only one point's security is breached.
- Incorporation of third-party components: Experts advise against using third-party components because they make the network more vulnerable to cyber-attack. These devices might be compromised with trojans, which could subsequently infect other network devices.
- Internet protocol utilization: Not all protocols are secure when it comes to data transmission. Some protocols use unencrypted data transport. As a result, they are prime targets for data theft through man-in-the-middle attacks.
- Insufficient personnel training: Any technology requires proper training to operate. Employees who are not well trained are more likely to fall for phishing scams.
- Maintenance: Even though the primary objective of maintenance is to keep things running smoothly, it can occasionally become a target for cyber-attacks. Operators frequently disable the security system for testing purposes while performing maintenance when it becomes easier for attackers to hit weak spots.

Depending on the factors that make microgrids vulnerable to cyber-attacks, the following attack points can be identified for a typical smart microgrid as shown in Fig. 4.

- Attack Point 1: The first type of attack is a physical attack on the hardware, such as tampering with PV modules, EV charging stations, wind turbines, combiner boxes, cables, or inverters. The most recent attacks have involved the theft and removal of PV modules to resell them [40].
- Attack Point 2: The second attack targets the inverter controller and algorithms and the plant supervisory system (for example, accessing and altering the inverter controller software or accessing the unit controller to shut down or destroy the plant). At any time, the plant monitoring and diagnostics system, internet-enabled

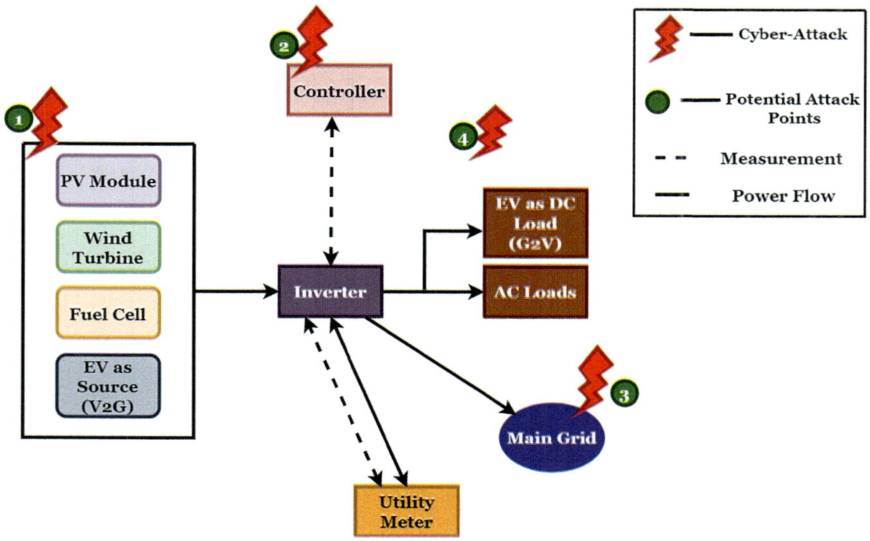

Fig. 4 Potential cyber-attack points in smart microgrid

communications, or the plant controller can be used to launch an attack on the inverter controls.

- Attack Point 3: Attacks targeting the grid can substantially influence the facility's functioning and overall safety (for example, faking energy demand and disconnecting the grid from the plant). This attack is identical to the second but spreads throughout the grid. By tripping breakers or causing low-voltage, high-voltage, or zero-voltage circumstances, hackers can disconnect inverters from the grid.
- Attack Point 4: Attacks spread across the supply chain (for example, faulty electronic components, poor analog or digital parts) are represented by attack number four. EV/PV inverters are high-tech electronic devices that make use of a variety of high-tech components. Malicious software could be hidden in these components, causing the inverter to malfunction.

B. Common Cyber Threats

Before discussing the detection and mitigation techniques, we explain some known cyber attacks. Active attacks and Passive attacks are two main types of cyber-attacks. Active attacks are more harmful than passive attacks because the attacker modifies or prevents the recipient from receiving the data, whereas passive attacks do not affect the data, and the attacker merely observes it. Some of these cyber disruptions are explained as follows:

(1) Eavesdropping Attack: In this type of passive attack, the data packets shared between the sender and the receiver are visible to the attacker.

(2) Traffic Analysis Attack: In this passive attack, the attackers constantly monitor and analyze traffic between the source and receiver.

(3) Masquerade Attack: In this type of active attack, the receiver continues to receive data from the attacker even though the sender is inactive.

(4) Replay Attack: This type of active attack, also referred to as playback attack, can repeat or delay the sensor data or control command to the distributed generators. In this attack, both the attacker and the sender send data to the receiver, making it difficult for the receiver to distinguish between genuine data from the sender and data routed through the attacker. The hackers save the data in the communication network first, then use it to fabricate sensor data by re-injecting it deliberately. This attack cannot be detected solely by watching sensor data or issuing a control command, but it can disrupt or deteriorate microgrid operations. Converter controller performance is harmed as a result of these discrepancies between recorded data and actual measurements.

(5) False Data Injection (FDI) Attack: In this type of active attack, instead of receiving data directly from the sender, the receiver receives data that the attacker has manipulated, but neither the sender nor the receiver is aware of the attacker's alteration. In several ways, the FDI attacks on measurement data and control orders are analogous to sensor and actuator problems. On the other hand, the faults are frequently thought to be random, independent events with a defined failure rate likelihood. On the other hand, adversaries can craft FDI attacks to cause the most harm possible while avoiding detection, perhaps leading to more significant consequences. In this instance, traditional fault-detection techniques would struggle to detect such sophisticated attacks [41–43]. For example, EV charging/discharging data is very widely used; therefore, precise measurement, processing, and analysis of the recorded data are critical. This charging/discharging data is collected by smart measuring devices installed at EV charging stations, opening a window for the FDI attack and thus causing severe damage to plug-in EVs.

(6) Denial-of-Service (DoS) Attack: A DoS attack is a common active IT attack that overloads a network with traffic, thus causing maximum possible deterioration of the system performance [44]. Sensor measurements and power references are inaccessible due to this type of attack. DoS attacks function by compromising a plant sensor and delaying measurement data, preventing the plant controller from obtaining inverter feedback on time. In this type of attack, the attacker goes against the data server by sending the server a large number of irrelevant requests, which the server then serves until all of its resources are used up. In addition, the sender/receiver seeks information from the server, but the request is refused because of a lack of resources. In terms of charging management systems and associated entities, attackers can target servers and prevent lawful EV users from sending valid requests. Important emergency vehicles (e.g., fire trucks, ambulances, and security vehicles) may be denied

charging due to Denial-of-Charge (a type of DoS), resulting in negative consequences for the various emergency services and society [45].

(7) Time-Delay Switch (TDS) Attack: Control systems frequently have time delays, which can affect their stability. Microgrids can contain inherent or intentional time delays. Microgrid communication channels are subjected to a TDS attack through the injection of time delays. Since the microgrid control system is highly dependent upon measurement data, these measured data are time-sensitive; therefore, a TDS might be catastrophic for the system. This work suggested by [46] proposes a Lyapunov-based method for real-time TDS attack estimation and detection. A neural-network-based detection technique is employed in [47] to identify and estimate TDS attacks quickly.

(8) Malware Injection via Electric Vehicle Supply Equipment (EVSE): EVSEs are vulnerable to malware injections due to their public availability, particularly at public charging stations. Payment information (debit/credit card information), personal information, charging time, payment amounts, and other sensitive information can be stolen from EVSEs infected with malware. The malware that infected the EVSE can potentially spread to a network of EVSEs. Malware implanted in EVSEs can spread to EVs, charging management systems, and power grids, causing harm to all parties [48]. The most effective strategy to combat these attacks is to do cybersecurity tests and assessments while deploying EVSEs.

(9) Stealthy Attack: The hacker's competence and professional understanding are essential for a stealthy attack. While remaining undiscovered, an attacker could have a continuous detrimental impact on the microgrid system's operation. Compared to the legacy power grid, an attacker can cause more significant harm to an IBR-based grid as remaining undiscovered for a more extended period. Reference [49] investigated stealthy attacks in a smart grid using a well-developed consensus-based protocol. Two machine learning approaches for detecting stealthy intrusions in a power grid were proposed by [50]. The length of time it takes to discover stealthy attacks on control systems is mostly determined by the complexity of the attack, which is based on the attackers' understanding of a power system model. Reference [51] examine how uncertainty in the power grid model affects the detection of stealthy attacks.

C. Emerging Technology and Cybersecurity Challenges

Integration of IBRs and EVs to smart grids require the utilization of other emerging technology at different scale of deployment. Industrial IoT (IIoT) facilitates the interconnection of the massive number of intelligent elements, such as loads, sensors, generators, protection relays, and grid operation components, in a smart grid [52]. IIoT enhances the grid efficiency and reliability in terms of serving customers and holistically coordinating facilities. IIoT provides end-to-end communication between remote access devices and the data center using low-power wide area networks relied on standard Internet Protocols (IPs) [53]. The common communication standard to automate the grid operation, specifically substation, is IEC 61850. A large deployment of internet connectivity in a grid depends on wireless communication, more specifically cellular technology, e.g., fifth-generation (5G) [54]. A cellular network is one of the pivotal technologies for enabling IoT, especially DER integration, for effective situational awareness and supervision. Considering the uncertain behavior of DERs and EVs energy exchange with the grid, wireless communication with low latency, noise immunity and traffic safety, and reliability, like 5G, is required to address smart grid concerns [55]. In addition, cloud technology is another critical emerging technology in the modern energy paradigm. Integrating intermittent energy sources and loads, like IBR and EVs, to grids necessitates the utilization of real-time functions and monitoring that is facilitated by cloud technology. Due to its pervasive application of the cloud in power systems, the security of this technology is vital for energy companies and stakeholders.

As discussed, all these technologies are interdependent, and a flaw or failure in one of them can cause disruptions to other technologies. For example, an issue with cloud computing service can disable the battery charging of electric vehicles at charging stations. Also, we need to ensure that these emerging technologies can be implemented in the current grid infrastructure that was not designed for these technologies. Therefore, to integrate DERs and EVs into a grid, we face the interdependency and interoperability concerns.

4 Detection and Mitigation Approaches

To outline cyber attack detection and mitigation methods, we need to learn the security objectives as discussed below:

A. Security Objectives of Cyber-Physical System (CPS)

The following security objectives must be met in order to assure the security [56] of a CPS:

(1) Availability: In a CPS, availability means providing service whenever required by preventing computation, control, and communication corruption caused by hardware failures, system upgrades, power outages, or attacks. It assures that each entity that uses the system's data, services, or resources may do so when needed.

(2) Reliability: In a CPS, reliability refers to the assurance that the data, transactions, and communications are authentic. In a CPS, the reliability checks all connected processes for uniqueness, such as monitoring, networking, computation, and actuation.

(3) Confidentiality: Confidentiality refers to the ability of the control system to prevent information from being revealed to unauthorized individuals or systems. Only

authorized users have access to the data as a result of this. In the microgrid network, there is much sensitive data flowing around. Client energy usage statistics, a customer identification number, and a list of appliances in use are all included in this data. An attacker can use this information to look into the customer's energy usage trends. If unauthorized individuals gain access to the data, an Internet Control Message Protocol (ICMP) flood attack can be conducted, and the reading interfered with, or altered [57]. As a result, utilities may suffer serious financial challenges, and customers may face exorbitant bills.

(4) Integrity: Modification of any resource or data is only permitted after authorization. A CPS must be capable of detecting any modifications introduced by unauthorized or malicious activity in the message being transmitted to ensure data integrity. It ensures that all data generated and shared during the system's functioning is correct and complete.

(5) Robustness: The degree to which a system can function properly and effectively even in faulty inputs, malfunctions, or disturbances is referred to as robustness of a CPS [58].

(6) Trustworthiness: In a CPS, trustworthiness refers to the system's ability to correctly complete system tasks under set operational and environmental conditions during a predetermined period. A security breach is referred to as a threat [59]. The entities which hamper normal grid operations are called attackers and to ensure smooth operation of microgrid, it must be protected from any kind of threats [60].

This section also discusses various cyber-attack detection and mitigation approaches that are classified as model-based cyber-attack and model-free techniques.

B. Model-Based Cyber-Attack Detection Strategies

Building a reliable attack detector is the first stage in constructing a secure CPS. Model-based cyber-attack detection methods compare actual system measurements with physics-based models that are supposed to imitate systems under no-fault situations to identify abnormalities by finding disparities between the modeled and actual performance [61]. A residual threshold is used to examine this inconsistency. The State-space model of a typical microgrid controller under attack can be derived as:

$$\begin{aligned} \dot{x} &= f_1(x, u_m) + d_1 \\ y &= f_2(x, u) + d_2 \\ u &= f_3(y_m) + d_3 \\ y_m &= y + \alpha_1 \\ u_m &= u + \alpha_2 \end{aligned} \tag{1}$$

Here, x, u, and y are the system state, input, and measurement variables. Nonlinear system dynamics, measurement equation, and control design equations are defined by f_1, f_2, f_3 respectively. d_1, d_2, d_3 represents disturbance signal; y_m, u_m defines manipulated input and output data; measurement information and the control decisions are defined by α_1 and α_2 respectively. This set of equations can be used to model the attacks discussed in Sect. 2.

(1) State Estimation Based Attack Detection Method (Weighted Least-Squares (WLS) Approach): For data measurement, the WLS methodology is an efficient and reliable attack detection tool. It is widely used, particularly in power systems and smart grids [62, 63]. The presence of a poor measurement is determined by comparing the produced measurement residual to a preset threshold. This idea can be elaborated by using the following standard WLS problem:

$$min\ x \equiv (y_m - f_2(x, u_m))^T W (y_m - f_2(x, u_m)) \tag{2}$$

$$r = y_m - f_2(\tilde{x}, u_m) \tag{3}$$

where, W is the weight matrix, and r is the residual. The estimated state $\tilde{x}$ based on measurement data y_m can be considered as a solution of Equation (2) [64]. An anomaly caused by cyber-attacks can be identified when the threshold is exceeded. The method is based on well-established power system state estimate approaches.

(2) Hypothesis Testing (Bayesian Detection Approach): The state-estimation method is well-established and works perfectly in most scenarios, but it contains some extreme loopholes that might prohibit them from working as intended. For instance, in Eqs. (2) and (3), this method will fail to converge if u_m is chosen in a way that makes r assign below the threshold.

The primary idea underlying Hypothesis Testing is to compute the conditional probability of a cyber-attack for two (or more) hypothetical scenarios; the attack is not present, or an adversary deliberates cyber-attacks with perfect system information [65]. The methods have been widely utilized to monitor a group of sensors [66], a subset is under attack, and a system monitor must find the corrupted sensors using all the incoming values. For example, Bayesian detection is used to identify deception attacks in sensor networks in [66, 67]. The performance limit of cooperative spectrum sensing is investigated in the face of Byzantine attacks, while the fusion center is harmed by false data, increasing the likelihood of faulty sensor output [68]. Reference [69] proposes a detector for both observable and unobservable events in Supervisory Control and Data Acquisition (SCADA) systems that are based on an enhanced likelihood ratio susceptible to colored Gaussian noise under false data injection attacks.

(3) Dynamic State Estimation Method: The primary idea behind utilizing dynamical state estimation methods is to determine the estimated system model and output y to compare with the received readings to detect cyber-attacks. In simple terms, a significantly different y and y_a means an anomaly is detected.

- Luenberger Observer Method: The Luenberger observer is a state estimation technique for a deterministic system with known control inputs and applies to a system with linear dynamics. A combination of matrix equalities and inequalities is solved for a continuous- or discrete-time dynamical system represented by state-space matrices, while providing asymptotic (or bounded) estimation error [70–72]. Observers frequently employ Matrix Solutions to these equations to estimate states and other dynamic quantities.
- $\mathcal{X}^2$ Detector based on Kalman filters: Instead of calculating the WLS, the Kalman filter residual characteristic is used in this approach, making it acceptable for usage in the presence of faulty or erroneous data. A residue-based detection approach is provided to detect deception attacks in a remote state estimation application where an estimator, equipped with a false data detector, is positioned at a distant location from the smart sensors and gets data from it [73]. This approach was also used to detect bias injection attacks for stochastic linear dynamical systems in [74]. An intrusion detection model was built using the $\mathcal{X}^2$ method feature selection and a multiclass Support Vector Machine (SVM) [75]. This method applies to nonlinear systems.

$$g = r^T Q^{-1} r \tag{4}$$

The $\mathcal{X}^2$ Detector calculates g by comparing the statistical characteristics of the obtained residual with the normal scenario. Where, Q is the covariance matrix of r.

(4) QFDI (Quasi-Fault Detection and Isolation) Approach: A well-known method that is frequently used in networked control systems (NCSs) is Quasi-fault detection and isolation (FDI). It entails keeping track of the system and determining the type, location, and occurrence of faults. This method has gained attention and has been used to check for external attacks in CPS applications. The CPS that had been the target of multiple attacks was treated as a descriptor system in [76], and the attacks were viewed as unidentified inputs that may impact both the state and the measurements. An Intelligent Generalized Predictive Controller (IGPC) which can detect both fault and cyber-attacks was developed in [77]. To identify defects and cyber-attacks in power grids a fault detection technique based on a geometric approach was proposed in [78].

C. Model-Free and Artificial-Intelligence (AI) Based Cyber-Attack Detection and Mitigation Strategies

Data-driven approaches that do not require physical models have piqued interest in smart grid applications in recent years. Data-driven approaches are utilized in power system areas to detect various cyber-attacks that misrepresent market and system operations. In general, most learning-based methods use data monitoring to discover system anomalies. The mismatch between fabricated data and estimated data is evaluated in terms of different features. Apart from detecting cyber-attacks, data-driven methods are utilized to differentiate between normal conditions, FDI, replay attacks, and physical problems.

Thousands of sensors are placed in the microgrid infrastructure as the traditional grid transitions to a smarter grid. These sensors continuously monitor the statuses of the grid components to which they are linked and generate massive amounts of data in the form of log files or time series data. Sensors in the microgrid network include irradiance sensors, module temperature sensors for PV sources, voltage monitor sensors, current monitor sensors, etc. The data from these sensors is saved on a server, and the data is pre-processed before being sent to the servers. The server can be either cloud or local. Posting data to a local server gives the maximum level of data security; nevertheless, it restricts the data's strength in discovering new patterns and gaining insights. When the data is kept on a cloud server, the user has more control over how the data is used because it can be accessed remotely and scraped to a workstation using a specific command, thus, making it more vulnerable to cyber disruptions. Data-driven techniques are more functional in cloud servers because they can train real-time data offline and implement it online. Different Machine Learning (ML) algorithms have been proven to be accurate in detecting these cyber intrusions as there is minimal detection error, and the error is independent of model uncertainty. ML, unlike rule-based techniques, detects intrusions based on historical data. This involves utilizing ML for automated vulnerability assessment, threat and risk prediction, and threat and risk intelligence collecting. The study of cryptographic protocols and lightweight cryptographic algorithms is another interesting topic for the security of microgrids. This is one of the fixes for the microgrid's communication protocol weaknesses. Data-driven techniques can be applied in supervised, semi-supervised, or unsupervised classifications.

(1) Supervised ML Methods: Data encryption is used to counter cyber-attack in a networked microgrid. Encryption involves taking some information and converting it into an unreadable form. The internet connection becomes encrypted when it is connected through a virtual private network (VPN), which means that if hackers attempted to intercept the data stream, all they would find is gibberish code. Encryption can be considered a type of secret code where a cipher is used to scramble the data and has a key that enables deciphering the message. The most endorsed VPN providers use Advanced Encryption Method (AES) with a 256-bit cipher, the strongest encryption standard currently in use. The bigger the cipher size, the harder it is to decode it. Banks and governments widely use data encryption [79]. A multilevel autoencoder model for detecting DoS attacks was developed in [80]. One input layer, one hidden layer, and one output layer make up an autoencoder. The model was trained with roughly 700 thousand packets of data and 49 features. Some features included source and destination IP and ports, source and destination jitters, record time, and attack category. The model was built using the UNSW-NB15 publically available data set. The autoencoder-based prediction model outperformed the random forest, decision tree, long short-term memory (LSTM), Lagrangian Support Vector Machine (LSVM), and k-nearest neighbor models.

(2) Deep Learning Methods: To identify power system disturbances, a combination of JRipper and Adaboost was created in [81]. Based on the data, the model produced three groups; natural disturbances, attacks, and no event. The researchers in [6] utilized an ensemble-based ML technique to detect an FDI attack. The model was tested on an IEEE 14 bus system. The results showed that unsupervised ensemble models outperformed individual models with the maximum accuracy of 73% when compared to linear regression, naive Bayes, decision tree, and SVM. Reference [82] uses a Multilayer Perceptron (MLP) to do a detailed investigation of the influence of FDI attacks on AI-based smart grids. The study's findings demonstrate that false data can impair the accuracy of ML algorithms by 15%, affecting vital smart grid decision-making. In the case of data poisoning, for example, if there is a disturbance and the model fails to foresee the disturbance owing to incorrect data, the grid can become unstable, resulting in catastrophic outcomes. A conditional Deep Belief Network (DBN) model for detecting FDI attacks for power theft in real-time is proposed in [83]. IEEE 118 and IEEE 300 bus systems were used to test the model, and the model's performance was compared to that of artificial neural networks and SVM in [83]. Microgrid digitization could benefit significantly from the application of edge computing, a unique computing paradigm [84]. The authors of [85] examined how cloud computing characteristics could be leveraged to improve security in the event of a DoS assault on the smart grid. A cloud-based firewall was proposed in [86] to protect the smart grid from DoS attacks. To recreate a DoS attack, the researchers generated 250 Gbps of data. The grid OpenFlow firewall has low latency, according to the simulation results. In [87], an attribute-based online/offline searchable encryption technique was presented to safeguard data access in the cloud for smart grid applications for authorized users. The authors of [88] presented a cloud-based secure home area network that is resistant to brute force, replay and capture, and other attacks. A smart grid security evaluation model based on a deep belief network (DBN) built of multiple RBMs, and a BP neural network was suggested in [89]. Security risks were assessed in five categories: policy and organizational risks, general technological risks, SaaS risks, PaaS risks, and IaaS risks.

Data authentication can be maintained by implementing SSL protocols. However, SSL protocols are subjected to DoS attacks; cryptographic algorithms can be applied to counterfeit the DoS attacks [90]. Network Intrusion Prevention System (NIPS) and Network Intrusion Detection System (NIDS) can be used in malware protection where the first one is used to prevent an attack from happening, and the latter is used to evaluate network traffic to safeguard the system against network-based attacks [91].

D. Emerging Technologies for Secure Design and Operation

Although cybersecurity concerns were raised earlier with some technologies integrating into smart grids, some of these technologies can be utilized to secure the integration of RESs and EVs into grids, as discussed in the followings.

(1) Digital Twin: A digital twin is a digital model of our system that replicates the principal physical components, cyber interfaces, and the entire grid behavior and process. Since smart grids have the combined nature of physical elements and cyber links with a large number of nodes and interactions, a digital twin can help the grid operators to analyze the grid functionality at different working modes in the presence of IBRs, and EVs [92]. Due to progress in Hardware-in-the-Loop (HIL) and Software-in-the-Loop (SIL) implementation of smart grids, we can design a grid equipped with DERs and IBRs with fewer cybersecurity risks since numerous grid conditions can be tested and verified using the HIL and SIL techniques [93]. A digital twin as a co-simulation can simulate the grid in a virtual environment, identify the cyber vulnerabilities, and verify the effectiveness of the proposed solutions to resolve the security flaws.

(2) Blockchain: Microgrids are changing to use big data for operations and services. Information and communication technologies (ICT) are a key component of this transformation, notably for the computing model that specifies how data analytics in a microgrid can be carried out. Due to its security, blockchain has recently become one of the most valuable technologies in various fields. The blockchain is a series of blocks containing each block's index, timestamp, previous hash, hash, and data. Because of the hashing, blockchain is thought to be secure. If someone attempts to modify the hash value of a block, the hash value of all prior blocks must be changed. Changing the hashes of all previous blocks becomes a computationally expensive process when numerous blocks are in the chain. To protect against FDI attacks, the authors in [94] developed a policy architecture based on blockchain for data interchange between autonomous system operations and under-running agents. The data layer, the detection layer, and the blockchain layer are the three layers in the model. The data layer is in charge of collecting data, which is then sent to the detection layer for community detection, and the blockchain layer secures the community detection and transaction record. The authors of [95] proposed a blockchain-based secure message transfer technique for smart meters and service providers as part of their research. On the smart meter side, the approach precludes FDI attacks. The smart meters initiate each transaction in this study, and the service provider is the master node. The transaction data is disseminated via the network and checked regularly through auditing and broadcasting transactions. Peer-to-peer (P2P) networks are used to connect service providers. Consensus verification is required before adding a new transaction/block, and the new block is only added after verification. Every transaction generates a key using the SHA-256 algorithm. The authors demonstrated in this study that data can be shared within a P2P service provider using a blockchain-based structure. The study [96] developed a decentralized security paradigm in the blockchain ecosystem based on the lightning network and smart contracts. Registration, scheduling, authentication, and billing are all phases of this approach. The authors of [68] presented a unique framework for grid-edge devices that combines integrated hardware security and a blockchain system to maintain a distributed cybersecurity technique that checks the provenance of messages sent to and from the devices.

(3) Software Defined Networking: To enhance network performance and monitoring, software-defined networking (SDN) technology offers a dynamic, resource-efficient, and programmable network setup. By separating the forwarding of network packets from the routing process and control activities, SDN centralizes the network intelligence in one network component. The performance, security, and safety of smart grid control systems can all be enhanced by SDN. SDN has intriguing applications for industrial control system (ICS) security, particularly for incident response. The ability to dynamically reconfigure the network following the discovery of a failure or a hacked device increases the control system's resilience and enables it to function even under challenging circumstances. Authors in [97] present a thorough analysis of SDN-based smart grid communication. Applications include microgrids, phasor measurement units, advanced metering infrastructures, and substation automation. Even though such an implementation might introduce a latency incompatible with the permitted latency in a microgrid context, SDN permits directly integrating a Network Intrusion Detection System (NIDS) within the SDN controller as shown in [98]. According to [99], microgrids could greatly benefit from self-healing network management, which entails quick switch resets and configurations to isolate suspicious devices from the microgrid and on-demand path formation for control commands to reduce the attacker's operation window.

5 Conclusion

Microgrids are superior to typical legacy power grids in terms of competence and productivity because they are more efficient than traditional power grids, utilize many renewable energy sources, and are ecologically beneficial. Additionally, the research delineated potential advantages and threats associated with the Microgrid. Overall, employing a microgrid needs a broad spectrum of security measures to combat cyber-attack challenges. Moreover, this study's conclusion emphasizes the significance of self-awareness regarding cyber-attacks on microgrids. The energy industry should be aware of the threats associated with the microgrid and take steps to reduce them by doing various risk assessments and case studies to offer an additional method of safeguarding the microgrid from various forms of cyberattack.

Microgrids are far more sensitive to cyber-attacks since they heavily require cyber systems and communication networks. Furthermore, since these power electronics-dominated microgrids have minimal inertia, cyber-attacks may impact their stability and operation more severely. After exploring different aspects of CPS in microgrids, this chapter has focused on cyber-attacks depending on data availability, integrity, and secrecy. Moreover, different types of model-based/model-free detection and mitigation techniques are discussed to address the potential cyber-attacks.

References

1. 2020 long-term reliability assessment. https://www.nerc.com/pa/RAPA/ra/Reliability%20 Assessments. Accessed 2022-07-10
2. Summary report on evs at scale and the U.S. electric power system. https://www.energy.gov/sites/prod/files/2019/12/f69/GITT%20ISATT. Accessed 2022-07-10
3. N. Pogaku, M. Prodanovic, T.C. Green, Modeling, analysis and testing of autonomous operation of an inverter-based microgrid. IEEE Trans. Power Electron. **22**(2), 613–625 (2007)
4. I.S. Bayram, I. Papapanagiotou, A survey on communication technologies and requirements for internet of electric vehicles. EURASIP J. Wirel. Commun. Netw. **2014**(1), 1–18 (2014)
5. N. Priyadharshini, S. Gomathy, M. Sabarimuthu, A review on microgrid architecture, cyber security threats and standards, in *Materials Today: Proceedings* (2020)
6. M. Ashrafuzzaman, S. Das, Y. Chakhchoukh, S. Shiva, F.T. Sheldon, Detecting stealthy false data injection attacks in the smart grid using ensemble-based machine learning. Comput. Sec. **97**, 101994 (2020)
7. S. Sahoo, T. Dragičević, F. Blaabjerg, Cyber security in control of grid-tied power electronic converters-challenges and vulnerabilities. IEEE J. Emerg. Select. Top. Power Electron. **9**(5), 5326–5340 (2019)
8. First-of-a-kind U.S. grid cyberattack hit wind, solar. https://www.wind-watch.org/news/2019/11/02/first-of-a-kind-u-s-grid-cyberattack-hit-wind-solar/
9. A. Bindra, Securing the power grid: Protecting smart grids and connected power systems from cyberattacks. IEEE Power Electron. Mag. **4**(3), 20–27 (2017)
10. E.L. McCoy, *Cyberterrorism* (Cavendish Square Publishing, LLC, 2018)
11. F. Nejabatkhah, Y.W. Li, H. Liang, R.R. Ahrabi, Cyber-security of smart microgrids: a survey. Energies **14**(1), 27 (2020)
12. H.F. Habib, C.R. Lashway, O.A. Mohammed, A review of communication failure impacts on adaptive microgrid protection schemes and the use of energy storage as a contingency. IEEE Trans. Ind. Appl. **54**(2), 1194–1207 (2017)
13. M.H. Cintuglu, O.A. Mohammed, K. Akkaya, A. Selcuk Uluagac, A survey on smart grid cyber-physical system testbeds. IEEE Commun. Surv. Tutor. **19**(1), 446–464 (2016)
14. Z. Li, M. Shahidehpour, F. Aminifar, Cybersecurity in distributed power systems. Proc. IEEE **105**(7), 1367–1388 (2017)
15. X. Shi, Y. Li, Y. Cao, Y. Tan, Cyber-physical electrical energy systems: challenges and issues. CSEE J. Power Energy Syst. **1**(2), 36–42 (2015)
16. G. Liang, J. Zhao, F. Luo, S.R. Weller, Z.Y. Dong, A review of false data injection attacks against modern power systems. IEEE Trans. Smart Grid **8**(4), 1630–1638 (2016)
17. J. Zhang, J. Ye, L. Guo, F. Li, W. Song, Vulnerability assessments for power-electronics-based smart grids, in *2020 IEEE Energy Conversion Congress and Exposition (ECCE)* (IEEE, 2020), pp. 1702–1707
18. J. Zhang, Q. Li, J. Ye, L. Guo, Cyber-physical security framework for photovoltaic farms, in *2020 IEEE CyberPELS (CyberPELS)* (IEEE, 2020), pp. 1–7
19. F. Li, R. Xie, B. Yang, L. Guo, P. Ma, J. Shi, J. Ye, W.Z. Song, *Detection and identification of cyber and physical attacks on distribution power grids with pvs: An online high-dimensional data-driven approach* (IEEE J. Emerg. Select. Top, Power Electron, 2019)
20. F. Li, Q. Li, J. Zhang, J. Kou, J. Ye, W.Z. Song, H.A. Mantooth, Detection and diagnosis of data integrity attacks in solar farms based on multilayer long short-term memory network. IEEE Trans. Power Electron. **36**(3), 2495–2498 (2020)
21. Y. Li, P. Zhang, L. Zhang, B. Wang, Active synchronous detection of deception attacks in microgrid control systems. IEEE Trans. Smart Grid **8**(1), 373–375 (2016)
22. O.A. Beg, T.T. Johnson, A. Davoudi, Detection of false-data injection attacks in cyber-physical dc microgrids. IEEE Trans. Ind. Inf. **13**(5), 2693–2703 (2017)
23. A. Sargolzaei, K.K. Yen, M.N. Abdelghani, S. Sargolzaei, B. Carbunar, Resilient design of networked control systems under time delay switch attacks, application in smart grid. IEEE Access **5**, 15901–15912 (2017)

24. S. Zuo, O.A. Beg, F.L. Lewis, A. Davoudi, Resilient networked ac microgrids under unbounded cyber attacks. IEEE Trans. Smart Grid **11**(5), 3785–3794 (2020)
25. S. Sahoo, T. Dragičević, F. Blaabjerg, Resilient operation of heterogeneous sources in cooperative dc microgrids. IEEE Trans. Power Electron. **35**(12), 12601–12605 (2020)
26. D. Shi, Yu. Pengfeng Lin, C.-C.C. Wang, X. Yan, P. Wang, Deception attack detection of isolated dc microgrids under consensus-based distributed voltage control architecture. IEEE J. Emerg. Select. Top. Circuits Syst. **11**(1), 155–167 (2021)
27. A. Chavez, C. Lai, N. Jacobs, S. Hossain-McKenzie, C.B. Jones, J Johnson, A. Summers, Hybrid intrusion detection system design for distributed energy resource systems, in *2019 IEEE CyberPELS (CyberPELS)* (IEEE, 2019), pp. 1–6
28. A.M. Kosek, Contextual anomaly detection for cyber-physical security in smart grids based on an artificial neural network model, in *2016 Joint Workshop on Cyber-Physical Security and Resilience in Smart Grids (CPSR-SG)* (IEEE, 2016), pp. 1–6
29. D.M. Shilay, K.G. Lorey, T. Weiz, T. Lovetty, Y. Cheng, Catching anomalous distributed photovoltaics: An edge-based multi-modal anomaly detection (2017). arXiv preprint arXiv:1709.08830
30. G.B. Gaggero, M. Rossi, P. Girdinio, M. Marchese, Detecting system fault/cyberattack within a photovoltaic system connected to the grid: a neural network-based solution. J. Sens. Actuator Netw. **9**(2), 20 (2020)
31. K. Fotiadou, T.H. Velivassaki, A. Voulkidis, D. Skias, C. De Santis, T. Zahariadis, Proactive critical energy infrastructure protection via deep feature learning. Energies **13**(10), 2622 (2020)
32. M.R. Khalghani, V. Verma, S.K. Solanki, J.M. Solanki, Resilient networked control of inverter-based microgrids against false data injections. Electronics **11**(5) (2022)
33. A. Bidram, A. Davoudi, Hierarchical structure of microgrids control system. IEEE Trans. Smart Grid **3**(4), 1963–1976 (2012)
34. L. Meng, E.R. Sanseverino, A. Luna, T. Dragicevic, J.C. Vasquez, J.M. Guerrero, Microgrid supervisory controllers and energy management systems: a literature review. Renew. Sustain. Energy Rev. **60**, 1263–1273 (2016)
35. M.R. Khalghani, J. Solanki, S.K. Solanki, A. Sargolzaei, Resilient and stochastic load frequency control of microgrids, in *2019 IEEE Power and Energy Society General Meeting (PESGM)* (2019), pp. 1–5
36. M.R. Khalghani, J. Solanki, S.K. Solanki, M.H. Khooban, A. Sargolzaei, Resilient frequency control design for microgrids under false data injection. IEEE Trans. Ind. Electron. **68**(3), 2151–2162 (2020)
37. X. Jin, W. Jianzhong, M. Yunfei, M. Wang, X. Xiandong, H. Jia, Hierarchical microgrid energy management in an office building. Appl. Energy **208**, 480–494 (2017)
38. F.G. Torres. *Advanced control of renewable energy microgrids with hybrid energy storage system*. PhD thesis, PHD thesis (Seville, Spain, 2015)
39. S. Tufail, I. Parvez, S. Batool, A. Sarwat, A survey on cybersecurity challenges, detection, and mitigation techniques for the smart grid. Energies **14**(18), 5894 (2021)
40. M. Avrutsky, R. Hadar, S. Arditi, Enhanced system and method for theft prevention in a solar power array during nonoperative periods, April 14 2015. US Patent 9,007,210
41. Z. Gao, C. Cecati, S.X. Ding, A survey of fault diagnosis and fault-tolerant techniques-part i: fault diagnosis with model-based and signal-based approaches. IEEE Trans. Ind. Electron. **62**(6), 3757–3767 (2015)
42. S. Yin, G. Wang, H. Gao, Data-driven process monitoring based on modified orthogonal projections to latent structures. IEEE Trans. Control Syst. Technol. **24**(4), 1480–1487 (2015)
43. G. Wang, S. Yin, Quality-related fault detection approach based on orthogonal signal correction and modified pls. IEEE Trans. Industr. Inf. **11**(2), 398–405 (2015)
44. K. Manandhar, X. Cao, H. Fei, Y. Liu, Detection of faults and attacks including false data injection attack in smart grid using kalman filter. IEEE Trans. Control Netw. Syst. **1**(4), 370–379 (2014)
45. J. Qin, M. Li, L. Shi, Yu. Xinghuo, Optimal denial-of-service attack scheduling with energy constraint over packet-dropping networks. IEEE Trans. Autom. Control **63**(6), 1648–1663 (2017)

46. M. Victorio, A. Sargolzaei, M.R. Khalghani, A secure control design for networked control systems with linear dynamics under a time-delay switch attack. Electronics **10**(3), 322 (2021)
47. A. Abbasspour, A. Sargolzaei, M. Victorio, N. Khoshavi, A neural network-based approach for detection of time delay switch attack on networked control systems. Proc. Comput. Sci. **168**, 279–288 (2020)
48. K. Harnett, B. Harris, D. Chin, G. Watson et al., Doe/dhs/dot volpe technical meeting on electric vehicle and charging station cybersecurity report. Technical report, John A. Volpe National Transportation Systems Center (US) (2018)
49. C. Zhao, J. He, P. Cheng, J. Chen, Analysis of consensus-based distributed economic dispatch under stealthy attacks. IEEE Trans. Industr. Electron. **64**(6), 5107–5117 (2016)
50. M. Esmalifalak, L. Liu, N. Nguyen, R. Zheng, Z. Han, Detecting stealthy false data injection using machine learning in smart grid. IEEE Syst. J. **11**(3), 1644–1652 (2014)
51. S. Harshbarger, M. Hosseinzadehtaher, B. Natarajan, E. Vasserman, M. Shadmand, G. Amariucai, (a little) ignorance is bliss: The effect of imperfect model information on stealthy attacks in power grids, in *2020 IEEE Kansas Power and Energy Conference (KPEC)* (IEEE, 2020), pp. 1–6
52. Z. Mahmood, *The Internet of Things in the Industrial Sector* (Springer, 2019)
53. S. Mumtaz, A. Alsohaily, Z. Pang, A. Rayes, K.F. Tsang, J. Rodriguez, Massive internet of things for industrial applications: Addressing wireless iiot connectivity challenges and ecosystem fragmentation. IEEE Ind. Electron. Mag. **11**(1), 28–33 (2017)
54. S. Al-Rubaye, E. Kadhum, Q. Ni, A. Anpalagan, Industrial internet of things driven by sdn platform for smart grid resiliency. IEEE Internet Things J. **6**(1), 267–277 (2019)
55. M. Garau, M. Anedda, C. Desogus, E. Ghiani, M. Murroni, G Celli, A 5g cellular technology for distributed monitoring and control in smart grid, in *2017 IEEE International Symposium on Broadband Multimedia Systems and Broadcasting (BMSB)* (2017), pp. 1–6
56. M.S. Mahmoud, Y. Xia, *Cloud Control Systems: Analysis, Design and Estimation* (Academic Press, 2020)
57. K. Shuaib, Z. Trabelsi, M. Abed-Hafez, A. Gaouda, M. Alahmad, Resiliency of smart power meters to common security attacks. Proc. Comput. Sci. **52**, 145–152 (2015)
58. N. Adam, Workshop on future directions in cyber-physical systems security, in *Report on workshop organized by Department of Homeland Security (DHS)* (2010)
59. R. Saltzman, A. Sharabani, Active man in the middle attacks, in *OWASP AU* (2009)
60. K.K. Fletcher, X. Liu, Security requirements analysis, specification, prioritization and policy development in cyber-physical systems, in *2011 Fifth International Conference on Secure Software Integration and Reliability Improvement-Companion* (IEEE, 2011), pp. 106–113
61. S. Tan, J.M. Guerrero, P. Xie, R. Han, J.C. Vasquez, Brief survey on attack detection methods for cyber-physical systems. IEEE Syst. J. **14**(4), 5329–5339 (2020)
62. G. Hug, J.A. Giampapa, Vulnerability assessment of ac state estimation with respect to false data injection cyber-attacks. IEEE Trans. Smart Grid **3**(3), 1362–1370 (2012)
63. L. Liu, M. Esmalifalak, Q. Ding, V.A. Emesih, Z. Han, Detecting false data injection attacks on power grid by sparse optimization. IEEE Trans. Smart Grid **5**(2), 612–621 (2014)
64. J. Liu, R. Singh, B.C. Pal, Distribution system state estimation with high penetration of demand response enabled loads. IEEE Trans. Power Syst. **36**(4), 3093–3104 (2021)
65. X. Ren, J. Yan, Y. Mo, Binary hypothesis testing with byzantine sensors: Fundamental tradeoff between security and efficiency. IEEE Trans. Signal Process. **66**(6), 1454–1468 (2018)
66. B. Kailkhura, Y.S. Han, S. Brahma, P.K. Varshney, Asymptotic analysis of distributed bayesian detection with byzantine data. IEEE Signal Process. Lett. **22**(5), 608–612 (2014)
67. B. Kailkhura, Y.S. Han, S. Brahma, P.K. Varshney, Distributed bayesian detection in the presence of byzantine data. IEEE Trans. Signal Process. **63**(19), 5250–5263 (2015)
68. A.S. Rawat, P. Anand, H. Chen, P.K. Varshney, Collaborative spectrum sensing in the presence of byzantine attacks in cognitive radio networks. IEEE Trans. Signal Process. **59**(2), 774–786 (2010)
69. B. Tang, J. Yan, S. Kay, H. He, Detection of false data injection attacks in smart grid under colored gaussian noise, in *2016 IEEE Conference on Communications and Network Security (CNS)* (IEEE, 2016), pp. 172–179

70. S.K. Park, L. Xu, *Data Assimilation for Atmospheric, Oceanic and Hydrologic Applications (Vol. II)* (Springer, 2013)
71. A. Radke, Z. Gao, A survey of state and disturbance observers for practitioners, in *2006 American Control Conference* (IEEE, 2006), p. 6
72. J. Qi, A.F. Taha, J. Wang, Comparing kalman filters and observers for power system dynamic state estimation with model uncertainty and malicious cyber attacks. IEEE Access **6**, 77155–77168 (2018)
73. Z. Guo, D. Shi, K.H. Johansson, L. Shi, Optimal linear cyber-attack on remote state estimation. IEEE Trans. Control Netw. Syst. **4**(1), 4–13 (2016)
74. J. Miloševič, T. Tanaka, H. Sandberg, K.H. Johansson, Analysis and mitigation of bias injection attacks against a kalman filter. IFAC-PapersOnLine **50**(1), 8393–8398 (2017)
75. I.S. Thaseen, C.A. Kumar, Intrusion detection model using fusion of chi-square feature selection and multi class svm. J. King Saud Univ.-Comput. Inf. Sci. **29**(4), 462–472 (2017)
76. F. Pasqualetti, F. Dörfler, F. Bullo, Attack detection and identification in cyber-physical systems. IEEE Trans. Autom. Control **58**(11), 2715–2729 (2013)
77. A.A. Yaseen, M. Bayart, Cyber-attack detection with fault accommodation based on intelligent generalized predictive control. IFAC-PapersOnLine **50**(1), 2601–2608 (2017)
78. H. Nishino, H. Ishii, Distributed detection of cyber attacks and faults for power systems. IFAC Proc. **47**(3), 11932–11937 (2014)
79. X. Li, X. Liang, L. Rongxing, X. Shen, X. Lin, H. Zhu, Securing smart grid: cyber attacks, countermeasures, and challenges. IEEE Commun. Mag. **50**(8), 38–45 (2012)
80. S. Ali, Y. Li, Learning multilevel auto-encoders for ddos attack detection in smart grid network. IEEE Access **7**, 108647–108659 (2019)
81. R.C. Borges Hink, J.M. Beaver, M.A. Buckner, T. Morris, U. Adhikari, S. Pan, Machine learning for power system disturbance and cyber-attack discrimination, in *2014 7th International symposium on resilient control systems (ISRCS)* (IEEE, 2014), pp. 1–8
82. D. Mukherjee, A novel strategy for locational detection of false data injection attack. Sustain. Energy, Grids Netw. **31**, 100702 (2022)
83. Y. He, G.J. Mendis, J. Wei, Real-time detection of false data injection attacks in smart grid: a deep learning-based intelligent mechanism. IEEE Trans. Smart Grid **8**(5), 2505–2516 (2017)
84. C. Feng, Y. Wang, Q. Chen, Y. Ding, G. Strbac, C. Kang, Smart grid encounters edge computing: Opportunities and applications. Adv. Appl. Energy **1**, 100006 (2021)
85. A. Califano, E. Dincelli, S. Goel, Using features of cloud computing to defend smart grid against ddos attacks, in *10th Annual Symposium on Information Assurance (Asia 15), ALBANY* (2015)
86. R.C. Diovu, J.T. Agee, A cloud-based openflow firewall for mitigation against ddos attacks in smart grid ami networks, in *2017 IEEE PES PowerAfrica* (IEEE, 2017), pp. 28–33
87. N. Eltayieb, R. Elhabob, A. Hassan, F. Li, An efficient attribute-based online/offline searchable encryption and its application in cloud-based reliable smart grid. J. Syst. Architect. **98**, 165–172 (2019)
88. B. Alohali, M. Merabti, K. Kifayat, A cloud of things (cot) based security for home area network (han) in the smart grid, in *2014 Eighth International Conference on Next Generation Mobile Apps, Services and Technologies* (IEEE, 2014, pp. 326–330
89. L. Chen, J. Liu, W. Ha, Cloud service security evaluation of smart grid using deep belief network. Int. J. Sens. Netw. **33**(2), 109–121 (2020)
90. F. Aloul, A.R. Al-Ali, R. Al-Dalky, M. Al-Mardini, W. El-Hajj, Smart grid security: threats, vulnerabilities and solutions. Int. J. Smart Grid Clean Energy **1**(1), 1–6 (2012)
91. L.A. Maglaras, J. Jiang, Ocsvm model combined with k-means recursive clustering for intrusion detection in scada systems, in *10th International Conference on Heterogeneous Networking for Quality, Reliability, Security and Robustness* (IEEE, 2014), pp. 133–134
92. A. Saad, S. Faddel, O. Mohammed, Iot-based digital twin for energy cyber-physical systems: design and implementation. Energies **13**(18), 4762 (2020)
93. V.H. Nguyen, Q.T. Tran, Y. Besanger, M. Jung, T.L. Nguyen, Digital twin integrated power-hardware-in-the-loop for the assessment of distributed renewable energy resources. Electr. Eng. **104**(2), 377–388 (2022)

94. M. Dehghani, M. Ghiasi, T. Niknam, A. Kavousi-Fard, M. Shasadeghi, N. Ghadimi, F. Taghizadeh-Hesary, Blockchain-based securing of data exchange in a power transmission system considering congestion management and social welfare. Sustainability **13**(1), 90 (2020)
95. H. Zhang, J. Wang, Y. Ding, Blockchain-based decentralized and secure keyless signature scheme for smart grid. Energy **180**, 955–967 (2019)
96. X. Huang, X. Cheng, P. Wang, H. Liu, Lnsc: a security model for electric vehicle and charging pile management based on blockchain ecosystem. IEEE Access **6**, 13565–13574 (2018)
97. M.H. Rehmani, A. Davy, B. Jennings, C. Assi, Software defined networks-based smart grid communication: a comprehensive survey. IEEE Commun. Surv. Tutor. **21**(3), 2637–2670 (2019)
98. A. Fausto, M. Marchese, Implementation details to reduce the latency of an sdn statistical fingerprint-based ids, in *2019 International Symposium on Advanced Electrical and Communication Technologies (ISAECT)* (IEEE, 2019), pp. 1–6
99. D. Jin, Z. Li, C. Hannon, C. Chen, J. Wang, M. Shahidehpour, C.W. Lee, Toward a cyber resilient and secure microgrid using software-defined networking. IEEE Trans. Smart Grid **8**(5), 2494–2504 (2017)

Improving Cybersecurity Situational Awareness in Smart Grid Environments

Thusitha Thilina Dayaratne, Fariha Tasmin Jaigirdar, Rumpa Dasgupta, Amin Sakzad, and Carsten Rudolph

Abstract Demand response (DR) and peer-to-peer (P2P) energy trading in smart grids use distributed architectures and multiple data sources to enable more consumer involvement. Given the reliance on and behind-the-meter data and the distributed and heterogeneous setups, these data and processes are prone to various cybersecurity attacks. Hence, identification of security risks and continuous situational awareness is essential to establish system trust and resilience. In such a multi-layered, distributed system, data origin and the steps for processing, modifying and aggregating data are highly significant. Data provenance denotes metadata describing data derivation throughout the different layers of the system. Tracking data provenance can provide valuable information on data history and lineage. However, while provenance generates metadata for data history, security-relevant information to estimate relevant risks are not addressed. This chapter emphasises the need for security-aware data provenance in residential DR and P2P energy trading. Based on the existing Prov-IoT model for security-aware provenance in the Internet of Things applications, we present a refined model with entities and metadata specific to smart grids and microgrids. This instantiation named *Prov-IoT-MG*, demonstrates the importance and necessity of security-aware provenance graphs for continuously estimating risks against man-in-the-middle, false data injection and load altering attacks. We

---

T. T. Dayaratne · F. T. Jaigirdar (✉) · R. Dasgupta · A. Sakzad · C. Rudolph
Faculty of IT, Monash University, Melbourne, Australia
e-mail: fariha.jaigirdar1@monash.edu

T. T. Dayaratne
e-mail: thusitha.dayaratne@monash.edu

R. Dasgupta
e-mail: rumpa.dasgupta@monash.edu

A. Sakzad
e-mail: amin.sakzad@monash.edu

C. Rudolph
e-mail: carsten.rudolph@monash.edu

H. Haes Alhelou et al. (eds.), *Power Systems Cybersecurity*, Power Systems,
https://doi.org/10.1007/978-3-031-20360-2_5

illustrate how Prov-IoT-MG graphs can be generated and evaluated at run-time and are useful in providing up-to-date information on active security controls and other security-relevant information. Finally, we discuss how these graphs help to improve the resilience of grid processes with higher situational awareness.

Keywords Smart grid security provenance · Cybersecurity awareness · Security metadata

1 Introduction

The integration of information and communication technology (ICT) and the penetration of distributed energy resources (DERs) across the traditional power grid enable smart grid solutions and related technologies such as Demand Response (DR) [9] and Peer-to-Peer (P2P) energy trading schemes [34]. DR and P2P energy trading schemes are expected to maintain the demand-supply balance across the grid and produce benefits for users and utility companies by reducing the cost. P2P energy trading allows prosumers and consumers to sell and buy their surplus and deficit energy to/from neighbours and other market participants. Further, peak demand reductions and efficient demand distribution potentially improve efficiency and reduce utilities' costs.

Besides the advantages, both schemes can open up entirely new sets of attack vectors and vulnerabilities that never existed with the traditional grid. For example, an adversary can launch Man-in-the-middle (MitM) attacks [19], False Data Injection Attacks (FDIAs) [24] or load-altering attacks (LAAs) [25] to manipulate the original data, inject false data and control signals through different system components such as DERs, Home Energy Management Systems (HEMSs), smart appliances, and communication channels. The effectiveness of DR and P2P energy trading schemes mainly depends on the precise, complete, and timely data generated from different sources and the secure propagation of data through different system components. Compromised data and control signals can move the system equilibrium away from the optimal solution where attackers can obtain financial benefits and disturb standard grid processes. Hence, maintaining data integrity and authenticity is essential. While system audits can validate the overall system security, they do not ensure that security controls are active at run-time or that systems run the newest software updates to provide the cybersecurity situational awareness. As many data sources are outside the perimeter of the grid operator or utility, approaches are needed to provide assurance for the security of data sources, aggregators and other devices processing data.

The very first step to secure these systems is to understand relevant risks, identify the possibility of intrusion or fault within system data propagation, and provide cybersecurity situational awareness. *Data provenance* provides a complete, structured view of data propagation within the system by recording the data lineage and processing information [20, 21]. A standardised provenance graph is defined by the PROV-DM data model [27]. While PROV-DM provides a structure to collect and report provenance information, it does not include security-related information. In prior work [22], we proposed a security-aware provenance model named Prov-IoT that includes security-related information at each step of the system's data propagation. It generates a comprehensive system's security view and creates an opportunity for identifying relevant system risks. Originally, this approach was developed to increase trust in health data in the Internet of Things (IoT) applications. In this context, malicious changes to data can have catastrophic consequences for individual patients.

Intuitively, as grid operations depend on data originating from devices outside the energy network operator's control perimeter, network operators should be interested in getting additional information on the security status of these devices. Similar issues exist for other entities in smart grid environments. However, smart grids are significantly more complex than health processes for individual patients. Thus, specific system analysis is required to identify processes suitable for security-aware provenance and evaluate the usability of the approach.

Various data provenance techniques have been suggested to ensure the origin and creation of data from IoT devices. Aman et al. [7] proposed a lightweight protocol that uses physical unclonable functions to ensure self-trust and data provenance in IoT systems. Olufowobi et al. [30] presented a data provenance model for IoT systems to guarantee the integrity and authenticity of data in IoT applications. Pagoda, a provenance-based intrusion detection system was proposed by Xie et al. [32] for big data environments. They considered the anomaly degree of a single provenance path and the whole provenance graph for precise and rapid detection in big data environments. Chia et al. [13] studied data provenance related attacks in a home energy monitoring network. Particularly, they focused on the attack through smart plugs collecting and providing energy usage data. F-Pro, a provenance-aware message authentication scheme for smart grid systems was introduced by Esiner et al. [18]. For a particular use-case, they presented that incorporating the source of messages, message delivery path, and message transformation route can thwart potential security threats. These approaches provide discussions and individual solutions for particular security issues in the smart grid and can provide building blocks for several security properties. However, none of the approaches explores attacks on a larger system level or the effect of compromised devices in the grid.

While a combination of securely configured devices without vulnerabilities and the use of suitable security controls can prevent many attacks, there is no guarantee that a secure status is always maintained. Moreover, network operators or market operators do not have full control over all devices. Hence, extending the scenarios in the analysis of security mechanisms, and providing and tracking security metadata in DR or P2P energy trading schemes should enable better security awareness for all participating entities. Refining and instantiating the Prov-IoT model, the result is a specific model for smart grid environments and microgrids (which we name as Prov-IoT-MG) that specifically identifies related entities, data and potential security-relevant metadata. Prov-IoT-MG should enable the development of processes in smart energy networks that generate suitable security-aware provenance graphs and support different entities in the system to improve their information about data flows and security controls in place. The analysis presented in this chapter shows that a security-aware provenance-based approach could be an effective way to ensure continuous situational awareness and progressive risk assessment in modern power systems. Also, DR and P2P play a vital role in reducing utility companies' service costs. Therefore, identifying and estimating risks in these parts of a smart grid system is crucial. This risk estimation should not only happen once or at the time of audits, it should be continuously updated.

In this chapter, using an idealised, but realistic, future smart grid scenario, we analyse the significance of a security-aware data provenance approach in the context of a grid-connected smart microgrid. We propose to add information on security controls, i.e, security metadata in multi-layer smart grid components, to enhance system trust from stakeholder perspectives. The key contributions of this chapter are as follows:

1. Analysis of potential benefits of security-aware provenance graphs in smart grid environments.
2. Designing security threats and security-aware provenance graphs (Prov-IoT-MG) for a smart microgrid system model to estimate associated risks.
3. Identify and propose potential security metadata for smart grid environments.

The rest of the chapter is organised as follows. We demonstrate a system model for a grid-connected smart microgrid in Sect. 2, identify possible security threats in Sect. 3, and design the Prov-IoT-MG in Sect. 4. Further, we illustrate how to apply Prov-IoT-MG to assess the associated risks in Sect. 5, discuss relevant concerns in Sect. 6 and conclude with possible future works in Sect. 7.

2 System Model

In this work, we analyse a common collective energy scheme in the smart grid context that comprises a residential DR scheme, which supports direct load control [12, 28] and a P2P energy market [23]. On a high level, the scheme consists of a set of connected microgrids and a grid operator/ distributed system operator (DSO) that act on behalf of the central grid as depicted in Fig. 1. Each microgrid consists of N households, a set of DERs and an independent market operator (MO). The MO manages and ensure the demand-supply balance of the microgrid. Further, the MO is the integration point between different microgrids and the central grid. We assume that the microgrid formation aims to obtain self-sufficiency among the connected community. Nevertheless, each microgrid connects to the central grid and neighbouring microgrids over the distribution network. The MO uses these connections in trading the energy between adjacent microgrids and the main grid.

Each household comprises a smart meter, a HEMS and a set of appliances. Further, some consumers possess DERs, such as solar panels and PV batteries. Houses that possess of DERs are in contact with a third-party service provider (DER SP). Further, every household is connected to a weather SP that provides weather forecasts. We

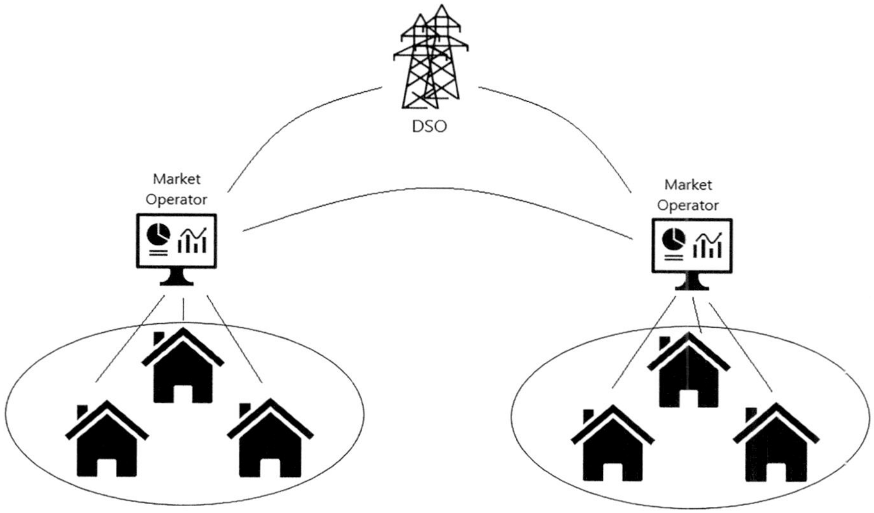

Fig. 1 High level view of the system

assume that the communication between home and DER SP and home and weather SP utilise HTTP connections. Figure 2 depicts the possible information/controls and power flows inside a smart home environment. Wireless protocols such as WiFi, Bluetooth, ZigBee, Z-Wave are commonly utilised for intra-house communication.

For the discussion on data provenance, we focus on data sources and data flows up to the grid/distributed system operator. The described system model is leveraged in implementing a smart energy trading scheme, as depicted in Fig. 3. Similar flows exist outside the DSO network. However, we have excluded them for brevity.

Upon installing DERs households, consumers are required to share the DER configuration information with the trusted DER SP. In this chapter, we denote the DER configuration information as DC_i where $i = 1, 2, \ldots, N$ with N being the total number of households. This configuration meta-information comprises the available DERs such as solar panel coordinates and angles, battery capacities, efficiency and charging/discharging rates. This information sharing process is a one-time operation where the information remains consistent unless there are some alterations. Thus, households can update this information if there is any change. All the appliances of a house are connected and potentially controlled by its HEMS. The house owner configures the earliest starting time, the latest finish time, the preferred start time, and the running duration for each appliance depending on their preferences. Similar to the DER metadata, configuring these parameters can also be considered as a one-time operation unless the preference changes or users want to add new appliances.

The Weather SP sends the weather forecast F of the following day to all the HEMSs before the day starts. Similarly, DER SP sends the corresponding DER's energy generation forecast e_i for individual households that possess DERs. Here, e_i

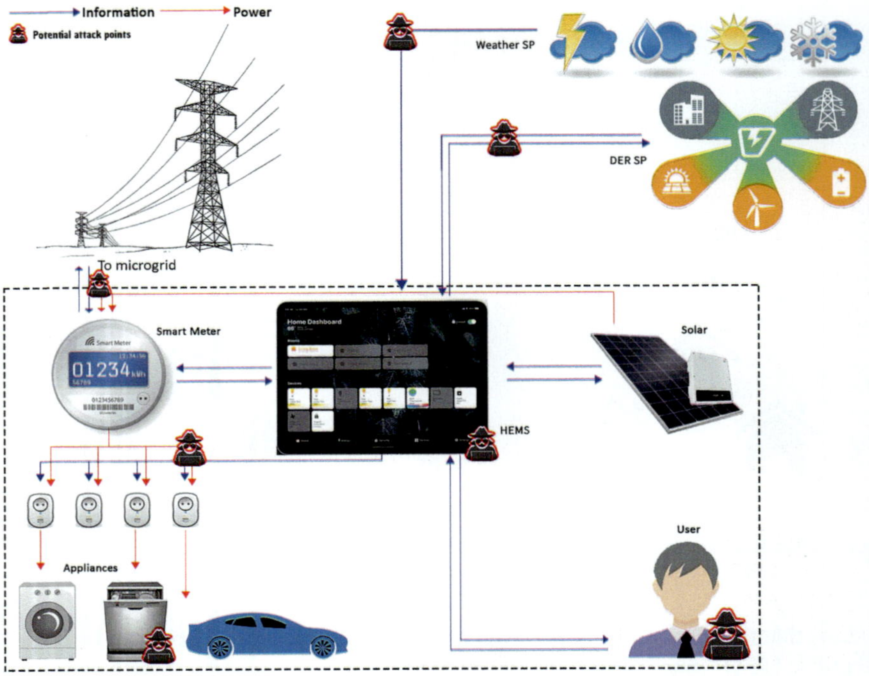

Fig. 2 Different information/power flows inside a house

is the DER's energy generation forecast of ith household. Upon receiving the F and e_i, HEMSs determine the optimal starting times of the connected appliances. HEMSs calculate excessive/deficient energy amounts using appliance schedules and send the forecasted energy excess/deficiency to the MO. This chapter defines the predicted excess or deficit energy of ith household as E_i. MO collects E_i from N households. Based on aggregated demand and supply $E = \sum_{i=1}^{N} E_i$ from all households and forecast energy price for that particular time slot, the MO determines the P2P market price of energy P_P^n, where n is the iteration number, and $n \geq 1$. Then, the MO triggers the P2P market by sending the P_P^n to all HEMSs. HEMS updates the E_i by adjusting schedules of shiftable appliances considering P_P^n. The updated E_i is again sent to the MO. MO upgrades P_P^n based on updated E_i of all households and sends the new price signal P_P^{n+1} to all HEMSs. This iterative process is stopped when the convergence criterion $P_P^{n+1} - P_P^n < \lambda$ is met, where λ is a small positive number to indicate the algorithm termination. Any existing surplus/deficient energy can be traded with the adjacent microgrid in the same manner. The remaining energy surplus or deficiency can sell or buy from the central grid operator at fixed prices. Note that the selling and buying energy price in the P2P market and adjacent microgrid is higher and lower, respectively than the grid. Hence, all HEMSs aim to trade the maximum energy in the P2P market and adjacent microgrid to maximize their benefit.

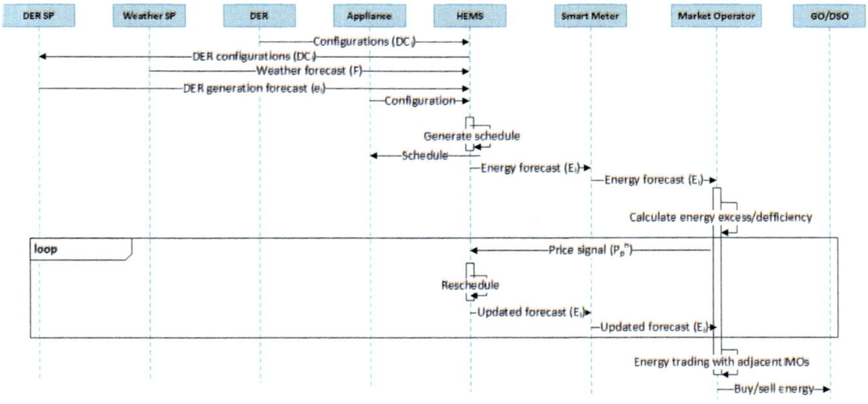

Fig. 3 High-level information flows in the system

Additionally, we assume that the MO has control over a set of smart appliances through connected HEMSs. MOs can issue control signals to turn on and off these appliances when the microgrid's demand-supply balance is jeopardised due to unforeseen errors. Possession of this capability allows the MO to instantly ensure grid safety by increasing or decreasing the demand of the grid. However, we denote that direct load control is considered as the last option, given the unavailability of sufficient demand from other microgrids or cannot export/store the energy excess.

3 Possible Security Threats

The efficiency and the effectiveness of the smart microgrid environment and its energy trading process (described in Sect. 2) rely on the accuracy and the timeliness of the propagated data and control signals. Possessing accurate, timely data can ensure that the system minimises the overall cost and satisfies the expectations of all its stakeholders [15]. However, the heterogeneous and distributed nature of the smart grid allows malicious individuals and organised groups to execute different cyberattacks targeting the accuracy of this data and controls. Several attack scenarios in the context of similar use cases have been documented [10, 11, 17]. Potential attack motivations can vary from financial benefit for attackers to disruption of the grid. Some of the potential attack points are depicted in Fig. 2. In this section, we analyse three types of attacks in the context of the described system model.

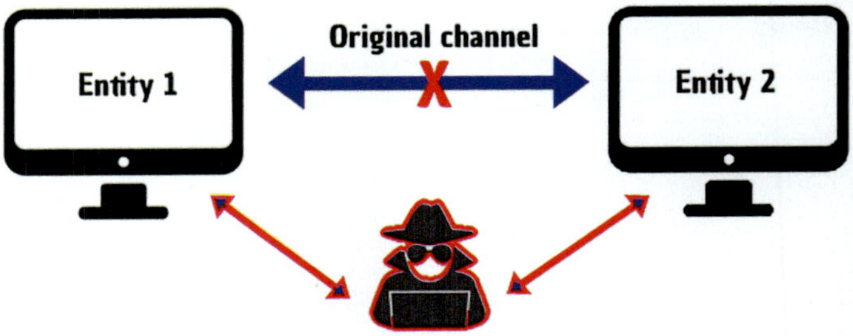

Fig. 4 Man-in-the-middle attack

3.1 Man-in-the-Middle (MitM) Attacks

MitM attack is a cyberattack where the attacker compromises the original communication channel between two entities to reside in the middle of the channel, as depicted in Fig. 4. This placement allows the adversary to eavesdrop and alter the communication between the entities, who believe to be communicating directly with each other.

MitM attacks are highly plausible in the depicted system model, given the use of wireless communication channels in local area networks (inside the house) and commonly adopted neighbourhood area networks in the microgrid environment. An attacker can execute a MitM attack to manipulate the DC_i sent by HEMSs. Manipulated DC_i can result in inaccurate generation forecasts that can produce excessive or deficient energy amounts as they do not reflect accurate information. Similarly, adversaries can exploit generated DER forecasts e_i sent by the DER SP using MitM attacks to construct incorrect DER predictions. For example, an attacker can manipulate the forecast to lower the predicted generation to influence the MO to borrow energy from adjacent microgrids. Malicious market participants can manipulate the market for their financial gain, and exploited excessive forecasts can result in insufficient power to serve the households. Adjacent MOs and potential competitive MOs can mount similar attacks either to gain financial advantages or to damage the reputation of the MO. MitM attacks on F sent by the weather SP can influence consumers to change their demand schedules. Some consumers might significantly deviate from their demand predictions during the day to re-mediate unforeseen or manipulated changes in weather forecasts. These changes can lead to sub-optimal solutions on DR objectives when the demand predictions are used with DR systems. Further, incorrect schedules can result in errors in the microgrid energy demand that influence erroneous trading decisions. Similarly, potential competing microgrids can use MitM attacks to manipulate the price signals to increase their profits.

3.2 False Data Injection Attacks (FDIAs)

In formal terms, a FDIA can be represented as

$$False\ Data,\ D_f = D_t + \delta$$

where D_t is the real data and δ is the injected false data. However, it is worth noting that, depending on the attack formulation, false data injection can occur in many other forms, including deletion of original data and replacement of existing data. Thus, false data can also be independent of actual data.

Adversaries can use FDIAs to inject inaccurate demand forecasts to obtain financial advantages (over others), which are impractical under normal circumstances. The attacks can be mounted through their HEMSs, compromised HEMSs at victim households, compromised smart meters or smart appliances. For example, an adversary user can forecast a significantly high demand, which they will not intend to consume. The incorrect forecast can influence other households to adjust their schedules to avoid unnecessary costs (due to foreseen high demand). Similarly, competent adversaries can use FDIAs via compromised HEMSs or collaboration among a set of houses to produce manipulated demand forecasts that can result in malicious outcomes or financial advantages. Further, advanced adversaries can compromise communication channels between smart meters and HEMSs, smart meters and the aggregator, or aggregator and MO to inject false data [14, 16, 31].

Another potential FDIA is to compromise the price signal P_P^n sent by the MO. HEMSs use this forecasted price signal to decide their optimal demand schedules. Thus, a manipulated P_P^n can result in sub-optimal demand schedules. Similarly, competitive consumers in the same microgrid can manipulate other households' energy trading bids (price, offering amounts, timeslots) to obtain financial advantages. Depending on the manipulation percentages, these attacks can also result in a cascade impact on the MO. Competitive market grid operators can also use FDIAs, similar to MitM attacks, on other operators to gain benefits or damage their reputation.

3.3 Load Altering Attacks (LAAs)

A load altering attack (LAA) is a cyber-physical attack against DR schemes, where it attempts to manipulate a set of unsecured controllable loads to damage the grid through circuit overflow or other adverse effects [8, 25]. LAAs on household appliances can be considered another potential attack under the depicted system model. Both internal (within the microgrid) and external entities (outside of the microgrid) can execute LAAs to disrupt the grid operation, increase the usage cost of a household, or obtain lower usage costs for the attacker.

An attacker can execute an LAA to manipulate an operating appliance via HEMS or compromise wireless communication protocols. A successful attack can result in the attacked appliance operating on a time slot that is not intended by the owner, which can potentially increase the owner's bill as well as cause an inconvenience. Several simultaneous LAAs can disrupt the demand-supply balance in the micro-grid as initial demand forecasts do not capture the LAA influence. Insufficient supply to cater for the required demand influences the MO to buy energy from adjacent microgrids or the central grid. Thus, competing MOs can utilise LAA on adjacent grids to sell their excess energy for a higher price.

An attacker inside the microgrid (a household) can leverage LAAs to reduce the demand for a particular time slot. E.g. an attacker can execute LAAs on the appliances that are scheduled to run on a time slot that the attacker expects to consume a large quantity of demand. As a result of the attack, the demand on the expected time slot will be lower and result in lower usage costs, given the price signal's dependency on aggregated demand. Similarly, an attacker can execute LAAs to operate appliances (compromised) on time slots, which can yield to higher bills.

Given the possibility of security compromises at different stages in the model, it is essential to ensure the accuracy and timeliness of the data transferred throughout the model, as inaccurate data can result in undesirable outcomes for most stakeholders. As best-practice segmentation and perimeter protection cannot prevent these attacks, a combination of security controls active at all steps in the process is required. Hence, along with data lineage, adding security-relevant information for devices and communication steps creates opportunities for a continuous data audit at each stage of data propagation. Thus, all stakeholders in the system can progressively update their risk estimation based on the provided security metadata.

4 Prov-IoT-MG: Evidence for Security Awareness and Opportunities

In this section, we first illustrate the structure of Prov-IoT-MG by describing different components in it. We also include examples of security metadata for the designed scenario. Then, we highlight the importance of security metadata in improving system security awareness.

4.1 Prov-IoT-MG Structure

The structure of Prov-IoT-MG requires understanding of data provenance and data provenance graph. Data provenance projects 'who-what-when-where' of a total data propagation summarising the history of ownership, the point of modification(s), and the actions performed on a data record [21, 26]. A data provenance graph [5, 22] includes various relationships among attributes in a data path and represents

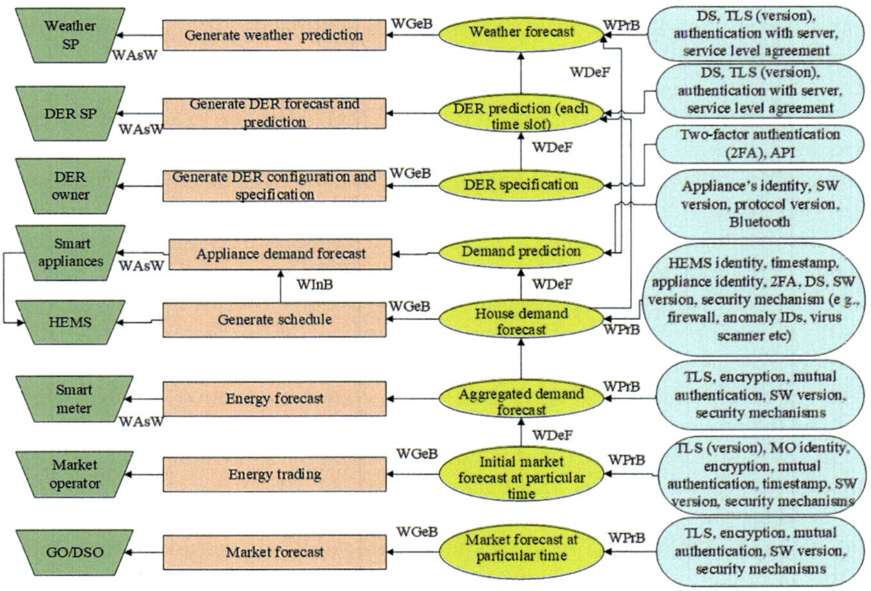

Fig. 5 A provenance graph with security metadata for the microgrid system (Prov-IoT-MG)

a structured data history. The PROV-DM data model structures the whole system in agents, activities and entities with their corresponding relations. Establishing a standard provenance graph for the system model is the initial step of achieving a security-aware provenance graph.

A security-aware provenance graph for a smart microgrid (Prov-IoT-MG) is presented in Fig. 5, which outlines the system provenance graph including step-wise security evidence. The green trapezoids show the actors responsible for certain activities in the scenario. These activities are represented as orange rectangles in the figure. Entities (yellow oval shapes in the figure) represent the data recorded, transferred and transformed within the system. The figure also projects the relevant relationships among different attributes in the provenance graph. For ease of presentation, we use the relationships as follows. WasGeneratedBy (WGeB), WasAssociatedWith (WAsW), WasInformedBy (WInB), WasDerivedFrom (WDeF) and WasProvedBy (WPrB).

It is important to understand that designing Prov-IoT-MG for specific grid architecture is vital. The original Prov-IoT model extends the PROV-DM model by including information on security-aware features (information on security controls) [22] with all processing steps of data propagation. The security metadata in the Prov-IoT model includes information on security controls, protocols used for data communication, information on encryption/authentication mechanisms, device identity, software versions, firewall, anti-malware mechanisms involved in the system etc. However, Prov-IoT is rather generic and has currently only been applied to the digital health context. Using Prov-IoT for smart grid scenarios requires identifying relevant entities, data and security metadata. Thus the Prov-IoT-MG model refines and instantiates Prov-IoT and its security metadata (light blue rounded rectangle shape) as depicted in Fig. 5. The example used to design the Prov-IoT-MG is specific to a grid-connected smart microgrid context, where MOs manage the demand-supply balance in their microgrids by enabling energy trading and DR coexisting and potentially connected with other microgrids.

While designing Prov-IoT-MG, we include relevant security metadata for each instance of the data process. For example, Weather SP (agent) generates the weather prediction (activity) and this process outcomes the weather forecast F (entity). Examples for security evidence during this process can be digital signatures (DS), used transport layer security (TLS) version used, authentication mechanism(s) used with the server, or any service level agreement (policy-based information) (security metadata). As the provenance graph documents this metadata, other entities can monitor this security information and due to better cybersecurity situational awareness can estimate the risk for F to be manipulated. Similarly, DER owners generate DER configuration and specification DC_i, where information on two-factor authentication (2FA) and/or application programming interface (API) are considered as required security information. Smart appliances in various HEMS generate appliance demand forecasts, where appliance's or HEMS's identity, software (SW) and protocol version used, and other security mechanisms used are considered as relevant security evidence. Accordingly, we include necessary security evidence for other processes included for various stakeholders in Fig. 5. Hence, documenting and reporting the presence or absence of security evidence presents the security status of the overall system.

We refer to Sect. 2 to elaborate more on the Fig. 5. As per our system model, the MO initiates the energy trading for a particular time slot based on forecasted aggregated energy demand and supply received from N smart meters. A smart meter receives the forecast from its HEMS. HEMSs use DER's energy generation forecast and schedules of smart appliances to calculate the demand and supply. The DER SP predicts the energy generation of a DER depending on its specification and the weather forecast. The Weather SP performs weather prediction activity and provides weather forecast F (entity) to the DER SP. During this process, DS, TLS version, authentication mechanism(s) with the server, or any service level agreement is useful to ensure the correctness of F. These security mechanisms are documented in the provenance graph to ensure that other agents can check the security metadata related to this process, given they provide relevant security evidence. After that, the DER SP receives the DER specification information DC_i from the DER owner. The DER owner can use 2FA and information on the API to validate the owner's identity and DC_i. This security evidence information is also added to the provenance graph. Now, the DER SP executes the DER energy generation forecast e_i and use DS, TLS version, authentication mechanism(s) with server, or any service level agreement to guarantee the accuracy of e_i. The provenance graph documents the accommodated security metadata. Following the above process, all other agents, such as smart appliances, HEMS, and smart meters, perform their respective activities to predict individual appliance energy demand, total household energy demand, and deficit or surplus energy information for that particular time slot. Relevant security metadata of each process are added to the provenance graph. Once the MO receives energy excess/dearth forecasts from N smart meters, the MO can monitor/check different security metadata related to each process of individual smart meters from the provenance graph. Based on documented security metadata, the MO can consume or discard data received from different smart meters. The risk assessment of different agents/entities based on documented security metadata of provenance graph is discussed in detail in Sect. 5.

4.2 Security Metadata Opportunities

Security metadata creates opportunities for system security awareness for the stakeholders to make an appropriate decision. Hence, adding security controls at the individual layer of the overall system is crucial. Moreover, particular standards for smart grid environments are vital attributes to be included as security metadata to project integrity and authenticity. With this aim, in this chapter, we discuss existing standards and propose to add those as potential security metadata.

Existing standards and frameworks covering some security aspects in the smart grid context are shown in Table 1. For example, IEC 62351 (standards for securing power system communications) requires private keys and role-based access control mechanisms in obtaining end-to-end security. Similarly, the OpenADR (Open Automated Demand Response) specification recommends using TLS in establishing secure channels between nodes for communication and the use of XML signatures to ensure non-repudiation [1]. Further, NIST has specified several high-level security requirements, including access control, identification, authentication and key management to ensure the cybersecurity of smart grids under NISTIR 7628 [2]. The IEEE 2030.5-2018 Standard for Smart Energy Profile Application Protocol [3] defines the application layer protocol that enables utility management of the end-user energy environment. In particular, it includes a set of security features, including device credentials, access control lists, and authorisation that are required in DR, management of distributed generation, and EVs. These standards specify several security controls that are potential candidates for security metadata to be included in the provenance graph. Thus, security-aware provenance graphs can also document compliance with standards with respect to security.

Remarkably, these standards and requirements are not rigidly or globally followed. For example, according to AEMO (Australian Energy Market Operator), there is no device-level cybersecurity standard for the Australian national energy market [4]. Thus, new standards, such as AS/NZ 4755, are being developed for DR capabilities and supporting technologies and DR enabled devices. We believe it is essential to adhere to or reuse existing standards and only supplement additional information

Table 1 Commonly used security standards/guidelines in smart grid

Name	Type	Focus	Authentication	Authorisation	Encryption	Signature	TLS
IEC-62351	Standard	System control operations and communication	✓	✓	✓	✓	✓
NISTIR-7628	Guideline	Overall smart grid components	✓	✓	✓	✓	✓
OpenADR	Standard	Demand Response	✓	✓	✓	✓	✓
IEEE 2030.5-2018	Standard	Smart energy profile application	✓	✓	✓	✓	✓

where necessary instead of developing novel frameworks. Thus, we have accommodated well-defined security standards and regulations while designing the security metadata.

5 Risk Assessment with Situational Awareness

The introduction of security metadata helps the relevant entities to continuously and progressively assess any associated risk of received data. The reason is steps of information flows can be modified to mitigate the potential impacts of the discussed attacks. Modifications in existing flows rely on entities capabilities to accurately determine the associated risks. Further, security metadata also enables entities to react to an increased risk or suspicious behaviour. For example, by checking optional redundant data sources or switching to a more resilient operational mode of the market and the microgrid, the relevant stakeholders can estimate risks and make appropriate decisions accordingly. This section discusses how Prov-IoT-MG can provide a basis for improving the assessment of associated risks under three main types of attacks.

5.1 Man-in-the-Middle Attacks

The depicted use case is susceptible to MitM attacks in several points of communications from both internal and external adversaries as discussed in Sect. 3. Initially, the DER SP can obtain the provenance information on the authentication mechanism (two-factor authentication (2FA)) used by HEMSs to initiate/update DC_i as these metadata are provided with Prov-IoT-MG. The DER SP can also check the HEMS software version through the meta-information. Given the availability of authentication information and software versions, the DER SP can assess the risk of received configurations being altered. The DER SP can ignore all requests from HEMSs with insufficient authentication mechanisms to ensure that the risk of receiving DC_i is the result of unauthorised access to the HEMS. It can also filter DC_i from HEMSs with outdated software versions as those can have security vulnerabilities that adversaries can use in altering the data. Similarly, if data is digitally signed, the DER SP can verify that DC_i are not modified during the communication between HEMSs and the DER SP.

HEMSs can use security metadata provided with provenance graphs based on Prov-IoT-MG to assess any associated risk of MitM considering the HEMSs' daily communications between the DER SP and the weather SP. In particular, we propose to include the digital signature and the security protocol used for transport as security metadata between HEMSs and both weather SP and DER SP. We denote that some of this information, such as the security protocol used, is available in standard communication. However, information on the protocol alone is insufficient. Security parameters, such as the negotiated cryptographic algorithms, key length, the mode of authentication, and the protocol versions, also need to be included in the meta-information provided with Prov-IoT-MG. For example, old TLS versions are vulnerable to various attacks. Also, TLS can be used without, with one-sided or with mutual authentication. Having access to all these metadata provides better risk assessments capabilities for HEMSs. Given the metadata, HEMSs can determine the risk of receiving e_i and F to be compromised. It can discard any potentially altered forecasts or use additional data sources when generating their demand schedules.

5.2 *False Data Injection Attacks*

Similar to the MitM attacks, the depicted use case comprises several points where adversaries can execute FDIAs. Prov-IoT-MG can be utilised to assess the risk of those FDIAs to mitigate the potential impact of attacks.

We propose to incorporate security metadata with all the communication between the smart appliances and HEMSs. This metadata includes appliances' identities, mode of communication (wired, Bluetooth, ZigBee, Z-Wave, etc.) software and protocol versions, and active security controls, such as intrusion detection systems. Given the availability of this metadata, HEMSs can estimate the risk of adversaries injecting false data into the systems through smart appliances' data/schedules. For example, some communication modes have a higher probability of being compromised. Similarly, smart appliances that possess outdated firmware versions have higher potential for FDIAs. HEMSs can either use correction approaches or some default data to mitigate potential attack impacts upon detecting any risk. Note that adversaries can inject false metadata, such as incorrect versions, via compromised devices. Hence, the PROV-IoT model cannot completely eliminate the threat. Additional security controls, such as anomaly detection in the HEMSs or TPM-based device attestation, can improve the overall security of the systems and can then be reported as security meta-data towards other entities.

Potentially, adversaries can also inject false data from HEMSs. Thus, the MO needs to assess the risk of FDIAs from HEMSs. We propose to include HEMS's software version, HEMS/smart meter identity, and associated security mechanisms (such as availability of IDS, firewall, and virus scanners) with the individual demand forecasts sent to the MO. The MO can use this metadata to assess the risk of received predictions and decide the optimal strategy. For example, the MO can ignore forecasts from HEMS, which have outdated software versions or do not possess anomaly detection capabilities. The MO can similarly reject data from HEMSs that do not enable 2FA or use outdated TLS versions for communication. This risk assessment can be further extended to each phase, such as initial forecast, commit forecast, and actual consumption. A similar risk assessment can be done considering the communication between MOs.

5.3 Load Altering Attacks

As discussed in Sect. 3, HEMS is one of the main entry points for LAAs, given that smart appliances are managed through the HEMS. Thus, HEMSs should ensure the authenticity of any received control signals prior to enact on them. MOs can include their identity, TLS version, and timestamp as security metadata when issuing command signals to connected HEMSs. HEMSs can utilise the metadata to verify the authenticity of the command. For example, a HEMS can ensure that the received control signal originated from the MO it is connected to and the specified timestamp has not expired. The HEMS can discard any control signal that it cannot verifies or not from the MO that is connected.

Specific to communication between HEMSs and appliances, HEMS should include its' identity, timestamp, and digital signature in the metadata. Smart appliances can use this metadata to ensure that the command signals originated from the HEMS and have not been compromised.

6 Discussion

In this chapter, we discuss the relevance of security metadata in improving cybersecurity situational awareness in a smart grid context. Most smart grid processes comprise distributed setups with several entities that exchange information to achieve individual and collective optimisation. Potential financial advantages for individuals can motivate the misbehaviours of stakeholders with competing objectives. Though existing standards provide security mechanisms for communication, privacy concerns prevent entities from accessing detailed information about other entities. Thus, no entity possesses complete knowledge of the entire data flow. Given this lack of control, each entity should assess the associated risk of incoming data before executing critical operational logic. However, existing security and communication standards do not provide any reliable mechanism to evaluate the associated risk of the propagated data. We suggest supplementing current protocols with additional security metadata.

We analyse the use of security-aware provenance models in a smart energy trading context of a microgrid environment. In particular, we depict how to leverage our refined provenance model Prov-IoT-MG for a better risk assessment considering three types of attacks, namely man-in-the-middle, false data injection and load altering attacks. We illustrate how standard security mechanisms and novel attributes can be documented and this information is forwarded to other entities. This provides better situational awareness and enables entities to assess the risk related to data flows. Summarised metadata can reduce the additional overhead. Each security control added to Prov-IoT-MG can extend the coverage of potential attacks.

However, it is worth mentioning that embedding the provenance metadata will not eliminate the risk. Instead, possessing the metadata enables each entity to evaluate the potential risk of incoming data. Upon the assessment, entities can decide the optimal strategy to continue. Additional security mechanisms can be embedded to improve the overall security of the process. For example, the MO can utilise an outlier detection method to compare incoming forecasts with historical demand values. Such capabilities allow MOs to add additional metadata to the provenance model, such as documenting suspicious data, the probability for it to be an anomaly and degree of variance, instead of just ignoring suspicious forecasts.

The security-aware provenance can also be used to validate system designs and add required security controls. One particularly problematic area is with devices such as DERs and smart inverters that operate in (potentially) hostile environments. Secure identification of these devices is essential but arduous. Currently, MAC addresses are used to identify devices in IP-based networks. This information can be easily changed and does not provide a secure identification. The use of cryptographic keys is also not ideal in device identification. Cryptographic keys can be transferred to other devices as they are often not securely bound to a particular device. Thus, additional mechanisms, such as hardware-based security or physical unclonable functions, are necessary to prevent malicious devices from accessing the network [6, 29, 33]. Furthermore, as mentioned above, the status of a device should be correlated with executable software running on the device and the device configuration. A compromised device often has additional executable code installed (e.g. Trojans, viruses, botnets, worms). Security controls that stop critical components from executing malicious code or regularly report the status of the component as security metadata can either prevent or enable fast reactions to attacks.

Finally, accounting and security information management can benefit from security-aware provenance. Dynamic energy tariffs, P2P trading and commitments for demand and supply always need to be associated with the exact date and time. External secure time-stamping services create additional dependencies, communication overhead and costs. Thus, if devices have hardware protected time-stamping mechanism, such as the one provided by the Trusted Platform Module (TPM), this time information can also be tracked via the provenance graph and bound to other security metadata. All stakeholders in the system can potentially benefit from this information and improve their security information management.

7 Conclusion and Future Work

Improving cybersecurity situational awareness of smart grid networks requires providing information on different steps in the data flows. Understanding the status of security controls at each step of data propagation empowers all stakeholders to establish continuous situational awareness and better risk perception for data and guide them (machine/human/system) in making relevant decisions. This study pinpoints the significance of security-aware data provenance information in smart grid environments. We have presented our system model in this chapter and identified the possible security threats. Then, the provenance graph for the considered system model is designed, and necessary security metadata is added to the provenance graph. Finally, we have demonstrated how the proposed security-aware data provenance graph helps risk estimation for the considered system model.

As the next step of this work, it is crucial to identify exact security controls to feed into data provenance graphs in real-life setup. Therefore, this study should motivate cross-disciplinary research so that cybersecurity experts and smart grid specialists work collectively for a system design enabling stronger resilience through better risk estimation.

References

1. OpenADR 2.0 Profile Specification. Technical report, OpenADR Alliance (2013)
2. Smart grid cybersecurity strategy architecture and high-level requirements. Technical report, The Smart Grid Interoperability Panel Smart Grid Cybersecurity Committee (2014). https://nvlpubs.nist.gov/nistpubs/ir/2014/NIST.IR.7628r1.pdf
3. IEEE standard for smart energy profile application protocol. IEEE Std 2030.5-2018 (Revision of IEEE Std 2030.5-2013) pp. 1–361 (2018). https://doi.org/10.1109/IEEESTD.2018.8608044
4. Electricity Rule Change Proposal. Technical report, Australian Energy Market Operator (2020)
5. U. Acar, P. Buneman, J. Cheney, J. Van Den Bussche, N. Kwasnikowska, S. Vansummeren, A graph model of data and workflow provenance, in *Procs. TAPP'10 workshop (Theory and Practice of Provenance)* (2010), p. 8
6. M.N. Aman, K.C. Chua, B. Sikdar, Mutual authentication in iot systems using physical unclonable functions. IEEE Internet Things J. **4**(5), 1327–1340 (2017)
7. M.N. Aman, K.C. Chua, B. Sikdar, Secure data provenance for the internet of things, in *Proceedings of the 3rd ACM International Workshop on IoT Privacy, Trust, and Security, IoTPTS '17* (Association for Computing Machinery, New York, 2017), pp. 11–14. https://doi.org/10.1145/3055245.3055255
8. S. Amini, F. Pasqualetti, H. Mohsenian-Rad, Dynamic load altering attacks against power system stability: attack models and protection schemes. IEEE Trans. Smart Grid **9**(4), 2862–2872 (2016)
9. V. Balijepalli, V. Pradhan, S. Khaparde, R. Shereef, Review of demand response under smart grid paradigm, in *2011 IEEE PES International Conference on Innovative Smart Grid*

Technologies-India, ISGT India 2011, vol. 1 (2011). https://doi.org/10.1109/ISET-India.2011.6145388
10. C. Barreto, A.A. Cárdenas, Impact of the market infrastructure on the security of smart grids. IEEE Trans. Industr. Inf. **15**(7), 4342–4351 (2019)
11. C. Barreto, H. Neema, X. Koutsoukos, Attacking electricity markets through iot devices. Computer **53**(5), 55–62 (2020)
12. C. Chen, J. Wang, S. Kishore, A distributed direct load control approach for large-scale residential demand response. IEEE Trans. Power Syst. **29**(5), 2219–2228 (2014)
13. M.H. Chia, S.L. Keoh, Z. Tang, Secure data provenance in home energy monitoring networks, in *Proceedings of the 3rd Annual Industrial Control System Security Workshop, ICSS 2017* (Association for Computing Machinery, New York, 2017), pp. 7–14. https://doi.org/10.1145/3174776.3174778
14. R. Dasgupta, A. Sakzad, C. Rudolph, Cyber attacks in transactive energy market-based microgrid systems. Energies **14**(4), 1137 (2021)
15. T. Dayaratne, C. Rudolph, A. Liebman, M. Salehi, Robust demand response for device scheduling under false data injection attacks in smart grids, in *2020 IEEE PES Innovative Smart Grid Technologies Europe (ISGT-Europe)* (IEEE, 2020), pp. 294–298
16. T. Dayaratne, C. Rudolph, A. Liebman, M. Salehi, We can pay less: coordinated false data injection attack against residential demand response in smart grids, in *Proceedings of the Eleventh ACM Conference on Data and Application Security and Privacy* (2021), pp. 41–52
17. T. Dayaratne, C. Rudolph, A. Liebman, M. Salehi, S. He, High impact false data injection attack against real-time pricing in smart grids, in *2019 IEEE PES Innovative Smart Grid Technologies Europe (ISGT-Europe)* (IEEE, 2019), pp. 1–5
18. E. Esiner, D. Mashima, B. Chen, Z. Kalbarczyk, D. Nicol, F-pro: a fast and flexible provenance-aware message authentication scheme for smart grid, in *2019 IEEE International Conference on Communications, Control, and Computing Technologies for Smart Grids (SmartGridComm)* (IEEE, 2019), pp. 1–7
19. L. Hadjidemetriou, G. Tertytchny, H. Karbouj, C. Charalambous, M.K. Michael, M. Sazos, M. Maniatakos, Demonstration of man in the middle attack on a feeder power factor correction unit, in *2020 IEEE PES Innovative Smart Grid Technologies Europe (ISGT-Europe)* (2020), pp. 126–130. https://doi.org/10.1109/ISGT-Europe47291.2020.9248779
20. X. Han, T. Pasquier, M. Seltzer, Provenance-based intrusion detection: opportunities and challenges, in *10th USENIX Workshop on the Theory and Practice of Provenance (TaPP 2018)* (2018)
21. F.T. Jaigirdar, C. Rudolph, C. Bain, Can i trust the data i see? in *Proceedings of the Australasian Computer Science Week Multiconference* (ACM, Sydney, 2019), pp. 1–10. https://doi.org/10.1145/3290688.3290731
22. F.T. Jaigirdar, C. Rudolph, C. Bain, Prov-IoT: a security-aware IoT provenance model, in *2020 IEEE 19th International Conference on Trust, Security and Privacy in Computing and Communications (TrustCom)* (IEEE, 2020), pp. 1360–1367
23. M. Khorasany, Y. Mishra, G. Ledwich, Hybrid trading scheme for peer-to-peer energy trading in transactive energy markets. IET Gener. Transm. Distrib. (2019). https://doi.org/10.1049/iet-gtd.2019.1233
24. G. Liang, J. Zhao, F. Luo, S. Weller, Z. Dong, A review of false data injection attacks against modern power systems. IEEE Trans. Smart Grid **8**, 1630–1638 (2017)
25. A.H. Mohsenian-Rad, A. Leon-Garcia, Distributed internet-based load altering attacks against smart power grids. IEEE Trans. Smart Grid **2**(4), 667–674 (2011)
26. Moreau, L., Clifford, B., Freire, J., Futrelle, J., Gil, Y., Groth, P., Kwasnikowska, N., Miles, S., Missier, P., Myers, J., Plale, B., Simmhan, Y., Stephan, E., Den Bussche, J.V.: The open provenance model core specification (v1.1). Future Gener. Comput. Syst. **27**(6), 743–756 (2011). https://doi.org/10.1016/j.future.2010.07.005
27. L. Moreau, P. Missier, Prov-dm: The prov data model, in *w3c Recommendation. World Wide Web Consortium* (2013)

28. H. Mortaji, S.H. Ow, M. Moghavvemi, H.A.F. Almurib, Load shedding and smart-direct load control using internet of things in smart grid demand response management. IEEE Trans. Ind. Appl. **53**(6), 5155–5163 (2017)
29. Y. Nozaki, M. Yoshikawa, Secret sharing schemes based secure authentication for physical unclonable function, in *2019 IEEE 4th International Conference on Computer and Communication Systems (ICCCS)* (2019), pp. 445–449. https://doi.org/10.1109/CCOMS.2019.8821698
30. H. Olufowobi, R. Engel, N. Baracaldo, L.A.D. Bathen, S. Tata, H. Ludwig, Data provenance model for internet of things (iot) systems, in *Service-Oriented Computing - ICSOC 2016 Workshops*. ed. by K. Drira, H. Wang, Q. Yu, Y. Wang, Y. Yan, F. Charoy, J. Mendling, M. Mohamed, Z. Wang, S. Bhiri (Springer International Publishing, Cham, 2017), pp.85–91
31. M. Pilz, F.B. Naeini, K. Grammont, C. Smagghe, M. Davis, J.C. Nebel, L. Al-Fagih, E. Pfluegel, Security attacks on smart grid scheduling and their defences: a game-theoretic approach. Int. J. Inf. Sec. (2019)
32. Y. Xie, D. Feng, Y. Hu, Y. Li, S. Sample, D. Long, Pagoda: A hybrid approach to enable efficient real-time provenance based intrusion detection in big data environments. IEEE Trans. Depend. Sec. Comput. **17**(6), 1283–1296 (2020). https://doi.org/10.1109/TDSC.2018.2867595
33. K. Yang, J. Park, M. Tehranipoor, S. Bhunia, Robust timing attack countermeasure on virtual hardware, in *2018 IEEE Computer Society Annual Symposium on VLSI (ISVLSI)* (2018), pp. 148–153. https://doi.org/10.1109/ISVLSI.2018.00036
34. C. Zhang, J. Wu, Y. Zhou, M. Cheng, C. Long, Peer-to-peer energy trading in a microgrid. Appl. Energy **220**, 1–12 (2018)

Hybrid Physics-Based and Data-Driven Mitigation Strategy for Automatic Generation Control Under Cyber Attack

Chunyu Chen, Junbo Zhao, Kaifeng Zhang, Yilong Liu, and Yang Chen

Abstract The fusion of information, data, control and electric power facilitates the electric cyber-physical system (ECPS). In the ECPS, though the high-level cyber-physical interaction and integration increase the flexibility and efficiency of power system operation, the accompanied cyber perils gradually endanger the system security. In fact, several energy sector-targeted attacks, including the infamous Ukraine power grid hack, have shown the power system vulnerability in cybersecurity incidents. Known as the critical power-balancing operation in real-time control systems, automatic generation control (AGC) is a typical ECPS application. Measurements from remote sensors may be manipulated by attackers when telemetered to the AGC center, thereby disrupting the balance of power and frequency stability. In this situation, cyber-attack- tolerant AGC plays an important role in the face of cybersecurity threats. To achieve the cyber-attack-tolerant AGC, some strategies including the physics-based and data-driven attack mitigation schemes have been employed. In this chapter, inspired by the fault detection, diagnosis and reconfiguration in fault-tolerant control theories, a hybrid physics-based and data-driven mitigation model is developed for AGC under false data injection attacks (FDIAs). A mathematical model is derived to reveal the causal relation between the FDIA signal and compromised AGC measurement data. Then, data-driven approaches are employed to establish the mapping between the compromised measurement data and power compensation. Finally, the compensation-based mitigation model is developed.

Keywords Electrical cyber-physical system · Automatic generation control · False data injection · Attack · Data-driven mitigation model

---

C. Chen (✉) · Y. Liu · Y. Chen
School of Electrical Engineering, China University of Mining and Technology, Xuzhou, China
e-mail: 5988@cumt.edu.cn

J. Zhao (✉)
Department of Electrical and Computer Engineering, University of Connecticut, Storrs, CT, USA
e-mail: junbo@uconn.edu

K. Zhang
School of Automation, Southeast University, Nanjing, China

H. Haes Alhelou et al. (eds.), *Power Systems Cybersecurity*, Power Systems,
https://doi.org/10.1007/978-3-031-20360-2_6

1 Introduction

The electric power system has evolved into a cyber-physical counterpart, owing to the ever-growing reinforcement of modern communication technologies. Innovative measurement technologies, such as the wide-area measurement system, fundamentally reshape the landscape of dynamic grid monitoring and operation by offering a variety of advanced metering devices and communication infrastructures. The coin of a cyber-physical system (CPS) has two sides. One side facilitates the information fusion for the improved decision-making on behalf of the grid operator; while the other side increases the susceptibility to cyber attacks, which may cause severe damage to the reliable grid operation. As is shown in Fig. 1, the architecture for the electrical-cyber-physical system consists of four main layers: the physical layer where the electrical components (e.g., generation units, substations, and transformers) are deployed to constitute the physical power grid; the data acquisition layer, which interfaces the physical layer to the cyber one, is responsible for collecting data of specific physical assets; the communication network layer takes charge of data packet forwarding. The application layer carries various power system applications that frequently exchange data or information with other layers. The integrated four-layer electrical CPS may suffer cybersecurity vulnerabilities at different locations. In the following, we briefly review cyber attacks against different targets or components across the cyber layer.

1.1 Cyber Attacks Against Advanced Metering Infrastructure

As one critical component in the advanced metering infrastructure (AMI), smart meters collect (even process) measurements of grid variables (e.g., power, voltage, and current) or device states (e.g., state of charge of an electric battery, reserve capacity of a generator). Healthy data measurements determine the well-functioning power system applications. Therefore, from the attacker's perspective, measurement tampering of smart meters is one essential cyber attack method. Researchers have investigated various smart meter-oriented cyber attacks. The remote terminal unit (RTU) in smart metering systems attracts considerable attention in the cyber attack study. For example, an RTU at a transform bank may suffer malware threats, which manipulate the state of circuit-breakers and overload the transformer [1]. To achieve the attack against RTUs, attackers usually manipulate meter readings or device states by exploiting the vulnerability of RTU hardware and software [2]. Recently, researchers have also begun to conduct experiments on various commercial-grade smart meters and measure their operation integration under cyber attack conditions. Both Internet control message protocol (ICMP) and pin flood can disrupt communication with smart meters [3]. Compared with a smart meter, the attack points of AMI significantly increase. For example, attackers can perform denial-of-service (DoS) attacks on any node and disrupt the traffic of AMI. Some researchers propose

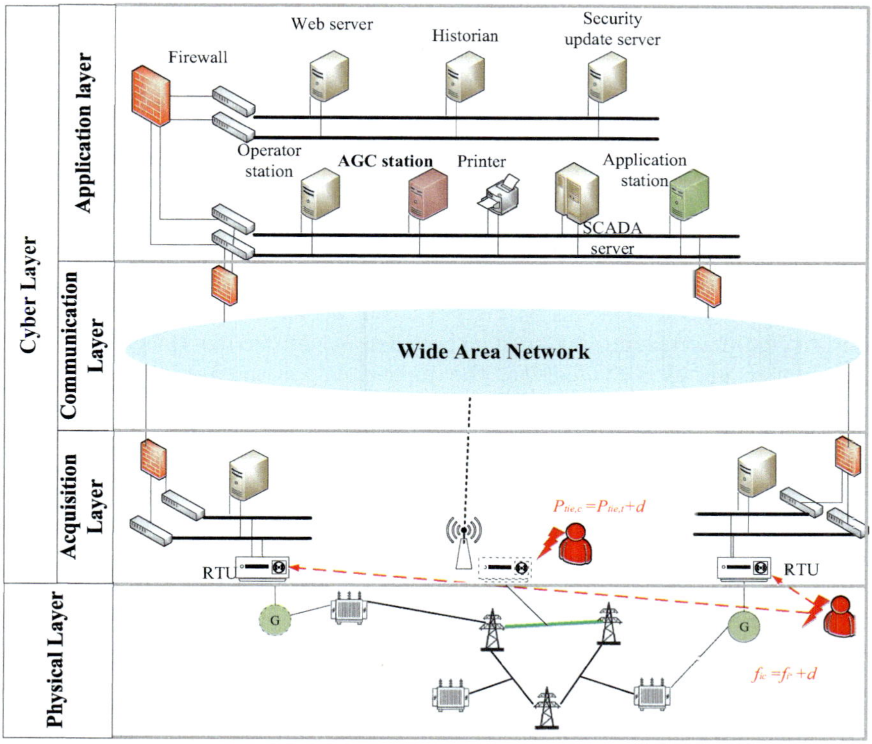

Fig. 1 The schematic diagram of four-layer electrical cyber-physical system

a concept called the cyber attack surface of AMI to evaluate the overall security exposure of AMI [4]. Based on the attack surface of AMI, defenders develop an evaluation framework, which can produce a quantitative information-based exposure metric to evaluate the completeness of implemented security mechanisms [5]. Researchers have studied various detection strategies for smart meters to reduce attack-induced costs and safety threats [6–9]. These detection strategies mainly adopt various anomaly detection algorithms such as clustering and deep learning. We often treat cyber attacks as low-probability events; hence clustering-based detection assumes that the intact data instances could form large dense clusters. Compromised data instances belong to outliers of these large dense clusters. K-means algorithm is one frequently-adopted unsupervised method for clustering analysis. Nevertheless, when the condition changes, the normal reconstruction errors in unsupervised methods may surpass the preset values. Hence, people turn to uncertainty-tolerant deep learning approaches.

1.2 Communication System-Oriented Cyber Attacks

A communication system acts as a bridge between applications and sensors. Fraught with telecommunication hardware and software, the communication system may also suffer cyber attacks if the hardware and software are not properly secured. A notable example is Internet hacktivism. Cyber attacks against communication systems concentrate on DoS and replay attacks. DoS attacks overload systems to make a network resource unavailable by flooding the target with superfluous requests. Replay attacks manipulate the communication data by replacing the original data packet with a fabricated one. As for a replay attack, attackers will hijack the sensors and repeat the intact readings for a certain period to avoid detection. Hence, the detection of replay attacks is challenging. The implementation and mitigation of replay attacks vary in different systems. Some researchers model the target as a discrete-time linear time-invariant Gaussian system and enhance the probability of detection by trading off the control performance [10]. In some network control systems, replay attacks may substitute the output with a delayed one. Though injecting authentication noise is an effective detection method, its accuracy is positively relevant to the magnitude of the noise. A spectral estimation method can abandon the injection of authentication noise, thus preserving the control performance [11]. We categorize DoS attacks into three types: logic attack, protocol attack, and bandwidth or flood attack. Flood attacks refer to the inundation of communication networks with superfluous packets, leading to a system crash [12]. The flooding methods mainly include pin flooding, smurf attack, and UDP flood. Rather than producing overburdened traffic, logic attacks rely on bugs or security holes of a device. Likewise, protocol attacks aim at the security holes of a specific protocol. To detect DoS attacks, researchers evaluate the following metrics: dropped package percentage, transaction duration, delay of request or response service, and divisional resources between communication channels. Usually, single metric-based methods cannot precisely identify a sophisticated DoS attack. Hence, it is more proper to use a set of DoS metrics. The power system spans extensive geographical regions. Besides, the widely-integrated distributed energy systems nurture the development of distributed networked systems. Therefore, the power system is more susceptible to distributed DoS or DDoS [13]. Basic security measures include the installation of firewalls and intrusion detection systems. Recent research on DDoS attacks begins to employ machine learning (ML) algorithms [14, 15], which can improve detection accuracy by learning the high-level features of packet flows relevant to DDoS.

1.3 Power System Application-Oriented Cyber Attacks

Ultimately, the goal of cyber attack boils down to the disruption of specific power system applications. From the perspective of power system operators, they are more concerned with application-oriented attacks rather than cyber attacks at other layers.

But it should be particularly mentioned, since the security level rises from the bottom physical layer to the top application layer, cyber attacks against smart meters or communication wares are usually the realization of indirect application-oriented attacks. Nevertheless, rather than focusing on the detail of the intrusion of a specific meter or telecommunication facility, application-oriented attacks take a macroscopic viewpoint on how cyber attacks affect the power system operational performance. Based on the type of power system applications, application-oriented attacks can be loosely categorized into: (1) SCADA-oriented attack, (2) state-estimation attack, (3) electricity market-oriented attack, and (4) power system control-oriented attack. We briefly review the latest progress in these areas.

SCADA-oriented attack: Supervisory control and data acquisition (SCADA) is a control system architecture comprising communication systems and human–machine interfaces. Attackers may gain unauthorized access to SCADA systems to disrupt the end devices [16]. In February, 2022, attackers infiltrated Florida's water plant's SCADA through a remote desktop application. The sharing of the same password for remote access and the lack of firewall protection are two main incentives in the post-event survey. This accident once again highlights how attackers can use everyday software to penetrate a critical infrastructure. Recently, some researchers develop SCADA testbeds to simulate various attacks and test SCADA's resilience [17]. Besides observing and evaluating the cyber attack influence, researchers also study how to detect SCADA-oriented attacks [18].

State estimation-oriented attack: Power system estimation is arguably the foundation of power system applications. The bad data detector (BDD) could be ineffective if attackers choose a specially structured injection vector [19]. Correspondingly, defenders turn to other detectors such as AC transmission systems [20]. In turn, from the attacker's perspective, defenders begin to consider other stealthy attack policies. Attack polices are usually goal-driven. In other words, attackers perform cyber attacks to achieve certain goals, e.g., the maximization of voltage deviation [21] or the minimization of the attack effort [22]. As with previous detection designs in SCADAs, ML also gain tremendous attention in the detection of cyber attacks against state estimation [23, 24].

Electricity market-oriented attack: In electricity markets, attackers may leverage the state estimation circumvention for financial rewards [25]. In other words, a profitable attack strategy is arguably the ultimate goal of attackers. The research verifies that by injecting undetectable errors into critical model parameters, the locational marginal prices in electricity markets can be significantly distorted, which leads to financial benefits for the attackers [26]. When the attacker has very limited knowledge about the topology and parametric information of targeted network, some researchers study a Monte Carlo-based attack strategy [27]. Besides, researchers consider the incomplete sensor information situation and design a two-stage attack policy [28].

Power system control-oriented attack: Voltage control applications are primary goals of cyber attacks [29]. Conventional voltage stability monitoring systems based on voltage stability indexes may fail to detect the instability issue under a smart attack.

Defenders can employ the multi-port equivalent circuit to calculate the Thevenin Equivalent (TE) parameters and the attack indicator, which can help reveal the attack on phasor measurement unit (PMU) data. As for the DoS attack on voltage control, self-organizing communication architectures are developed in [30]. Most of our previous work centers on frequency control under cyber attacks [31–34]. We categorize our research into three stages: stage I is the detection stage where we use ML to screen out false data injection attacks (FDIAs) [31]; stage II is the parameter tuning stage where we tune critical control parameters to attenuate the influence of time-delay attack [32]; and stage III is the controller reconfiguration stage where we reconfigure the conventional automatic generation control (AGC) using the estimation and compensation schemes [33, 34].

This chapter is the refinement of our previous work on compensation-based controller reconfiguration [34]. The improvement includes the complete physics analysis of FDIA against AGC using a detailed multi-machine classic model and the in-depth analysis of different key simulation setups in case studies. We desire to have this chapter serve as a step-stone in the area of attack-tolerant frequency control for large-scale complex power systems. The remainder of the chapter is as follows: Sect. 2 gives the background of AGC under FDIA and the detailed multi-machine classic model-based AGC system; Sect. 3 presents the hybrid physics-based and data-driven mitigation strategy; Sect. 4 tests the proposed strategy via IEEE benchmark systems; finally, we end this chapter by giving conclusions in Sect. 5.

2 Automatic Generation Control Under False Data Injection Attacks

Before presenting the compromised AGC system and its corresponding model, we first briefly review its intact counterpart. Researchers use the system frequency response (SFR) model extensively to simulate the frequency behavior of the bulk system. The SFR model averages multiple generating units into a single machine via dynamic equivalencing [35]. The SFR model ignores the slow thermal system dynamics of the boiler and ultra-fast electric generator response, both of which are beyond the interested midrange frequencies. Hence, supposing the system is dominated by the steam engine-driven thermal units, the resulting SFR model comprises the governor server motors, steam turbine, and inertia.

2.1 Automatic Generation Control Based on Detailed Multi-machine Classical Model

IN this chapter, instead of relying on the SFR model, we use the detailed multi-machine classical model (MMCM), which can reflect the real-world power systems.

The MMCM mainly contains the dynamic generator model, interface model, and network model. The first takes the form of a first-order differential equation. The latter two can be described by algebraic constraints. Since the generator in this chapter is a three-phase synchronous machine, the most direct component concerning the system frequency is the rotor speed ω_i:

$$\frac{d\delta_i}{dt} = \omega_i - \omega_s \tag{1}$$

Further, in the dynamic generator model, we use the swing of the rotor of a synchronous machine, which is described by a second-order differential equation, to describe the relationship between the rotor swing and power exchange:

$$\frac{2H}{\omega_s}\frac{d^2\delta_i}{dt^2} = P_{mi} - P_{ei} \tag{2}$$

where the mechanical power P_{mi} in the output of the turbine governor control system:

$$\begin{aligned} \dot{P}_{mi} &= \frac{1}{T_{ti}}(-P_{mi} + u_{gi}) \\ \dot{u}_{gi} &= \frac{1}{T_{gi}}\left(-u_{gi} - \frac{(\omega_i - \omega_i)}{R_i} + P_i^r\right) \end{aligned} \tag{3}$$

The electric power P_{ei} is written by:

$$P_{ei} = E'_{qi}I_{qi} + E'_{di}I_{di} - (x'_{di} - x'_{qi})I_{di}I_{qi} \tag{4}$$

To obtain the dq-axis voltages or currents, we use the direct-quadrature-zero transformation to transfer the three-phase quantities into a two-axis rotating dq0 reference frame. By ignoring the ultrafast subtransient process, we define the dynamics of the two-axis voltage behind the transient reactance as:

$$\begin{aligned} T'_{doi}\frac{dE'_{qi}}{dt} &= -E'_{qi} - (x_{di} - x'_{di})I_{di} + E_{fdi} \quad i = 1, \ldots, m \\ T'_{qoi}\frac{dE'_{di}}{dt} &= -E'_{di} - (x_{qi} - x'_{qi})I_{qi} \quad i = 1, \ldots, m \end{aligned} \tag{5}$$

where the excitation voltage E_{fdi} can be controlled by proportional controller:

$$\dot{E}_{fi} = -\frac{E_{fi}}{T_{ei}} + \frac{K_{ei}}{T_{ei}}(V_{refi} - V_i) \tag{6}$$

The interface model describes the transformation from the internal to the terminal voltage:

$$E'_{di} - V_i \sin(\delta_i - \theta_i) - r_{si} I_{di} + x'_{qi} I_{qi} = 0$$
$$E'_{qi} - V_i \cos(\delta_i - \theta_i) - r_{si} I_{qi} - x'_{di} I_{di} = 0 \quad (7)$$

The definition of parameters and variables from (1)–(7) can be found in [36] and not repeated herein.

Supposing all loads are modelled as constant impedances, the network model in the rectangular coordinate system can be written by:

$$\begin{bmatrix} \boldsymbol{V}_g \\ \boldsymbol{V}_p \\ \boldsymbol{V}_l \end{bmatrix} = \begin{bmatrix} \boldsymbol{Z}_{gg} & \boldsymbol{Z}_{gp} & \boldsymbol{Z}_{gl} \\ \boldsymbol{Z}_{pg} & \boldsymbol{Z}_{pp} & \boldsymbol{Z}_{pl} \\ \boldsymbol{Z}_{lg} & \boldsymbol{Z}_{lp} & \boldsymbol{Z}_{ll} \end{bmatrix} \begin{bmatrix} \boldsymbol{I}_g \\ \boldsymbol{I}_p \\ \boldsymbol{I}_l \end{bmatrix} \quad (8)$$

where $\boldsymbol{V}_g = \begin{bmatrix} V_1 & \cdots & V_m \end{bmatrix}^T$ represents the vector of terminal voltages of m generators; $\boldsymbol{V}_p = \begin{bmatrix} V_{p1} & \cdots & V_{pn} \end{bmatrix}^T$ represents the vector of voltage at p tie-lines. $\boldsymbol{I}_g = \begin{bmatrix} I_1 & \cdots & I_m \end{bmatrix}^T$ represents the vector of currents injected into the network from m generators; $\boldsymbol{I}_p = \begin{bmatrix} I_1 & \cdots & I_n \end{bmatrix}^T$ represents the vector of currents injected into the network from n tie-lines. $\boldsymbol{V}_l = \begin{bmatrix} V_{l1} & \cdots & V_{lk} \end{bmatrix}^T$ represents the vector of voltage at k load buses; $\boldsymbol{I}_l = \begin{bmatrix} I_{l1} & \cdots & I_{lk} \end{bmatrix}^T$ represents the vector of currents injected into the k load buses. $\boldsymbol{Z_{gg}}$, $\boldsymbol{Z_{gp}}$, $\boldsymbol{Z_{gl}}$, $\boldsymbol{Z_{pg}}$, $\boldsymbol{Z_{pp}}$, $\boldsymbol{Z_{pl}}$, $\boldsymbol{Z_{lg}}$, $\boldsymbol{Z_{lp}}$, and $\boldsymbol{Z_{ll}}$ are the corresponding impedance matrices. Combining (1)–(8) we can write the complete MMCM as a set of differential–algebraic equations (DAEs):

$$\dot{\boldsymbol{x}} = \boldsymbol{f}(\boldsymbol{x}, \boldsymbol{u}, \boldsymbol{y})$$
$$0 = \boldsymbol{g}(\boldsymbol{x}, \boldsymbol{u}, \boldsymbol{y}) \quad (9)$$

where

$\boldsymbol{x} = \begin{bmatrix} \delta_1 & \omega_1 & E'_{d1} & E'_{q1} & P_{m1} & u_{g1} & E_{f1} & \cdots & \delta_m & \omega_m & E'_{dm} & E'_{qm} & P_{mm} & u_{gm} & E_{fm} \end{bmatrix}^T$; $\boldsymbol{u} = \begin{bmatrix} P_1^r & V_{ref1} & \cdots & P_m^r & V_{refm} & I_{p1} & \cdots & I_{pn} & I_{l1} & \cdots & I_{lk} \end{bmatrix}^T$; $\boldsymbol{y} = \begin{bmatrix} \boldsymbol{V}_p^T & \boldsymbol{V}_l^T \end{bmatrix}^T$. Since AGC mainly considers the frequency stability under small disturbances, we can use the small signal model by linearizing (9) around the equilibria (nominal states). The small signal model is expressed by:

$$\Delta\dot{x} = A\Delta x + B\Delta u + C\Delta y$$
$$0 = D\Delta x + E\Delta u + F\Delta y \quad (10)$$

Based on the center-of-inertia (CoI) frequency, we can write the area control error (ACE) as:

$$\Delta y_c = \beta \frac{\sum_i H_i \Delta\omega_i}{\sum_i H_i} + \sum_j \Delta P_{tie,j} \quad (11)$$

Note that the controlled output contains the tie-line power deviation terms, we treat them as extended states, and thus the differential equations in (10) can be rewritten by:

$$\Delta\dot{\overline{x}} = \overline{A}\Delta\overline{x} + \overline{B}\Delta u \tag{12}$$

where $\Delta\overline{\boldsymbol{x}} = \begin{bmatrix} \Delta\boldsymbol{x}^T & \Delta P_{tie,1} & \cdots & \Delta P_{tie,n} \end{bmatrix}^T$, $\overline{A} = diag(A, I_{n\times n})$, $\overline{B} = \begin{bmatrix} B^T & 0 \end{bmatrix}^T$. Combining (11) and (12), we can obtain the complete AGC control system model:

$$\begin{aligned} \Delta\dot{\overline{x}} &= \overline{A}\Delta\overline{x} + \overline{B}_c\Delta u_c + \overline{B}_d\Delta u_d \\ \Delta y_c &= C_c\Delta\overline{x} \end{aligned} \tag{13}$$

where $\Delta\boldsymbol{u}_c$ represents the vector of the control input; while $\Delta\boldsymbol{u}_d$ represents the vector of the exogenous disturbance. Based on (13), we can design controllers to stabilize Δy_c under $\Delta\boldsymbol{u}_d$. Take the PI controller as an example, the control input $\Delta\boldsymbol{u}_c$ can be expressed by:

$$\Delta u_{ci} = k_p\Delta y_c + k_i\int\Delta y_c dt \tag{14}$$

2.2 *Power Imbalance-Oriented False Data Injection Attacks*

With the aid of the AGC control system (1)–(14), we reveal the operating principle of FDIA against AGC. It should be mentioned that the computer system-oriented cyber intrusions, for example, cybercriminals' exploitation of hardware backdoors to infiltrate smart meters or interfacing devices on the data acquisition layer, are not the focus of this chapter. Instead, we concentrate on the compromised power systems application (i.e., AGC) and study the working mechanism of FDIA targeting specific attack goals. The FDIA in this case is equivalent to the data manipulation (injected false data) added to the true data:

$$x_c = x_t + d \tag{15}$$

where x_c denotes the compromised state information uploaded to the SCADA system; x_t denotes the true information before cybercriminal incidents occur; and d denotes the injected false data. As is shown in (15), the FDIA is demonstrated or modelled as a math formula (summation), which is sufficient to characterize the intrusion process for the goal-oriented FDIA working principle study. The concrete implementation, for example, that the attacker bypasses security protection of RTUs to achieve FDIA in (15), is out of scope for this research.

Usually, the security level is directly proportional to the importance. The greater the importance of the protégé is, the higher its security level is. In the hierarchical security architecture for smart grids, the increasing importance is commonly accompanied by the rise of grid layers. Specifically, the power system application on the top layer receives the highest protection from various security measures, including firewalls, data encryption, and security patches. In contrast, subject to the installation and implementation costs, the bottom layer where RTUs (PMUs) collect data from physical components usually lacks sufficient protection and is thus susceptible to cyber attacks. In terms of AGC, RTUs are generally classified into two categories. The first one serves the purpose of measuring system frequency and tie-line power. The second one directly interfaces the governor (of the synchronous machine) and dispatch its respective AGC order. Since the second category is usually the inherent component of the highly secure power plant monitoring and control system, attackers may barely gain its write permissions and manipulate the AGC order. Hence, most of the existing research studies have focused on measurement tampering. Most of them use the center-of-inertia (CoI) frequency, which is substituted by ω_{COI} for convenience of analysis. Unlike the tie-line power measurement, the CoI frequency (speed) ω_{COI} is simultaneously dependent upon multiple generator values, which can offer backup protection via measurement redundancy. Take the FDIA as an example, there will always exists a difference between the compromised generator frequency measurement and the remaining intact ones. And the difference is the exact injected false data. Therefore, we can achieve the detection by comparing the difference with the predetermined threshold. To avoid detection, the attacker may simultaneously corrupt all frequency measurements by enforcing the same attack laws, e.g., the same injected false data in the FDIA. Besides the frequency measurement tampering, the attacker may tamper with the tie-line power measurement. Both lead to incorrect ACE determination, thus causing power imbalance and deteriorated frequency quality. Figure 1 demonstrates the complete procedure of these two attack scenarios under the framework of a four-layer electrical-cyber-physical system. Whatever the attack scenario is, the compromised ACE can always be written by the sum of the true ACE and the change:

$$ACE_c = ACE_t + \varepsilon \tag{16}$$

Based on the A1/A2 and B1/B2 control performance criteria, ACE of a specific control area should return to zero. Then (16) can be transformed into:

$$\Delta\omega_{COI} = -\frac{\Delta P_{tie} + \varepsilon}{\beta} \tag{17}$$

To evaluate the steady-state CoI frequency, we aggregate all interconnected control areas into one integrated system and write the total ACE as the sum of compromised and intact ACEs:

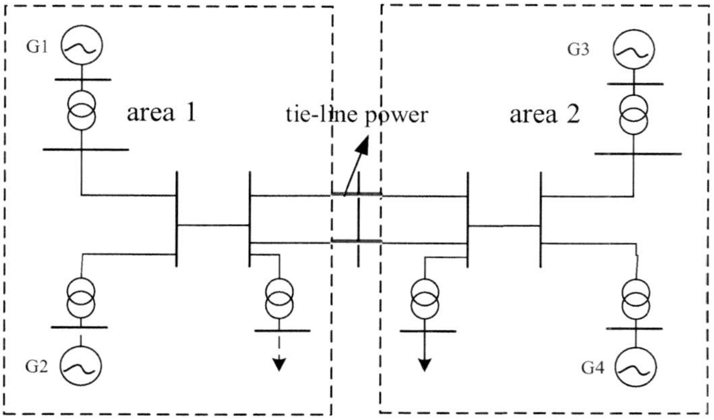

Fig. 2 The schematic diagram of false data injection attack against tie-line power measurement of Kundur's two-area system

$$ACE_{tot} = \sum_{i \in N_c} ACE_{ic} + \sum_{j \notin N_c} ACE_{jt} \tag{18}$$

where N_c represents the set of compromised areas. For a pair of interconnected areas i and j, we have:

$$\Delta P_{tie,i} = -\Delta P_{tie,j} \tag{19}$$

By substituting (19) into (18), it follows that:

$$ACE_{tot} = \sum_{k} \beta_k \Delta\omega_{COI,k} + \sum_{i \in N_c} \varepsilon_i \tag{20}$$

where $\Delta\omega_{COI,k}$ represents the CoI frequency of control area k; β_k represents the bias coefficient of control area k; ε_i represents the compromised ACE change of the control area i. Since ACE_{tot} returns to zero under the A1/A2 and B1/B2 control performance criteria, considering the homogeneity of the steady-state $\Delta\omega_{COI,k}$, we can finally obtain the steady-state $\Delta\omega_{COI,k} = -\sum_{i \in N_c} \varepsilon_i / \sum_k \beta_k$. Under the condition that β_k is fixed for each control area k, the attacker can increase the injected false data d_i to elevate ε_i, thus deteriorating the frequency quality by increasing $\Delta\omega_{COI,k}$.

We use Kundur's system in Fig. 2 to demonstrate the compromised AGC system. We assume the attacker tampers the tie-line power measurement between control area 1 and 2. There is no backup sensor (RTU) and the defender cannot immediately detect and switch to other healthy backups. Consequently, the system would experience abnormalities such as the irregular frequency excursions. Figure 3 shows the machine speed and COI frequency deviations considering the tie-line power measurement tampering. FDIAs will cause persistent frequency deviation and is dangerous to the

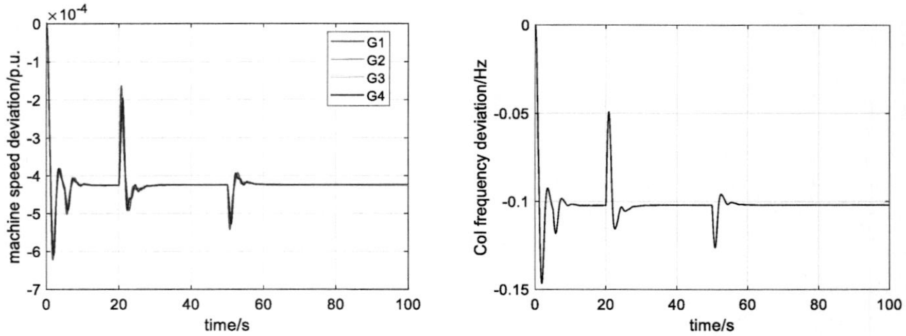

(a) Machine speed deviations (b) CoI frequency deviations

Fig. 3 Machine speed and CoI frequency deviations under tie-line power measurement tampering of Kundur's System

power system stability. Therefore, we must design mitigation strategy to attenuate the influence of FDIAs.

3 Hybrid Physics-Based and Data-Driven Mitigation Strategy

IN this section, we present the hybrid physics-based and data-driven mitigation strategy for AGC under FDIAs. First, we study the compensation-based mitigation from the perspective of model-based control. Then, we expand on the limitation of model-based compensation in complex power systems, and introduce the conception and realization of data-driven mitigation strategy.

3.1 Physics Analysis of Compensation-Based Mitigation Strategy

Section 2 describes the working principle of compromised AGC using a detailed MCMM, which can be written by:

$$\begin{aligned} \Delta\dot{\overline{x}} &= \overline{A}\Delta\overline{x} + \overline{B}_c\Delta u_c + \overline{B}_d\Delta u_d \\ \Delta y_c &= C_c\Delta\overline{x} \end{aligned} \tag{21}$$

Hence, we can formulate the cyber-attack mitigation problem as:

Problem Definition Consider the linear time-invariant system (21), the control objective is to design the control input $\Delta \boldsymbol{u}_c$, such that Δy_c is asymptotically stable.

Since the compromised ACE change ε is equivalent to a sensor fault, we can use a fault-tolerant controller for cybersecurity risk mitigation. We can categorize fault-tolerant control system (FTCS) into active and passive FTCS. The former reconfigures the controller using the real-time fault diagnosis information; the latter does not reconfigure the controller in response to real-time faults by pre-considering all the faulty and normal operating conditions at the design stage. The passive FTCS stresses the robustness of the control system, which can adapt to various faulty conditions. Compared with the active FTCS, the passive FTCS usually discards complex control structures since it accommodates the overall fault conditions rather than every individual fault condition. Hence, the passive FTCS can guarantee improved simplicity in practice compared with its active counterpart. Nevertheless, the passive FTCS barely achieves the optimal fault suppression for any particular faulty condition, which may be essential for high control performance requirements. In fact, most of the robust control algorithm-based passive FTCSs only accomplish the suboptimal performance. For example, the robust H-infinity method tolerates a specific range of uncertainties by satisfying prescribed control performance constraints:

$$\left\| H_{\varepsilon}^{e}(s) \right\|_{\infty} \leq \gamma \tag{22}$$

where $H_{\varepsilon}^{e}(s)$ represents the transfer function between the uncertainties (the compromised ACE change) and the control output. As can be seen from (22), though the gain from the input to the output is attenuated by the prescribed γ, Δy_c is still dependent upon ε. In other words, the robust control performance may deteriorate under intensive FDIAs with big injected false data. Furthermore, the compromised ACE change ε would always shade the true ACE, such that we cannot distinguish the true frequency values. To overcome this inherent problem of passive FTCS, we should actively estimate and compensate ε. This design belongs to the active FTCS. Specifically, we use the estimated $\hat{\varepsilon}$ to correct the compromised output as $\Delta y_c - \hat{\varepsilon}$, which can be used to generate the control input $\Delta \boldsymbol{u}_c$.

The active FTCS has received tremendous attention via model-based methods, among which the active disturbance rejection control (ADRC) method gains popularity for actively estimating and compensating uncertainties. One of the well-established uncertainty estimation strategies is the sliding mode observer, which maintains the sliding motion of the estimation error dynamics e_x upon the prescribed sliding surface:

$$s = \left(\frac{d}{dt} + \delta \right)^{n-1} e_x \tag{23}$$

where n denotes the dimension of the output state; δ denotes a strictly positive number. The compromised ACE change ε is equivalent to the noise, then we can reformulate the output using a dynamic filter and treat it as an extended state, such that the reformulated system can filter out ε by moving it to the input channel:

$$\begin{aligned} \Delta x_e &= A_e \Delta x_e + B_e \Delta u_c + F_e \varepsilon \\ \Delta y_c &= C_e \Delta x_e \end{aligned} \tag{24}$$

where $\boldsymbol{x}_e$ is the state vector considering the extended state. A_e, B_e, F_e and C_e are the corresponding matrices. Then we can design a sliding mode observer based on (24):

$$\Delta \dot{z} = A_e \Delta z + B_e \Delta u_c - G_1(C_e z - \Delta y_c) + G_2 \eta \tag{25}$$

where z is the vector of observed or estimated states; G_1 is the gain vector. The discontinuous term η is defined by:

$$\eta = \begin{cases} -\rho \| C_e F_e \| \frac{P_0 e_y}{\| P_0 e_y \|} & if \quad e_y = 0, \\ 0 & otherwise \end{cases} \tag{26}$$

If the design matrices (e.g., C_e, F_e, A_e) satisfy certain conditions, we can prove that the estimation errors converge to zero and obtain the corresponding reconstructed $\hat{\varepsilon}$. The details are presented in [37] and not repeated herein.

The aforementioned observer-based active FTCS belongs to model-based control and extensively exploit the physical dynamics of MCMM for the observer design. Nevertheless, in the context of the bulk power systems, it may be difficult to employ the model-based active FTCS directly due to the following reasons:

The complexity of system model: Multiple generators, including the steam and hydraulic turbine-driven synchronous generators and power electronics interfaced renewable generators, aggravate the system complexity. Mathematically, this enhanced complexity corresponds to the increased system size, order, and nonlinearity, which is usually at odds with the purpose of model-based control for simple structured systems.

The uncertainty of the system model: Integrating a high share of unpredictable renewable energy and plug-in electric vehicles, along with unplanned generator tripping, relay failures, and other emergency situations, expose the modern power system to more uncertainty issues. In turn, the dynamic system model can be uncertain rather than deterministic. And we cannot precisely recover the system dynamics via mathematical modeling. Without the reliable physical system model, it is difficult to design model-based control strategies.

Considering the degrading applicability of model-based active FTCS in face of complex power systems, we resort to model-free active FTCS. Unlike its model-based counterpart, the model-free version is independent of the system model. Hence it refrains from the curse of system model complexity or uncertainty. Inspired by the sliding observer-based sensor fault reconstruction, we can see that the key to FDIA mitigation strategy is the estimation of the compromised ACE change. With this in mind, we should design a model-free controller (observer) considering the high-order AGC system.

A model-free controller in its extreme form is completely data-driven. In other words, we treat the system as a black box and achieve the control laws by learning the causal relationship between the input and output of the black box. A famous example is the reinforcement learning (RL), which models the control law as a trained projector (a tabular form or a neural network) by maximizing the prescribed rewards based on the observation of the black box output at specific states. Though the idea of learning-based control is promising and implementable, its practical application faces challenges of heavy data and computational burdens. Since the key of FDIA mitigation is the estimation of compromised ACE change, we design the estimation-oriented regression method rather than the control input-oriented RL method. The regression is irrelevant with the decision making which incorporates the machine learning and dynamic optimization over multiple learning episodes, thus greatly reducing the computational complexities.

3.2 *Causal Relationship Between AGC Variables of Interest*

IN summary, we call the proposed model-free FDIA-tolerant AGC scheme as the physics-based and data-driven mitigation strategy. In other words, to achieve the regression-based compromised ACE change estimation, we first build up the regression model via physics analysis. Second, we design the data-driven estimation scheme using the established regression model. In the following, we expand on these two aspects respectively.

We use the physics-based regression model analysis to infer the causal relationship between the independent and dependent variables (i.e., ε). Suppose the tie-line power measurement of control area i suffers the FDIA with the injected false data ε. The compromised aggregated AGC system can be expressed by:

$$\begin{aligned} \dot{x} &= Ax + Bu + Fd \\ ACE_c &= cx + \varepsilon \end{aligned} \tag{27}$$

where ACE_c represents the compromised ACE of the whole system. $\boldsymbol{x}$ represents the vector of system state deviations from which Δ is removed for brevity; $\boldsymbol{u}$ represents the vector of reference power adjustments; $\boldsymbol{d}$ represents the vector of aggregated disturbances. We assume that the influence of $\boldsymbol{d}$ is far less than ε. Then we can remove the $\boldsymbol{Fd}$ term. Meanwhile, let $u = \begin{bmatrix} u_1 \cdots u_m \end{bmatrix}$; then we can calculate u_i under the integral controller as:

$$u_i = \alpha_i k_i \int ACE_c dt \tag{28}$$

where α_i represents the allocation coefficient; k_i represents the integral control coefficient. Substituting (28) into (27) leads to:

$$\dot{x} = Ax + BD\int (cx + \varepsilon)dt \tag{29}$$

where $D = \begin{bmatrix}\alpha_1 k_1 \cdots \alpha_m k_m\end{bmatrix}^T$; let x_e be the extended state $x_e = \int cxdt$, we reformulate (29) by:

$$\underbrace{\begin{bmatrix}\dot{\boldsymbol{x}} \\ \dot{x}_e\end{bmatrix}}_{\dot{\overline{\boldsymbol{x}}}} = \underbrace{\begin{bmatrix}A & BD \\ \boldsymbol{c} & 0\end{bmatrix}}_{\overline{A}} \underbrace{\begin{bmatrix}\boldsymbol{x} \\ x_e\end{bmatrix}}_{\overline{\boldsymbol{x}}} + \underbrace{\begin{bmatrix}\boldsymbol{1} \\ 0\end{bmatrix}}_{\overline{B}} \boldsymbol{\eta} \tag{30}$$

where $\boldsymbol{\eta} = BD\int \varepsilon dt$. From (30) we can see that the false data-related term $\boldsymbol{\eta}$ affects the state dynamics. Furthermore, considering the CoI frequency deviation as the measured output y, the complete state space model is:

$$\begin{aligned} \dot{\overline{\boldsymbol{x}}} &= \overline{A}\overline{\boldsymbol{x}} + \overline{B}\boldsymbol{\eta} \\ y &= h\overline{\boldsymbol{x}} \end{aligned} \tag{31}$$

From (31) we can see that there exists a causal relationship between $\boldsymbol{\eta}$ and y. In other words, we can perform the regression analysis to reconstruct the mapping from y to $\boldsymbol{\eta}$. By discretization, the discrete-time state-space model of (31) is:

$$\begin{aligned} \overline{\boldsymbol{x}}[k+1] &= \overline{A}_d\overline{\boldsymbol{x}}[k] + \overline{B}_d\boldsymbol{\eta}[k] \\ y[k] &= h_d\overline{\boldsymbol{x}}[k] \end{aligned} \tag{32}$$

Take the z-transform of both sides of (32):

$$\begin{aligned} z\overline{\boldsymbol{X}}(z) &= \overline{A}_d\overline{\boldsymbol{X}}(z) + \overline{B}_d\boldsymbol{\eta}(z) \\ Y(z) &= h_d\overline{\boldsymbol{X}}(z) \end{aligned} \tag{33}$$

The transfer function of (33) is:

$$Y(z)/\boldsymbol{\eta}(z) = h_d\left(zI - \overline{A}_d\right)^{-1}\overline{B}_d \tag{34}$$

Ultimately, we can express the causal relationship as a set of difference equations:

$$y[k] + a_1 y[k-1] + \text{L} = \eta_1[k] + b_{11}\eta_1[k-1] + \text{L} + \eta_2[k] + b_{12}\eta_2[k-1] + \text{L} \tag{35}$$

From (35) we know that it is difficult to calculate $\boldsymbol{\eta}$, especially when we may have no precise information of the coefficients in (35), i.e., a_1, a_2,..., b_{11}, b_{21}, ..., b_{12},

b_{22}, ...Though the physics-based analysis cannot compute ε directly, it reveals the causal relationship between the compromised ACE change ε and the CoI frequency deviation y. It means we may perform the regression analysis upon ε and y, and use the regression model to estimate ε.

3.3 Regression-Based Estimation and Compensation-Based Mitigation Strategy

Figure 4 shows the sequence diagram of the complete regression-based estimation and compensation-based mitigation process. After the FDIA occurs at T_f, it takes t_{dec} to achieve the detection, followed by the data collection and regression-based estimation time t_{est}. Since the regression-based estimation takes place instantaneously, the data collection takes up most of t_{est}. Ultimately, we implement the compensation at T_2. Detection in stage I is not the emphasis of this chapter, but it is essential to stage II (estimation of ε). Besides discerning the presence of FDIAs, the detection in stage I determines how quickly the defender can start to collect data y. Slow detection compresses the allowable post-detection reaction time T_3- T_1, which would affect the accuracy of regression-based estimation (we will explain this phenomenon in the following). Therefore, the rapidity of detection contributes to the enhanced safety of time margin and squeezes more time to take mitigation measures. Specifically, we can adopt prediction-based simultaneous change-point and outlier detection algorithms, which can render the detection time negligible.

Before presenting the regression-based estimation method in stage II, we define the compromised ACE change ε as the constant signal. We model the static constant signal as a point value and use the sequence-to-point regression model to estimate ε.

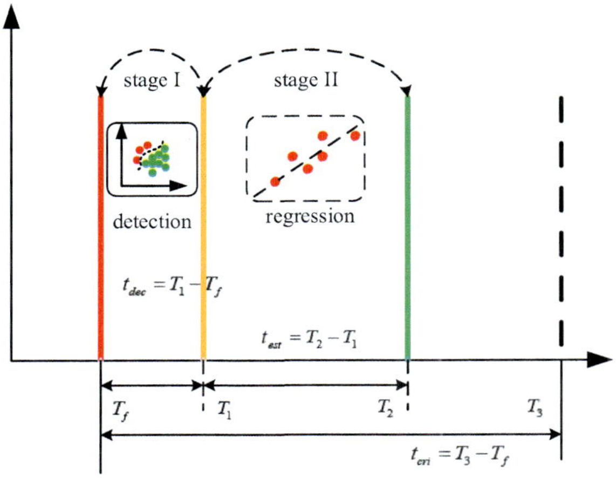

Fig. 4 The sequence diagram of the complete regression-based estimation and compensation-based mitigation process

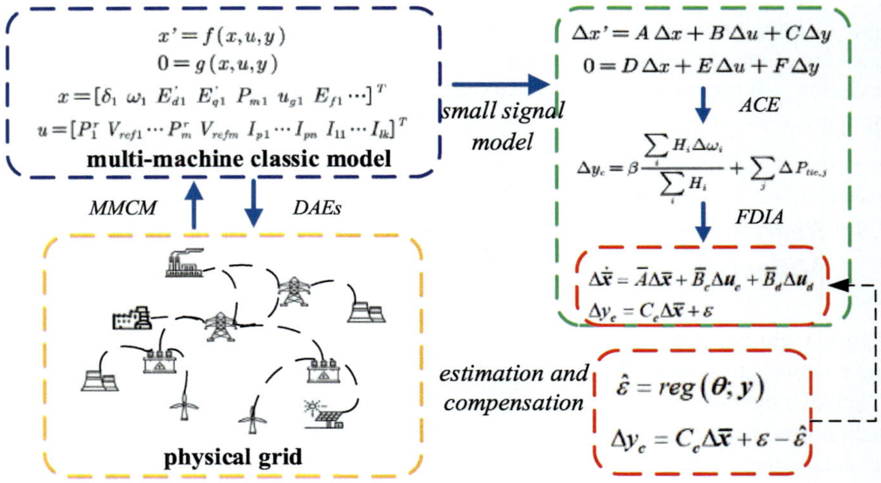

Fig. 5 The schematic diagram of hybrid physics-based and data-driven mitigation strategy

Sequence-to-point regression: The independent variable is y, while the dependent variable is ε, and the regression model is:

$$\hat{\varepsilon} = reg(\boldsymbol{\theta}; \boldsymbol{y}) \tag{36}$$

where $\boldsymbol{\theta}$ represents the vector of parameters; $\boldsymbol{y}$ represents the vector of CoI frequency deviation at different time points $\left[\, y(1T) \ \cdots \ y(qT) \,\right]$. Algorithm 1 summarizes the procedures of regression-based estimation in stage II. After estimation we can directly compensate the compromised ACE change at T_2. The complete design process is summarized in Fig. 5.

Algorithm 1 Regression-based Estimation of Compromised ACE Change

1 **input:** The set of injected false data $\boldsymbol{D}$
2 **output:** The regression model $reg(\boldsymbol{\theta}; \boldsymbol{y})$
3 **init**: sampling period T, the total number of sampling points q, the simulation model $\boldsymbol{M}$
4 $\boldsymbol{D}_{train}, \boldsymbol{D}_{test} \overset{split}{\leftarrow} \boldsymbol{D}$
5 $\boldsymbol{\varepsilon}_{train}, \boldsymbol{y}_{train,0} \leftarrow simulate(\boldsymbol{M}, \boldsymbol{D}_{train}), \boldsymbol{\varepsilon}_{test}, \boldsymbol{y}_{test,0} \leftarrow simulate(\boldsymbol{M}, \boldsymbol{D}_{test})$
6 $\boldsymbol{y}_{train} \leftarrow sample(\boldsymbol{y}_{train,0}, T, q), \boldsymbol{y}_{test} \leftarrow sample(\boldsymbol{y}_{test,0}, T, q)$
7 Train the regression model: $\min\|\boldsymbol{\varepsilon}_{train} - reg(\boldsymbol{\theta}; \boldsymbol{y}_{train})\|$
8 Test the regression model:$\mathrm{RMSE}(\boldsymbol{\varepsilon}_{test} - reg(\boldsymbol{\theta}; \boldsymbol{y}_{test}))$

4 Case Studies

In this section, we use the IEEE benchmark systems to test the proposed mitigation strategy. The details of the benchmark systems, including Kundur's and 68-bus systems, can be found in [34] and not repeated herein. First, we study the influence of different input data upon the regression-based estimation. Second, we perform the regression-based estimation using different regression methods. Finally, we test the compensation-based mitigation strategy using the regression-based estimation.

4.1 *Input Data Analysis*

Once we determine the regression method, the remaining most important determinant is the input data $\boldsymbol{y} = \left[y(1T) \cdots y(qT) \right]$, which is dependent upon the number of sampling points q and sampling time T. Also, the detection time $t_{dec} = T_1 - T_f$ in Fig. 6, which determines when we start to collect the input data, would also affects the data volume once the data collection time $T_2 - T_1$ is given. In this section, to study the influence of input data upon the regression performance, we test Algorithm 1 using the IEEE benchmark system under different input data compositions.

In practice, the sensors can record the real-time frequency measurement and compute $\boldsymbol{y}$. Based on (17) we know that the FDIA will cause persistent frequency deviations. The operating codes regulate the allowable duration period for a specific frequency deviation value. Generally, the wider the frequency deviates, the shorter the allowable time is. In other words, it means that we need to reduce q for large $\boldsymbol{y}$. Moreover, whatever $\boldsymbol{y}$ is, shorter observation period (i.e., smaller qT) means we

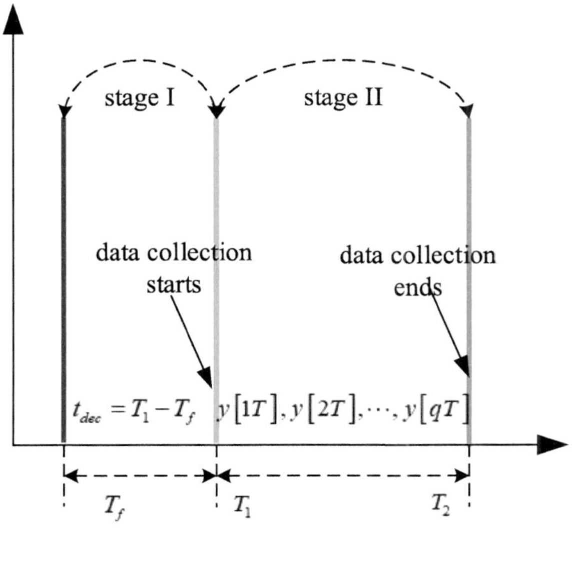

Fig. 6 The diagram of sequential detection and regression process

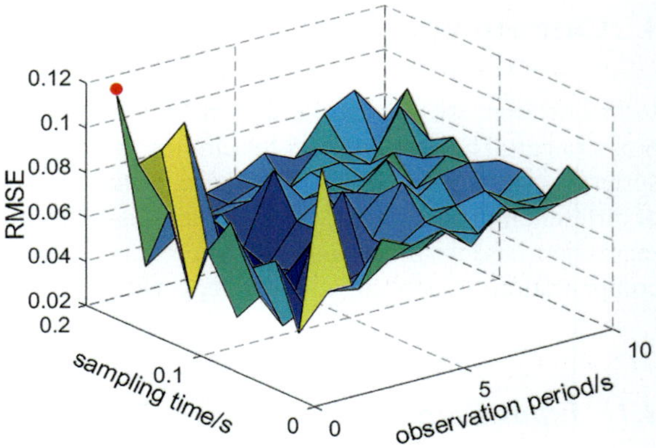

Fig. 7 Root mean squared errors under different sampling time and observation periods

could reduce the reaction time below the critical one, thus preventing the system from transferring to the unrecoverable states beyond T_3. To further demonstrate the influence of sampling period T and the number q of sampling points upon the regression performance, we use Algorithm 1 for different pairs of (T, q). The simulation model $\boldsymbol{M}$ is the MCMM of the Kundur's system. Besides, we choose the linear regression model for $reg(\boldsymbol{\theta}; \boldsymbol{y})$ considering the linearized small signal model of the compromised system. We generate 1000 random samples. We use 700 of them to train the linear regression model and 300 of them to test the performance and calculate the root mean square error (RMSE) as shown in Fig. 7.

From Fig. 7 we know that the estimation errors, defined by RMSE of the test instances, generally decrease with prolonged observation periods. As for the same observation period, a smaller sampling interval T usually means smaller estimation errors. This phenomenon makes sense since longer observation periods with high sampling rates offer more regression-relevant information on the dynamic trajectory of $\boldsymbol{y}$. Consequently, the shortest observation period (1 s) with the largest sampling interval (0.2 s) has the largest RMSE, as shown by the red dot in Fig. 7.

When the FDIA occurs at T_f, we need to save the detection time $t_{dec} = T_1 - T_f$ for the data collection and regression-based estimation time $t_{est} = T_2 - T_1$. Rather than neglecting t_{dec} in the prediction-based detection strategies, we consider the ex-post analysis-based detection where the detector needs compromised system responses or features after the occurrence of FDIA. Usually, as with the regression-based estimation, this ex-post analysis-based detection requires the data collection over a period of time. Therefore, we need to ascertain to what extent the regression can tolerate the detection-caused reaction delay. In other words, we should evaluate the allowable $t_{est} = T_2 - T_1$ for a fixed T_2 by varying the length of $t_{dec} = T_1 - T_f$. Let $T_2 = 7, 8, 9, 10s$, we use Algorithm 1 for different pairs of (T, t_{dec}), and Fig. 8 shows the results.

From Fig. 8 we can see that the estimation accuracy does not significantly deteriorate as $t_{est} = T_2 - T_1$ decreases. Take Fig. 8a as an example, with the increase of detection time t_{det} from 1 to 5 s, t_{est} decreases from 6 to 2 s. The estimation errors

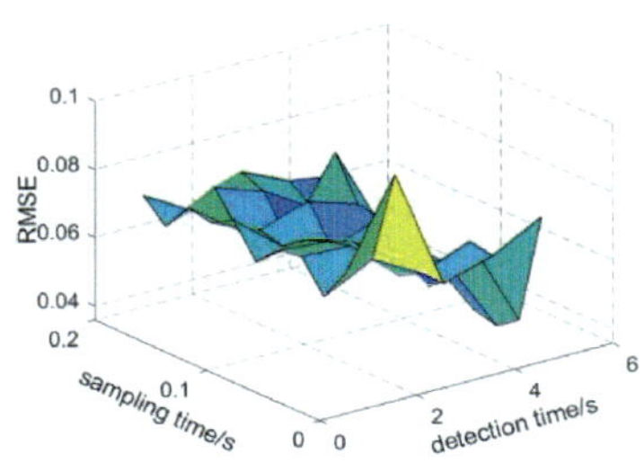

(a) Estimation errors under different pairs of $(T,\ t_{dec})$ and $T_2 = 7s$

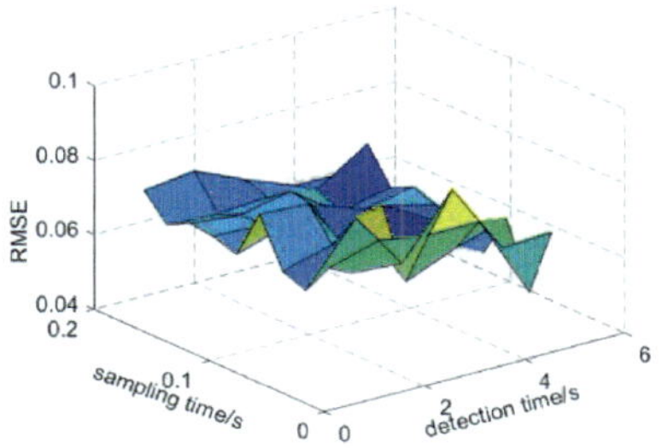

(b) Estimation errors under different pairs of $(T,\ t_{dec})$ and $T_2 = 8s$

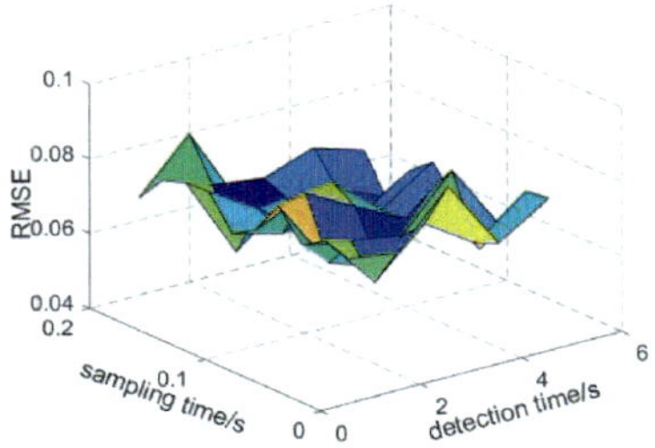

(c) Estimation errors under different pairs of $(T,\ t_{dec})$ and $T_2 = 9s$

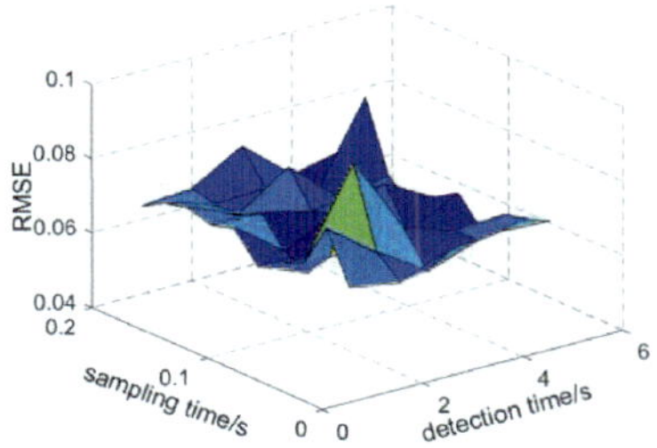

(d) Estimation errors under different pairs of $(T,\ t_{dec})$ and $T_2 = 10s$

Fig. 8 Root mean squared errors under different input data composition

under the decreasing t_{est} fall within the range between 0.06 and 0.08. We can get similar conclusions for other T_2 values. One plausible explanation of the insensitivity to detection-caused latency is that ε is highly dependent upon the steady-state y_s. More specifically, an injective function $y_s = -\varepsilon / \sum_k \beta_k$, as derived previously in Sect. 2, can quantify the relationship between ε and y_s. That is to say, we can reconstruct ε as long as we know y_s. From Fig. 9 we can observe that the settling time of the compromised CoI frequency excursions is around the range of 5 s to 7.5 s, as shown in the black rectangular window. It means that we can preserve the information of y_s as long as the data collector can capture the data points from the latter half of the trajectory. In other words, the trajectory during detection-caused latency mainly contains the non-steady y_{ns}, which is irrelevant with ε. Therefore, the accuracy maintains an inflexible level however long t_{det} is.

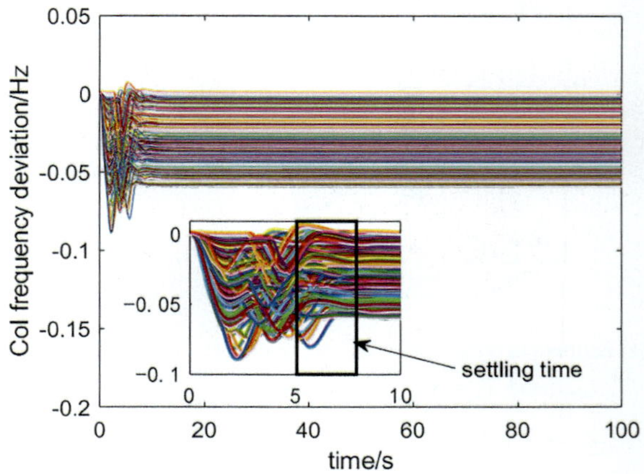

Fig. 9 Dynamic CoI frequency deviations under different FDIAs

4.2 Regression Method Analysis

In this section, we test the performance of SoTA regression methods regarding the estimation of ε. We choose three of the most popular regression methods: (1) least squares-based linear regression, (2) error margin constraint-based support vector regression (SVR), and (3) uncertainty-based Gaussian process regression (GPR). In the linear regression model, the objective is to minimize the sum of deterministic squared errors. Rather than minimizing the estimation errors, SVR treats the error term as a constraint. Moreover, with the kernel function-based transformation, SVR can handle the potential nonlinear relation between ε and y. As with SVR, GPR is a nonparametric regression method. GPR considers the model uncertainty and outputs a distribution rather than the point. We use the three regression methods for the following four scenarios. scenario 1: $t_{det} = 1$ s, $t_{est} = 9$ s, T $= 0.02$ s; scenario 2: $t_{det} = 5$ s, $t_{est} = 2$ s, T $= 0.2$ s; scenario 3: $t_{det} = 3s$, $t_{est} = 5$ s, T $= 0.1$ s; scenario 4: $t_{det} = 2$ s, $t_{est} = 6$ s, T $= 0.08$ s. Among the four scenarios, 1 and 2 generate two polarizing data compositions. The former has the largest volume of data with the shortest detection-caused latency t_{det}, data collection period t_{est}, and the smallest sampling time T; the latter has the opposite results. Scenarios 3 and 4 fall between. We randomly generate 1000 samples (ε and $\boldsymbol{y}$) for each scenario. We use 700 of them to train the regression model, while the remaining 300 samples are used to check the performance by evaluating the estimation errors. Figure 10 shows the distribution of errors in different scenarios using the three regression methods.

From Fig. 10 we learn that the errors under GPR are far less than those under LR or SVR, which validates the adaptability of GPR in face of uncertain (random) attack scenarios. It indicates that regression methods that focus on error minimization or constraint preservation of the known (training) data, may not have equal performance for the unknown (test) data.

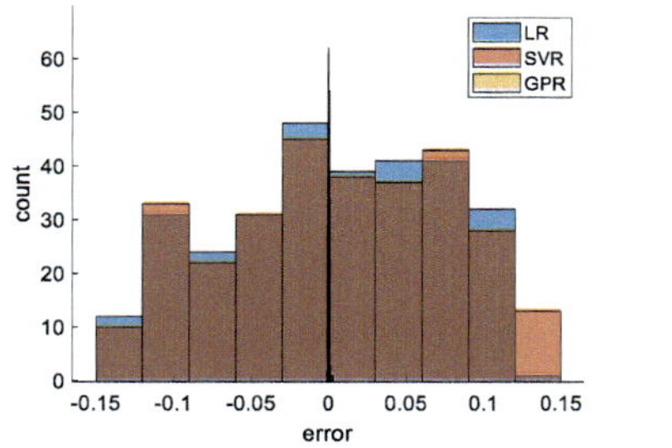

(a) Estimation errors using different regression methods at scenario 1

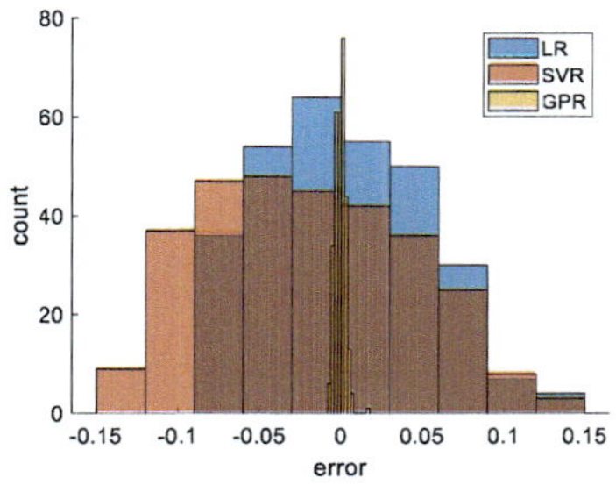

(b) Estimation errors using different regression methods at scenario 2

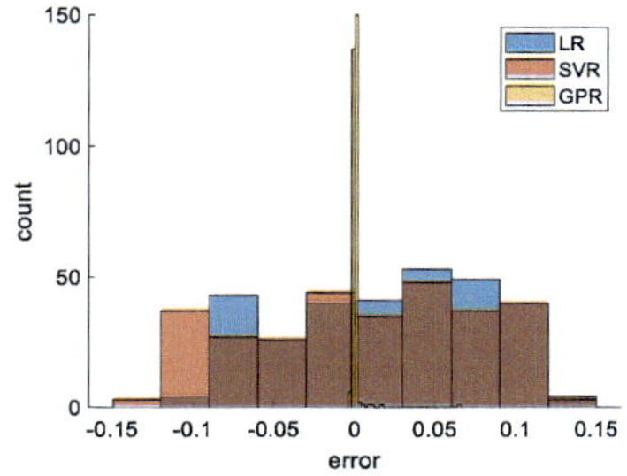

(c) Estimation errors using different regression methods at scenario 3

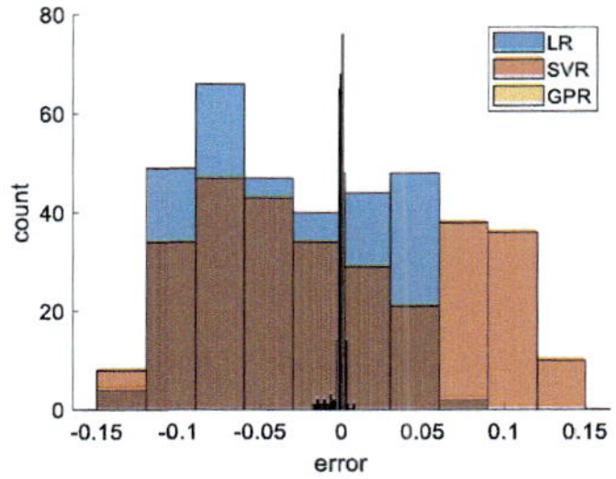

(d) Estimation errors using different regression methods at scenario 4

Fig. 10 Histograms of estimation errors under different scenarios and regression methods

4.3 Compensation-Based Mitigation Analysis

We can perform the compensation-based mitigation strategy using the estimation of ε. The steady-state CoI frequency deviation after compensation is $y_s = -(\varepsilon - \hat{\varepsilon})/\sum_k \beta_k$. If the estimation $\hat{\varepsilon}$ is close to ε enough, we can attenuate y_s to the utmost. We simulate the complete mitigation process using Kundur's and the IEEE 68-bus system. For convenience, we suppose that the FIDA occurs at 0 s. We use the trained regression models in scenarios 2 and 3 to estimate and compensate ε. Figure 11 shows the trajectory of the CoI frequency before and after compensation.

From Fig. 11 we note that the proposed compensation strategy works well for both benchmark systems. The left half of the red dotted line in each subfigure represents the reaction period (detection and data collection time), and the right half corresponds to the mitigation or recovery period. We do not consider the extra correction (e.g., load shedding) in the data collection period. Hence the CoI frequency keeps deviating before settling at the steady-state value. After compensation, the frequency deviation significantly decreases. All three regression methods guarantee acceptable frequency errors. In order to improve the frequency quality, we can reduce the detection time and data collection time to lower the compromised frequency deviation during the reaction period. Also, we can tune the control parameter to reduce the settling time,

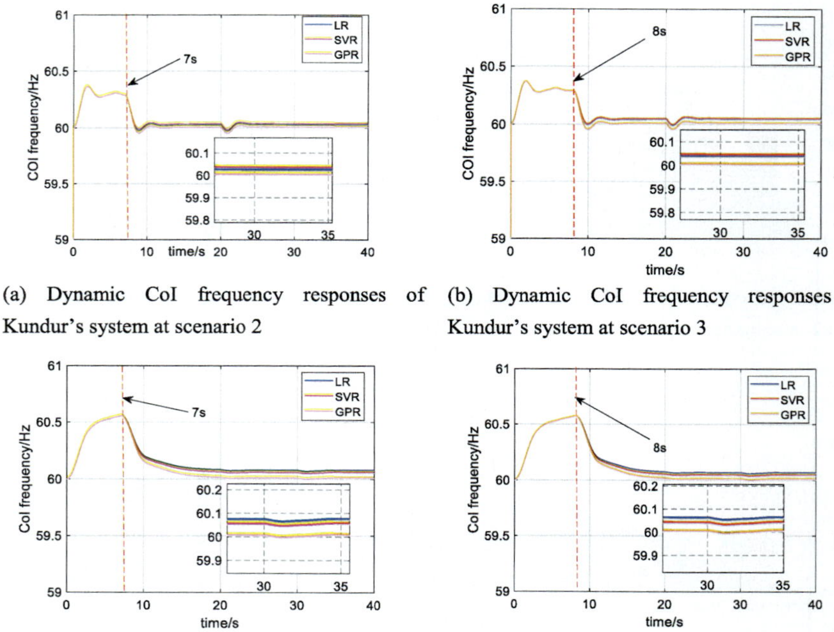

(a) Dynamic CoI frequency responses of Kundur's system at scenario 2

(b) Dynamic CoI frequency responses of Kundur's system at scenario 3

(c) Dynamic CoI frequency responses of IEEE 68-bus system at scenario 2

(d) Dynamic CoI frequency responses of IEEE 68-bus system at scenario 3

Fig. 11 Dynamic CoI frequency responses under different scenarios using the proposed mitigation strategy

thus capturing the steady-state frequency deviation value over a short period, which can also prevent the system from operating at long-lasting under-frequency or over-frequency states.

5 Conclusions

This chapter proposes a hybrid physics-based and data-driven mitigation strategy for the compromised AGC under cyber attack. We first perform the physics-based analysis based on a detailed multi-machine classical model and deduce the causal relationship between the variables of interest, i.e., the compromised frequency deviation and the ACE change. Then we use the regression to reconstruct the ACE change and compensate for it in the controlled output. From the simulation studies, we prove the effectiveness of the proposed mitigation strategy.

We mainly treat injected false data as the constant signal in this chapter. In practice, the attacker may opt for other types of attack signals. Hence it remains a question of how to develop a versatile mitigation scheme that can tolerate FDIAs featuring various signal types. We have made an exploratory study on how to mitigate periodic FDIAs [34]. The studied periodic FDIAs therein are confined to sinusoids. Nevertheless, periodic signals can take on many forms, which we do not address in our previous work. The key to predicting the complete periodic signals is the prediction over one cycle. When the sinusoid over the prediction horizon contains the peak point, we can reconstruct a sinusoid if we can predict the future fragment over a quarter of a cycle. However, when the periodic signals contain multiple modes, it requires the prediction over a complete period. The longer the period is, the poorer the regression performance may be. Besides the periodic signals, attackers may choose irregular stair-like signals, thus aggravating the issue. We can barely completely filter out these irregularly abnormal signals as we do to the random noise. In this situation, unilateral regression-based estimation is no longer useful. Instead, we may return to the conventional passive attack-tolerant control methods to free the reconfigured controller of the burden of actively predicting attack signals. The trade-off between active compensation and passive tolerance of the attack signals thus becomes an interesting topic in our future research.

References

1. P. Top, E. Banks, P.D. Barnes, et al., Simulation of a rtu cyber attack on a transformer bank, in *IEEE Power & Energy Society General Meeting* (IEEE, 2017), pp. 1–5
2. A. Parizad, C. Hatziadoniu, A laboratory set-up for cyber attacks simulation using protocol analyzer and RTU hardware applying semi-supervised detection algorithm, in *2021 IEEE Texas Power and Energy Conference (TPEC)* (IEEE, 2021), pp. 1–6
3. S. Kumar, H. Kumar, G.R. Gunnam, Security integrity of data collection from smart electric meter under a cyber attack, in *2019 2nd International Conference on Data Intelligence and Security (ICDIS)* (IEEE, 2019), pp. 9–13
4. J.C. Foreman, D. Gurugubelli, Identifying the cyber attack surface of the advanced metering infrastructure. Electr. J. **28**(1), 94–103 (2015)
5. A. Hahn, M. Govindarasu, Cyber attack exposure evaluation framework for the smart grid. IEEE Trans. Smart Grid **2**(4), 835–843 (2011)
6. M. Raciti, S. Nadjm-Tehrani, Embedded cyber-physical anomaly detection in smart meters[M]//Critical information infrastructures security. (Springer, Berlin, Heidelberg, 2013) pp. 34–45
7. C.C. Sun, D.J.S. Cardenas, A. Hahn et al., Intrusion detection for cybersecurity of smart meters. IEEE Trans. Smart Grid **12**(1), 612–622 (2020)
8. M. Ismail, M.F. Shaaban, M. Naidu et al., Deep learning detection of electricity theft cyber-attacks in renewable distributed generation. IEEE Trans. Smart Grid **11**(4), 3428–3437 (2020)
9. A. Kavousi-Fard, W. Su, T. Jin, A machine-learning-based cyber attack detection model for wireless sensor networks in microgrids. IEEE Trans. Ind. Inf. **17**(1), 650–658 (2020)
10. Y. Mo, B. Sinopoli, Secure control against replay attacks, in *2009 47th annual Allerton Conference on Communication, Control, and Computing (Allerton)* (IEEE, 2009), pp. 911–918
11. B. Tang, L.D. Alvergue, G. Gu, Secure networked control systems against replay attacks without injecting authentication noise, in *2015 American Control Conference (ACC)* (IEEE, 2015), pp. 6028–6033
12. M. Bogdanoski, T. Suminoski, A. Risteski, Analysis of the SYN flood DoS attack. Int. J. Comput. Netw. Inf. Secur. (IJCNIS) **5**(8), 1–11 (2013)

13. S.T. Zargar, J. Joshi, D. Tipper, A survey of defense mechanisms against distributed denial of service (DDoS) flooding attacks. IEEE Commun. Surv. Tutor. **15**(4), 2046–2069 (2013)
14. G.S. Kushwah, V. Ranga, Voting extreme learning machine based distributed denial of service attack detection in cloud computing. J. Inf. Secur. Appl. **53**, 102532 (2020)
15. M. Asad, M. Asim, T. Javed et al., Deepdetect: detection of distributed denial of service attacks using deep learning. Comput. J. **63**(7), 983–994 (2020)
16. N.R. Rodofile, K. Radke, E. Foo, Extending the cyber-attack landscape for SCADA-based critical infrastructure. Int. J. Crit. Infrastruct. Prot. **25**, 14–35 (2019)
17. A. Hahn, A. Ashok, S. Sridhar et al., Cyber-physical security testbeds: Architecture, application, and evaluation for smart grid. IEEE Trans. Smart Grid **4**(2), 847–855 (2013)
18. M. Kalech, Cyber-attack detection in SCADA systems using temporal pattern recognition techniques. Comput. Secur. **84**, 225–238 (2019)
19. Y. Liu, P. Ning, M.K. Reiter, False data injection attacks against state estimation in electric power grids. ACM Trans. Inf. Syst. Secur. **14**(1), 1–33 (2011)
20. B. Li, G. Xiao, R. Lu et al., On feasibility and limitations of detecting false data injection attacks on power grid state estimation using D-FACTS devices. IEEE Trans. Ind. Inf. **16**(2), 854–864 (2019)
21. M. Jin, J. Lavaei, K.H. Johansson, Power grid AC-based state estimation: Vulnerability analysis against cyber attacks. IEEE Trans. Autom. Control **64**(5), 1784–1799 (2018)
22. H. Wang, J. Ruan, G. Wang et al., Deep learning-based interval state estimation of AC smart grids against sparse cyber attacks. IEEE Trans. Ind. Inf. **14**(11), 4766–4778 (2018)
23. J.Q. James, Y. Hou, V.O.K. Li, Online false data injection attack detection with wavelet transform and deep neural networks. IEEE Trans. Ind. Inf. **14**(7), 3271–3280 (2018)
24. C. Chen, Y. Wang, M. Cui, et al., Data-driven detection of stealthy false data injection attack against power system state estimation. IEEE Trans. Ind. Inf. (2022)
25. L. Xie, Y. Mo, B. Sinopoli, Integrity data attacks in power market operations. IEEE Trans. Smart Grid **2**(4), 659–666 (2011)
26. H. Xu, Y. Lin, X. Zhang et al., Power system parameter attack for financial profits in electricity markets. IEEE Trans. Smart Grid **11**(4), 3438–3446 (2020)
27. H. Badrsimaei, R.A. Hooshmand, S. Nobakhtian, Stealthy and profitable data injection attack on real time electricity market with network model uncertainties. Electric Power Syst. Res. **205**, 107742 (2022)
28. P.K. Jena, S. Ghosh, E. Koley et al., Design of AC state estimation based cyber-physical attack for disrupting electricity market operation under limited sensor information. Electric Power Syst. Res. **205**, 107732 (2022)
29. M. Ghafouri, M. Au, M. Kassouf et al., Detection and mitigation of cyber attacks on voltage stability monitoring of smart grids. IEEE Trans. Smart Grid **11**(6), 5227–5238 (2020)
30. C. Cameron, C. Patsios, P.C. Taylor et al., Using self-organizing architectures to mitigate the impacts of denial-of-service attacks on voltage control schemes. IEEE Trans. Smart Grid **10**(3), 3010–3019 (2018)
31. C. Chen, K. Zhang, K. Yuan et al., Novel detection scheme design considering cyber attacks on load frequency control. IEEE Trans. Ind. Inf. **14**(5), 1932–1941 (2017)
32. C. Chen, Y. Chen, K. Zhang et al., System redundancy enhancement of secondary frequency control under latency attacks. IEEE Trans. Smart Grid **12**(1), 647–658 (2020)
33. C. Chen, M. Cui, X. Fang et al., Load altering attack-tolerant defense strategy for load frequency control system. Appl. Energy **280**, 116015 (2020)
34. C. Chen, Y. Chen, J. Zhao et al., Data-driven resilient automatic generation control against false data injection attacks. IEEE Trans. Ind. Inf. **17**(12), 8092–8101 (2021)
35. P.M. Anderson, M. Mirheydar, A low-order system frequency response model. IEEE Trans. Power Syst. **5**(3), 720–729 (1990)
36. P. Kundur, N.J. Balu, M.G. Lauby, *Power system stability and control* (McGraw-hill, New York, 1994)
37. C. Edwards, C.P. Tan, Sensor fault tolerant control using sliding mode observers. Control. Eng. Pract. **14**(8), 897–908 (2006)

Data-Driven Cyber-Resilient Control of Wide Area Power Systems

Yasin Asadi, Malihe Maghfoori Farsangi, Ali Moradi Amani, Hassan Haes Alhelou, Seyed Mehran Dibaji, and Ehsan Bijami

Abstract With the uptake of advanced communication technologies, monitoring and operation of power systems have been experiencing a paradigm shift. Many centralized algorithms can now be implemented in a distributed scheme. For example, local decisions can make faster actions resulting in a resilient power grid. Despite these advantages, these communication networks have made power networks vulnerable to cyber-attacks. Therefore, only attack-resilient algorithms can be reliably implemented using these technologies. In this chapter, a Data-Driven Cyber-Resilient Control (DDCRC) method is proposed for the frequency stability of wide area power grids. In the proposed method, the automatic generation control signals are generated without using a predefined model such that the frequency stability of the power system in the presence of Deception Attacks (DA) is guaranteed. Simulation results on a three-area power grid show the efficiency and superiority of the proposed method.

Keywords Data driven control · Automatic generation control · Deception attack

Y. Asadi (✉) · M. M. Farsangi · E. Bijami
Shahid Bahonar University of Kerman, Kerman, Iran
e-mail: yasinasadi@eng.uk.ac.ir

M. M. Farsangi
e-mail: mmaghfoori@uk.ac.ir

E. Bijami
e-mail: e.bijami@eng.uk.ac.ir

A. M. Amani
School of Engineering, RMIT University, Melbourne, VIC, Australia
e-mail: ali.moradiamani@rmit.edu.au

H. H. Alhelou (✉)
Department of Electrical and Computer Systems Engineering, Monash University, Clayton, VIC 3800, Australia
e-mail: Hassan.haesalhelou@monash.edu

S. M. Dibaji
Massachusetts Institute of Technology (MIT), Cambridge, MA, USA
e-mail: dibaji@mit.edu

H. Haes Alhelou et al. (eds.), *Power Systems Cybersecurity*, Power Systems,
https://doi.org/10.1007/978-3-031-20360-2_7

1 Introduction

Nowadays, Information Communication (IC) and Data-Driven (DD) technologies have been largely applied to improve the monitoring and control performance of critical infrastructures [1]. Thanks to these technologies, the implementation of advanced Artificial Intelligence (AI) and Machine Learning (ML) technics have been facilitated to compensate for the lack of precise models of these complex systems by using recorded data. However, the dependence of these systems on communication networks makes them increasingly susceptible to cyber-attacks which may cause serious damage and disruptions. A cyber-attack is technically defined as intentional failures and disruptions caused by unauthorized access to Cyber-Physical Systems (CPSs), which may indeed impact their performance and stability [2, 3].

Wide area power systems are among the important CPSs that support our modern lifestyle. In these systems, individual power areas, which are possibly interconnected through a control network, shape a wide area interconnected power system. Research on these systems in both academia and industry has become mature over decades, and several solutions and technologies have been developed to support them. However, with the paradigm shift in power systems, from traditional large-scale centralized generation units to distributed renewable-based generation (see Fig. 1), new challenges have emerged that force researchers to revisit the old studies. In other words, intermittence and unpredictability of renewable resources and new technologies, such as electric vehicles, as well as the transformation of power distribution grids from "a network of consumers" to "a network of prosumers, i.e. consumers/generator units", have made power systems a very complex environment where many of the previously designed algorithms do not necessarily work. Despite these fundamental changes, consumers still expect to receive stable and reliable electricity.

In spite of the desire to employ more renewable resources in power grids, several technical issues should be addressed to keep the reliability of the grid. Modeling and control of new power grids is such a complex and sophisticated activity that traditional approaches may fail [4]. In addition, several advanced distributed control algorithms should be developed to mitigate undesired effects of the uncertain and almost unpredictable behavior of renewable resources [5]. This, indeed, means a huge amount of communication over control and monitoring networks. With the increase in the knowledge of cyber-attackers about the model and dynamics of power systems, cyber-attacks on power systems have become very destructive. Recently, in order to deal with these problems, some cyber-resilient control approaches have been proposed to maintain the performance of the power grids.

1.1 Cyber-Resilient Control

The cyber-resilience of the control system is the ability of the controller to preserve an acceptable performance in presence of a major cyber-physical disruption. In

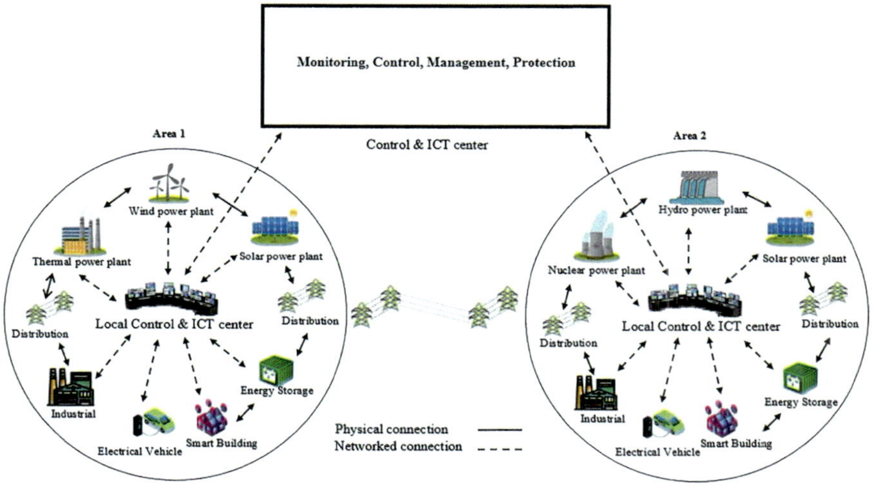

Fig. 1 The new structure of integrated power grids

recent years, several resilience methods have been proposed to guarantee stability and performance of the system when attackers can bypass IT security mechanisms. For example, in [6] two approaches to make the system secure against stealth attacks are proposed which are the optimal allocation of countermeasures to secure selected signals, and augmenting the measurement vector with extra secure data.

Generally, a cyber-resilient strategy includes the following two parts [7]. First, the resilience of the system against the attack is estimated and the possibility of increasing that by parameter tuning is studied. Then, an efficient algorithm is applied to estimate the state in presence of attacks. Using this view, predictive control approaches have been proposed to deal with deception attacks [8, 11]. Many algorithms proposed in the control community for Networked Control Systems (NCS) can be also extended to the cyber-attack condition, see e.g., attack detection and control approach proposed in [9–14]. For example, [9, 10] propose a state filtering strategy for NCS that have been presented in the case that the control signal to the plant is lost due to the attack. In [14], the online estimation approach is obtained by extending a modified version of the Unknown Input Kalman filter (UIKF) for the treatment of permanent unknown inputs.

1.2 Cyber-Resilient Control of Power Systems

With the increased number of cyber-attacks in recent years, research activities in both academia and power systems industry have been focused on attack detection and mitigation algorithms [15]. For example, it has been shown that Denial-of-Service (DoS), False Data Injection (FDI) and Resonant Attacks (RA) can destabilize power networks [16–18]. In DoS attacks, the attacker creates artificial traffic on the network to add delay in communications and prevent packets from arriving on time. In FDI attacks, the attacker maliciously changes the packets sent to the network such that a false value of the sensor measurement or control command reaches the destination. Based on these reports and advances in resilient controller design, power network researchers tend to design cyber-resilient control systems. For example, in [19] an event-triggered automatic generation controller has been designed for power system. In fact, with this resilient design, the frequency stability of the interconnected power network is guaranteed in the presence of DoS and DA attacks. In [20], an automatic generation controller has been designed based on model based adaptive method to maintain the frequency stability of the power grid by considering the minimum safe communication channels. In [21–23], to improve the performance of power grids in the presence of DA, unknown input estimators are extended. In [24, 25], using attack detection algorithms, designer first tried to prevent an FDI attack in the power grids. In fact, in this type of solutions, an attempt has been made to prevent the instability of the power network after detecting the attack by using intelligent mechanisms. In [26, 27] a virtual resilience layer has been designed to prevent instability in the DC and AC power grid after a cyber-attack on the control signals of secondary control level.

Despite the efforts mentioned in the previous paragraphs, in [28] proved that if the attacker knows the dynamics of the power system and its control mechanism, can successfully apply its scenario and unstable the power grid. This means that the attacker, knowing the dynamics of the system, tries to make malicious changes in the dynamics of the system. A very effective solution to solve this problem is to use DD methods along with adaptive control mechanisms. The main idea of these methods is the non-dependence of the controller design on the system model. In fact, in these methods, the control signal is updated based on the input and out data received from the controlled system. Therefore, if the system dynamics is changed by the attacker, the controller has the ability to follow the changes and produces the control signal in a way that weakens the effect of the attacker's destructive behavior.

Based on this view, in the last few years, two DD controllers have been introduced to combat cyber-attacks [29, 30]. The results of these two studies have shown that DD design can be a very effective mechanism for the stability of power systems in the presence of cyber-attacks. Due to the best of authors' knowledge, the DD method for wide area power systems in the presence of cyber-attacks and saturation phenomenon is not presented. This shortcoming motivated us to present a data-driven cyber-resilient controller for wide area power systems that is resilient to saturation and cyber-attacks in this chapter. In addition, for the first time, a DA on the AGC signal in wide area power systems is modeled. The presented model is a complete model in which the attacker's behavior is modeled as an unknown function. In the following, after explaining the different types of DD controllers and stating different issues in this field, the proposed DDCRC method will be presented.

2 Data-Driven Control

One of the earliest contributions to Data-Driven Control (DDC) was proposed by Ziegler and Nichols [31] where a tuning technic for PID parameters using the process data was introduced. In recent years, many DDC methods based on time series analysis and learning-based approaches have been proposed, such as DD optimal control [32], unfalsified control [33], Iterative Feedback Tuning (IFT) [34], Virtual Reference Feedback Tuning (VRFT) [35] and Neural Network-based control [36]. These techniques have their own assumptions about the system. For example, IFT and VRFT assume that the control structure is predetermined, while data-enabled predictive control [37] is developed only for linear systems. In general, DDC approaches can be categorized into online and off-line methods. Off-line methods, such as Data-Driven Model Predictive Control (DDMPC) [38], need a pre-measured I/O dataset of the system for the controller design. However, online techniques, such as Model Free Adaptive Control (MFAC) [39], can perform simultaneous modeling and control actions while the control loop is running. From another perspective, DD techniques can be 'indirect', i.e., they need to first extract a model of the system using I/O data and then design a controller for it, or 'direct' where the intermediate modeling step is avoided [40]. Interestingly, it is shown that direct approaches perform closer to optimal for larger I/O datasets, while indirect ones work still sub-optimally [40]. It emphasizes that choosing the proper DDC approach heavily depends on the characteristics of the system under control.

MFAC is known as one of the most effective online DD based algorithms for unmodeled nonlinear systems [41]. In this method, a sequential modeling-control approach is used where a linear dynamic model of the system is first extracted from the I/O data using which, control signals are then generated [42]. It means that there is no need for a training dataset or a pre-define controller structure [43]. Furthermore, it has a computationally simple structure which makes it suitable for real-time applications. Also, unlike intelligent control approaches such as fuzzy and neural networks, it needs no training process. In recent years, this method is proposed to solve many industrial problems, for example in [44] the issue of saturation is investigated for output saturated nonlinear systems with a data-based MFAC solution. In [45], two model-less adaptive control algorithms, according to Dynamic Linearization (DL) concept, are suggested for a Single Input and Single Output (SISO) nonlinear dynamical systems. The authors in [46] develop a PID-MFAC scheme with a new frame and extend it to partial and full forms. In addition, the quantized DDC structure is introduced for nonlinear systems with physical output constrained in [47, 48]. The predictive algorithm of MFAC approach, which combines the advantages of predictive control and MFAC, are developed in [49, 50] for unknown nonlinear systems. Since this method is proposed for unknown nonlinear systems and has a very simple control structure, it can be suitable for interconnected wide area power systems. In this chapter, with the development of this DD method for nonlinear systems in the presence of input saturation phenomenon, a DDCRC method for Automatic Generation Control (AGC) of wide area power systems is presented.

2.1 Data-Driven Cyber-Resilient Automatic Generation Control of Power Systems

Due to the fact that frequency is a very important parameter in the network, frequency stability has become a very important research topic in power networks. Typically, three levels of primary, secondary, and tertiary control are used for frequency stability in power systems. In the first level, the governor-based primary control is used to quickly converge the imbalance error between load and generation to zero. In the second level, in order to keep the network frequency at the nominal value and also to control the power exchange between different areas, an AGC is used. The third-control level is also used to reduce the generation costs. Thanks to ICT, these control levels are connected to each other using a network. However, this network makes the maintenance, installation, supervision and management of the grid easier, it can make the power system vulnerable to cyber-attacks.

Figure 2 shows the common control structure of *i*th area of interconnected power grid. In this control structure, T_{gi} and T_{chi}, are presented for modeling the time constant of governors and turbines. Also, D_i and M_i are used to show the load damping effect and the inertia constant. Furthermore, the stiffness factor of the tie_line between areas, power exchange among areas and power disturbance are shown by T_{ij}, $\Delta P_{tie\text{-}i}$ and ΔP_{di}, respectively. Furthermore, in order to implement of primary control loop, $1/R_i$ is used.

In this control structure, the AGC calculates the Area Control Error (ACE) based on the received information from the other areas. Then, it sends the reference signals of primary control loop vial network. Due to the importance of the information sent from the AGC, this network can be a very good target for cyber attackers.

A. **The Deception Attack Scenario**

False Data Injection (FDI) or Deception Attack (DA) is one important type of cyber-attacks on cyber-physical power grids [51, 52]. In this kind of attack, the attackers maliciously change the measured signals [53] or control commands [52]. Totally, based on the past studies, DA can be modeled as follows [53],

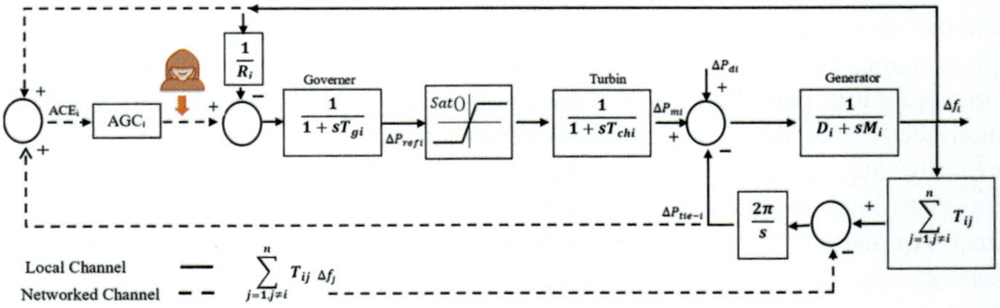

Fig. 2 A common AGC structure in power systems

$$\varphi_a(\tau) = \varphi(\tau) + \chi(\tau) \tag{1}$$

In model (1), $\varphi(\tau)$, $\varphi_a(\tau)$ and τ are normal data, corrupted data and sampling time. Based on the different definitions of the $\chi(\tau)$, the following different types of DA are defined [52],

Random attack: $\chi(\tau) = rand(\tau) * \varphi(\tau)$
Ramp attack: $\chi(\tau) = \tau * \varphi(\tau)$
Pulse attack: $\chi(\tau) = \Theta * \varphi(\tau)\tau_1 < \tau < \tau_2$
Scaling attack: $\chi(\tau) = \Theta * \varphi(\tau)$
here $\Theta \in$ R.

Recently, the attackers try to implement more complex behavior rather than the 4 models of pulse, random, scale and ramp [28]. Indeed, the attackers try to improve their knowledge about the target and then perform a complex and clever attack [28]. Therefore, we have to use a more precious and comprehensive model for deception attacks. In this chapter a new model is used for DA as follow:

$$\boldsymbol{u}_a(\tau) = \Gamma(\boldsymbol{u}(\tau)) \tag{2}$$

Here, $\boldsymbol{u}_a(\tau) = [u_{1a}(\tau),\ldots,u_{ma}(\tau)]^T$ and $\boldsymbol{u}(\tau) = [u_1(\tau), \ldots, u_m(\tau)]^T$ describe the control commands of AGC, after and before the DA, as well as Γ represents the attacker's behavior. One of the advantages of proposed model (2) is that it includes all type of model (1) and is more general than model (1).

Nowadays, thanks to the excellence in cyber-attack detection methods, if the attackers do a random behavior, their intrusion can be easily detected [28].

In the following, to make the proposed model more accurate, the following assumption is made.

Assumption 1 $\frac{\partial \Gamma(\boldsymbol{u}(\tau))}{\partial u(\tau)}$ is available in each time sample. In the other word, attackers are looking for slow changes to avoid detection.

B. Modeling of the AGC Problem under the Deception Attack

Assume that the AGC problem can be described by the following model,

$$\boldsymbol{y}(\tau+1) = \Psi\big(\boldsymbol{y}(\tau), \ldots, \boldsymbol{y}(\tau-n_y), \boldsymbol{u}(\tau), \ldots, \boldsymbol{u}(\tau-n_u)\big) \tag{3}$$

in (3), considering β_i as a frequency bias factor, the output vector $y(\tau)$ involves $y_i(\tau) = \Delta P_{tie\text{-}i}(\tau) + \beta_i \Delta\omega_i(\tau)$. $n_u \in Z^+$ and $n_y \in Z^+$ are unknown orders. Moreover, the following assumptions are considered on Ψ (.) which is unknown function.

Assumption 2 $\frac{\partial \Psi(\tau)}{\partial u_i(\tau)}$ $\forall s \in [\tau, \tau\ -n_u]$ and $i = [1, \ldots, m]$ is available.

Assumption 3 The global Lipschitz condition is satisfied by System (3), that is,

$$\|\Delta \boldsymbol{y}(\tau+1)\|_2 \leq L\|\Delta \boldsymbol{u}(\tau)\|_2;\ \forall\tau$$

here L is Lipschitz constant, $\|L\|_2 = \sqrt{L^T L}$, $\|L\|_2 > 0$ and $\Delta \boldsymbol{u}(\tau) = [\Delta u_1(\tau), \ldots, \Delta u_m(\tau)]^T$ for which, $\Delta u_i(\tau) = u_i(\tau) - u_i(\tau - 1)$ for $i = 1, \ldots, m$.

Assumptions 2 and 3 are two logical assumptions in the data-driven controller literature [43–45]. For example, based on assumption 3, the limited energy changes at the input cause limited energy changes at the output.

Lemma 1 ([44]) *If Assumption* 1–3 *are met for system* (3), *power system* (3) *can be written to the following form.*

$$\Delta \boldsymbol{y}(\tau + 1) = \psi(\tau)\Delta \boldsymbol{u}(\tau) \tag{4}$$

where $\psi(\tau) = \begin{bmatrix} \omega_{11}(\tau) & \omega_{12}(\tau) & \cdots & \omega_{1m}(\tau) \\ \omega_{21}(\tau) & \omega_{22}(\tau) & \cdots & \omega_{2m}(\tau) \\ \vdots & \vdots & \vdots & \vdots \\ \omega_{m1}(\tau) & \omega_{m2}(\tau) & \cdots & \omega_{mm}(\tau) \end{bmatrix}$, $\|\psi(\tau)\|_2 < C$, $\omega_{ij}(\tau) = \frac{\Delta y_i(\tau+1)}{\Delta u_j(\tau)}$, and $C \in R^+$.

According to (2), one of the behaviors that an attacker can perform is saturation. This means that, the primary level controller may also become saturated due to a malicious increase in the AGC signal. For this purpose, by defining the saturation functions in the following forms, we try to suggest an automatic generation controller that is resistant to this phenomenon.

The saturated control signal can be approximated as follows:

$$u_{is}(\tau) \triangleq \boldsymbol{U}_i tanh(u_{ia}(\tau)) = \begin{cases} -\boldsymbol{U}_i & u_{ia}(\tau) < -\boldsymbol{U}_i \\ u_{ia}(\tau) & -\boldsymbol{U}_i \le u_{ia}(\tau) \le \boldsymbol{U}_i \\ \boldsymbol{U}_i & u_{ia}(\tau) > \boldsymbol{U}_i \end{cases} \tag{5}$$

here $\boldsymbol{U}_i \in \mathbb{R}$. Considering the definition of input saturation (5) and the attack model (2), the AGC problem is converted to:

$$\boldsymbol{y}(\tau + 1) = \Psi_s\left(\boldsymbol{y}(\tau), \ldots, \boldsymbol{y}\left(\tau - n_y\right), \boldsymbol{u}(\tau), \ldots, \boldsymbol{u}(\tau - n_u)\right) \tag{6}$$

where, the effect of DA on AGC problem of wide area power system is shown in the Ψ_s. In the following, the expression of Proposition 1 shows that the dynamic linearization is still correct after the DA.

Proposition 1 *Considering Assumptions* 1, 2 and 3, *the AGC problem* (6) *can be written as:*

$$\Delta \boldsymbol{y}(\tau + 1) = \psi_s(\tau)\Delta \boldsymbol{u}(\tau) \tag{7}$$

$$here, \| \psi_s(\tau)\|_2 \leq \overline{b} \in \mathbb{R}, \psi_s(\tau) = \begin{bmatrix} \omega_{s11}(\tau) & \omega_{s12}(\tau) & \cdots & \omega_{s1m}(\tau) \\ \omega_{s21}(\tau) & \omega_{s22}(\tau) & \cdots & \omega_{s2m}(\tau) \\ \vdots & \vdots & \vdots & \vdots \\ \omega_{sm1}(\tau) & \omega_{sm2}(\tau) & \cdots & \omega_{smm}(\tau) \end{bmatrix} \in \mathbb{R}^{m \times m}.$$

Proof Considering the Eq. (2), $\frac{\Psi_s(\boldsymbol{y}(\tau),\boldsymbol{u}(\tau))}{\partial \boldsymbol{u}(\tau)}$ can be written as:

$$\frac{\partial \Psi_s(\boldsymbol{y}(\tau), \boldsymbol{u}(\tau))}{\partial \boldsymbol{u}(\tau)} = \frac{\partial \Psi(\boldsymbol{y}(\tau), \Gamma(\boldsymbol{u}(\tau)))}{\partial \Gamma(\boldsymbol{u}(\tau))} \cdot \frac{\partial (\Gamma(\boldsymbol{u}(\tau))}{\partial \boldsymbol{u}(\tau)}$$

According to the Assumptions 1 and 2, this partial Derivative is always available. Furthermore,

We know that the output energy of the generator is always less than the amount of input energy. Thus,

$$\|\Delta \boldsymbol{y}(\tau + 1)\|_2 \leq \overline{b} \|\Delta \boldsymbol{u}(\tau)\|_2$$

where $\overline{b}$ is a constant. ■

3 DDCRC Design and Stability Analysis

Suppose the following control input criterion function for AGC problem (6):

$$J(u(\tau)) = (\Re(\tau + 1) - \boldsymbol{y}(\tau + 1))^T (\Re(\tau + 1) - \boldsymbol{y}(\tau + 1)) + (\Delta \boldsymbol{u}(\tau))^T \lambda (\Delta \boldsymbol{u}(\tau)) \tag{8}$$

here $\Re(\tau) \in$ and $\lambda > 0$ are the desired vector and penalty factor, respectively. Based on the optimal theory and after of doing a little simplification the following control vector is obtained [44]:

$$\Delta \boldsymbol{u}(\tau) = \delta \frac{\psi_s^T(\tau)(\Re(\tau + 1) - \boldsymbol{y}(\tau))}{\lambda + \|\psi_s(\tau)\|_2^2} \tag{9}$$

here, $0 < \delta \leq 1$.

Due to the possibility of input saturation, the following control vector is applied to the power system:

$$\begin{aligned} u(\tau + 1) &= u(\tau) + \Delta u(\tau + 1) \; ; \quad \forall i > 0 if \|u_i(\tau + 1)\| \leq U_i \\ u(\tau + 1) &= u(\tau) + \varrho(\tau)\Delta u(\tau + 1) ; \; if \exists i > 0 : \|u_i(\tau + 1)\| > \boldsymbol{U}_i \end{aligned} \tag{10}$$

here, $i = 1, \ldots, m$,

$$\varrho(\tau) \triangleq \begin{bmatrix} \varrho_{11} & 0 & \cdots & 0 \\ 0 & \varrho_{22} & \cdots & 0 \\ \vdots & \vdots & \ddots & \vdots \\ 0 & 0 & \cdots & \varrho_{mm} \end{bmatrix}$$

where, $0 \le \varrho_{ii} \le 1$. Indeed, this parameter indicates the effect of saturation phenomenon on the control signal ith.

Remark 5 As $\psi_s(\tau)$ is required to generate a control signal at any given time. Therefore, in the following, an iterative method for estimating this matrix is presented. For this purpose, consider the following cost function:

$$\begin{aligned} J(\psi_s(\tau)) &= \left(\Delta \boldsymbol{y}(\tau) - \widehat{\psi}_s(\tau)\Delta \boldsymbol{u}(\tau-1)\right)^T \left(\Delta \boldsymbol{y}(\tau) - \widehat{\psi}_s(\tau)\Delta \boldsymbol{u}(\tau-1)\right) \\ &+ \left(\widehat{\psi}_s(\tau) - \widehat{\psi}_s(\tau-1)\right)^T \mu\left(\widehat{\psi}_s(\tau) - \widehat{\psi}_s(\tau-1)\right) \end{aligned} \tag{11}$$

here, $0 < \mu \le 1$. Based on the optimal theory and after of doing a little simplification the following estimation is obtained:

$$\widehat{\psi}_s(\tau) = \widehat{\psi}_s(\tau-1) + \frac{\eta\left[\Delta \boldsymbol{y}(\tau) - \widehat{\psi}_s(\tau-1)\Delta \boldsymbol{u}(\tau-1)\right]\Delta \boldsymbol{u}(\tau-1)^T}{\mu + \|\Delta \boldsymbol{u}(\tau-1)\|_2^2} \tag{12}$$

here $\widehat{\psi}_s(\boldsymbol{\tau}) = [\widehat{\psi}_{s_{ij}}(\boldsymbol{\tau})]_{m\times m}$. In order to improve the estimation algorithm (12), the following reset mechanism is used to make the estimation more accurately [44]:

$$\widehat{\psi}_{ii}(\tau) = \widehat{\psi}_{ii}(1) if \left|\widehat{\psi}_{ii}(\tau)\right| < b_2 or\ sign(\widehat{\psi}_{ii}(\tau)) \neq sign(\widehat{\psi}_{ii}(1))$$

$$\widehat{\psi}_{ij}(\tau) = \widehat{\psi}_{ij}(1) if \left|\widehat{\psi}_{ij}(\tau)\right| > b_1 or sign(\widehat{\psi}_{ij}(\tau)) \neq sign(\widehat{\psi}_{ij}(1))$$

$\widehat{\psi}_{s_{ii}}(1)$ and $\widehat{\psi}_{s_{ij}}(1)$ are initial values of $\widehat{\psi}_{ii}(\tau)$ and $\widehat{\psi}_{ij}(\tau)$, respectively.

Proof of stability

Before starting to prove the stability, the following lemma is provided to make the stability analysis easier.

Lemma 2 *Based on the saturation function* (5) *and the control vector* (10), *it can be concluded that:*

$$\|\varrho(\tau)\|_2 \le 1$$

Proof Since $|\varrho_{ii}| \le 1$ and $\|\varrho(\kappa)\|_2 = max\ (\varrho_{ii})$ therefore it can be concluded that. $\|\varrho(\tau)\|_2 \le 1$.

In the following the stability of the proposed method for AGC in presence of cyber attack and input saturation is evaluated. ■

Theorem 1 *If the Assumptions* 1, 2 and 3 *are satisfied for wide area interconnected power system,* $\mathfrak{R}(\tau+1) = \mathfrak{R}(\tau)$ *and the proposed method in* (9)–(12) *is used to the AGC problem, there is a* $\lambda_{min} > 0$ *such that for any* $\lambda > \lambda_{min}$ $lim_{\tau\to\infty}\mathfrak{R}(\tau+1)$-$\boldsymbol{y}(\tau) = 0$.

Proof Please consider, the error vector as follows:

$$\epsilon(\tau) = \mathfrak{R}(\tau) - \boldsymbol{y}(\tau) \tag{13}$$

according to Lyapunov Function $\upsilon(\tau) = \epsilon(\tau)^T\epsilon(\tau)$, it can be concluded that:

$$\Delta\upsilon(\tau+1) = \epsilon^T(\tau+1)\epsilon(\tau+1) - \epsilon^T(\tau)\epsilon(\tau) \tag{14}$$

based on $\mathfrak{R}(\tau+1) = \mathfrak{R}(\tau)$, Eqs. (7) and (14) can be written as,

$$\epsilon(\tau+1) = \mathfrak{R}(\tau+1) - \boldsymbol{y}(\tau) - \psi_s(\tau)\delta\varrho(\tau)\frac{\widehat{\psi}_s^T(\tau)(\mathfrak{R}(\tau+1) - \boldsymbol{y}(\tau))}{\lambda + \|\widehat{\psi}_s(\tau)\|_2^2}$$

after simplification

$$\epsilon(\tau+1) = \mathcal{H}(\tau)\epsilon(\tau) \tag{15}$$

where, $\mathcal{H}(\tau) \triangleq \left[I_m - \psi_s(\tau)\delta\varrho(\tau)\frac{\widehat{\psi}_s^T(\tau)}{\lambda+\|\widehat{\psi}_s(\tau)\|_2^2}\right]$. Therefore, Eq. (13) becomes

$$\Delta\upsilon(\tau+1) = \epsilon^T(\tau)\mathcal{H}^T(\tau)\mathcal{H}(\tau)\epsilon(\tau) - \epsilon^T(\tau)\epsilon(\tau)$$

$$= \epsilon^T(\tau)[\mathcal{H}^T(\tau)\mathcal{H}(\tau) - I_m]\epsilon(\tau) \tag{16}$$

to prove the stability, it is sufficient to prove that $\|\mathcal{H}(\tau)\|_2 < 1$. Based on Proposition 1 and Lemma 2, It is easy to see that there is a $\lambda_{\min} > 0$ that:

$$\begin{aligned} &\|I_m - \psi_s(\tau)\delta\varrho(\tau)\frac{\widehat{\psi}_s^T(\tau)}{\lambda+\|\widehat{\psi}_s(\tau)\|_2^2}\|_2 \le 1 \\ &\|\psi_s(\tau)\delta\varrho(\tau)\frac{\widehat{\psi}_s^T(\tau)}{\lambda+\|\widehat{\psi}_s(\tau)\|_2^2}\|_2 \le \|\delta\|_2\|\psi_s(\tau)\|_2\|\varrho(\tau)\|_2\|\frac{1}{2\sqrt{\lambda_{\min}}}\|_2 \end{aligned} \tag{17}$$

thus:

$$\|\psi_s(\tau)\delta\varrho(\tau)\frac{\widehat{\psi}_s^T(\tau)}{\lambda + \|\widehat{\psi}_s(\tau)\|_2^2}\|_2 \leq \|\frac{\overline{b}\delta}{2\sqrt{\lambda_{\min}}}\|_2 < 1 \tag{18}$$

■

Theorem 2 **If Assumptions** 1, 2 *and* 3 *are satisfied for the wide area interconnected power systems* (6), $\|\Delta\,\mathfrak{R}(\tau)\|_2 < \gamma$ *and the proposed method is applied for this system, there exists a* $\gamma \in \mathbb{R}^+$ *such that*:

$$\|\epsilon(\tau+1)\|_2 \leq d^\tau\|\epsilon(1)\|_2 + \gamma \tag{19}$$

here, $0 < d \leq 1$.

Proof If the error vector is considered as $\Delta\,\mathfrak{R}(\tau) = \mathfrak{R}(\tau+1) - \mathfrak{R}(\tau)$, Eq. (13) can be written as,

$$\begin{aligned}\|\epsilon(\tau+1)\|_2 &= \|\mathfrak{R}(\tau+1) - y(\tau) - \psi_s(\tau)\Delta u(\tau)\|_2 \\ &= \|\mathfrak{R}(\tau) - y(\tau) - \psi_s(\tau)\Delta u(\tau) + \Delta\mathfrak{R}(\tau)\|_2\end{aligned} \tag{20}$$

Based on the Theorem 1 and Eq. (15) we can concluded that:

$$\begin{aligned}\|\epsilon(\tau+1)\|_2 &\leq d^\tau\|\epsilon(1)\|_2 + \|\Delta\mathfrak{R}(\tau)\|_2 \\ &\leq d^\tau\|\epsilon(1)\|_2 + \gamma\end{aligned} \tag{21}$$

4 Simulation

Figure 3 shows a three-area connected power system in which there is the ability to transfer and receive power between areas via tie lines. Considering $x_i = [\Delta P_{tie\text{-}i}, \Delta P_{vi}, \Delta P_{mi}, \Delta\omega_i]^T$, the dynamic of each area can be described by the following model [54],

$$X_i(\tau+1) = A_{ii}X_i(\tau) + \sum_{j=1, j\neq i}^{N} A_{ij}X_j(\tau) + B_i u_i(\tau)$$

here, $\Delta P_{tie\text{-}1}$, ΔP_{vi}, ΔP_{mi}, $\Delta\omega_i$ are the deviation of power exchanged over the tie line, the deviation of mechanical powers, the deviation electrical power and the frequency deviation, respectively. In this chapter of the book, when generating a control signal, A_{ii}, A_{ij}, B_i and B_{wi} are assumed to be unknown. Moreover, $y_i(\tau) = \beta_i\Delta\omega_i(\tau) + \Delta P_{tie\text{-}i,}$ and $u_i = \Delta P_{refi}$ are the measured vector of ith area and control command which is received by the ith AGC. The parameters used in the simulation are shown in Table 1.

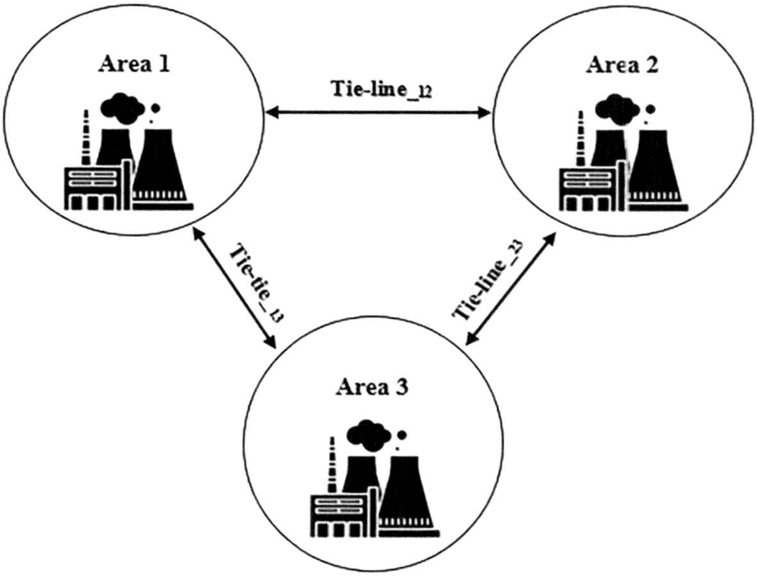

Fig. 3 The three-area interconnected power system

Table 1 The parameters of three area interconnected systems [54]

Symbol	Area1	Area2	Area3
T_{chi}	0.3	0.4	0.35
R_i	0.05	0.05	0.05
β_i	21	21.5	21.8
T_{gi}	0.1	0.17	0.2
M_i	10	12	12
D_i	1	1.5	1.8

In this section, in order to evaluate the performance of the proposed method, DDCRC method is implemented along with the Resilient Output Feedback Controller (ROFC), proposed in [54] method on three-area inter connected power system showed in Fig. 3 and the results obtained from both methods have been compared with each other.

In simulation, in order to measure the efficiency of the proposed method in the event of abnormal conditions, the following two scenarios are considered.

Scenario 1. In the presence of $\pm$ 0.1 p. u step load change and 10% parameter variations, for $t > 2$ s a deception attacks with the profile $\Gamma(\boldsymbol{u}_i(\tau)) = \alpha_i \times Sin(\boldsymbol{u}_i(\tau))$ is applied on AGC control signal in all area.

Scenario 2. In addition to the conditions expressed in Scenario 1, governor saturation is considered between -0.3 and 0.3 in all areas.

Based on [54], the parameters of ROFC controller are considered as $K_1 = [-0.0470, -0.0102]$, $K_2 = [-0.0504, -0.0089]$ and $K_3 = [-0.0530, -0.0089]$.

Also, the basic conditions of the proposed controller are considered as follows.

$$\eta = \mu = \delta = 1, \lambda = 0.1$$
$$\widehat{\psi}(1) = \widehat{\psi}(2) = \widehat{\psi}(3) = 1000\begin{bmatrix} 1 & 0.1 & 0.1 \\ 0.1 & 1 & 0.1 \\ 0.1 & 0.1 & 1 \end{bmatrix};$$

Figure 4 demonstrates the performance of the proposed DDCRC in presence of a step load change at T = 1, a DA at T = 2 s and the variations of model's parameters. Figure 4a and b show that mechanical and electrical powers of power's area. These figures show that in the presence of the unstable scenario 1, the proposed method has performed better than the ROFC method in the convergence of variation of electrical and mechanical power. Figure 4c also shows that the proposed method is able to convergence the frequency variation in the all areas to zero in the. However, the ROFC method in the third area is not able to convergence the frequency variation. The reason for the superiority of the proposed method is that the controller produces the control signal appropriate to each moment of the control system without any dependence on the system model and according to the online identification.

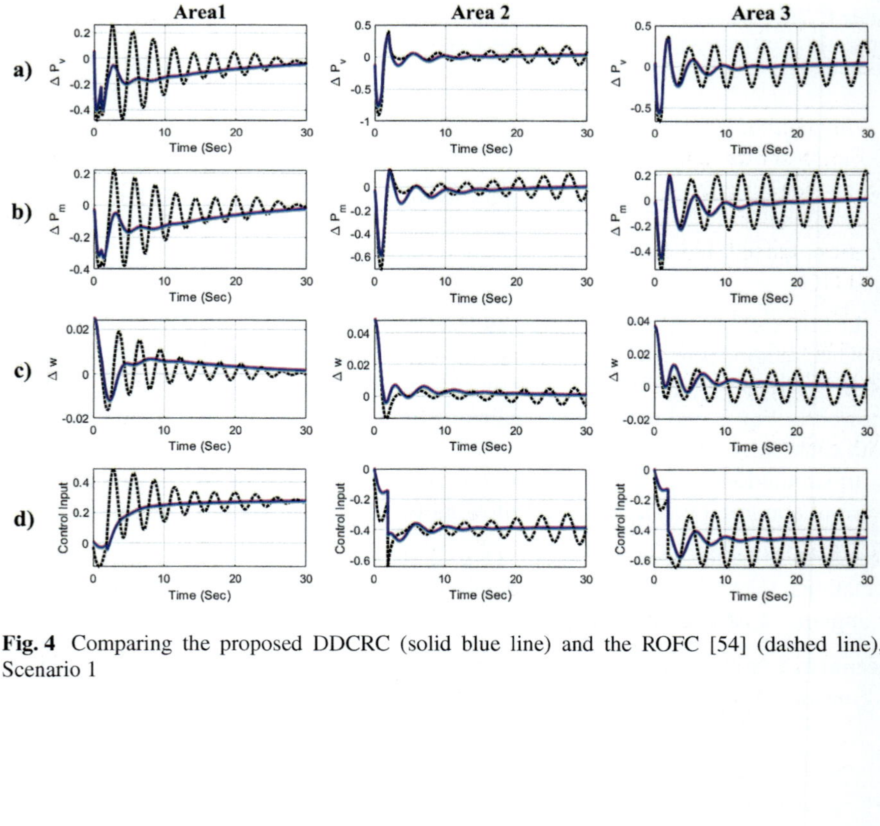

Fig. 4 Comparing the proposed DDCRC (solid blue line) and the ROFC [54] (dashed line), Scenario 1

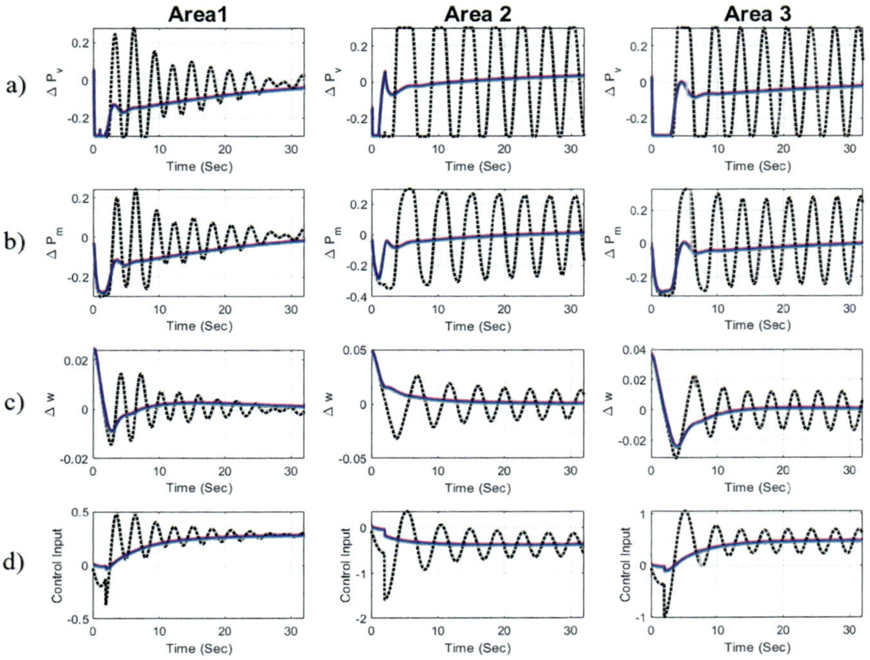

Fig. 5 Comparing the proposed DDCRC (solid blue line) and the ROFC [54] (dashed line), Scenario 2

In Scenario 2, the control signals are modified so that governor saturation occurs. Figure 5 shows that the proposed method performed better than the second method when the saturation limit occurred. The reason for this advantage is that the proposed controller is not saturated under the influence of nonlinear behavior due to its DD feature. Indeed, thanks to online estimation and control updates based on Eqs. (10) and (12).

The simulation results showed that the proposed method without using the model information succeeded in keeping the power system stable in the presence of saturation phenomenon and cyber-attacks. The reason for this is that since the method is model free, it has more resistance to load changes, cyber-attacks and saturated nonlinear behavior than the model-based methods such as ROFC.

5 Conclusion

In This chapter, a DDCAR method is proposed for wide-area power system subject to cyber-attack. To this end, firstly a new model of DA to the wide-area interconnected power systems is presented. The valuable point in the proposed model is that the behavior of attacker is model by unknown nonlinear function, this means that the attacker can make any changes to the data. Secondly, to deal with this type of attack, a DD control method based on dynamic model is proposed. The dynamic model is updated using input/output signals of the grid, even under attack. Based on this model, an efficient adaptive control signal is generated, and the stability of the power system is guaranteed. In simulation section, the proposed controller is applied to the AGC problem in a sample three-area interconnected power system when a DA resulted in the governor saturation. Simulations are performed in severe scenarios when parameters of the system varied, and a step load change happened in addition to an attack. The obtained results showed that the proposed controller could tolerate the attack without the necessity of any pre-defined model. It worked more reliable than its model-based peer.

References

1. G. Bedi, G.K. Venayagamoorthy, R. Singh, R.R. Brooks, K.-C. Wang, Review of Internet of Things (IoT) in electric power and energy systems. IEEE Internet Things J. **5**(2), 847–870 (2018)
2. M. Tan, Z. Song, X. Zhang, Robust leader-following consensus of cyber–physical systems with cyber-attack via sampled-data control. ISA Trans. **109**, 61–71 (2021)
3. D. Ye, S. Luo, A co-design methodology for cyber-physical systems under actuator fault and cyber-attack. J. Franklin Inst. **356**(4), 1856–1879 (2019)
4. S.A.G.K. Abadi, S.I. Habibi, T. Khalili, A. Bidram, A model predictive control strategy for performance improvement of hybrid energy storage systems in DC microgrids. IEEE Access **10**, 25400–25421 (2022)
5. S.A.G.K. Abadi, A. Bidram, A distributed rule-based power management strategy in a photovoltaic/hybrid energy storage based on an active compensation filtering technique. IET Renew. Power Gener. **15**(15), 3688–3703 (2021)
6. S.D. Bopardikar A. Speranzon, On analysis and design of stealth-resilient control systems, in *Resilient Control Systems (ISRCS), 6th International Symposium on Resilient Control Systems (ISRCS) 2013* vol. 48, no. 53, pp. 13–15. (Aug 2013)
7. H. Fawzi, P. Tabuada, S. Diggavi, Secure estimation and control for cyber-physical systems under adversarial attacks. IEEE Trans. Autom. Control **59**(6), 1454–1467 (2014)
8. Z. Pang, G. Liu, Design and implementation of secure networked predictive control systems under deception attacks. IEEE Trans. Control Syst. Technol. **20**(5), 1334–1342 (2012)
9. J.Y. Keller, D. Sauter, K. Chabir, State filtering for discrete-time stochastic linear systems subject to random cyber attacks and losses of measurements, in *2012 20th Mediterranean Conference on Control & Automation (MED)* (3–6 July 2012) , pp 935,940
10. S. Deshmukh, B. Natarajan, A. Pahwa, State estimation over a Lossy network in spatially distributed cyber-physical systems. IEEE Trans. Signal Process. **62**(15), 3911–3923 (1 Aug 2014)
11. A. Rosich, H. Voos, Li. Yumei, M. Darouach, A model predictive approach for cyber-attack detection and mitigation in control systems, in *52nd Annual Conference on Decision and Control (CDC)* (IEEE, 2013). (Dec 2013), pp. 6621–6626, 10–13
12. A. Teixeira, H. Sandberg, K.H. Johansson, Networked control systems under cyber attacks with applications to power networks, in *American Control Conference (ACC), 2010* (June 30 2010–July 2 2010), pp. 3690–3696.
13. A.F. Taha, A. Elmahdi, J.H. Panchal, D. Sun, Unknown input observer design and analysis for networked control systems. Int. J. Control **88**(5), 920–934 (2015)

14. A. Ahmadi, Y. Asadi, A.M. Amani, M. Jalili, X. Yu, Resilient model predictive adaptive control of networked Z-source inverters using GMDH. IEEE Trans. Smart Grid. https://doi.org/10.1109/TSG.2022.3174250
15. A.M. Amani, M. Jalili, Power grids as complex networks: resilience and reliability analysis. IEEE Access (2021)
16. K. Wang, M. Du, S. Maharjan, Y. Sun, Strategic honeypot game model for distributed denial of service attacks in the smart grid. IEEE Trans. Smart Grid **8**(5), 2474–2482 (2017)
17. X. Liu, Z. Li, Z. Li, Optimal protection strategy against false data injection attacks in power systems. IEEE Trans. Smart Grid **8**(4), 1802–1810 (2016)
18. Y. Wu, Z. Wei, J. Weng, X. Li, R.H. Deng, Resonance attacks on load frequency control of smart grids. IEEE Trans. Smart Grid **9**(5), 4490–4502 (2017)
19. J. Liu, Y. Gu, L. Zha, Y. Liu, J. Cao, Event-triggered h∞ load frequency control for multiarea power systems under hybrid cyber attacks. IEEE Trans. Syst. Man Cybern. Syst. **49**(8), 1665–1678 (2019)
20. C. Peng, J. Zhang, H. Yan, Adaptive event-triggering H∞ load frequency control for network-based power systems. IEEE Trans. Ind. Electron. **65**(2), 1685–1694 (2017)
21. C. Wu, Z. Hu, J. Liu, L. Wu, Secure estimation for cyber-physical systems via sliding mode. IEEE Trans. Cybern. **48**(12), 3420–3431 (2018)
22. M.A. Sid, S. Chitraganti, K. Chabir, Medium access scheduling for input reconstruction under deception attacks. J. Franklin Inst. **354**(9), 3678–3689 (2017)
23. J.Y. Keller, K. Chabir, D. Sauter, Input reconstruction for networked control systems subject to deception attacks and data losses on control signals. Int. J. Syst. Sci. **47**(4), 814–820 (2016)
24. X. He, X. Liu, P. Li, Coordinated false data injection attacks in AGC system and its countermeasure. IEEE Access **8**, 194640–194651 (2020)
25. M. Khalaf, A. Youssef, E. El-Saadany, Joint detection and mitigation of false data injection attacks in AGC systems. IEEE Trans. Smart Grid **10**(5), 4985–4995 (2018)
26. S. Zuo, T. Altun, F.L. Lewis, A. Davoudi, Distributed resilient secondary control of DC microgrids against unbounded attacks. IEEE Trans. Smart Grid **11**(5), 3850–3859 (2020)
27. S. Zuo, O.A. Beg, F.L. Lewis, A. Davoudi, Resilient networked AC microgrids under unbounded cyber attacks. IEEE Trans. Smart Grid **11**(5), 3785–3794 (2020)
28. G. Liang, J. Zhao, F. Luo, S.R. Weller, Z.Y. Dong, A review of false data injection attacks against modern power systems. IEEE Trans. Smart Grid **8**(4), 1630–1638 (2016)
29. X. Huang, D. Zhai, J. Dong, Adaptive integral sliding-mode control strategy of data-driven cyber-physical systems against a class of actuator attacks. IET Control Theory Appl. **12**(10), 1440–1447 (2018)
30. X. Qiu, Y. Wang, X. Xie, H. Zhang, Resilient model-free adaptive control for cyber-physical systems against jamming attack. Neurocomputing **413**, 422–430 (2020)
31. C.C. Hang, K.J. Åström, W.K. Ho, Refinements of the Ziegler-Nichols tuning formula. Control Theory Appl. **138**(2), 111–118 (1991)
32. D. Selvi, D. Piga, G. Battistelli, A. Bemporad, Optimal direct data-driven control with stability guarantees. Eur. J. Control. **1**(59), 175–187 (2021)
33. T. Wonghong, S. Engell, Automatic controller tuning via unfalsified control. J. Process Control **22**(10), 2008–2025 (2012)
34. C. Tian, Y. Peng, Data-driven iterative tuning based active disturbance rejection control for piezoelectric nano-positioners. Mechatronics **65**, 102321 (2021)
35. C.L. Remes, R.F. Binz, J.V. Flores, L. Campestrini, Virtual reference feedback tuning applied to cascade control. IET Control Theory Appl. **14**(20), 3738–3746
36. E. Aggelogiannaki, H. Sarimveis, Nonlinear model predictive control for distributed parameter systems using data driven artificial neural network models. Comput. Chem. Eng. **32**(6), 1225–1237
37. A. Allibhoy, J. Cortés, Data-based receding horizon control of linear network systems. IEEE Control Syst. Lett. **5**(4), 1207–1212
38. J. Berberich, J. Köhler, M.A. Müller, F. Allgöwer, Data-driven model predictive control with stability and robustness guarantees. IEEE Trans. Autom. Control **66**(4), 1702–1717

39. Z. Hou, W. Huang, The model-free learning adaptive control of a class of SISO nonlinear systems, in *Proceedings of the 1997 American Control Conference (Cat. No. 97CH36041)*, vol. 1, (1997), pp. 343–344
40. V. Krishnan, F.P Asqualetti, On direct vs indirect data-driven predictive control, in *2021 60th IEEE Conference on Decision and Control (CDC)* (IEEE, 2021), pp. 736–741
41. Y. Asadi, M. Maghfoori, E. Bijami, The design of data-driven adaptive predictive controller for a class of unknown no-linear system featuring output saturation. J. Control **15**(3), 23–33 (2021)
42. Z. Hou, Y. Zhu, Controller-dynamic-linearization-based model free adaptive control for discrete-time nonlinear systems. IEEE Trans. Ind. Inf. **9**(4), 2301–2309 (2013)
43. Y. Asadi, A. Ahmadi, S. Mohammadi, A.M. Amani, M. Marzband, B. Mohammadi-Ivatloo, Data-driven model-free adaptive control of Z-source inverters. Sensors **21**(22), 7438–7444 (2021)
44. Y. Asadi, M.M. Farsangi, E. Bijami, A.M. Amani, K.Y. Lee, Data-driven adaptive control of wide-area nonlinear systems with input and output saturation: a power system application. Int. J. Electr. Power Energy Syst. **133**, 107225 (2021)
45. Z. Hou, C. Ronghu, G. Huijun, An overview of dynamic-linearization-based data-driven control and applications. IEEE Trans. Ind. Electron. **64**(5), 4076–4090 (2017)
46. Z. Hou, J. Shangtai, Model-free adaptive control for a class of nonlinear discrete-time systems based on the partial form linearization, in *IFAC Proceedings* **41**(2), 3509–3514 (2008)
47. X. Bu, Z. Hou, Q. Yu, Y. Yang, Quantized data driven iterative learning control for a class of nonlinear systems with sensor saturation. IEEE Trans. Syst. Man Cybern. Syst. **50**(12), 5119–5129 (2018)
48. Z. Hou, L. Ting, Constrained model free adaptive predictive perimeter control and route guidance for multi-region Urban traffic systems. IEEE Trans. Intell. Transp. Syst. (2020)
49. Y. Wang, H. Mingdong, Model-free adaptive integral terminal sliding mode predictive control for a class of discrete-time nonlinear systems. ISA Trans. **93**, 209–217 (2019)
50. Y. Guo, Z. Hou, S. Liu, S. Jin, Data-driven model-free adaptive predictive control for a class of MIMO nonlinear discrete-time systems with stability analysis. IEEE Access **7**, 102852–102866 (2019)
51. J. Yang, A controllable false data injection attack for a cyber physical system. IEEE Access **9**, 6721–6728 (2021)
52. S.M. Dibaji, M. Pirani, D.B. Flamholz, A.M. Annaswamy, K.H. Johansson, A. Chakrabortty, A systems and control perspective of CPS security. Annu. Rev. Control. **47**, 394–411 (2019)
53. A. Ahmadi, M. Nabipour, S. Taheri, B. Mohammadi-Ivatloo, V. Vahidinasab, A new false data injection attack detection model for cyberattack resilient energy forecasting. IEEE Trans. Ind. Inf. (2022)
54. X. Zhou, Z. Gu, F. Yang, Resilient event-triggered output feedback control for load frequency control systems subject to cyber attacks. IEEE Access 58951–58958 (2019)

Cyberattack-Resilient Control in Multi-area Power Generation

Zhijian Hu, Rong Su, Shichao Liu, Zeyuan Xu, and Kai Zhang

Abstract While automatic generation control of multi-area power systems has better dynamic performance, it intrinsically shares the distributed control and communication frameworks and is thus inevitably vulnerable to cyber attacks. To better explore the merits of multi-area power systems, and at the same time, guarantee the security for power infrastructures and the safety for operators, countermeasures to cyber attacks must be taken. In this chapter, one new cyberattack-resilient control method suitable for automatic power generation is systematically revealed. Different from the existing techniques, this chapter presents a completely novel, unexplored resilient control framework for power generation systems under cyber attacks. This framework is equipped with two security layers. Specifically, the first security layer is about off-line tolerant control gains seeking, and the second security layer concerns on-line cyberattack detection. In the first security layer, different categories of cyber attack models can be presumed by specific concerns, with which the composite power system model is derived. Leveraging the Lyapunov method, the tolerant control gains with expected performance indexes can be sought. In the second security layer, the received signals from neighboring areas are carefully examined by the real-time cyberattack detection procedure before executing control actions, to protect the communication network from cyber attacks. With the proposed cyberattack-resilient control method embedded in each control unit, the secure operation of multi-area power systems are well promoted.

Z. Hu · R. Su
School of Electrical and Electronic Engineering, Nanyang Technological University, Singapore 639798, Singapore

S. Liu
Department of Electronics, Carleton University, Ottawa, ON K1S5B6, Canada

Z. Xu
Department of Control Science and Engineering, Harbin Institute of Technology, Harbin 150001, People's Republic of China

K. Zhang (✉)
Key Laboratory of Ultra-Precision Intelligent Instrumentation, Harbin Institute of Technology, Harbin 150001, People's Republic of China
e-mail: kaizhang0116@163.com

H. Haes Alhelou et al. (eds.), *Power Systems Cybersecurity*, Power Systems,
https://doi.org/10.1007/978-3-031-20360-2_8

Keywords Multi-area power system · Cyberattack-resilient control · Off-line tolerant control gains seeking · On-line cyberattack detection

1 Introduction

As a typical kind of cyber-physical energy systems, the modern power systems play the critical role both in environment and economy. A general modern power system usually contains five parts, that is, generation, transmission, transformation, distribution, and consumption. In this chapter, we mainly focus on the power generation, namely automatic generation control (AGC). AGC has been widely adopted in power systems to keep the active power balance between the power generation and power consumption. For a multi-area power system, its physical architecture usually contains several geographic regions, such as multiple provinces or cities. As a result, the multi-area power generation system usually shares distributed control framework for the sake of enhancing its dynamic performance, by cooperatively generating anticipated power output of each area [1–5]. Compared with the traditional centralized or decentralized control modes, distributed AGC utilizes the tie-line power signals to assist other areas in generating the anticipated power, and thus has greater practical significance in multi-area power generation systems.

Though distributed AGC framework for multi-area power systems has better dynamic performance compared with its counterparts, its real-time implementation in practical power systems is not straightforward. The successful implementation of distributed AGC heavily relies on the effective data transmission between different areas [6–9]. Only when the received measurements transmitting from neighboring areas are accurate and timely can the local area obtain the better dynamic performance. To enhance the flexibility of the communication network and simultaneously reduce the construction cost, modern power systems usually adopt the wireless sensor network (WSN) to transmit signals between different areas [10]. That is, the phasor measurement units (PMUs), which are used for data transmission installed at the control centers, must receive/send state measurements from/to their neighboring control centers via the wireless communication network. While the wireless communication network facilitates the data transmission and is more economic, it is much more vulnerable to cyber attacks (see Fig. 1 as an illustration).

In the last decades, cyber attacks happened from time to time. A typical example is that the Iran nuclear power plant was decimated under Stuxnet attacks in 2010 [11]. Also in 2015, a power distribution company was attacked in Ukraine, which led to a three-hour outage at seven substations [12]. These incidents have heightened people's alert of cyber attacks, and have promoted the improvement and innovation of cyberattack-resilient control in power systems, especially in AGC. From the attacker's point of view, two main sorts of cyber attacks, i.e., denial-of-service (DoS) attacks and false data injection (FDI) attacks, may be chosen in order to realize the conspiracy of destroying the national power grid and hindering national economic growth. DoS attacks are also known as jamming attacks [13, 14], which mainly refer

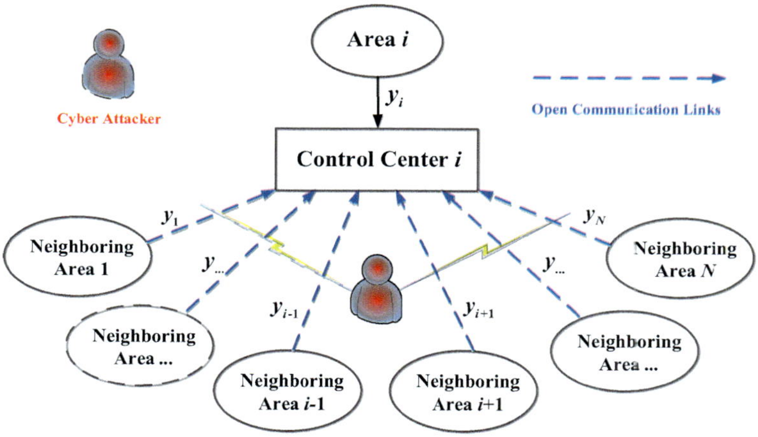

Fig. 1 An illustration of cyber attacks

to blocking the data transmitting from one terminal to another in wireless communication link, and are usually launched by tampering the communication protocol. The destruction of DoS attacks to power systems is usually reflected in preventing data exchange among different control areas. FDI attacks emphasize the false information injected into the data packet during transmission, which means the received signals are not exactly what they are. The harm of FDI attacks is destroying the authenticity of the data [15, 16].

In response to cyber attacks on power systems, many scholars have carried out the frontier exploration on the initiation and defense of cyber attacks, and a series of effective methods and strategies have been revealed in the field of AGC [17–24]. In [18], one rule-based fallback control approach was presented to enhance the security of islanded microgrids with the consideration of DoS attacks, in which the extra control loop was adopted to cooperate with the distributed resources. Zhou et al. discussed both DoS attacks and FDI attacks on communication links, local controller and primary controller, and the respective impacts on optimal operating points for microgrids. Literature [21] firstly analyzed the characteristics of statistical data of state vectors without cyber attacks, and then proposed an on-line χ^2 detection strategy based on Kalman filter and particle swarm optimization. Hu et al. adopted the credibility evaluation into the distributed AGC, which mitigated the impacts of FDI attacks on frequency regulation dynamics [22].

While these works have certain efficacy in promoting the security operation for power systems, their research focuses are excessively simplex, leading to a limited effectiveness of security control under cyber attacks. Encouraged by the above discussions, this chapter aims at presenting a systematic and unexplored resilient control framework for power generation systems under cyber attacks. The framework includes two hierarchical security control layers, that is, off-line tolerant control gains seeking and on-line cyberattack detection. In the first security layer, we presuppose one category of mathematical models to depict the specific attacking behaviors. By integrating the attack model into the power system model, and with the Lyapunov method, the tolerant control gains with expected performance indexes (e.g., $\mathcal{H}_\infty$ robustness) can be sought. It is not difficult to see that, the control effects of the first security layer heavily rely on the presumed cyber attack models. When the practical attack behaviors deviate from the presumed models, the derived tolerant control gains can hardly guarantee the secure operation of the power system. To deal with the imperfection of the first security layer, an on-line cyberattack detection mechanism is further equipped as the second security layer. In the second security layer, the real-time cyberattack detection unit is set up at each control center, which is responsible for examining the received signals from neighboring areas before executing control actions. Once the received signals are constantly judged as the compromised ones, the control center can take emergency measures (e.g., alarm and stop operating).

The remainder of this chapter is arranged as below. Section 2 elaborates the off-line tolerant control gains seeking method. Section 3 designs the on-line cyberattack detection mechanism. Section 4 validates the effectiveness of Sects. 2 and 3. Section 5 sums up the chapter.

2 Off-Line Tolerant Control Gains Seeking Method

2.1 Modeling of Multi-area Power Generation System

Before modeling the power generation system, we firstly discuss the communication network of multi-area power systems. Usually, one phasor measurement unit (PMU) is installed in each area, which is responsible for broadcasting the local state measurements to its neighboring areas, and at the same time, receiving the neighboring state information from the neighboring PMUs [25]. All data exchange is completed by the well-designed communication topology, which is usually designed based on the physical structure of the power generation system. To make potential readers more clear with the data transmissions among areas of the power system, we hereby introduce the graph theory to illustrate the communication topology of multi-area power systems.

In the graph theory, $\mathbb{G} = (\mathbb{V}, \mathbb{E}, \mathbb{L})$ is the mathematical formulation of a directed graph, which is adopted to describe the communication topology in this chapter. $\mathbb{V} = \{1, 2, \ldots, n\}$ corresponds to the label of different areas. $\mathbb{E} \subseteq \mathbb{V} \times \mathbb{V}$ depicts the communication links. $\mathbb{L} = (l_{ij})_{n\times n}$ represents adjacency matrices corresponding to communication links. $l_{ij} > 0$ indicates there are data transmissions between the ith PMU and the jth PMU. Area j is usually defined as area i's neighbor if $l_{ij} = 1$, based on which there is $\mathbb{N}_i \triangleq \{j \in \mathbb{V} | (i, j) \in \mathbb{E}\}$, which means the ith PMU can receive state measurements from its neighbors $j \in \mathbb{N}_i$ according to the communication topology.

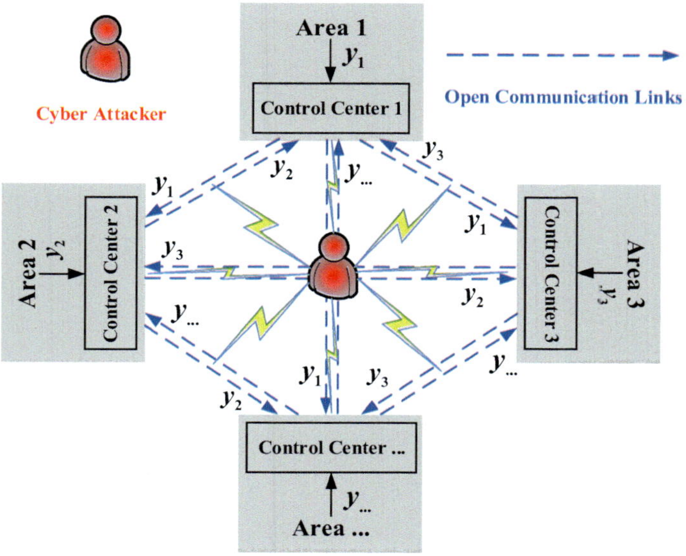

Fig. 2 The architecture of a 4-area power system under cyber attacks

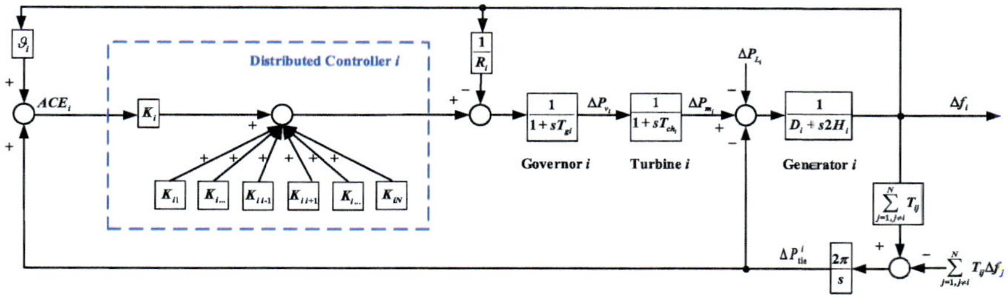

Fig. 3 Framework of AGC of area i

As an illustration, the architecture of a four-area power system is represented in Fig. 2. The generator is chosen as the non-reheat turbine, and each control area is equivalent as a single-generator system to facilitate the controller design. The AGC dynamics of each area contain the following five parts. Figure 3 draws the framework of AGC of area i.

The turbine dynamic is

$$\Delta \dot{P}_{m_i} = -\frac{1}{T_{ch_i}} \Delta P_{m_i} + \frac{1}{T_{ch_i}} \Delta P_{v_i} \tag{1}$$

The governor dynamic is

$$\Delta \dot{P}_{v_i} = -\frac{1}{R_i T_{g_i}}\Delta f_i - \frac{1}{T_{g_i}}\Delta P_{v_i} + \frac{1}{T_{g_i}}\Delta P_{c_i} \tag{2}$$

The load-generation dynamic is

$$\Delta \dot{f}_i = -\frac{D_i}{2H_i}\Delta f_i + \frac{1}{2H_i}\Delta P_{m_i} - \frac{1}{2H_i}\Delta P_{\text{tie}}^i - \frac{1}{2H_i}\Delta P_{L_i} \tag{3}$$

The tie-line power flow dynamic is

$$\Delta \dot{P}_{\text{tie}}^i = \sum_{j=1, j\neq i}^{N} 2\pi T_{ij}(\Delta f_i - \Delta f_j) \tag{4}$$

The ACE signal can be described as

$$ACE_i = \vartheta_i \Delta f_i + \Delta P_{\text{tie}}^i \tag{5}$$

The parameters in above five equations, as well as their physical significance, are given in Table 1.

The compact form of (1)–(5) in state-space equation is

Table 1 Parameters of area i

Δf_i	Deviation of frequency
ΔP_{m_i}	Deviation of generator mechanical power
ΔP_{v_i}	Deviation of turbine value position
ΔP_{c_i}	Load reference set-point deviation
ΔP_{tie}^i	Net tie-line active power flow
ΔP_{L_i}	Load disturbance
ACE_i	Area control error
T_{ch_i}	Time constant of turbine
T_{g_i}	Time constant of governor
R_i	Speed drop
H_i	Equivalent inertia constant of generator
D_i	Equivalent damping coefficient of generator
T_{ij}	Tie-line synchronizing coefficient between area i and j
ϑ_i	Frequency bias constant $\vartheta_i = 1/R_i + D_i$
N	Quantity of areas in a multi-area power system

$$\begin{cases} \dot{x}_i = \mathcal{A}_i x_i + \sum\limits_{j=1, j\neq i}^{N} \mathcal{A}_{ij} x_j + \mathcal{B}_i u_i + \mathcal{F}_i v_i, \\ y_i = \mathcal{C}_i x_i, \\ z_i = \mathcal{D}_i x_i, \end{cases} \tag{6}$$

with x_i denoting the state vector, u_i denoting the control input, y_i denoting measured output, and z_i denoting any interested signal.

$$x_i = \begin{bmatrix} \Delta f_i & \Delta P_{m_i} & \Delta P_{v_i} & \Delta P^i_{\text{tie}} & \int ACE_i \end{bmatrix}^T,$$

$$\mathcal{A}_i = \begin{bmatrix} \frac{-D_i}{2H_i} & \frac{1}{2H_i} & 0 & \frac{-1}{2H_i} & 0 \\ 0 & \frac{-1}{T_{ch_i}} & \frac{1}{T_{ch_i}} & 0 & 0 \\ \frac{-1}{R_i T_{g_i}} & 0 & \frac{-1}{T_{g_i}} & 0 & 0 \\ \sum\limits_{j=1, j\neq i}^{N} 2\pi T_{ij} & 0 & 0 & 0 & 0 \\ \vartheta_i & 0 & 0 & 1 & 0 \end{bmatrix},$$

$$\mathcal{A}_{ij} = \begin{bmatrix} 0 & 0\,0\,0\,0 \\ 0 & 0\,0\,0\,0 \\ 0 & 0\,0\,0\,0 \\ -2\pi T_{ij} & 0\,0\,0\,0 \\ 0 & 0\,0\,0\,0 \end{bmatrix}, \quad \mathcal{C}_i = \begin{bmatrix} \vartheta_i & 0 \\ 0 & 0 \\ 0 & 0 \\ 1 & 0 \\ 0 & 1 \end{bmatrix}^T,$$

$$\mathcal{B}_i = \begin{bmatrix} 0 & 0 & \frac{1}{T_{g_i}} & 0 & 0 \end{bmatrix}^T, \quad \mathcal{F}_i = \begin{bmatrix} \frac{-1}{2H_i} & 0 & 0 & 0 & 0 \end{bmatrix}^T,$$

$$\mathcal{D}_i = \begin{bmatrix} 1\,0\,0\,0\,0 \end{bmatrix}, \quad v_i = \Delta P_{L_i}, \quad u_i = \Delta P_{c_i}.$$

Under distributed framework, each area exchanges the active power via tie-lines with each other. Since the active power is measured and transmitted by digital devices (e.g., remote telemetry units (RTUs) and PMUs), we thereby derive the discrete-time state-space model to facilitate the execution of distributed AGC.

The discrete-time state-space model of area i is

$$\begin{cases} x_i(k+1) = A_i x_i(k) + B_i u_i(k) + \sum\limits_{j=1, j\neq i}^{N} A_{ij} x_j(k) + F_i v_i(k), \\ y_i(k) = C_i x(k), \\ z_i(k) = D_i x(k). \end{cases} \tag{7}$$

with $A_i = e^{\mathcal{A}_i h}$, $B_i = \int_0^h e^{\mathcal{A}_i \tau} \mathcal{B}_i d\tau$, $A_{ij} = e^{\mathcal{A}_{ij} h}$, $C_i = \mathcal{C}_i$, $F_i = \int_0^h e^{\mathcal{A}_i \tau} \mathcal{F}_i d\tau$, $D_i = \mathcal{D}_i$, and h indicating the sampling period.

2.2 Distributed Automatic Generation Controller Design Under DoS Attacks

To deal with the distributed AGC, output feedback control is more preferable than state feedback control considering the availability of the state signals. Therefore, in this chapter, the output feedback control method is adopted. To achieve a better dynamic performance of AGC, the information of the neighboring states is reasonable to be included in the local controller design. For area i, the controller is formulated as

$$u_i(k) = K_i y_i(k) + \sum_{j=1, j\neq i}^{N} l_{ij} K_{ij} y_j(k). \tag{8}$$

with K_i and K_{ij} representing the distributed controller gains corresponding to local area i and its neighboring area j.

However, the above representation is not practical. In realities, $y_i(k)$ should be firstly measured by PMUs, and afterwards be sent to its neighbors through the wireless communication network. During the measurements transmitting from one area to another, the adversaries are likely to congest open communication links by tampering with communication protocol, and thus prevent the neighboring measurements from reaching the local control center. As a result, the received signals at the local control center are no longer the ones sending from its neighbors. To guarantee the secure operation of the power systems, we take the underlying DoS attacks into account, and then we aim to design a tolerant controller to reduce the impacts of DoS attacks on AGC.

For a concise description, in this section, we employ the Bernoulli variable $\beta_{ij}(k)$ to depict the DoS attacks. In the practical circumstance, at the control center of area i, the received measurements sending from its neighbor j is changed to

$$\widetilde{y}_j(k) = \beta_{ij}(k) y_j(k), \tag{9}$$

where

$$\text{Prob}\{\beta_{ij}(k) = 1\} = \bar{\beta}_{ij}, \quad \text{Prob}\{\beta_{ij}(k) = 0\} = 1 - \bar{\beta}_{ij},$$

with $\bar{\beta}_{ij} \in [0, 1]$ signifying the probability that the open communication link is not attacked.

So far, we have finished designing the distributed automatic generation controller under DoS attacks. The closed-loop dynamics under cyber attacks are deduced as

$$\begin{aligned} x_i(k+1) &= (A_i + B_i K_i C_i) x_i(k) + F_i v_i(k) \\ &\quad + \sum_{j=1, j\neq i}^{N} l_{ij} (A_{ij} + \beta_{ij}(k) B_i K_{ij} C_j) x_j(k). \end{aligned} \tag{10}$$

Expected performance indexes are critical factors to be considered when designing the distributed automatic generation controller, such as stability, reliability and robustness. In this chapter, as an illustration, we discuss the $\mathcal{H}_\infty$ robustness performance in the mean square asymptotic stability sense. The definition of the $\mathcal{H}_\infty$ robustness performance is given in [22].

2.3 Off-Line Tolerant Control Gains Seeking

Theorem 1 *Given the DoS attack probabilities on open communication links $\bar{\beta}_{ij}$, AGC system (10) is mean square asymptotically stable with Δf_i signals meeting the prescribed $\mathcal{H}_\infty$ performance level γ, if there exist matrices $P_i > 0$, such that for $i = 1, 2, \ldots, N$,*

$$\begin{bmatrix} \Phi_1 & \Phi_2 \\ * & \Phi_3 \end{bmatrix} < 0, \tag{11}$$

where,

$$\begin{aligned}
\hat{A} &= \tilde{A} + \tilde{B}\tilde{K}\tilde{C} + \tilde{B}S, \ \tilde{B} = diag_N\{\bar{B}_i\}, \\
S &= \breve{K}\tilde{C} \odot ((\bar{\beta}\bar{L}) \otimes 1_{2\times 5}), \\
\tilde{A} &= diag_N\,\{\bar{A}_i\} \odot (L \otimes 1_{2\times 5}), \\
\bar{A}_i &= col_N^i\{[A_{i1}, \ldots, A_i, \ldots, A_{iN}]\}, \\
\tilde{P} &= \frac{1}{N} diag_N\{P\}, \ P = diag_N\{P_i\}, \\
\breve{K} &= diag_N\{\breve{K}_i\}, \ \breve{K}_i = col\,^i_N\{[K_{i1}, K_{i2}, \ldots, K_{iN}]\}, \\
\bar{B}_i &= diag_N^i\{B_i\}, \bar{B} = diag\,\{I_N \otimes B_1, \ldots, I_N \otimes B_N\}, \\
R &= diag_N\{\sum_{j=1}^{N}(\beta_{ij} - \bar{\beta}_{ij})vec_N^j\{l_{ij}K_{ij}C_j\}\}, \\
\tilde{K} &= I_N \otimes diag_N\,\{K_i\}, \ \tilde{C} = diag_N\,\{\bar{C}\}, \\
\tilde{C}_i &= diag_N^i\{\hat{C}\}, \ \hat{C} = diag_N\{C_i\}, \\
U &= \bar{B}\breve{K}\breve{C} \odot ((\Upsilon\bar{L}) \otimes 1_{2\times 5}), \\
\bar{L} &= diag_N\{l_i\}, \ l_i = \ diag_N\{l_{ij}\}, \\
\breve{C} &= diag_N\{\hat{C}\}, \ \bar{C} = diag_N\{C_i\}, \\
\Upsilon &= diag_N\{\sigma_i\}, \ \sigma_i = \ diag_N\{\sigma_{ij}\}, \\
\bar{\beta} &= diag_N\{\beta_i\}, \ \beta_i = \ diag_N\{\beta_{ij}\}, \\
\sigma_{ij} &= \sqrt{\bar{\beta}_{ij}(1 - \bar{\beta}_{ij})}, \ K_{ii} = 0, \ L = (l_{ij})_{N\times N},
\end{aligned}$$

$$\begin{aligned}\check{K} &= diag_N\{\hat{K}_i\},\ \hat{K}_i = diag_N\{K_{ij}\},\\ \bar{P} &= diag\ \{I_N \otimes P_1, \ldots, I_N \otimes P_N\},\\ \Phi_1 &= (\hat{A})^T \bar{P}\hat{A} + U^T \bar{P} U + \tilde{D}^T \tilde{D} - \tilde{P},\\ \Phi_2 &= (\hat{A})^T \bar{P}\tilde{F},\ \Phi_3 = \tilde{F}^T \bar{P}\tilde{F},\\ \tilde{D} &= diag_N\{\bar{D}_i\},\ \bar{D}_i = diag_N^i\{D_i\},\\ \tilde{F} &= diag_N\{\bar{F}_i\},\ \bar{F}_i = diag_N^i\{F_i\},\end{aligned}$$

where, $\odot$ signifies matrix dot product operation, $\text{diag}_N^i\{B_i\}$ indicates only the ith block of N has nonzero value, $\otimes$ represents the Kronecker product.

Proof 1 For AGC system (10), this chapter designs the Lyapunov function as

$$V(k) = \sum_{i=1}^{N} x_i^T(k) P_i x_i(k)$$

Moreover, one obtains

$$\begin{aligned}\Delta V(k) &= \mathcal{E}\{V(k+1)\} - V(k)\\ &= \mathcal{E}\Big\{\sum_{i=1}^{N} x_i^T(k+1) P_i x_i(k+1)\Big\} - \sum_{i=1}^{N} x_i^T(k) P_i x_i(k)\\ &= \mathcal{E}\Big[(\hat{A} + \tilde{B}W + \bar{B}R)\tilde{x}(k) + \tilde{F}\tilde{v}(k)\Big]^T \bar{P}\\ &\quad\times\Big[(\hat{A} + \tilde{B}W + \bar{B}R)\tilde{x}(k) + \tilde{F}\tilde{v}(k)\Big]\\ &\quad-\tilde{x}^T(k)\tilde{P}\tilde{x}(k)\\ &= \zeta^T(k)\Phi\zeta(k),\end{aligned}$$

in which

$$\begin{aligned}\tilde{x}(k) &= 1_{N\times 1} \otimes \text{col}_N\{x_i(k)\},\\ \tilde{v}(k) &= 1_{N\times 1} \otimes \text{col}_N\{v_i(k)\},\\ \zeta(k) &= [\tilde{x}^T(k), \tilde{v}^T(k)]^T.\end{aligned}$$

When the load disturbances are zero,

$$\Delta V(k) \leqslant \tilde{x}^T(k)\Phi_1\tilde{x}(k) \tag{12}$$

From Theorem 1 we know that, $\Delta V(k)$ is always below zero for any nonzero states $\tilde{x}(k) \neq 0$. Therefore, AGC system (10) satisfies the definition of mean square asymptotic stability.

Further, when there are some load disturbances in (10), we define one new index

$$\mathcal{J}(k) = \mathcal{E}\{V(k+1)\} - V(k) + \tilde{z}^T(k)\tilde{z}(k) \\ -\gamma^2\tilde{v}(k)^T(k)\tilde{v}(k). \quad (13)$$

where $\tilde{z}(k) = 1_{N\times 1} \otimes \mathrm{col}_N\{z_i(k)\}$.

By accumulating the expectation of (13) , one can obtain

$$\begin{aligned}\bar{\mathcal{J}} &= \mathcal{E}\Big\{\sum_{k=0}^{\infty}\mathcal{J}(k)\Big\} \\ &= \mathcal{E}\{V(\infty)\} + \mathcal{E}\Big\{\sum_{k=0}^{\infty}\tilde{z}^T(k)\tilde{z}(k)\Big\} - V(0) \\ &\quad -\gamma^2\sum_{k=0}^{\infty}\tilde{v}^T(k)\tilde{v}(k).\end{aligned}$$

From Theorem 1, we conclude that $\bar{\mathcal{J}}$ is always below zero. And based on $\mathcal{E}\{V(\infty)\} \geqslant 0$ and $V(0) = 0$, $\|\tilde{z}\|^2_{\mathcal{E}_2} < \gamma^2\|\tilde{v}\|^2_2$ is always satisfied when nonzero load disturbances meeting $v_i \in l_2[0, \infty)$. Hence, the proof of Theorem 1 is completed.

In Theorem 1, we assume the cyberattack model as DoS attacks, and analyze the existing condition of the distributed automatic generation controller under DoS attacks. However, with the current form of the sufficient condition (11), the distributed controller gains K_i and K_{ij} are coupled with the Lyapunov matrices P_i, so we cannot derive the values of K_i and K_{ij}. Thus, we further propose Theorem 2, in which the cone complementarity linearization technique is employed to select out the values of the tolerant controller gains under DoS attacks in the standard LMI toolbox.

Theorem 2 *Given the DoS attack probabilities on open communication links $\bar{\beta}_{ij}$, AGC system (10) is mean square asymptotically stable with Δf_i signals meeting the prescribed $\mathcal{H}_\infty$ performance level γ, if there exist matrices $P_i > 0$, $X_i > 0$, such that for $i = 1, 2, \ldots, N$,*

$$\begin{bmatrix}\Omega_1 & \Omega_2 \\ * & \Omega_3\end{bmatrix} < 0 \quad (14)$$

$$P_i X_i = I \quad (15)$$

where,

$$\begin{aligned}\Omega_{1pq} &= diag\,\{-I_N \otimes X_1, \ldots, -I_N \otimes X_N\}, \\ \Omega_{2pq} &= \begin{bmatrix}(\hat{A}^{pq})^T & (\hat{B}^{pq})^T & (U^p)^T \\ \tilde{E}^T & 0 & 0\end{bmatrix}^T, \\ \Omega_{3pq} &= diag\,\{\tilde{D}^T\tilde{D} - \tilde{P}, -\gamma^2 I\}.\end{aligned}$$

Proof 2 Using Schur complement to (11), and define $X_i = P_i^{-1}$, we can arrive at (14). Therefore, Theorem 2 is established. The detailed implementation can refer to [26].

Up to now, the distributed automatic generator controller gains with a certain tolerance to DoS attacks can be obtained, which is the first security layer of the resilient control framework we proposed in this chapter. The design process of tolerant control gains seeking is off-line, and the function of the obtained tolerant control gains is to initialize the operation of power systems.

3 On-Line Cyberattack Detection Mechanism

In Sect. 2, we have systematically discussed the tolerant control gains seeking method. Before deriving the tolerant control gains, we need to firstly presume the cyber attack model, and then incorporate it into the power system model, based on which we can finally obtain the tolerant control gains with the specific performance indexes satisfied. As a result, the efficacy of the designed tolerant controller relies to a certain extent on the authenticity of the presumed cyber attack model. When the presumed model deviates largely from the actual attacking behavior, the efficacy of the tolerant controller will reduce. Inspired by this fact, this section will comprehensively investigate the on-line cyberattack detection mechanisms, aiming at finding the underlying cyber attacks by detecting the transmitted data, which can be seen as an effective complement to the first security layer.

3.1 Two Typical Cyberattack Detection Mechanisms

3.1.1 Intrusion-Detector-Dependent Cyberattack Detection

Faced with the underlying cyber attacks in AGC, some scholars tend to distinguish attacking behaviors by predicting the load and by comparing the predicted values and real-time measurements. For example, [27] analyzed the impact of data loss caused by cyber attacks on AGC, and proposed a model-based abnormal behavior detection and attack mitigation algorithm. Some scholars aim to judge the attack behaviors in AGC by checking the consistency of predicted frequencies and measured ones, such as in [28], authors presumed adversaries can arbitrarily design attacking model by eavesdropping frequency signal from PMUs, and exploited a fast optimal attack detection algorithm to examine, identify, and mitigate the attack from the defender side, even though adversaries are launching an optimal control sequence. However, in practical power systems, the real-time value of the load is almost unknown. The above two attack detection schemes designed by predicting the load have limited feasibility with the concern of accuracy. Besides, these schemes ignore the impacts of environmental noise on the detection results, let alone the countermeasures of false alarm rate increasing.

To further enhance the robustness and accuracy of cyber attack detections, intrusion detector dependent attack detection strategies have aroused extensive attention in the field of AGC [29–38]. To reduce the impact of stochastic load disturbances on detection accuracy, Ameli et al. presented a stochastic unknown input estimators based attack detection method, to distinguish the compromised measurements by FDI attacks from the healthy data [29]. This method skillfully used the physical characteristics of the power system, and let intrusion detection and security defense work in parallel, which makes the attack detection mechanism independent of load changes, and thus leads to a higher detection accuracy. On the other side, this method well minimized the impact of load disturbances on states estimation, by designing the gains of stochastic unknown input estimators with the optimization techniques. Liu et al. focused on DoS attacks in microgrids and proposed an intrusion detector based AGC approach [32]. In this scheme, the detected DoS attack choice is used to dynamically adjust the control gains, rather than the actual attack choice. Thus, the presented control scheme is more robust to possible false alarms since it has the ability to maintain the asymptotic stability for power systems under high false detection rates.

χ^2 is also seen as a typical kind of intrusion detectors to detect the cyber attacks in power systems. The χ^2 test is a classical hypothesis test method, which was first proposed by a British mathematician called Karl Pearson in 1900. Its basic purpose is to test whether the distribution series of actual values and of theoretical values are within a reasonable range, that is, to quantitatively describe the difference between observed values and expected values. The standard formula of χ^2 test is

$$\chi^2 = \Sigma \frac{(O - E)^2}{E} \tag{16}$$

where O represents a sequence of observed values, and E represents a sequence of expected values.

The value of χ^2 reflects the degree that the observed values conform to the expected ones. A smaller χ^2 value indicates a more reliable sequence of the observation, and vice versa. Hence, in practical applications of χ^2 detection, the probabilities that the measurements are in the line with the actual values can be calculated by checking up the χ^2 test table. In recent decades, χ^2 detector based intrusion detection approaches have been widely explored in the field of power systems cybersecurity. As an illustration, we hereby briefly introduce the idea of χ^2 intrusion detection in multi-area AGC [35].

Taking the cyber attacks into consideration, the real-time measurements transmitted from area j to area i can be represented by

$$\tilde{y}_j(k) = y_j(k) + G_j g_j(k) \tag{17}$$

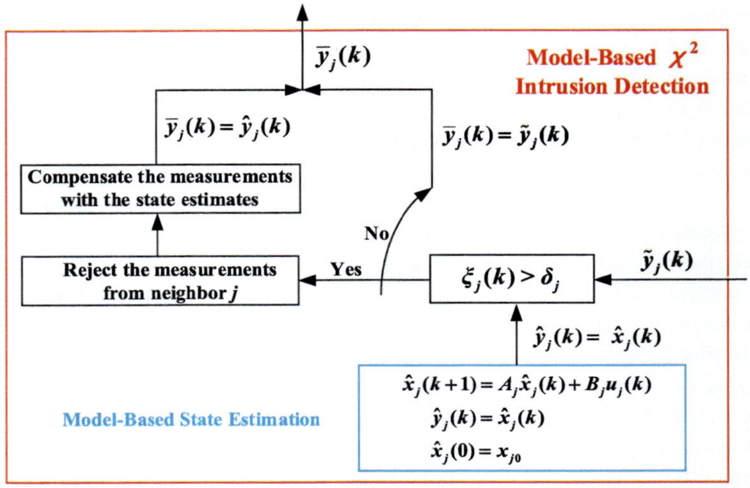

Fig. 4 Model-based χ^2 intrusion detection [35]

where $y_j(k)$ stands for the true value, $g_j(k)$ signifies the injected false information by attackers, and G_j indicates a column vector composed by 0 or 1, with 0 representing the communication link is attacked while 1 representing the communication link is healthy.

From the perspective of power systems security, a χ^2 intrusion detection unit may be set up at the control center of each area, to examine the states transmitted from the neighbors, see Fig. 4.

Considering the load variations in power industries generally obey the normal distributions [39], we formulate

$$\xi_j(k) = \sum_{l=k-\Gamma+1}^{k} [\hat{y}_j(l) - \tilde{y}_j(l)]^T [\hat{y}_j(l) - \tilde{y}_j(l)], k \geq \Gamma. \tag{18}$$

Therefore, $\xi_j(k)$ obeys χ^2 distribution, and its freedom degree is $(5 \times \Gamma)$, where Γ is the sliding window, and $\hat{y}_j(l)$ is the estimated states based on the decentralized model.

The χ^2 detector at time k can be formulated as

$$\xi_j(k) \underset{H_1}{\overset{H_0}{\lesseqgtr}} \delta_j \tag{19}$$

with δ_j representing the false alarm probability, which can be specifically designed by the given power system. The H_0 hypothesis usually supposes there are no differences between the transmitted and the estimated states. Once the H_0 hypothesis no longer holds, also say the H_1 hypothesis is satisfied, an alarm will be instantly triggered to report the underlying cyber attacks to the grid operator.

3.1.2 Credibility-Based Cyberattack Detection

Section 3.1.1 retrospects the intrusion-detector-dependent cyberattack detection approaches, and some achievements in the applications of power systems. This section will from another perspective, discuss the credibility-based cyberattack detections. The basic idea of the credibility-based cyberattack detection is to distinguish the attacking behavior by setting up the credibility metrics [40–45]. In recent years, the credibilities of open communication links are usually defined by β distribution function. A general idea of β-based credibility scheme is to describe the extent of security by judging the behaviors of the transmitted data. β-based credibility scheme adopts time scaling factors, which are determined by the historical transmitted data. According to the time scaling factors, values of the credibility of each communication link can be calculated. By referring to the credibility value, the pattern of data sharing and collaborative perception among the whole communication network can be determined. Finally, an optimal global decision can be selected out by the data fusion algorithm at the data fusion center.

β-based credibility scheme utilizes the β probability density function to depict the security level of communication links [43]. This probability density function provides mathematic foundations for the combination of feedback control and expressing confidence levels. Josang et al. in [43] designed several kinds of β-based credibility mechanisms for different specific instances, and validated their feasibility and effectiveness, respectively.

To evaluate and classify the transmitted data, credibility metrics are reasonable to be set up [44, 45]. In [44], an optimal credibility-based intrusion detection strategy was constructed to distinguish the underlying FDI attacks, in which the credibility metric is formulated by assessing the consistency of the received measurements and the fused data of whole communication network. The formulated credibility metric in [44] is expressed as

$$r_i(t) = \begin{cases} r_i(t-1)+1, & v_i \neq v \\ r_i(t-1), & v_i = v \end{cases} \tag{20}$$

where $r_i(t)$ represents the credibility metric of sensor node i at sampling instant t, which is used to count the number of false signals by time t, v_i represents the received signals from sensor node i, v represents the fused data of whole communication network.

Zeng et al. set up the credibility metric based on the difference of the data transmitted between contiguous sampling instants [45]. When the difference is smaller than the detection threshold, the transmitted signal is judged as credible, and vice versa. The validations of [45] is upon the distributed networked control in the application

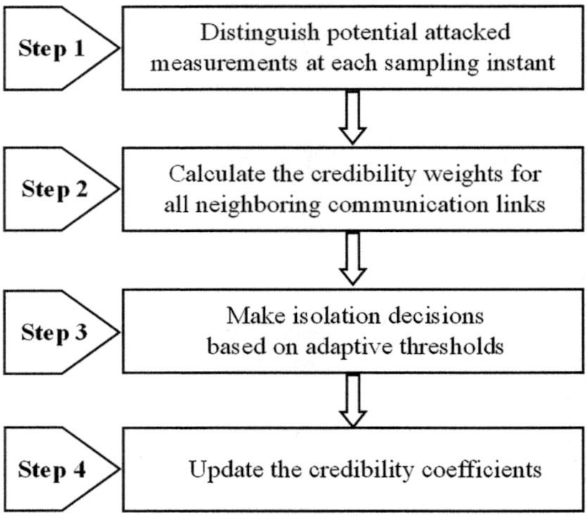

Fig. 5 Flow diagram of the credibility-based AGC [22]

of multiple unmanned aerial systems. As for the applications of the credibility-based cyberattack detection in power systems, a detailed discussion will be developed in the next section.

3.2 *Implementation of Credibility-Based AGC Under Cyber Attacks*

As is discussed in Sect. 3.1.2, the credibility-based cyberattack detection approach has broad applications. This section will focus on its specific implementation in multi-area power generation systems. When implementing the distributed AGC, the local area needs to collect the signals of tie-line power and frequency from its neighbors. These signals are transmitted from neighboring PMUs to the local PMU via the communication network, e.g., cognitive radio networks or WiMAX, and have the possibilities of suffering cyberattack threats.

In this section, a credibility-based AGC strategy will be systematically revealed. The presented credibility-based AGC strategy mainly contains four steps, see Fig. 5.

The first step is to distinguish the abnormal behavior of each received signal. Under the ideal case, the transmitted signal from one neighbor is decreased as the time goes by. Therefore, we set up a credibility metric to distinguish the attacked measurement by comparing the gradient variation of the measurements at adjacent sampling instants [46]. At control area i, a buffer should be installed to record the historical measurements. The credibility metric is then formulated as

$$S_{ij}(k) = \begin{cases} S_{ij}(k-1)+1, & \text{if } \varphi(y_j(k)) - \varphi(y_j(k-1)) \leqslant T(k); \\ S_{ij}(k-1), & \text{if } \varphi(y_j(k)) - \varphi(y_j(k-1)) > T(k). \end{cases} \tag{21}$$

with $S_{ij}(k)$ indicating the number of abnormal observations by time k, $\varphi(y_j(k))$ representing a function of measurements, $T(k)$ denoting a threshold function to be designed, and L standing for the maximal storage capacity of the buffer. When $\varphi(y_j(k)) - \varphi(y_j(k-1)) > T(k)$, the received signal will be judged as abnormal, and vice versa.

The second step concerns the credibility weight calculation. Based on the credibility metric, a sequence of credible and incredible detection results will be made for each communication link. A local credibility manager is thus established to update the credibility weights of each communication link. The proposed credibility weight calculation method improves the traditional β distribution function [43] by integrating the calculation period. The formula for calculating the credibility weight is designed as

$$Cre_{ij}(k) = \begin{cases} \dfrac{\theta_0(S_{ij}(k) - S_{ij}(k-M_{ij})) + \theta_1}{\theta_0 M_{ij} + \theta_2} & \text{if } k > M_{ij}; \\ \dfrac{\theta_0 S_{ij}(k) + \theta_1}{\theta_0 k + \theta_2} & \text{if } k \leqslant M_{ij}. \end{cases} \tag{22}$$

where, M_{ij} is the calculation period used to determine credibility weights, θ_0, θ_1 and θ_2 are three constants used to adjust the changing speed of credibility weights with the constraint $\theta_1 \leqslant \theta_2$.

The third step is to design an adaptive-threshold-based isolation decision. Based on the credibility weights calculated in step 2, we aim for designing thresholds to determine when to isolate the related communication links to ensure the security operation of power systems. Since accurately and timely isolating the attacked communication links is vital to both power infrastructures and grid operators, we propose the adaptive threshold functions. Once the calculated credibility weight is smaller than the adaptive threshold, isolate the corresponding communication link immediately, which means the information transmitting from this neighbor will no longer be used. The isolation decision is made by

$$Cre_{ij}(k) \leqslant Cre_{ij}^{th}(k) \tag{23}$$

with the adaptive threshold function

$$Cre_{ij}^{th}(k) = \frac{\gamma_1 k + \gamma_0}{\gamma_2 k + \gamma_0} \tag{24}$$

where γ_0, γ_1 and γ_2 are three constants to be selected according to specific systems.

For a clear demonstration, the specific implementation of the credibility-based AGC is constructed in Fig. 6.

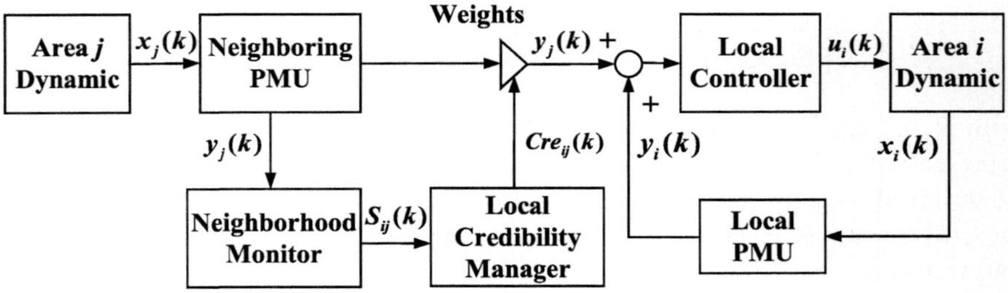

Fig. 6 Specific implementation of credibility-based AGC [22]

The last step is to update the credibility weights. Based on the steps 1–3, the real-time credibility weights will be updated. As a result, the closed-loop dynamic of AGC is formulated by

$$\begin{aligned} x_i(k+1) = {} & (A_i + B_i K_i C_i)x_i(k) + F_i \Delta P_{L_i}(k) \\ & + \sum_{j=1, j\neq i}^{4} Cre_{ij}(k) l_{ij}(A_{ij} + B_i K_{ij} C_j)x_j(k), \end{aligned} \tag{25}$$

where

$$Cre_{ij}(k) = \begin{cases} Cre_{ij}(k), & \text{if } \; Cre_{ij}(k) > Cre_{ij}^{th}(k); \\ 0, & \text{if } \; Cre_{ij}(k) \leqslant Cre_{ij}^{th}(k). \end{cases} \tag{26}$$

4 Validations of the Cyberattack-Resilient AGC

4.1 System Structure and Parameters

Figure 7 shows the system structure used for validations of the presented cyberattack-resilient AGC framework. Table 2 gives the involved parameters. The sampling period is selected as $0.02s$. Load disturbances are simulated as $v_i(k) = 0.008e^{-0.01k}\sin(\pi/20k)$. $\mathcal{H}_\infty$ robustness index is prescribed as $\gamma = 1$.

4.2 Validations of the First Security Layer

To validate the effectiveness of the first security layer, the off-line tolerant control gains seeking method, we compared the frequency dynamics of the traditional controller (without the consideration of DoS attacks) and the proposed tolerant controller (with the consideration of DoS attacks). Two different DoS attack probabilities are tested, that is, a) The DoS attack probability is 0.4 of each communication link; b) The DoS attack probability is 0.8 of each communication link.

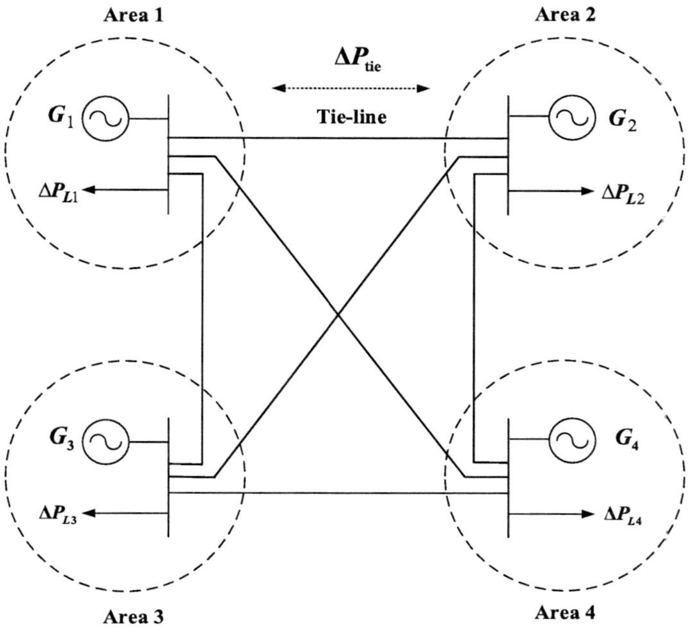

Fig. 7 Interconnected four-area power system

Table 2 System metrics

Area 1	Area 2	Area 3	Area 4
$D_1 = 1$	$D_2 = 1.5$	$D_3 = 3$	$D_4 = 2$
$2H_1 = 10$	$2H_2 = 12$	$2H_3 = 15$	$2H_4 = 18$
$T_{ch_1} = 1.7$	$T_{ch_2} = 1.2$	$T_{ch_3} = 0.9$	$T_{ch_4} = 0.8$
$T_{g_1} = 0.7$	$T_{g_2} = 0.9$	$T_{g_3} = 1.5$	$T_{g_4} = 0.7$
$R_1 = 0.05$	$R_2 = 0.06$	$R_3 = 0.08$	$R_4 = 0.1$
$T_{12} = 0.15$	$T_{21} = 0.15$	$T_{31} = 0.2$	$T_{41} = 0.3$
$T_{13} = 0.2$	$T_{23} = 0.25$	$T_{32} = 0.25$	$T_{42} = 0.35$
$T_{14} = 0.3$	$T_{24} = 0.35$	$T_{34} = 0.4$	$T_{43} = 0.4$

Figure 8 demonstrates the dynamic performance of Δf_1. From Fig. 8a we can see that, under the DoS attack probability 0.4, the traditional controller behaves a larger overshoot and a longer settling time compared with the tolerant controller we proposed in this chapter, though both the two controllers can maintain the system stability. From Fig. 8b we observe that, with the DoS attack probability increasing to 0.8, the traditional controller can hardly guarantee the system stability, while the proposed tolerant controller can still stabilize the system quickly.

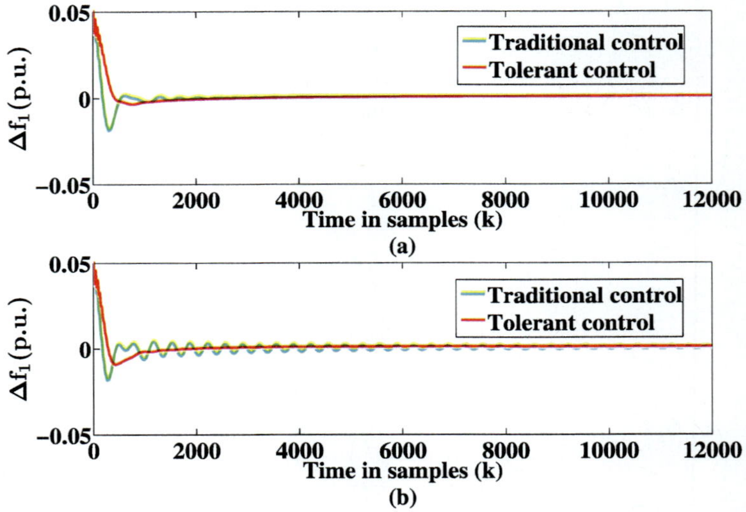

Fig. 8 Dynamic performance of Δf_1 under different DoS attacks. **a** DoS attack probability is 0.4. **b** DoS attack probability is 0.8

4.3 Validations of the Second Security Layer

In order to test the efficacy of the second security layer, the on-line cyberattack detection mechanism, two kinds of the FDI attacks are considered. We assume the FDI attacks are launched on the communication links from area 2 to area 1, and from area 3 to area 2. Specifically,

Case 1:

$$\tilde{y}_2(k) = C_2x_2(k) + [1;\ 1] * 4k, \quad \text{from} \quad k = 4000,$$
$$\tilde{y}_3(k) = C_3x_3(k) + [1;\ 1] * 8k, \quad \text{from} \quad k = 8000.$$

Case 2:

$$\tilde{y}_2(k) = C_2x_2(k) + [1;\ 1] * 8k, \quad \text{from} \quad k = 4000,$$
$$\tilde{y}_3(k) = C_3x_3(k) + [1;\ 1] * 16k, \quad \text{from} \quad k = 8000.$$

Note that the controller gains used in this section are derived from the first security layer, that is, the off-line tolerant control gains under the DoS attack probability 0.4. From $k = 0$ to $k = 4000$, we still assume the system is suffering DoS attacks with the probability 0.4. The parameters in the on-line credibility-based intrusion detection mechanism are designed as below.

In step 1, the $ACE_j(k)$ signal is selected as the reference. The credibility metric (21) is designed as

$$S_{ij}(k) = \begin{cases} S_{ij}(k-1)+1 & \text{if} \\ & ACE_i(k) - ACE_i(k-1) \leqslant 0 \\ & \text{when} \quad ACE_i(k) > 0 \quad \text{or} \\ & ACE_i(k) - ACE_i(k-1) > 0 \\ & \text{when} \quad ACE_i(k) < 0; \\ S_{ij}(k-1) & \text{otherwise;} \\ S_{ij}(0) = 0. \end{cases} \tag{27}$$

In step 2, $\theta_0 = 1$, $\theta_1 = 2$, $\theta_2 = 3$, $M_{ij} = 200$.
In step 3, the adaptive threshold function is designed as

$$Cre_{ij}^{th}(k) = \frac{5000 + 0.4k}{5000 + k}. \tag{28}$$

Figure 9 compares the frequency deviation dynamics with and without the proposed credibility-based cyberattack detection mechanism under Case 1 and Case 2, respectively. It can be seen that without the intrusion detection mechanism, the dynamic of the frequency deviation becomes unstable under both Case 1 and Case 2, which means the attackers will be constantly destroying the system. And the attacks with the larger amplitude will lead a greater divergence of the frequency dynamic. While with the proposed on-line cyberattack detection mechanism, the dynamic of the frequency deviation shows a short period of divergence at first, and then quickly returns to the stable states. The results illustrate that the proposed cyberattack detection mechanism has the ability to isolate the compromised communication links

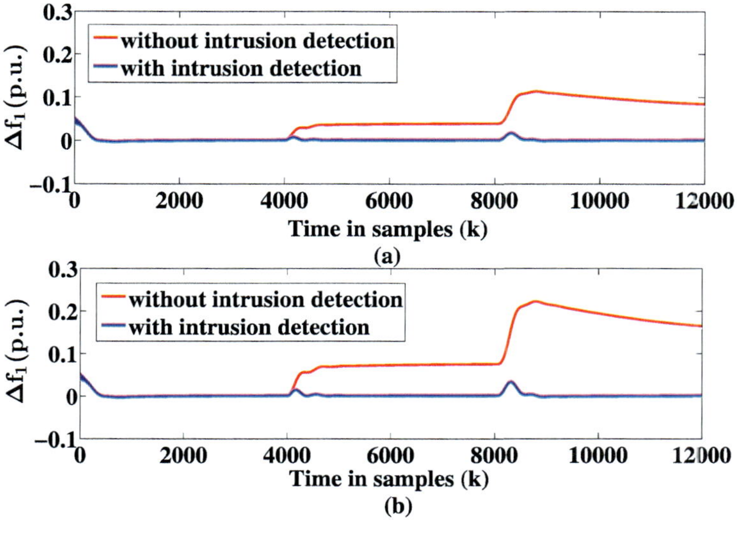

Fig. 9 Dynamic performance of Δf_1 under different FDI attacks. **a** Case 1. **b** Case 2

when the credibility weights decrease to the predefined thresholds, and to guarantee the secure operation of the power generation system.

Note that a short period of divergence of the system dynamic is reasonable and allowable because the detection time is needed before making the isolation decision for the on-line cyberattack detection mechanism. The detection time, which corresponds to the degree of divergence, is to do with the threshold function $Cre_{ij}^{th}(k)$ and the calculation period of credibility weight M_{ij}. A larger threshold function usually needs a longer detection time, and thus leading a greater divergence. Once the threshold function $Cre_{ij}^{th}(k)$ is determined, the detection time will be positively correlated with M_{ij}, indicating that the larger M_{ij} will cause the greater divergence and vice versa. Nevertheless, a smaller M_{ij} not always brings a better detection effectiveness. When M_{ij} decreases, the rate of making false isolation decision will increase. That is to say, the credibility-based intrusion detection mechanism will be more sensitive to load variations. In such cases, some stronger load disturbances may be mistaken as the cyber attacks. Therefore, when employing the credibility-based intrusion detection mechanism proposed in this chapter to an actual power system, delicately designing both $Cre_{ij}^{th}(k)$ and M_{ij} based on the required security level, is crucial and deserves pondering.

5 Summary

A systematic cyberattck-resilient control framework with two security layers for multi-area power generation systems is proposed in this chapter. The first security layer concerns the off-line tolerant control gains seeking, and the second security layer concerns the on-line credibility-based cyberattack detection. Simulations are conducted under two typical categories of cyber attacks, DoS attacks and FDI attacks. The simulation results show that, under different DoS attack probabilities, the off-line tolerant controller can maintain the stable operation of the system, while under FDI attacks with different amplitudes, the on-line credibility-based cyberattack detection mechanism can distinguish and isolate the attacked communication links and finally keep the system stable.

References

1. H.H. Alhelou, A.Y. Abdelaziz, P. Siano, *Wide Area Power Systems Stability, Protection, and Security* (Springer, 2020)
2. L. Xiaoqing, Yu. Xinghuo, J. Lai, Y. Wang, J.M. Guerrero, A novel distributed secondary coordination control approach for islanded microgrids. IEEE Trans. Smart Grid **9**(4), 2726–2740 (2018)
3. Z. Hu, S. Liu, W. Luo, L. Wu, Resilient distributed fuzzy load frequency regulation for power systems under cross-layer random denial-of-service attacks. IEEE Trans. Cybern. (2020)

4. K. Zhang, B. Zhou, G. Duan, Event-triggered and self-triggered control of discrete-time systems with input constraints. IEEE Trans. Syst. Man Cybern. Syst. **52**(3), 1948–1957 (2022)
5. H.H. Alhelou, M.E.H. Golshan, P. Siano, Frequency response models and control in smart power systems with high penetration of renewable energy sources. Comput. Electr. Eng. **96**, 107477 (2021)
6. Y. Xu, Z.Y. Dong, J.H. Zhao, P. Zhang, K.P. Wong, A reliable intelligent system for real-time dynamic security assessment of power systems. IEEE Trans. Power Syst. **27**(3), 1253–1263 (2012)
7. C. Peng, J. Li, M. Fei, Resilient event-triggering H_∞ load frequency control for multi-area power systems with energy-limited dos attacks. IEEE Trans. Power Syst. **32**(5), 4110–4118 (2016)
8. P. Zhang, H. Xue, S. Gao, J. Zhang, Distributed adaptive consensus tracking control for multi-agent system with communication constraints. IEEE Trans. Parallel Distrib. Syst. **32**(6), 1293–1306 (2021)
9. B. Wang, R. Su, A distributed platoon control framework for connected automated vehicles in an urban traffic network. IEEE Trans. Control Netw. Syst. 1–11 (2022)
10. L. Huang, W. Junfeng, Y. Mo, L. Shi, Joint sensor and actuator placement for infinite-horizon lqg control. IEEE Trans. Autom. Control **67**(1), 398–405 (2022)
11. J.P. Farwell, R. Rohozinski, Stuxnet and the future of cyber war. Survival **53**(1), 23–40 (2011)
12. Defense Use Case. Analysis of the cyber attack on the Ukrainian power grid, in *Electricity Information Sharing and Analysis Center (E-ISAC)*, vol. 388 (2016)
13. A. Garnaev, Y. Liu, W. Trappe, Anti-jamming strategy versus a low-power jamming attack when intelligence of adversary's attack type is unknown. IEEE Trans. Signal Inf. Process. Netw. **2**(1), 49–56 (2016)
14. R. Ma, P. Shi, W. Ligang, Dissipativity-based sliding-mode control of cyber-physical systems under denial-of-service attacks. IEEE Trans. Cybern. **51**(5), 2306–2318 (2021)
15. A.S. Musleh, G. Chen, Z.Y. Dong, A survey on the detection algorithms for false data injection attacks in smart grids. IEEE Trans. Smart Grid **11**(3), 2218–2234 (2020)
16. Y. Mo, B. Sinopoli, False data injection attacks in control systems, in *Preprints of the 1st Workshop on Secure Control Systems* (2010), pp. 1–6
17. H.H. Alhelou, P. Cuffe, A dynamic-state-estimator-based tolerance control method against cyberattack and erroneous measured data for power systems. IEEE Trans. Industr. Inf. **18**(7), 4990–4999 (2022)
18. M. Chlela, D. Mascarella, G. Joós, M. Kassouf, Fallback control for isochronous energy storage systems in autonomous microgrids under denial-of-service cyber-attacks. IEEE Trans. Smart Grid **9**(5), 4702–4711 (2018)
19. R. Ma, P. Shi, W. Ligang, Sparse false injection attacks reconstruction via descriptor sliding mode observers. IEEE Trans. Autom. Control **66**(11), 5369–5376 (2021)
20. Q. Zhou, M. Shahidehpour, A. Alabdulwahab, A. Abusorrah, A cyber-attack resilient distributed control strategy in islanded microgrids. IEEE Trans. Smart Grid **11**(5), 3690–3701 (2020)
21. R. Chen, X. Li, H. Zhong, M. Fei, A novel online detection method of data injection attack against dynamic state estimation in smart grid. Neurocomputing **344**, 73–81 (2019)
22. H. Zhijian, S. Liu, W. Luo, W. Ligang, Credibility-based secure distributed load frequency control for power systems under false data injection attacks. IET Gener. Trans. Distrib. **14**(17), 3498–3507 (2020)
23. D. Bose, A.P. Jayasumana, A reputation-based method for detection of attacks in virtual coordinate based wireless sensor networks, in *2015 IEEE 40th Conference on Local Computer Networks (LCN)* (2015), pp. 486–489
24. X. Zeyuan, D. Wang, G. Yi, H. Zhijian, A novel tracking control approach of amplitude signals for vibratory gyroscopes suppressing high-frequency disturbance. Measurement **195**, 110981 (2022)
25. H. Zhijian, S. Liu, L. Yang, W. Ligang, Distributed fuzzy filtering for load frequency control of non-linear interconnected power systems under cyber-physical attacks. IET Control Theory Appl. **14**(4), 527–538 (2020)

26. J. Liu, W. Chengwei, Z. Wang, W. Ligang, Reliable filter design for sensor networks using type-2 fuzzy framework. IEEE Trans. Industr. Inf. **13**(4), 1742–1752 (2017)
27. S. Sridhar, M. Govindarasu, Model-based attack detection and mitigation for automatic generation control. IEEE Trans. Smart Grid **5**(2), 580–591 (2014)
28. R. Tan, H.H. Nguyen, E.Y.S. Foo, D.K.Y. Yau, Z. Kalbarczyk, R.K. Iyer, H.B. Gooi, Modeling and mitigating impact of false data injection attacks on automatic generation control. IEEE Trans. Inf. Forens. Sec. **12**(7), 1609–1624 (2017)
29. A. Ameli, A. Hooshyar, A.H. Yazdavar, E.F. El-Saadany, A. Youssef, Attack detection for load frequency control systems using stochastic unknown input estimators. IEEE Trans. Inf. Forens. Sec. **13**(10), 2575–2590 (2018)
30. W. Bi, K. Zhang, Y. Li, K. Yuan, Y. Wang, Detection scheme against cyber-physical attacks on load frequency control based on dynamic characteristics analysis. IEEE Syst. J. **13**(3), 2859–2868 (2019)
31. H. Zhijian, S. Liu, W. Ligang, Credibility-based distributed frequency estimation for plug-in electric vehicles participating in load frequency control. Int. J. Electr. Power Energy Syst. **130**, 106997 (2021)
32. S. Liu, P. Siano, X. Wang, Intrusion-detector-dependent frequency regulation for microgrids under denial-of-service attacks. IEEE Syst. J. **14**(2), 2593–2596 (2020)
33. T. Huang, B. Satchidanandan, P.R. Kumar, L. Xie, An online detection framework for cyber attacks on automatic generation control. IEEE Trans. Power Syst. **33**(6), 6816–6827 (2018)
34. X. Zeyuan, B. Xi, G. Yi, D. Wang, A novel model for fully closed-loop system of hemispherical resonator gyroscope under force-to-rebalance mode. IEEE Trans. Instrum. Meas. **69**(12), 9918–9930 (2020)
35. H. Zhijian, S. Liu, W. Luo, W. Ligang, Intrusion-detector-dependent distributed economic model predictive control for load frequency regulation with pevs under cyber attacks. IEEE Trans. Circuits Syst. I Regul. Pap. **68**(9), 3857–3868 (2021)
36. T.S.L.V. Ayyarao, I. Ravi Kiran, A two-stage kalman filter for cyber-attack detection in automatic generation control system. J. Modern Power Syst. Clean Energy 1–10 (2021)
37. L.-Y. Lu, H.J. Liu, H. Zhu, Chia-Chi Chu. Intrusion detection in distributed frequency control of isolated microgrids. IEEE Trans. Smart Grid **10**(6), 6502–6515 (2019)
38. S. Yan, W. Sun, X. Yu, H. Gao, Adaptive sensor fault accommodation for vehicle active suspensions via partial measurement information. IEEE Trans. Cybern. 1–12 (2021)
39. T. Zhang, W. Sheng, X. Song, X. Meng, C. Shi, Probabilistic modelling and simulation of stochastic load for power system studies, in *2013 UKSim 15th International Conference on Computer Modelling and Simulation* (2013), pp. 519–524
40. P. Bai, X. Zhang, F. Ye, Reputation-based beta reputation system against ssdf attack in cognitive radio networks, in *2017 Progress in Electromagnetics Research Symposium - Fall (PIERS - FALL)* (2017), pp. 792–799
41. K. Gerrigagoitia, R. Uribeetxeberria, U. Zurutuza, I. Arenaza, Reputation-based intrusion detection system for wireless sensor networks, in *2012 Complexity in Engineering (COMPENG). Proceedings* (2012), pp. 1–5
42. H. Alzaid, M. Alfaraj, S. Ries, A. Jøsang, M. Albabtain, A. Abuhaimed, Reputation-based trust systems for wireless sensor networks: a comprehensive review, in *IFIP International Conference on Trust Management* (Springer, 2013), pp. 66–82
43. A. Josang, R. Ismail, The beta reputation system, in *Proceedings of the 15th Bled Electronic Commerce Conference*, vol. 5 (2002), pp. 2502–2511
44. P. Wang, C. Chen, S. Zhu, L. Lyu, W. Zhang, X. Guan, An optimal reputation-based detection against ssdf attacks in industrial cognitive radio network, in *2017 13th IEEE International Conference on Control Automation (ICCA)* (2017), pp. 729–734
45. W. Zeng, M.-Y. Chow, A reputation-based secure distributed control methodology in d-ncs. IEEE Trans. Industr. Electron. **61**(11), 6294–6303 (2014)
46. Y. Wang, T. Wang, X. Yang, J. Yang, Gradient descent-barzilai borwein-based neural network tracking control for nonlinear systems with unknown dynamics. IEEE Trans. Neural Netw. Learn. Syst. 1–11 (2021)

Cyber-Security of Protection System in Power Grids—Part 1: Vulnerabilities and Counter-Measures

Shahin Riahinia, Amir Ameli, Mohsen Ghafouri, and Abdulsalam Yassine

Abstract The recent trend to expand the use of Information Technology (IT) in power networks has made the electric grid potentially vulnerable to cyber-attacks. Protection systems are among the most critical cyber-vulnerable components, as they directly affect the integrity and stability of power systems. On this basis, this chapter reviews and investigates the cyber-vulnerabilities of protection systems in power networks. These vulnerabilities can be categorized into three main groups: (i) vulnerabilities of substations, (ii) vulnerabilities of Communication-Assisted Protection (CAP) schemes, and (iii) vulnerabilities of Wide-Area Protection (WAP) schemes. This chapter also discusses the techniques that can be used to detect cyber-attacks against protection schemes and differentiate them from faults.

Keywords Communication-assisted protection · Cyber-attacks · Cyber-security · Substations · Wide-area protection · Intrusion detection systems

---

S. Riahinia · A. Ameli (✉) · A. Yassine
Department of Electrical Engineering, Lakehead University, Thunder Bay, ON P7B 5E1, Canada
e-mail: aameli@lakeheadu.ca

S. Riahinia
e-mail: sriahini@lakeheadu.ca

A. Yassine
e-mail: ayassine@lakeheadu.ca

M. Ghafouri
Concordia University, 1455 De Maisonneuve Blvd. West, EV 7.640, Montreal, QC H3G 1M8, Canada
e-mail: mohsen.ghafouri@concordia.ca

H. Haes Alhelou et al. (eds.), *Power Systems Cybersecurity*, Power Systems,
https://doi.org/10.1007/978-3-031-20360-2_9

1 Introduction

Power networks are among the most important critical infrastructures of any society since the operation of other infrastructures is dependent upon energy supplied by these systems. Recently, there has been a trend toward embedding information and communication technologies (ICTs) in power networks to enhance their performance, meet future needs of energy systems, and improve consumer-centricity. Such a shift toward a more modern, efficient, and accurate system results in the development of smart grids. However, modernization of the grid has also exposed smart grids to a variety of cyber-attacks, in which, adversaries exploit the vulnerabilities of the cyber layer and/or utilize Industrial Electronic Devices (IEDs)—e.g., protective relays—to sabotage the underlying physical layer, initiate instability, or create a blackout. Among different systems in a smart grid, the protection system is considered the most critical cyber-vulnerable, because protection schemes preserve the grid's integrity and help to stabilize it following physical disturbances. Indeed, the U.S. Department of Homeland Security has introduced digital protective relays, i.e., the main components of protection systems, as critical targets for cyber-attacks [1]. Additionally, the U.S. Department of Energy confirmed the consequences of such threats against protection systems by experimentally exploiting the Aurora vulnerability, and launching an attack based on it to destroy the grid's physical components [2]. Therefore, cyber-security of protection schemes in various domains of smart grids is of paramount importance and should be investigated thoroughly.

In addition to being destructive, cyber-attacks against protection systems are generally stealthy, with a long gap from several months to several years between their launch and detection. Evidences from previous high-profile cyber-attacks against energy infrastructure and other associated critical industrial control systems show that during this gap attackers try to gain information about when and where they can create maximum damage [3]. For instance, BlackEnergy 3 malware propagated without notice for six months before switching off 30 substations [4]. The long gap between initiation and detection of cyber-attacks proves the stealthiness of these cyber-attacks against energy infrastructure, which makes them technically challenging to detect.

Protection schemes are often deployed in substations to protect power system components, such as transmission lines. The high integration of instrumentation and interconnectedness in the automation systems of substations, however, make these power grid facilities vulnerable to cyber-attacks. Studies, e.g., [5, 6], show substations' vulnerability to cyber-intrusions. Such vulnerabilities were also proven in the real world when cyber-attacks impaired the Ukrainian power system in 2015 and 2016 [3]. Thus, securing substations against cyber-attacks is the first step toward the cyber-resiliency of protection systems. Although securing these facilities is necessary to prevent cyber-attacks against protection schemes, it is not adequate, since Communication-Assisted Protection (CAP) schemes, e.g., Line Current Differential Relays (LCDRs), are dependent on the data received from devices/relays installed outside of substations. Recently, technical developments in communication technology and measurement synchronization have facilitated the design and deployment of CAP schemes to protect transmission lines more reliably. The reliance of CAP schemes on communicated data, however, makes them even more susceptible to cyber-attacks, since these schemes can be targeted by breaking into their external communication channel as well. Additionally, Wide Area Protection (WAP) system is utilized to complement conventional protection schemes and to avoid their malfunction during different conditions. WAP schemes take protective actions after collecting time-synchronized measurements from various remote locations in the network [7]. The reliance of WAP schemes on communicated data makes them vulnerable to cyber-attacks that can create country-wide blackouts. As a result, protecting WAP schemes against cyber-attacks is crucial for maintaining the integrity of power grids.

2 Substations

Substations are parts of generation, transmission, and distribution systems, in which the voltage level is transformed from one level to another. Substations contain two groups of equipment: (i) power system components, such as switching devices, transformers, reactors, capacitors, and voltage regulators, which are used to operate bulk power systems and deliver energy to stakeholders, and (ii) equipment used for monitoring, control, and protection of the components in the first group and measuring their associated parameters. Protection devices are categorized in the second group. Generally, substations are unattended, relying on an automated system that is remotely supervised and controlled through the supervisory control and data acquisition (SCADA) system. Thus substations are integrated with advanced intelligent electronic devices (IEDs)—such as relays and controllers—with complex functionalities, accurate time synchronizing, and high-speed communication infrastructure, which make them vulnerable to cyber-attacks [8, 9]. On this basis, this section is focused on the vulnerabilities of substations. It first elaborates on the automation system of a substation and discusses the communication protocols and standards used within a substation to investigate the existing cyber-vulnerabilities. Afterwards, it discusses different types of media and topologies used for communicating with other substations and/or control centres, to have a clearer idea about the vulnerabilities that can stem from outside of substations.

2.1 Substation Automation

A substation automation system is a set of hardware and software components that are used to remotely monitor and operate an electrical system. To improve the system's overall efficiency and production, a substation automation system automates some repetitive and error-prone processes.

Traditional substations consist of mechanical relays and meters, which barely supported recording and had no methods of communication. Fault recorders captured data mostly in the form of paper charts, making reading and analyzing of the data difficult. Thus, repair or troubleshooting in traditional substations was costly and time-consuming.

With the advancement of microprocessor technology, digital protection and control devices became more sophisticated. New IEDs may receive and record data on a variety of system parameters, process it in a fraction of a second using advanced processors, and make judgments in abnormal conditions [10]. Additionally, modern substation devices may store information in their internal storage for a period of time and then send it to third-party software packages for additional analysis. IEDs may now communicate with a local or distant user via a variety of communication media, as will be discussed in Sect. 3. This gives operators more freedom in terms of how and when they process data. Additionally, to make the operation of substations easier, the SCADA system was developed to collect data from a variety of IEDs in a substation using various communication methods, and control and monitor them using different visualization technologies. Following this modernization, implementation of higher-level protocols that simplify configuration, commissioning, and testing is necessary.

2.1.1 IEC 61850 Standard

IEC 61850, which is a standard that provides security guidelines for substation communication protocols, is quickly becoming the most promising and widely-used standard for substation communication. IEC 61850 describes the data models, communication protocols, and services. Figure 1 describes the three layers of a substation communication system as categorized by IEC 61850, the substation layer, the bay layer, and the process layer. These layers are associated with the three main functions of monitoring, remote operation, and control and protection, which are necessary for the substation automation system. The process level consists of different analog equipment, such as transformers, Current Transformers (CT), Voltage Transformers (VT), Circuit-Breakers (CB), switch-gears, and surge arresters. The process layer uses Merging Units (MU) to perform Analog-to-Digital (A/D) conversion and transfer the resulting information to the bay layer, which employs protection and control embedded devices, such as Digital Fault Recorders (DFRs) and IEDs [9]. These devices control the process level devices. The station level collects information from the bay level equipment for real-time monitoring and analysis. The upcoming interconnection between the substations and control centres is performed from the station level.

IEC 61850, which comprehensively describes the standard communication protocols for substation automation systems, offers numerous advantages, as follows:

- Flexibility: the standard considers different performance requirements in order to support different services;

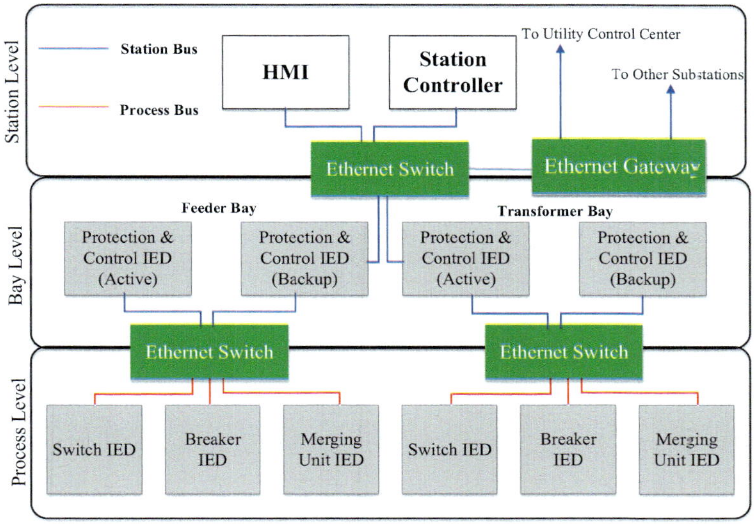

Fig. 1 Substation architecture in three levels: process, bay, and station level

- Configurability: a standard language is defined for substation configuration in order to designate different services offered by all the equipment;
- Interoperability: equipment made by various vendors are able to provide their services; which results in improvement of configuration, time, and reliability issues.

Communication infrastructure plays a key role in the protection applications. Optical fiber Ethernet communication offers a platform for protection devices to exchange information, such as voltage and current Sampled Values (SVs), as well as Generic Object Oriented Substation Event (GOOSE) messages, which include, information such as status, tripping, and blocking. On the other hand, wide area communication system has facilitated the exchange of information between faraway locations. IEC 61850 defines five different types of communication services which are shown in Fig. 2. As can be seen in this figure, massages can be divided into the following two main categories:

1. messages with strict time requirements are considered as real-time traffic, which is exchanged in the process and bay levels;
2. messages with low or no time requirements, which are transferred at the station level.

The popularity of IEC 61850 may be linked to two primary factors: the simplicity of connecting through Ethernet rather than older hardwired systems, and the defined message forms that assure compatibility. SVs and GOOSE messages are two kinds of defined real-time messages in the IEC 61850 standard, which are shared through Ethernet, using Local Area Networks (LANs) [7]. The communication model of the IEC 61850 standard is shown in Fig. 2.

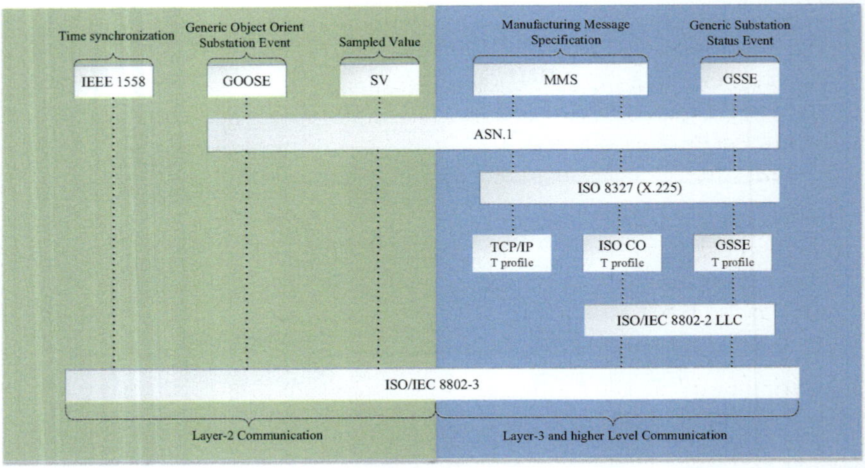

Fig. 2 IEC 61850 communication models

2.2 Vulnerabilities of Substation Automation Systems

From the security point of view, power plants and substations are among the most important and critical facilities in power networks [11–15]. Therefore, the more these facilities are automated and digitized, the more the quantity of potential threats and cyber-attacks will be. Communication protocols utilized in substations and their path of digitization and standardization have been defined in IEC 61850. However, at the same time, this standard suffers from some vulnerabilities that expose substations to cyber-attacks, since it does not focus on the security and regulation of modernized power systems. Later, IEC 62351 was developed to improve the security and reliability of automation systems in substations. In the literature, there are several attack models against IEC 61850 standard that take advantage of IEDs for malicious purposes [16, 17]. Some of these models are specific to substation automation systems, whereas other types are renowned threats already found in Ethernet-based networks. These attacks include, but are not limited to,

- Denial of Service Attack (DoSA): IEDs are virtually disconnected from the network, so they cannot send or receive Ethernet frames;
- Spoofing: IEDs can be physically or virtually spoofed to send erroneous commands or status events to other devices in the network;
- Unauthorized access: IEDs are accessed by an attacker to modify their configuration or settings, or to obtain sensitive data that can compromise components or the entire substation;
- Eavesdropping: critical data are intercepted to obtain information about devices or the substation;

- Man-In-The-Middle (MITM) attack: communications between IEDs and monitoring systems are intercepted and modified by an attacker to pretend that the attacker is one of the end nodes of the communication;
- Replay attack: it is a more particular type of MITM attack, in which an attacker fakes an event by intercepting the legitimate data and sending data recorded during that event.

As demonstrated in the literature, cyber-attacks against substations can considerably impact the operation of substations and/or the entire power grid. It is shown that for instance in [18, 19], cyber-attacks against substations and protective relays can result, in the worst case, in instability or blackout. The most common impacts of cyber-attacks against substations are as follows [20]:

- Interfering of monitoring system: IEDs are prevented from receiving data from MUs or other IEDs, or data is manipulated to mislead the monitoring system;
- Undesirable control and protection actions: protection and control equipment receive wrongful signals, or they are prevented from receiving the information, which results in an undesirable control and protection action;
- Network interruption: substation components become unable to exchange data with each other, which endangers the integrity of the system and the supply of energy;
- Synchronization tampering: control centre receives false time-stamped data, which may result in wrong control or protection decisions.

There are two primary categories of studies focused on the cyber-security of substations, both of which are in their early stages. The first category concentrates on attack modelling, risk assessment, and optimal techniques to allocate security mechanisms in substations [21–23]. The second category, on the other hand, focuses on improving substations from cyber security point of view [6, 23, 23–25]. In [6], host- and network-based anomaly detection systems are presented to detect simulated attacks in substations. In [24], a temporal-event-construction-based anomaly detection method has been developed to detect attacks in substations. The authors of [25] present a mathematical model to distinguish between malicious incidents of alter-and-hide attacks and disturbances. A graph colouring approach, which employs game theory, is presented in [23] to obtain the best distribution of security mechanism that reduces substation vulnerabilities. In [26], a novel type of firewall has been implemented to prevent attackers from accessing the substation. However, this type of solution is not adequate for protecting substations against different types of cyber-attacks.

One important point that must be considered when cyber-security improvement techniques are developed for substations is that data transfer speed should not be affected, especially when time requirements have strict limitations. For example, for those applications that have time requirements lower than 4 milliseconds, e.g., GOOSE and SV messages in substations, encryption is not recommended. Table 1 presents time constraints for IEC 61850 process level messages.

Table 1 Time constraints for IEC 61850 process level messages [27]

Application	Message type	Time requirement
Fast messages	GOOSE	$\leq$3 ms
Fast messages	GSSE	$\leq$10 ms
Raw data	SV	$\leq$3 ms
Medium-speed messages	MMS	$\leq$100 ms
Low-speed messages	MMS	$\leq$500 ms
Time synchronization	IEEE 1588	>500 ms
File transfer	MMS	>500 ms

3 CAP Schemes

CAP schemes were developed to deal with transient stability problems, since they can significantly reduce the clearing time of faults compared to conventional protection schemes. Thus, the purpose of CAP schemes, which are also called pilot protection schemes, is to provide high-speed fault clearing from two or more terminals of an entire protected line [28]. This objective is achieved by communicating the data between the terminals through communication media. The following subsections first elaborate on commonly used communication media for power system applications, and then introduce different types of CAP schemes.

3.1 Commonly Used Communication Media in Power Systems

There are various communication media and topologies that can be used in power networks. These media could be mainly classified into wired and wireless. Wired communication media often provide wide bandwidth, and high reliability [29]. Fiber optics and power line carrier (PLC) are the most commonly-used technologies that transfer data through a physical medium [30]. However, this category suffers from some limitations due to their physical connections, such as difficult installation and high costs. Although wired communication is more reliable, wireless communication becomes increasingly more prevalent due to the technological advancements in communication systems. The most commonly-used wireless communication media are cellular, microwave, and satellite. Selecting an appropriate communication medium for an application depends on many factors, such as cost, the required reliability, the number of endpoints and channels, and the communication distance. A comparison between wired and wireless communication is provided in Table 2. Additionally, Table 3 categorizes different communication media used for synchrophasor data transfer or SCADA systems. In the following, the communication media described

Table 2 Comparison between wired and wireless communication media

	Mobility	Cost	Expansion	Remote access capacity	Cyber vulnerability
Wired	Limited by the length of cable	High	Difficult	Weak	High
Wireless	Not limited to physical location	Low (easy deployment)	Easy	Strong	Low (depending on deployed security mechanism)

Table 3 Communication methods for synchrophasor and SCADA data transfer

Wired communication	Power Line Communication (PLC)
	Leased line
	Optical fiber communication
Wireless communication	Cellular communication
	Microwave-based communication
	Radio-based communication
	Satellite communication

in Table 3 are discussed in more detail, and their advantages and disadvantages are explained.

1. PLC: Transmission Power Line Communication (PLC) exchanges data through a physical conductor used for transfer of electric power [31], as shown in Fig. 3 [30]. As a result, there is no need to install new cables to transfer the data. Such a characteristic makes the PLC technology the easiest and most economical way to communicate data. However, PLC suffers from various limitations, such as (i) low bandwidth, (ii) high noise pollution, and (iii) loss of communication during faults and physical disturbances, which might affect the performance of protection relays. Additionally, PLC suffers from cyber-security issues arising from the nature of power lines [32]. Power cables are not twisted and use no shielding, thus producing a fair amount of Electromagnetic Interference (EMI), which can be received via radio receivers easily. As a result, PLC communication requires a strong encryption technique to prevent the eavesdropping of critical data.
2. Leased lines: A leased line is a telecommunication circuit between two or more nodes used to ensure continuous data flow from one to another point of a communication system. This communication medium is similar to PLC, while the main difference is that it is installed exclusively for transferring data between the pre-defined locations. Leased lines are typically used for SCADA communications. Despite the effectiveness of this communication method, its cost is high for long distances and it is not the best solution for multi-channel appli-

cations. Eavesdropping on leased lines can be done by intercepting or tapping into communication signals. Additionally, this communication medium mainly suffers from DoSAs rather than interception due to the highly specialized and often proprietary data that they carry [33].

3. Fiber optic: Fiber optic is a flexible fiber made of glass, which transmits information as light pulses along the glass fiber. This medium utilizes visible light or infra-red rays to transmit the data between two nodes, while providing high bandwidth, especially for long distances. The typical structure of fiber optic communication technology is shown in Fig. 4. Compared to the PLC, a fiber optic cable has a higher bandwidth, lower attenuation, better reliability, and negligible interference [30]. However, due to their physical constraints, fiber optic cables suffer from the following disadvantages: their installation and maintenance costs are high, they are difficult to be spliced or curved, and they are highly susceptible and can be easily damaged. From the security point of view, fiber is the most secure medium available for carrying sensitive data, since it is very difficult to tap fiber without being detected, and it does not radiate electromagnetic energy to intercept.
4. Cellular: Cellular communication is the most popular wireless communication since its infrastructure is widely implemented all around the world and it is one of the most economical ones among other methods. The typical configuration of cellular communication is shown in Fig. 5. It contains several cellular stations, Mobile Switching Centres (MSCs), which are connected to the cellular network [34]. The most common vulnerability of this communication medium is its shared nature. Because of its non-deterministic latencies and limited service dependability, cellular communication is not reliable for most protection applications, which require a permanent and reliable high-speed connection between the sender and receiver. Furthermore, cellular communication is vulnerable to EMI and noise injection, which overshadows its reliability. All wireless communication media, including cellular, pose serious cyber-security issues since the signals can be captured by nearby devices. Therefore, eavesdropping is a very probable attack if efficient authentication and encryption mechanisms are not in place.
5. Microwave: A microwave communication channel consists of a line-of-sight wireless communication technology that provides high-speed connections to transfer the data. The process of microwave-based communication is similar to fiber-optic communication, except it is wireless. The typical structure of microwave communication technology is depicted in Fig. 6. Microwave is cost-effective in terms of installation, and it provides very wide bandwidth, as well as high capacity and data rate [30]. However, it suffers from a few disadvantages, including high maintenance cost and requiring a clear line of sight with multi-path propagation. Microwave communication is vulnerable to interception, due to free-space line of sight wave propagation. Thus, the signals can be eavesdropped or interfered with using an antenna and a decoder, making this method of communication exposed to cyber-attacks and somehow unreliable.

This issue is of greater importance for protection applications since they require high reliability in exchanging data.

6. Radio system: A radio system, as can be seen in Fig. 7, is a technology for communicating signals through radio waves, which are electromagnetic waves of frequency between 30 hertz and 300 gigahertz [30]. They can be produced by a radio transmitter connected to an antenna that generates the waves, and can be received by another antenna connected to a radio receiver. Radio is useful for communicating with inaccessible areas. Nevertheless, this method of communication suffers from several disadvantages, such as having a high potential for noise contamination and possible interference with adjacent channels, EMI that can interrupt data transfer, and being prone to spoofing, jamming, and eavesdropping attacks. Due to these drawbacks, radio systems are somehow unsuitable for protection applications that require high reliability.
7. Satellite: Satellite communication is a technique in which radio telecommunication signals are transmitted and strengthened through a transponder. Thus, a communication channel is created between a source transmitter and a receiver at different locations on Earth. Compared to other communication methods, this medium provides uninterrupted communication, since its equipment is in space [30]. The typical structure of the satellite technology is illustrated in Fig. 8. Satellite communication depends on the location of the receiver antenna and has less control over data transmissions. In addition to its high cost, the rate of delay in sending and receiving information is high. Satellite communication can be negatively affected by different weather and environmental conditions, such as rain, storm, solar flares, and cosmic radiation in space. These drawbacks make satellite communication unreliable for protection applications that need a high-speed connection with minimal delay. From cyber-security point of view, satellite communication is susceptible to jamming and spoofing [35]. Jamming can take place by disrupting the radio frequency signal with electronic attacks. Spoofing can occur by sending fake signals to the receiver and pretending to have valid information.
8. Spread spectrum radio: The spread-spectrum technique is a communication method in which a signal generated with a specific frequency is spread in the frequency domain, and as a result, the signal gains a wider bandwidth. The block diagram of a typical Spread spectrum radio communication system is depicted in Fig. 9. As can be seen in this figure, the actual signal data is integrated with a sequence of bits (modulation) that spreads the signal over a wide bandwidth that is typically stretched from 20 to 30 MHz. Then, the receiver demodulates the signal using the same frequency spreading generator to recreate the original signal [36]. The advantages of this communication technique include its higher security against eavesdropping, and its higher resistance to natural interference and noise since this method employs a reliable encoding/decoding approach [37]. Similar to other wireless communication techniques, spread spectrum radio is prone to EMI, radio signal interference, and jamming.

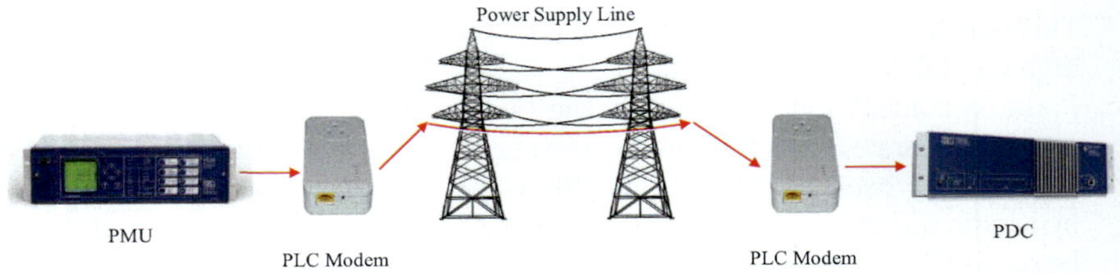

Fig. 3 PLC structure in power systems

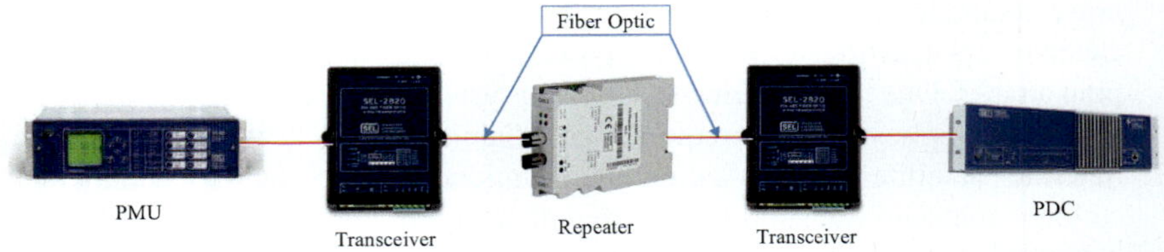

Fig. 4 Structure of fiber optic communication for transmitting data from a Phasor Measurement Units (PMUs) to a Phasor Data Concentrator (PDC)

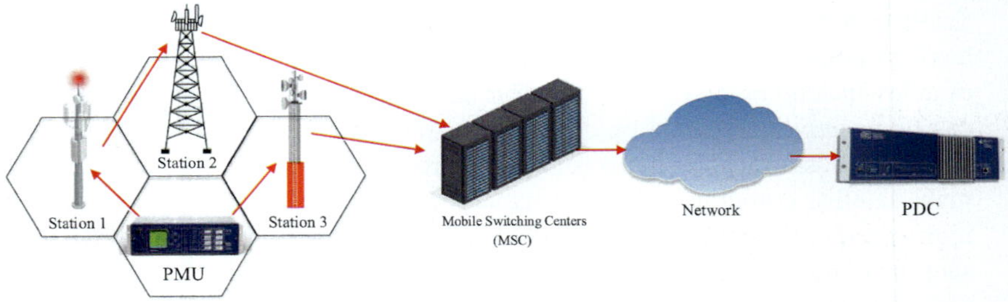

Fig. 5 The typical structure of cellular communication

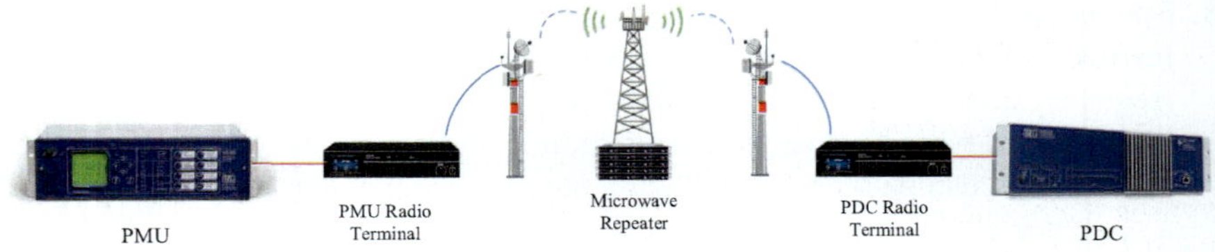

Fig. 6 The structure of microwave communication technology

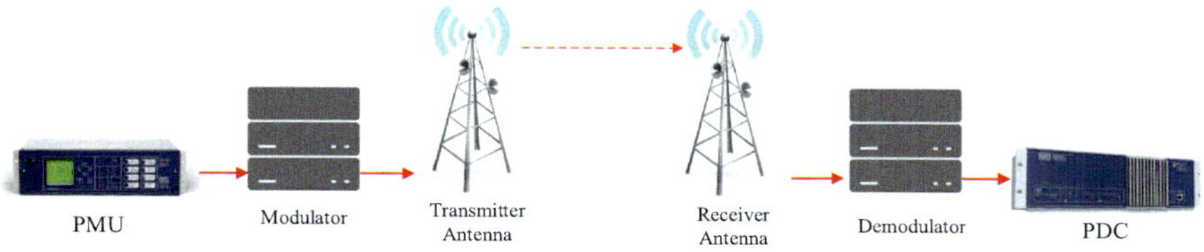

Fig. 7 The structure of radio communication system

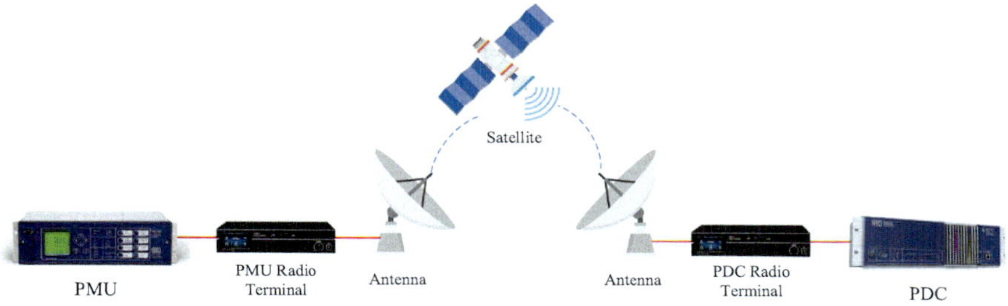

Fig. 8 The structure of satellite communication technology

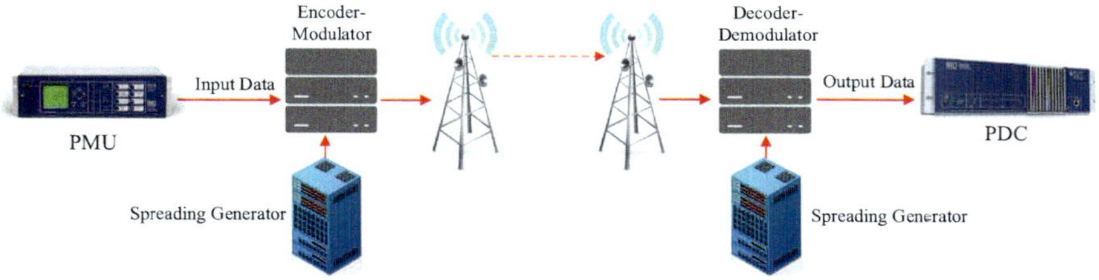

Fig. 9 The structure of spread spectrum radio communication

3.2 Types of CAP Schemes

CAP schemes are often divided into three groups: directional comparison, differential, and phase comparison schemes. Directional comparison schemes—which are used for phase and ground distance elements, as well as for zero- and negative-sequence directional overcurrent elements—include Permissive Over-reaching Transfer Trip (POTT), Permissive Under-reaching Transfer Trip (PUTT), Direct Transfer Trip (DTT), Direct Under-reaching Transfer Trip (DUTT), Directional Comparison Blocking (DCB), and Directional Comparison Unblocking (DCUB) [38]. In these schemes, permissive means that the permission to trip is shared with other relays, and blocking means that other relays are prevented from tripping (block). On the other hand, the differential scheme, which is used by Line Current Differential Relays (LCDRs), detects faults by comparing the currents measured at all line terminals

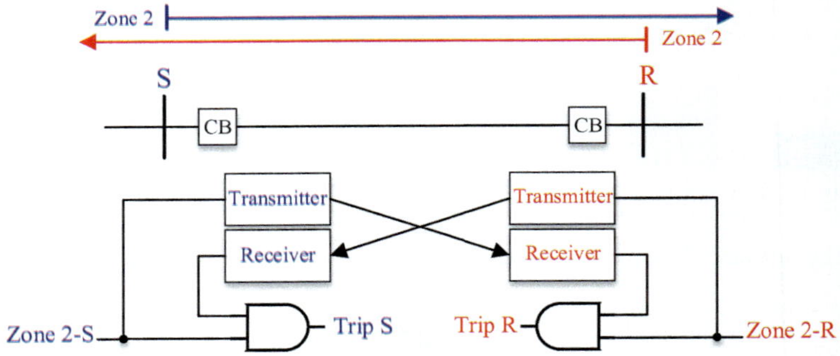

Fig. 10 Simplified logic and single line diagram of POTT scheme

[19]. Phase comparison scheme, which is an extension of differential scheme, monitors the phase angle of the current at each terminal of a line and transmits data to other terminals via a communication channel [39]. Thanks to these CAP schemes, faults on the 100% of a line can be cleared in a timely manner, so improving the stability of the system. The following subsections elaborate on the above-mentioned CAP schemes.

3.2.1 Permissive Over-Reaching Transfer Trip (POTT)

The POTT scheme, depicted in Fig. 10, uses an overreaching Zone 2 element to trip the local circuit breaker and send a permissive trip signal to the remote end. If the remote Zone 2 element finds a fault, the remote relay trips the breaker whenever it gets the permissive signal [40]. The POTT scheme is suitable to be used with all digital teleprotection equipment applied over direct and multiplexed fiber-optic and radio systems. This scheme is inherently tolerant of propagation delays and channel asymmetry. POTT communications schemes need to provide high security and high speed (4–8 ms is typical).

3.2.2 Permissive Under-Reaching Transfer Trip (PUTT)

In the PUTT scheme, when a relay detects a fault in its zone 1, it trips the line and sends a signal to the remote relay via a communication channel to let it know that the fault has happened on the line [40]. As Fig. 11 shows, if the remote relay also detects a fault in its zone 2, it commands the circuit breaker to trip the line.

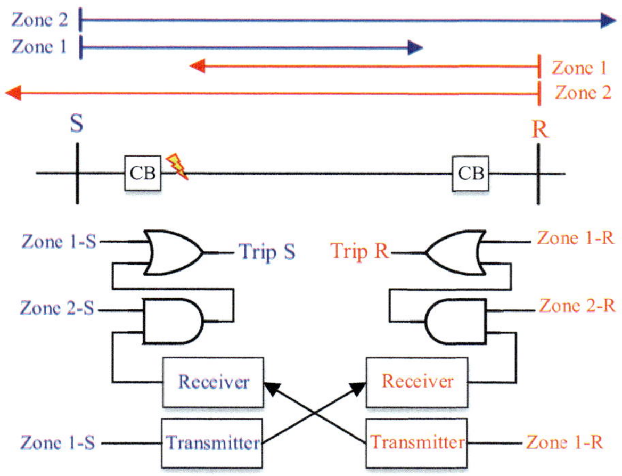

Fig. 11 Simplified logic and single line diagram of PUTT scheme

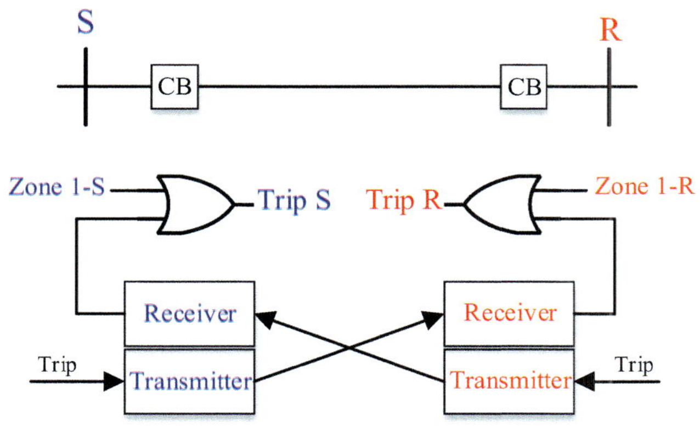

Fig. 12 Simplified logic and single line diagram of DTT scheme

3.2.3 Direct Transfer Trip (DTT)

DDT is a protection method in which a trip signal is sent from one side of the line to the remote side, and the circuit breaker of the remote side is tripped. A similar mechanism is implemented on the other side of the line as well [40]. The logical and single-line diagram of DTT method is shown in Fig. 12.

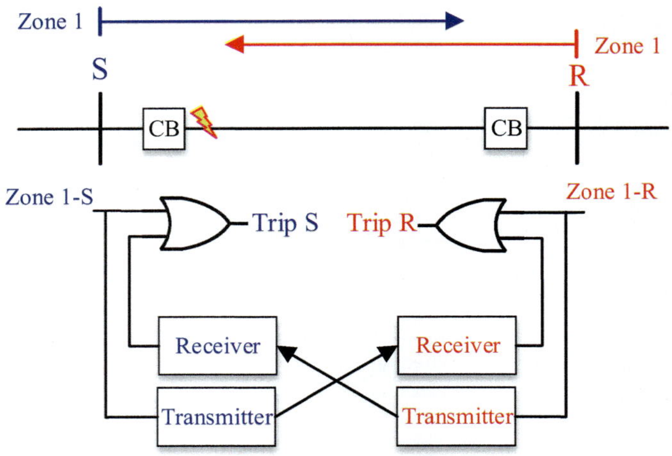

Fig. 13 Simplified logic and single line diagram of DUTT scheme

3.2.4 Direct Under-Reaching Transfer Trip (DUTT)

In the DUTT scheme, when zone 1 of a relay detects a fault, it trips the line and sends a trip signal to the remote circuit breaker as well. Without using this scheme, it takes about 20 cycles for the remote relay to trip the breaker for faults occurring in zone 2, but implementing the DUTT decreases this time to a few cycles (for communication delays) [40]. The simplified logic and single line diagram of DUTT scheme are depicted in Fig. 13.

3.2.5 Directional Comparison Blocking (DCB)

Unlike POTT and DCUB schemes, which send a trip signal when they detect a fault in the forward direction, DCB scheme sends a blocking signal when it detects a fault in the opposite direction [40]. The DCB scheme uses an instantaneous reverse-looking element for Zone 3 (as shown in Fig. 14) to send the blocking signal to the remote relay.

3.2.6 Directional Comparison Unblocking (DCUB)

The DCUB scheme works in the same way as the POTT scheme does, but with one difference: it uses an additional logic to trip the line for a short period of time (typically 150 ms) during a communications failure caused by noise or any other disturbance. To this aim, the DCUB scheme uses an additional signal called "unblock", as shown in Fig. 15 [40].

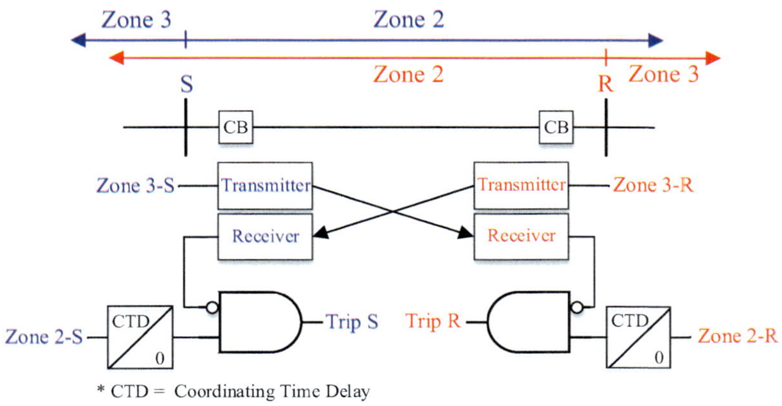

Fig. 14 The simplified logic and single line diagram of DCB scheme

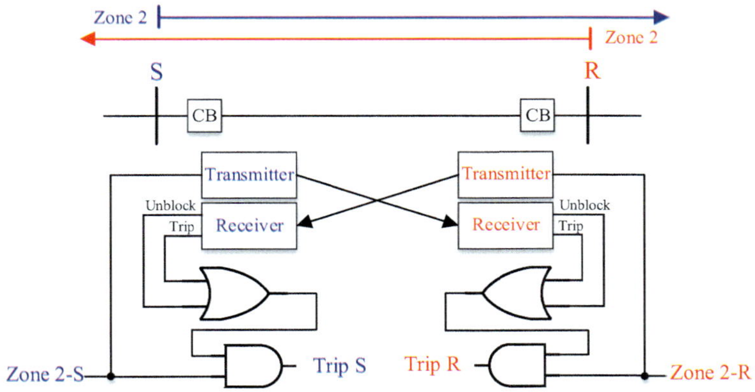

Fig. 15 The simplified logic and single line diagram of the DCUB scheme

3.3 Line Current Differential Relay (LCDR)

LCDRs are protective devices that protect transmission lines by comparing the difference between the in-going and out-going currents of a line with a predefined threshold. Ideally, the difference has to be zero in a normal situation. In practice, however, there might be a small differential current, since there are sources of error, such as measurement errors, CT ratio mismatch, or the charging currents of lines [41]. To protect a line, an LCDR shares its current measurements with other terminals using communication media. LCDRs can be categorized into two main types: Phasor-based LCDR, and SV-based LCDRs. Phasor-based LCDRs transfer current phasors. On the contrary, SV-based LCDRs, exchange the SVs of currents. SV-based LCDRs have the following advantages over phasor-based ones:

- they transmit less data (approximately 50%), since they communicate instantaneous values rather than phase angle and magnitude of phasors,
- they send information with a higher rate compared to phasor-based LCDRs, which transfer data every 4 ms [42],
- they do not rely on the Global Positioning System (GPS), the SV-based relays can be synchronized by measuring the communication delay and then interpolating the transmitted information, and
- they can extract more information from the communicated SVs, such as harmonics or rate of change of currents, thus enabling LCDRs to function better [41].

3.4 Phase Comparison Scheme

Phase comparison scheme monitors the current direction at one terminal of a protected line and sends this information to the other terminal using a communication channel. There is no fault in the protection zone if the two currents are almost in phase. In other words, a fault has happened if these two currents are almost 180° out of phase. If current flows into the line from both terminals, each terminal compares the local and distant current directions and trips the line if they are out of phase. The communication channel is generally on/off, sending only when the overcurrent detector's thresholds are exceeded. Because this scheme functions exclusively based on the line current, it is not affected by power fluctuations or out-of-phase circumstances in the phase comparison. In addition, no potential transformers are required. The phase comparison approach also has the benefit of not requiring a voltage to create directionality or distance measurements.

3.5 Vulnerabilities of CAP Schemes to Cyber-Attacks

As explained in Sect. 3.2, one of the most important requirements of CAP schemes is a reliable and high-speed communication medium to share the information with remote relays. As a result, these schemes are vulnerable to any cyber-attack that compromises the integrity of the communication system for CAP schemes. Such cyber-attacks can disrupt the communication between relays, affect the bandwidth and data transfer rate, or inject false information instead of the authentic data, so interrupting the normal operation of CAP schemes or misleading them into making wrong decisions.

Additionally, some of the CAP schemes, such as LCDRs, require time synchronization. Thus, the performance of these schemes depends highly on the integrity of the synchronization mechanism as well [39]. An attacker can exploit this vulnerability by spoofing the GPS signal and manipulating the time-tag of measurements [43]. It has been proven in the literature that manipulating the time tags of measurements is equivalent to changing the measurements' phase angle [44].

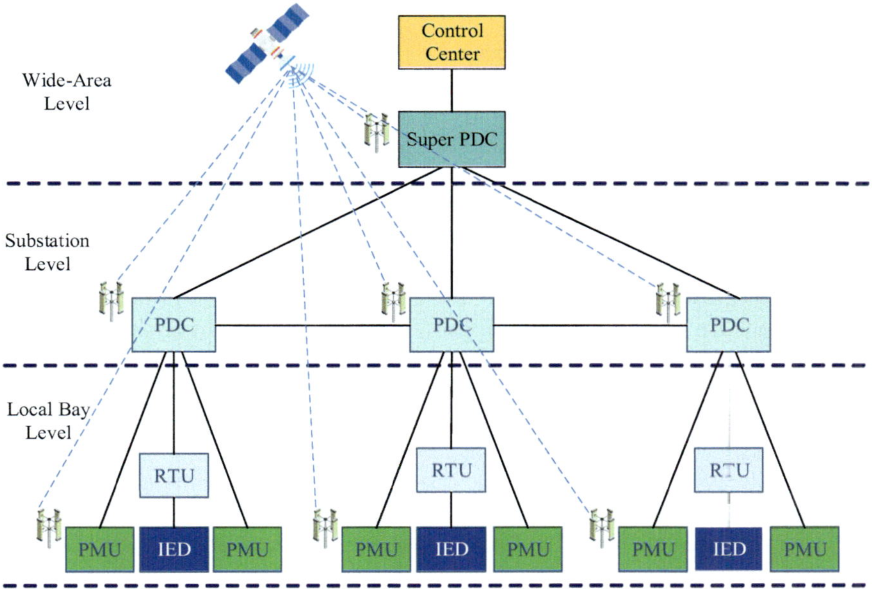

Fig. 16 Architecture of WAP system

4 WAP Schemes

WAP schemes use system-level information gathered over a wide geographical area to take a series of rapid actions and make important decisions to protect the system from the spread of disturbances [45]. In addition, these schemes are utilized to complement conventional protection schemes and to avoid their malfunction during different conditions. With the advancement of communication technologies and time synchronization mechanisms, control centres can gather synchronized real-time measurements—such as system voltages, angles, and active and reactive power flows—captured by IEDs and PMUs installed at various locations in the network. WAP schemes use wide-area measurements for system monitoring and state estimation, event recording, analysis of system/load characteristics, controlling and protecting the grid [46, 47]. As shown in Fig. 16, the WAP system is comprised of three different levels: local bay, substation, and wide-area. More information about these levels and the general architecture of WAP system can be found in [48]. WAP schemes include remedial action [49], fault localization [50], backup protection [51], emergency protection [52], and adaptive protection [53].

- Remedial action schemes: These schemes detect unusual conditions in power systems, and react to them with quick, corrective, and automatic measures, such as load/generator shedding, generation rescheduling, and line tripping. In fact, remedial action schemes reduce the implications of disturbances, prevent cascading failures, and ensure power system reliability. These schemes include out-of-step relaying, under-frequency and under-voltage load shedding, direct load and generation tripping, and controlling reactive power components [54].

- Fault localization scheme: Locating a fault on transmission lines is often a difficult task due to their extensive and complex construction. It is critical to accurately locate faults on lines since this accelerates the repair process and reduces Energy Not Served (ENS) costs. Fault localization scheme uses wide-area measurements to accurately locate faults in power networks.
- Backup protection scheme: In a conventional power system, relays' settings and locations are coordinated such that each one protects a certain protection zone. These zones are intentionally overlapped, with certain backup relays to guarantee that all faults are detected. In practice, however, such a local backup protection plan may fail under-stressed operation circumstances. This issue can be addressed by using a wide-area backup protection scheme, which leverages PMU data and communication between relays and the control centre.
- Emergency protection scheme: Some disturbances, such as faults and overloading, heavily affect the stability and operation of power systems. Such disturbances, in the worst case, can result in cascading outages or system-wide blackouts. The emergency protection scheme can predict such events and take precautionary measures to reduce their impacts.
- Adaptive protection scheme: In practice, the settings of protective relays may not be optimal for all situations, and these settings must be modified for different operating conditions and system topologies. Adaptive protection takes care of this situation by automatically adjusting the settings of relays based on the information received from PMUs and IEDs [55]. Among the above-mentioned WAP schemes, adaptive protection is one of the most critical ones from the cyber-security point of view, since it directly affects the settings of protective relays. Given that the settings of relays are critical for proper operation of protection systems, any cyber-attack against this scheme can compromise the stability and reliable operation of power systems.

4.1 Vulnerabilities of WAP Schemes

This section describes the potential vulnerabilities of different components of the WAP system. These vulnerabilities stem from PMUs, PDCs, gateways, communication links, and historian databases.

- Vulnerabilities of PMUs: The correctness of data received from a PMU depends on the accuracy of time synchronization mechanism, communication link between the PMU and PDC, and security of its firmware. The GPS signal, which is used for time synchronization, lacks authentication and encryption mechanisms [56], and it can be jammed or spoofed [57]. A PMU's data can be also lost or modified if the communication link between the PMU and PDC is compromised. It's worth noting that traditional methods, such as basic cryptographic techniques and hash functions, may be used to secure the integrity and confidentiality of PMU data. However, these techniques are ineffective if the attacker accesses the PMU firmware. In such cases, for instance, FDIAs be launched to mislead WAP schemes [58].

- Vulnerabilities of PDCs: Since the communication links of PDCs are not encrypted, they are subject to a variety of cyber-attacks. If a PDC is compromised or misconfigured, the authentic information can be replaced by incomplete or wrongful data, so the accuracy and validity of WAP schemes will be negatively affected. Furthermore, the functions that PDCs use to time-align data are sensitive to delays. Thus, an adversary can manipulate the timestamps of a PMU's measurements to drop the measurements of some other uncompromised PMUs. More details about the vulnerabilities of PDCs can be found in [59, 60].
- Vulnerabilities of communication links: WAP system can utilize the communication media discussed in Sect. 3.1. These communication channels can be jammed or slowed down by targeting network components such as routers or by performing a flooding attack on the network. Other kinds of attacks against communication networks include Main-in-the-Middle (MITM) attacks, False Data Injection Attacks (FDIAs), eavesdropping, and delay attacks. These attacks are explained in more detail in Sect. 5.
- Vulnerabilities of historian Database: If network interfaces allow for unauthorized access, an attacker can alter or remove measurements in the wide-area historian data. As a result, critical data for monitoring and controlling applications would be modified or lost. Additionally, such data can be used for replay attacks.

5 Taxonomy of Cyber-Attacks Against Protection Systems

Generally speaking, cyber-attacks can have a detrimental impact on at least one of three major aspects of the system security, namely, Confidentiality, Integrity, and/or Availability, often referred to as CIA triad [61]. These three aspects are discussed in this section along with attacks that target each one.

- Confidentiality indicates that only a certain number of authorized persons should have access to information required for the operation of the system. In other words, the system should be equipped with certain tools that prevent unauthorized access to important and effective information about the system. This security aspect is crucial for protection applications since they continuously receive real-time measurements, which can be used by adversaries to launch attacks.
- Integrity of a system means that its data should not be manipulated by unauthorized persons or devices, or by authorized users outside of their scope of work. If the integrity of a system is compromised, the system's operation may be severely affected, and might deviate from normal. For instance, FDIAs and malicious firmware interference are among the most well-known integrity attacks. System

integrity is especially important for protection purposes since these applications can immediately and severely affect the topology and operation of the power grid.
- Availability of a system means that its resources should be always available to authorized users. If an attacker compromises the availability of a system, he/she can prevent legitimate users from accessing the system and/or its vital information; as a result, the system may become temporarily or permanently unreachable. DoSA is one of the most important threats to the availability of a system. This attack can be launched either through the communication system, for instance by sending a huge amount of packets to a system to prevent it from processing the actual data, or physically. Threats against the availability of protection systems are of paramount importance since they result in the inaccessibility of protection devices, and consequently in sustaining faults or disconnecting a large portion of the grid.

5.1 Attacks Against the Protection System

This subsection introduces the important attacks that can target the protection system of a power grid.

5.1.1 Physical Damage or Device Capture

Unattended protection IEDs can be at risk of physical damages, e.g., sabotaging communication links, firmware manipulation, tampering attacks, or information extraction by using available communication ports [62]. Besides the physical damage, an attacker can create unwanted operation of protection IEDs, e.g., by changing their setting, establishing a physical connection to the device aiming at inserting malicious codes or manipulating firmware through I/O ports of the relay. In such a case, for instance, the inserted malicious code can be malware or downgrading of the firmware to a previous version with known vulnerabilities. Ultimately, the attacker can use the malware or the known vulnerabilities to take the control of protection device and cause their malfunction. Moreover, with the physical access to the device, the adversary can gather various information about the network, e.g., connections, relay features, and measured data, and use this information for crafting future attacks. In addition, since several protection schemes require constant communications with IEDs, an adversary can penetrate those communication links, capture the control of a device, and present it as a legitimate node in the system. It should be noted that most of the protection devices installed in power plants and substations are well-guarded; thus, this kind of attack is less likely to occur.

5.1.2 Denial of Service Attacks (DoSAs)

DoSAs can happen physically, or through the software or network. In the case of software damage, an attack targets protective relays through the vulnerabilities of its software, whereas if the communication becomes available, the attack has targeted the network. DoSAs often occur by (i) sending too many requests over the service host that cause the buffer to overfill the relay connections; (ii) physical deletion of the device; and (iii) the interruption of communication between the protection devices when the data exchange is necessary. The protection devices which are not up-to-date are more possible to be compromised due to the vulnerabilities identified in the previous version of their firmware. As a result, the chance of being infected by viruses, which are designed to send floods of requests and congest the communication bandwidth, is high. A harmful update can potentially lead to the same results [63].

5.1.3 Node Jamming Attack

Node jamming attacks occur when an adversary obscures the communication links by interfering signals, e.g., radio frequency signals, or when the attackers reduce the signal-to-interference-plus-noise ratio (SINR). This specific form of attack prevents the information from being transmitted to relays; as a result, makes them unreachable or uncontrollable. Moreover, this kind of attack may result in unavailability of time-sensitive data [64].

5.1.4 False Data Injection Attack (FDIA)

FDIAs occur when the integrity of data is affected by injecting false data. This attack can be launched by sabotaging sensors at the physical layer or by manipulating the information at the network layer. Due to the sensitivity of protection devices to false data, the entire protection scheme and consequently the grid can be affected by even a small injection of false commands or measurements. Therefore, the attackers try to launch such attacks by injecting a minimum amount of false data to damage the grid, while the attack remains stealthy [64]. FDIAs are more likely to occur in WAP schemes, since IEEE 37.188, which is the common protocol in wide area systems, lacks the necessary security mechanisms to verify the authenticity of the data. It is worth mentioning that FDIAs can be launched by inserting malicious software into IEDs such that the transferred data packets, which include false information, appear normal from the perspective of communication system monitoring applications. As a result, detection and prevention of these attacks are often complicated even in the presence of network monitoring and encryption techniques [65].

5.1.5 Eavesdropping

Eavesdropping, which is also called network snooping or sniffing, is a type of attack in which attackers gather critical information from the communication system to increase their information about the system [66]. The leaked information could then be used to target a component in the protection system by cyber-attacks. The data gathering process often occurs in vulnerable networks and aims to access the information exchanged between two communication devices, e.g., between relays and the control centre in the WAP system. This attack is one of the most common attacks when wireless communication media are used.

5.1.6 Man In The Middle (MITM) Attack

MITM attack occurs when the communication route is replaced by a third person or a device so that the adversary takes the control of transmitted data packets. In this attack, exchange of information, such as measurements and trip commands, takes place through the attackers and not the authorized devices. As a result, the adversaries will be able to inject malicious information into the system at any time [67]. MITM attacks are likely to happen when the system has no or weak deployed encryption mechanisms or lacks communication monitoring systems.

5.1.7 Spoofing

Spoofing is a type of attack that occurs when an adversary pretends to be an authorized part of the system, e.g., GPS clock or relays, and sends data to mislead the operator or other devices to mislead them into taking wrong decisions or force them to reveal sensitive information. This happens when the attacker hides her/his identity and impersonates a trusted device. This attack often happens in the network layer and communication network. For instance, in GPS spoofing, an adversary manipulates the timing of the system by impersonating a GPS transmitter [68]. This type of attack can also lead to leakage of information and can be used to design more destructive attacks.

5.1.8 Time Synchronization Attacks (TSAs)

TSA is a group of intrusions that target the time synchronization mechanism. Any other type of attack that targets the authenticity of time synchronization, such as FDIA and spoofing against GPS, is categorized in this group. Given that some protection schemes, such as differential scheme for transmission lines, are highly dependent on time-synchronization, a TSA against these schemes can result in their malfunction. This issue has been proven in Sect. 2.1 of Chap. 10. for LCDRs.

5.1.9 Dictionary/Brute-Force Attacks

In a dictionary attack, adversaries send, for example to a protective relay, a huge number of known credentials in order to guess its passwords [69]. When authentication measures are poor, an attack is viable, and it gets simpler when factory-set credentials are still in use and not updated [70]. As a result, failing to update users' credentials [71] and implementing inadequate privacy standards [72] might allow an adversary to achieve high-level access and manage the system following a brute force attack. Furthermore, a dictionary attack is successful when log-in attempts and the user credentials of connected persons or devices to a protection IED are not logged, or when the same credential is used by multiple devices in a protection system.

5.1.10 Reverse Engineering Attacks

By reverse engineering an IED's source code and firmware, adversaries can obtain sensitive information about it or the entire system. Attackers can use this technique to find sensitive information left by software programmers, e.g., hard-coded credentials, faults, or system data, and use it to launch attacks in the future. The information obtained by this attack can also be used to create and deploy malicious software [73].

5.1.11 Unauthorized Access

This attack frequently results in concerns with privacy and confidentiality [71]. To acquire access over network devices in protection systems, implant malicious code in IEDs, or eavesdrop on information transferred over the network, attackers may leverage open ports, user elevated rights, or default passwords. Furthermore, insecure network security protocols might expose private system meta-data. Adversaries can leverage network data's location and temporal information to assess users' states [70].

6 Cyber-Attack Detection and Prevention for Power System Protection Applications

This section elaborates on cyber-attack prevention and detection methods for protection applications in power grids.

6.1 Preventing Cyber-Attacks

Preventing cyber-attacks requires defending against a variety of vulnerabilities inside the cyber and physical layers of protection schemes. For each specific attack, its corresponding countermeasures must be deployed to deter the attacker from exploiting vulnerabilities or weaknesses that should be exploited for launching that attack. On this basis, this section discusses different techniques that can be used to prevent cyber-attacks in protection systems.

6.1.1 Edge Protection

As previously discussed, a major family of threats against protection systems targets edge devices, such as relays, IEDs, or MUs, through firmware manipulation and malware, to name a few. Therefore, there is always a possibility for an attacker to take advantage of vulnerabilities inside the edge devices/nodes, take their control, and affect the physical grid [74]. The security of these edge devices, however, can be strengthened using several techniques such as Trojan activation, circuit modification, securing firmware, and cryptographic strategies.

- Trojan activation: Trojans are malware that hide in a protection device aiming at creating a malfunction. Trojans, often change heat distribution inside the relay, for instance, while launching DoS or brute force attacks [75]. To detect this malware, a Trojan activation approach compares the heat distribution of the relay with the patterns during its normal operation [76].
- Circuit modification: Having a mechanism to modify the circuit of components in the system following detection of cyber-attacks is a prevention technique used to stop the adversary from launching the next stage of an attack. These modifications include installing sleep/kill or self-destruct mechanisms, temporarily deactivating the component [77] or delaying its operation [78] until the security experts investigate the problem. For instance, when a backup relay is compromised by illegitimate access or tampered with, the sleep/kill or self-destruct mechanism can disable the device, so preventing the attack from taking the control of the relay, and stopping the attacker from tripping the components. It is clear that depending on the target components and schemes, some or all of these circuit modification methods might not be feasible.
- Secure firmware updates: One way to keep a system safe is to securely update its firmware into a new version in order to prevent attacks. In this process, an authenticated server informs the devices that a new update is available. Then, each node with the new firmware announces the update confirmation to the neighbour devices. If a device has out of date firmware, it will request the update from the legitimate source [79].
- Cryptographic strategies: Lightweight cryptography techniques and hash-based methods are effective solutions to ensure the security of the protection schemes. The hash-based methods can operate in a locked or unlocked state, where in the

former the device operates only when it receives the correct hash. In the latter state, however, the operation is not dependent on the hash [80]. Additionally, the lightweight cryptography techniques can be used to keep the data secure from intrusions, and at the same time, respect the requirement of low computation, which is an important criterion for real-time applications.

6.1.2 Securing Communication Media

As discussed earlier, secure and resilient communication is a key part of safe information exchange in protection applications. The most important communication securing techniques include cryptographic strategies and role-based authorization [81].

- Cryptographic strategies: These techniques are crucial for maintaining the security of information exchange in protection applications, to ensure that the data is not being leaked or eavesdropped on by an adversary [82]. It should be noted that not all the signals in a protection scheme can be encrypted due to the limited computation capacity of devices and the requirements of fast-rate communications for timely fault identification in power grids.
- Role-based authorization: A role-based authorization mechanism can be used to verify requests and messages sent from various sources. In such a case, only authorized sources can send/receive information, and consequently, the attacker's access remains prohibited [83].

6.2 Intrusion Detection Systems (IDSs) for Protection Applications

IDS can be used to secure power system protection applications by identifying malicious activities and distinguishing them from attack-free operations. These systems often use the system model or the network behaviour to detect intrusions. IDSs can be divided into two main categories: (i) if the IDS benefits from predefined rules or behaviour of the system during attack-free operation and/or intrusions, it is referred to as traditional IDS, and (ii) if the IDS leverages state-of-the-art Machine Learning methods to detect known and unknown cyber-attacks, it is called Data-driven IDS [84]. These two categories are discussed in the following.

6.2.1 Traditional IDSs

Traditionally, IDSs are used to monitor and analyze the network and system measurements searching for intrusions, attack signatures, abnormal traffic, and anomalies. Broadly speaking, these IDSs can be categorized into signature-based and anomaly detection techniques. Signatures are a set of patterns that are defined based on the system rules defined for proper and secure operation of the system and to monitor and protect it against possible attacks [85]. Signature-based IDSs keep the record of system behaviour during previous attacks and compare it with the system measurements to find similarities that represent the same threat. In fact, these IDSs search for similar patterns compared to a set of pre-defined threats to identify the intrusions, similar to an anti-virus. These techniques, however, suffer from several drawbacks, such as the possibility of missing new attacks and lack of generalization. Anomaly-based IDSs, on the other hand, keep the record of natural behaviour of the system and devices during different valid operation points and identify the intrusions by monitoring the deviations in the system behaviour. If the deviation is large, it means an attack or a disturbance has occurred. In some situations, distinguishing between attacks and natural disturbances is a challenge for anomaly-based techniques. For instance, anomaly-based IDSs can monitor message emanations and the rate of data transfer between devices and identify the presence of attackers when these parameters exceed a predefined threshold. As an example, Sect. 4 of Chap. 10 presents an anomaly-based IDS for LCDRs.

6.2.2 Machine-Learning-Based Methods

With the development of Artificial Intelligent (AI) and Machine Learning (ML) technologies, a new type of IDSs has been developed to overcome the inherent problems of model-based IDSs. ML-based IDSs gather data from power network components, and use it to learn the behaviour of the system under attack-free conditions, as well as during faults or attacks, to identify the attack based on the features extracted from the data. ML-based IDSs can be categorized into supervised, unsupervised, semi-supervised, and reinforcement learning methods [84].

- Supervised ML methods: These methods often require a substantial amount of labelled data for their training phase. Supervised ML methods for classification assign the input to one of the categories used in the training phase, for instance, attack-free condition, attack, or fault categories. Artificial Neural Network (ANN), Support Vector Machine (SVM), and K-Nearest Neighbour (KNN) are examples of classifier models. Regression, on the other hand, predicts the output based on the input. The data or components of protection schemes can be examined using these methods to identify malicious activities based on previously-observed occurrences in the training dataset [86].
  - ANN: This method is a set of neuron layers that have the capacity to estimate a function by mapping a set of input features (or data) to the classified events, e.g., fault or attack. In an ANN, each layer consists of neurons, each of which calculates its output as a combination of a weight factor with the inner product of its corresponding layer inputs. This output is then followed by an activation function. The layers formed by these neurons are often referred to as Perceptron layers. Multi-Layer Perceptron (MLP) networks, as a family of ANNs, have the

capability to statically map inputs to outputs. When these MLPs are augmented with other learning processes, they acquire extra features. Some examples of ANN are Recurrent Neural Network (RNN) [87], Long Short Term Memory (LSTM) [88], and Gated Recurrent Units [89]. Additionally, in recent years, deep learning techniques—which are composed of MLPs and Convolutional Neural Networks (CNN) with a large number of layers— have received a surge of attention due to their capability to identify cyber-attacks [90]. Section 3 of Chap. 10 employs ANN to develop an IDS for LCDRs.
  - Support Vector Machines (SVMs): In SVM technique, a hyperplane is utilized to separate the under-attack and attack-free data. As a result, it can be used to identify intrusions. When SVM is combined with kernels (i.e., mathematical functions that add to the dimensions of the data), its performance improves.
  - K-Nearest neighbour (KNN): KNN is a non-parametric supervised ML-based method that classifies or predicts the data based on proximity. This method works based on the idea that comparable points can be discovered close together. In KNN method, the Euclidian distance of specific data with other points is used to determine the class of that data with respect to the neighbours. Then, k-minimum distances are utilized to find the most probable classes. This approach can also help to estimate the unavailable information of the lost nodes. Additionally, in combination with other ML techniques, KNN can detect anomalies or outlier data [91].

- Unsupervised ML methods: These methods, e.g., Principal Component Analysis (PCA) and K-means Clustering [86], can be used to extract useful information and hidden patterns without labelling the data.

  - PCA: This technique is used to change the basis of the data using the calculated principal components, which represent the most important dimensions of the data. In fact, PCA is a technique used to reduce the dimension of the data by keeping only the important components. This technique is particularly beneficial for the substation protection systems, due to existence of massive amount of data and the capability of PCA to be integrated into other ML techniques [92].
  - K-means clustering is a technique that separates the data into k clusters and assigns the nearest mean to each one. In this technique, K is an arbitrary hyper-parameter; nonetheless, the model continuously iterates until it obtains the best centroids. Because this approach is label-free, it is a good option for protection applications that do not involve labelled datasets [93].

- Semi-supervised ML methods: These methods use a combination of labelled and unlabelled data, i.e., a small quantity of labelled and a big number of unlabelled. In other words, semi-supervised methods are examples of weak supervision. These methods are particularly important when the learning procedure should be accurate and at the same time, the cost of data labelling is high. For instance, combining clustering and classification techniques results in a semi-supervised learning method, where clustering algorithms—which are unsupervised—are used for grouping data based on their similarities. Then, by identifying the most relevant samples in the

data set, the unlabelled data may be labelled based on their clusters. This data can then be used to train a supervised ML model for classification.
- Reinforcement learning (RL): This group of techniques takes the advantage of both supervised and unsupervised learning, as its functionality does not depend on the data that is previously labelled; instead, it calculates the final output through gradual improvements using trial and error [94]. This technique learns the optimum actions by interacting with the environment. Markov Decision Process (MDP) [95] is often used to represent the environment by describing a number of states and a set of corresponding actions. As a result, in each state, the RL model has several actions to pick from, and it will be rewarded or penalized based on its state and the chosen action. State changes are typically probabilistic, occur in reaction to the activities, and are connected to a transition matrix that represents the likelihood of shifting from one state to another in response to each action. RL is especially effective when the borders between regular and malicious traffic vary based on attack strategies and parameters.

7 Conclusion

In this chapter, cyber-vulnerable protection schemes were divided into three major groups: (i) those that are implemented in substations, (ii) CAP schemes, and (iii) WAP schemes. In order to analyze the cyber-vulnerabilities of protection schemes in substations, first, the automation system, communication protocols, media, and existing cyber vulnerabilities of substations were discussed. Deployment of the IEC 61850 standard, different types of attacks against substation automation systems, and potential impacts of cyber-attacks against substations were also discussed in this chapter. Then, CAP schemes—including directional comparison, differential, and phase comparison schemes—were elaborated on, and their vulnerabilities to cyber-attacks were explained in detail. Afterward, the major applications of WAP schemes and their vulnerabilities to cyber-attacks were detailed. Additionally, this chapter explained different types of attacks that can target the protection system of a power grid. The taxonomy of attack detection and prevention for power system protection applications was another topic discussed in this chapter. The companion chapter provides two case studies for detecting attacks against LCDRs, one using a model-based IDS and another one using ML.

Acknowledgements This research is financially supported by the Public Safety Canada under grants PSIMS 22168.

References

1. Department of Homeland Security, Cyber resilience and response, 2018 public-private analytic exchange program, Technical Report, 2018. https://www.dhs.gov/sites/default/files/publications/2018AEPCyberResilienceandResponse.pdf
2. J. Meserve, Mouse click could plunge city into darkness, experts say, CNN. com, vol. 27 (2007)
3. D.E. Whitehead, K. Owens, D. Gammel, J. Smith, Ukraine cyber-induced power outage: analysis and practical mitigation strategies, in *2017 70th Annual Conference for Protective Relay Engineers (CPRE)* (2017), pp. 1–8
4. R. Khan, P. Maynard, K. McLaughlin, D. Laverty, S. Sezer, Threat analysis of blackenergy malware for synchrophasor based real-time control and monitoring in smart grid, in *4th International Symposium for ICS and SCADA Cyber Security Research 2016*, vol. 4 (2016), pp. 53–63
5. C. Jiwen, L. Shanmei, Cyber security vulnerability assessment for smart substations. IEEE PES Asia-Pacific Power and Energy Engineering Conference (APPEEC) **2016**, 1368–1373 (2016)
6. J. Hong, C.-C. Liu, M. Govindarasu, Integrated anomaly detection for cyber security of the substations. IEEE Trans. Smart Grid **5**(4), 1643–1653 (2014)
7. M. Adamiak, A. Apostolov, M. Begovic, C. Henville, K. Martin, G. Michel, A. Phadke, J. Thorp, Wide area protection-technology and infrastructures. IEEE Trans. Power Delivery **21**(2), 601–609 (2006)
8. G. Allen, R. Cheung, Integration of protection, control and monitoring functions for transmission and distribution substations. IEEE Trans. Power Delivery **13**(1), 96–101 (1998)
9. Y. Zheng, D. Wang, Z. Zhou, T. Cao, Hierarchical protection control system of smart substations. J. Modern Power Syst. Clean Energy **2**(3), 282–288 (2014)
10. W.G. on Centralized Substation Protection, I. P. S. R. C. Control, Advancements in centralized protection and control within a substation. IEEE Trans. Power Delivery **31**(4), 1945–1952 (2016)
11. S. Adepu, N.K. Kandasamy, J. Zhou, A. Mathur, Attacks on smart grid: power supply interruption and malicious power generation. Int. J. Inf. Secur. **19**(2), 189–211 (2020)
12. T. Nguyen, S. Wang, M. Alhazmi, M. Nazemi, A. Estebsari, P. Dehghanian, Electric power grid resilience to cyber adversaries: state of the art. IEEE Access **8**, 87 592–87 608 (2020)
13. V.S. Kumar, J. Prasad, R. Samikannu, A critical review of cyber security and cyber terrorism-threats to critical infrastructure in the energy sector. Int. J. Crit. Infrastruct. **14**(2), 101–119 (2018)
14. T. Bartman, K. Carson, Securing communications for scada and critical industrial systems, in *2016 69th Annual Conference for Protective Relay Engineers (CPRE)* (2016), pp. 1–10
15. Y. Zhang, L. Wang, Y. Xiang, C.-W. Ten, Power system reliability evaluation with scada cybersecurity considerations. IEEE Trans. Smart Grid **6**(4), 1707–1721 (2015)
16. K. Kimani, V. Oduol, K. Langat, Cyber security challenges for iot-based smart grid networks. Int. J. Crit. Infrastruct. Prot. **25**, 36–49 (2019)
17. D. Mashima, Securing smart-grid infrastructure against emerging threats, in *Solving Urban Infrastructure Problems Using Smart City Technologies* (Elsevier, 2021), pp. 359–382
18. A. Ameli, A. Hooshyar, E.F. El-Saadany, A.M. Youssef, An intrusion detection method for line current differential relays. IEEE Trans. Inf. Forensics Secur. **15**, 329–344 (2020)
19. A. Ameli, A. Hooshyar, E.F. El-Saadany, Development of a cyber-resilient line current differential relay. IEEE Trans. Industr. Inf. **15**(1), 305–318 (2019)
20. Y. Yang, H.T. Jiang, K. McLaughlin, L. Gao, Y. Yuan, W. Huang, S. Sezer, Cybersecurity test-bed for iec 61850 based smart substations. IEEE Power Energy Soc. Gener. Meeting **2015**, 1–5 (2015)
21. Y. Chen, J. Hong, C.-C. Liu, Modeling of intrusion and defense for assessment of cyber security at power substations. IEEE Trans. Smart Grid **9**(4), 2541–2552 (2018)
22. M. Bahrami, M. Fotuhi-Firuzabad, H. Farzin, Reliability evaluation of power grids considering integrity attacks against substation protective IEDs. IEEE Trans. Industr. Inf. **16**(2), 1035–1044 (2020)

23. M. Touhiduzzaman, A. Hahn, A.K. Srivastava, A diversity-based substation cyber defense strategy utilizing coloring games. IEEE Trans. Smart Grid **10**(5), 5405–5415 (2019)
24. C.-W. Ten, J. Hong, C.-C. Liu, Anomaly detection for cybersecurity of the substations. IEEE Trans. Smart Grid **2**(4), 865–873 (2011)
25. C. Wang, C.-W. Ten, Y. Hou, A. Ginter, Cyber inference system for substation anomalies against alter-and-hide attacks. IEEE Trans. Power Syst. **32**(2), 896–909 (2017)
26. N. Moreira, E. Molina, J. Lázaro, E. Jacob, A. Astarloa, Cyber-security in substation automation systems. Renew. Sustain. Energy Rev. **54**, 1552–1562 (2016)
27. M. Rodríguez, J. Lázaro, U. Bidarte, J. Jiménez, A. Astarloa, A fixed-latency architecture to secure goose and sampled value messages in substation systems. IEEE Access **9**, 51 646–51 658 (2021)
28. IEEE Guide for Protective Relay Applications to Transmission Lines, *IEEE Std C37.113-2015 (Revision of IEEE Std C37.113-1999)* (2016), pp. 1–141
29. R. Leelaruji, L. Vanfretti, Power system protective relaying: basic concepts, industrial-grade devices, and communication mechanisms (2011)
30. H.H. Alhelou, A.Y. Abdelaziz, P. Siano, *Wide Area Power Systems Stability, Protection, and Security* (Springer, 2020)
31. S.H. Horowitz, A.G. Phadke, *Power System Relaying* (Wiley, 2014)
32. W. Liu, H. Widmer, P. Raffin, Broadband plc access systems and field deployment in european power line networks. IEEE Commun. Mag. **41**(5), 114–118 (2003)
33. W.M. Karwisch Jr, Confidentiality of information on dedicated leased lines (2003)
34. B. Appasani, D.K. Mohanta, A review on synchrophasor communication system: communication technologies, standards and applications. Protect. Control Modern Power Syst. **3**(1), 1–17 (2018)
35. J. Rooker, *Satellite vulnerabilities* (Technical Report, Marine Corps Command and Staff Coll Quantico VA, 2008)
36. S. Leader, *Telecommunications handbook for transportation professionals: the basics of telecommunications* (United States. Federal Highway Administration, Technical Report, 2004)
37. K. Feher, *Wireless Digital Communications: Modulation and Spread Spectrum Applications* (Prentice-Hall, Inc., 1995)
38. B. Kasztenny, M.V. Mynam, N. Fischer, A. Guzmán, Permissive or blocking pilot protection schemes? how to have it both ways, in *proceedings of the 47th Annual Western Protective Relay Conference, Spokane, WA* (2020)
39. S.H. Horowitz, A.G. Phadke, *Power System Relaying* (Wiley, 2014)
40. W.A. Elmore, *Protective Relaying: Theory and Applications*, vol. 1 (CRC Press, 2003)
41. H. Miller, J. Burger, N. Fischer, B. Kasztenny, Modern line current differential protection solutions, in *2010 63rd Annual Conference for Protective Relay Engineers* (2010), pp. 1–25
42. D. Costello, *Event Analysis Tutorial* (Schweitzer Engineering Laboratories, 2013)
43. B. Kasztenny, N. Fischer, K. Fodero, A. Zvarych, Communications and data synchronization for line current differential schemes, in *Proceedings of the 38th Annual Western Protective Relay Conference, Spokane, WA* (2011), pp. 1–4
44. Z. Zhang, S. Gong, A.D. Dimitrovski, H. Li, Time synchronization attack in smart grid: impact and analysis. IEEE Trans. Smart Grid **4**(1), 87–98 (2013)
45. A. Guzman, D.A. Tziouvaras, E.O. Schweitzer, K. Martin, Local- and wide-area network protection systems improve power system reliability, in Power Systems Conference: Advanced Metering, Protection, Control, Communication, and Distributed. Resources **2006**, 174–181 (2006)
46. Z. Bo, J. Jayasinge, R. Hewett, R. Aggarwal, A. Johns, New scheme for monitoring and protection of power transmission system based on global positioning system, in *The 1996 31 st Universities Power Engineering Conference. Part 1(of 3)* (1996), pp. 25–28
47. Z. Bo, G. Weller, T. Lomas, M. Redfern, Positional protection of transmission systems using global positioning system. IEEE Trans. Power Delivery **15**(4), 1163–1168 (2000)
48. V. Ashok, A. Yadav, A.Y. Abdelaziz, A comprehensive review on wide-area protection, control and monitoring systems, in *Wide Area Power Systems Stability, Protection, and Security* (2021), pp. 1–43

49. V.K. Singh, M. Govindarasu, Decision tree based anomaly detection for remedial action scheme in smart grid using pmu data. IEEE Power Energy Society General Meeting (PESGM) **2018**, 1–5 (2018)
50. J.-A. Jiang, J.-Z. Yang, Y.-H. Lin, C.-W. Liu, J.-C. Ma, An adaptive pmu based fault detection/location technique for transmission lines. i. theory and algorithms. IEEE Trans. Power Delivery **15**(2), 486–493 (2000)
51. J. Zare, F. Aminifar, M. Sanaye-Pasand, Synchrophasor-based wide-area backup protection scheme with data requirement analysis. IEEE Trans. Power Delivery **30**(3), 1410–1419 (2015)
52. M. Begovic, D. Novosel, D. Karlsson, C. Henville, G. Michel, Wide-area protection and emergency control. Proc. IEEE **93**(5), 876–891 (2005)
53. G. Rockefeller, C. Wagner, J. Linders, K. Hicks, D. Rizy, Adaptive transmission relaying concepts for improved performance. IEEE Trans. Power Delivery **3**(4), 1446–1458 (1988)
54. J. De La Ree, V. Centeno, J.S. Thorp, A.G. Phadke, Synchronized phasor measurement applications in power systems. IEEE Trans. Smart Grid **1**(1), 20–27 (2010)
55. X. Yin, Z. Zhang, Z. Li, X. Qi, W. Cao, Q. Guo, The research and the development of the wide area relaying protection based on fault element identification. Protect. Control Modern Power Syst. **1**(1), 1–13 (2016)
56. B. Moussa, M. Debbabi, C. Assi, Security assessment of time synchronization mechanisms for the smart grid. IEEE Commun. Surv. Tutor. **18**(3), 1952–1973 (2016)
57. M.S. Almas, L. Vanfretti, R.S. Singh, G.M. Jonsdottir, Vulnerability of synchrophasor-based wampac applications' to time synchronization spoofing. IEEE Trans. Smart Grid **9**(5), 4601–4612 (2018)
58. R. Kateb, P. Akaber, M.H.K. Tushar, A. Albarakati, M. Debbabi, C. Assi, Enhancing wams communication network against delay attacks. IEEE Trans. Smart Grid **10**(3), 2738–2751 (2019)
59. C. Blask, A. Lee et al., *Nescor electric sector failure scenarios and impact analyses - version 3*, Technical report (2015)
60. B. Moussa, A. Al-Barakati, M. Kassouf, M. Debbabi, C. Assi, Exploiting the vulnerability of relative data alignment in phasor data concentrators to time synchronization attacks. IEEE Trans. Smart Grid **11**(3), 2541–2551 (2020)
61. J.P. Anderson, *Computer security technology planning study* (Technical Report, Anderson (James P) and CO Fort Washington PA Fort Washington, 1972)
62. X. Liu, C. Qian, W.G. Hatcher, H. Xu, W. Liao, W. Yu, Secure internet of things (iot)-based smart-world critical infrastructures: survey, case study and research opportunities. IEEE Access **7**, 79 523–79 544 (2019)
63. G. Loukas, G. Öke, Protection against denial of service attacks: a survey. Comput. J. **53**(7), 1020–1037 (2010)
64. X. Yang, X. Zhang, J. Lin, W. Yu, X. Fu, W. Zhao, Data integrity attacks against the distributed real-time pricing in the smart grid, in *IEEE 35th International Performance Computing and Communications Conference (IPCCC)*, vol. 2016 (IEEE, 2016), pp. 1–8
65. S. Li, Y. Yilmaz, X. Wang, Quickest detection of false data injection attack in wide-area smart grids. IEEE Trans. Smart Grid **6**(6), 2725–2735 (2015)
66. K. Hamlen, M. Kantarcioglu, L. Khan, B. Thuraisingham, Security issues for cloud computing. Int. J. Inf. Secur. Privacy (IJISP) **4**(2), 36–48 (2010)
67. F. Aliyu, T. Sheltami, E.M. Shakshuki, A detection and prevention technique for man in the middle attack in fog computing. Proc. Comput. Sci. **141**, 24–31 (2018)
68. S. Bhamidipati, K. J. Kim, H. Sun, P.V. Orlik, Wide-area gps time monitoring against spoofing using belief propagation, in *2019 16th Annual IEEE International Conference on Sensing, Communication, and Networking (SECON)* (IEEE, 2019), pp. 1–8
69. M. Alsaleh, M. Mannan, P.C. van Oorschot, Revisiting defenses against large-scale online password guessing attacks. IEEE Trans. Dependable Secure Comput. **9**(1), 128–141 (2012)
70. N. Neshenko, E. Bou-Harb, J. Crichigno, G. Kaddoum, N. Ghani, Demystifying iot security: an exhaustive survey on iot vulnerabilities and a first empirical look on internet-scale iot exploitations. IEEE Commun. Surv. Tutor. **21**(3), 2702–2733 (2019)

71. W. Iqbal, H. Abbas, M. Daneshmand, B. Rauf, Y.A. Bangash, An in-depth analysis of iot security requirements, challenges, and their countermeasures via software-defined security. IEEE Internet Things J. **7**(10), 10 250–10 276 (2020)
72. J.P. Walters, Z. Liang, W. Shi, V. Chaudhary, Wireless sensor network security: a survey: security in distributed, grid, and pervasive computing ed. by Y. Xiao. Auerbach Publications
73. S. Berger, O. Bürger, M. Röglinger, Attacks on the industrial internet of things-development of a multi-layer taxonomy. Comput. Secur. **93**, 101790 (2020)
74. Y. Xiao, Y. Jia, C. Liu, X. Cheng, J. Yu, W. Lv, Edge computing security: state of the art and challenges. Proc. IEEE **107**(8), 1608–1631 (2019)
75. A.N. Nowroz, K. Hu, F. Koushanfar, S. Reda, Novel techniques for high-sensitivity hardware trojan detection using thermal and power maps. IEEE Trans. Comput. Aided Des. Integr. Circuits Syst. **33**(12), 1792–1805 (2014)
76. M. Tehranipoor, F. Koushanfar, A survey of hardware trojan taxonomy and detection. IEEE Design Test Comput. **27**(1), 10–25 (2010)
77. M. Zhang, N.K. Jha, Finfet-based power management for improved dpa resistance with low overhead. ACM J. Emerg. Technol. Comput. Syst. (JETC) **7**(3), 1–16 (2011)
78. Y. Zhou, D. Feng, Side-channel attacks: ten years after its publication and the impacts on cryptographic module security testing. Cryptology ePrint Archive (2005)
79. N.S. Mtetwa, P. Tarwireyi, A.M. Abu-Mahfouz, M.O. Adigun, Secure firmware updates in the internet of things: a survey, in *International Multidisciplinary Information Technology and Engineering Conference (IMITEC)*, vol. 2019 (IEEE, 2019), pp. 1–7
80. J. Daemen, V. Rijmen, The advanced encryption standard process, in *The Design of Rijndael* (Springer, 2002), pp. 1–8
81. C. Peng, H. Sun, M. Yang, Y.-L. Wang, A survey on security communication and control for smart grids under malicious cyber attacks. IEEE Trans. Syst. Man Cybern. Syst. **49**(8), 1554–1569 (2019)
82. B. Alohali, K. Kifayat, Q. Shi, W. Hurst, A survey on cryptography key management schemes for smart grid. J. Comput. Sci. Appl. Sci. Educat. **3**(3A), 27–39 (2015)
83. S. Misra, A. Vaish, Reputation-based role assignment for role-based access control in wireless sensor networks. Comput. Commun. **34**(3), 281–294 (2011)
84. V.K. Singh, M. Govindarasu, Cyber kill chain-based hybrid intrusion detection system for smart grid, in *Wide Area Power Systems Stability, Protection, and Security* (Springer, 2021), pp. 571–599
85. R.G. Bace, P. Mell et al., Intrusion detection systems (2001)
86. S.H. Haji, S.Y. Ameen, Attack and anomaly detection in iot networks using machine learning techniques: a review. Asian J. Res. Comput. Sci. **9**(2), 30–46 (2021)
87. T. Mikolov, M. Karafiát, L. Burget, J. Cernockỳ, S. Khudanpur, Recurrent neural network based language model. Interspeech **2**(3), 1045–1048 (2010). Makuhari
88. S. Hochreiter, J. Schmidhuber, Long short-term memory. Neural Comput. **9**(8), 1735–1780 (1997)
89. J. Chung, C. Gulcehre, K. Cho, Y. Bengio, Empirical evaluation of gated recurrent neural networks on sequence modeling (2014), arXiv preprint arXiv:1412.3555
90. S. Lawrence, C. Giles, A.C. Tsoi, A. Back, Face recognition: a convolutional neural-network approach. IEEE Trans. Neural Netw. **8**(1), 98–113 (1997)
91. H.H. Pajouh, R. Javidan, R. Khayami, A. Dehghantanha, K.-K.R. Choo, A two-layer dimension reduction and two-tier classification model for anomaly-based intrusion detection in iot backbone networks. IEEE Trans. Emerg. Top. Comput. **7**(2), 314–323 (2019)
92. J. Zhang, Machine learning with feature selection using principal component analysis for malware detection: a case study (2019), arXiv preprint arXiv:1902.03639
93. S.M. Tahsien, H. Karimipour, P. Spachos, Machine learning based solutions for security of internet of things (iot): a survey. J. Netw. Comput. Appl. **161**, 102630 (2020)

94. C. Wu, J. Shi, Y. Yang, W. Li, Enhancing machine learning based malware detection model by reinforcement learning, in *Proceedings of the 8th International Conference on Communication and Network Security* (2018), pp. 74–78
95. G. Kalnoor et al., *Markov decision process based model for performance analysis an intrusion detection system in iot networks*. J. Telecommun. Inf. Technol. (2021)

Cyber-Security of Protection System in Power Grids—Part 2: Case Studies on Securing Line Current Differential Relays

Shahin Riahinia, Amir Ameli, Mohsen Ghafouri, and Abdulsalam Yassine

Abstract This chapter is the second part of a two-part chapter that thoroughly discusses cyber-vulnerabilities of protection systems in power networks. Part I comprehensively explained different cyber-attacks and the taxonomy of attack detection for power system protection applications. In this Part, as an example, the working principle and cyber-security of Line Current Differential Relays (LCDRs) are studied. Additionally, False Data Injection Attacks (FDIAs) and Time Synchronization Attacks (TSAs) against LCDRs are formulated in this chapter. The reasons for focusing on LCDRs are that (i) this type of relay is highly vulnerable to cyber-attacks due to its dependence on communication media and the Global Positioning System (GPS) for collecting time-synchronized remote measurements from all terminals of the line in its protection zone, and (ii) LCDRs are widely used for protecting major transmission lines carrying giga-watts of power. Finally, this chapter elaborates on two types of methods, i.e., a Machine Learning (ML)-based and a model-based, to address the cyber-security problem of this family of Communication-Assisted Protection (CAP) schemes.

Keywords Communication-assisted protection · Cyber-attacks · Cyber-security · Line current differential relays · Intrusion detection systems

S. Riahinia · A. Ameli (✉) · A. Yassine
Department of Electrical Engineering, Lakehead University, Thunder Bay, ON P7B 5E1, Canada
e-mail: aameli@lakeheadu.ca

S. Riahinia
e-mail: sriahini@lakeheadu.ca

A. Yassine
e-mail: ayassine@lakeheadu.ca

M. Ghafouri
1455 De Maisonneuve Blvd. West, EV 7.640, Montreal, QC H3G 1M8, Canada
e-mail: mohsen.ghafouri@concordia.ca

H. Haes Alhelou et al. (eds.), *Power Systems Cybersecurity*, Power Systems,
https://doi.org/10.1007/978-3-031-20360-2_10

1 Introduction

Line Current Differential Relays (LCDRs) are Communication-Assisted Protection (CAP) devices that are widely used in power system protection applications. This type of relay compares the currents measured at all terminals of a line to detect faults. The first and foremost requirement of LCDRs to detect faults is a quick, secure, and reliable communication medium to exchange the current measurements of all terminals [1]. In order to exchange data measurements between LCDRs, various types of communication media can be used, among which dedicated fiber optic links and multiplexed channels are commonly utilized. Dedicated fiber optic communication provides a fast and secure way of exchanging information between LCDRs. Additionally, since fiber optics provides a symmetrical point-to-point channel, no external reference signal, such as a Global Positioning System (GPS) signal, is needed for time synchronization [2]. Multiplexed channels, on the other hand, are economic and efficient communication links for LCDRs, and provide the possibility of wireless communication with remote and hard-to-reach areas [3]. The availability of multiplexed channels, however, depends on the availability of devices and physical channels that carry the data. Additionally, due to signal conversion and buffering, multiplexed channels experience larger delays than dedicated point-to-point fiber. Another weakness of the multiplexed channels is its dependence on an external time reference, like GPS, due to the non-symmetrical nature of this communication medium [2]. The protocol that LCDRs use to communicate, i.e., IEEE 61850, is also vulnerable to cyber-attacks [4, 5]. As a result, the dependence of LCDRs on the communication system and/or the GPS expose this type of relay to cyber-attacks. Denial of Service Attacks (DoSAs), False Data Injection Attacks (FDIAs), and Time Synchronization Attacks (TSAs) are some of the most important cyber-threats that can target LCDRs.

DoSAs halt the transmission of data between LCDRs in different substations, resulting in unavailability of LCDRs [6]. However, since the majority of commercial LCDRs, e.g., the ones in [7–9], are equipped with backup protection elements, such as overcurrent and distance elements, the line remains protected during DoSAs. Additionally, some LCDRs are equipped with redundant communication medium [9]. Thus, DoSAs do not seriously impact LCDRs. FDIAs, on the other hand, can overshadow the security and protection of power systems by intercepting legitimate data and injecting wrong information [10]. This type of attack is of paramount importance since it can result in maloperation of LCDRs. Another type of intrusion against LCDRs is the TSA, which compromises the time-synchronization of measurements by spoofing the GPS signal. This type of attack can mislead LCDRs into tripping healthy lines. More information about the vulnerabilities of LCDRs can be found in [11, 12].

The above discussion demonstrates that the cyber-security of LCDRs is a crucial problem to tackle. On this basis, Sect. 2 elaborates on the working principle of LCDRs, and formulates FDIAs and TSAs against this family of CAP schemes. Sections 3 and 4, respectively, develop a Machine-Learning (ML)-based Intrusion

Detection System (IDS) and an anomaly-based IDS for LCDRs. The performance of these IDSs is evaluated on the New-England 10-generator 39-bus grid system using PSCAD/EMTDC software. Finally, Sect. 5 concludes this chapter.

2 Working Principle of LCDRs

LCDRs operate based on Kirchhoff's current law, where a meaningful difference between the input and output currents of a line (for all line terminals) indicates a fault. To accurately calculate this difference, time-synchronized current measurements from all terminals of the line are required [13]. The relay picks up and trips the line if the trajectory of the relay moves into the trip zone of the differential-restraining plane illustrated in Fig. 1 [7]. In this figure, differential (I_d) and restraining (I_r) currents are, respectively, defined as the sum of all currents going into the line, and the sum of the magnitudes of all terminals' currents. Additionally, I_b, I_{d0}, k_1, and k_2, are, respectively, the cross over threshold, minimum operating current, and the slopes of the first and second characteristic lines. The differential and restraining currents can be expressed as:

$$I_d\,[k] = |I_1\,[k] + I_2\,[k] + \cdots + I_n\,[k]| \tag{1}$$

$$I_r\,[k] = |I_1\,[k]| + |I_2\,[k]| + \cdots + |I_n\,[k]| \tag{2}$$

where the number of line terminals is depicted by N, and $I_n\,[k] = |I_n|\angle\theta_n$ denotes the current phasor of the nth $\in N$ terminal of a line at time-step k.

An LCDR issues a trip command when the relay's trajectory moves into the operating zone, i.e., $I_d \geqslant I_{op}$, where, I_{op} depends on restraining current and can be calculated as:

$$I_{op}\,(I_r) = \begin{cases} I_{d0} + k_1 \times I_r & I_r \leqslant I_b \\ I_{d0} + k_1 \times I_b + k_2\,(I_r - I_b) & I_r \geqslant I_b \end{cases} \tag{3}$$

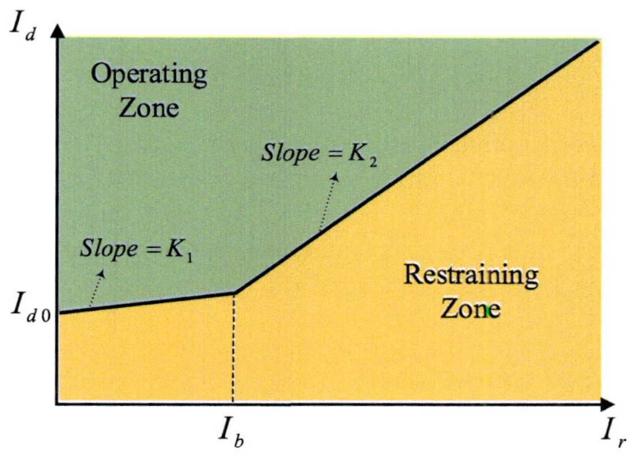

Fig. 1 The characteristic of LCDRs [7]

2.1 FDIAs and TSAs Against LCDRs

The correct operation of LCDRs requires accurate and time-synchronized measurements from all terminals of a line. As a result, if adversaries modify the phase angle or amplitude of current measurements at remote terminals, they can shift the operating point into the trip zone and cause the line to be tripped. Similarly, when the time stamp of remote measurements is manipulated by adversaries, the LCDR calculates a wrong phase angle for manipulated measurements, and consequently, it picks up [14].

When an LCDR operates normally, the value of differential current is very small, i.e., ideally zero. Hence, based on (1), for any terminal n $(1 \leqslant n \leqslant N)$, (4) holds:

$$\sum_{i\in\{1,\cdots,N\}-\{n\}} (|I_i| \angle\theta_i) = -\,(|I_n| \angle\theta_n) \tag{4}$$

If an attack manipulates the measurements by (i) changing their phase angle from θ_n to $\theta_n + \Delta\theta_n$ and (ii) their magnitudes from $|I_n|$ to $|I_n| + |\Delta I_n|$, the differential current I_d is shifted from zero to a value that can be calculated using:

$$I_d = \left| \sum_{i\in\{1,\cdots,N\}-\{n\}} (|I_i| \angle\theta_i) + ((|I_n| + |\Delta I_n|) \angle(\theta_n + \Delta\theta_n)) \right| \tag{5}$$

By substituting (4) into (5), (6) can be obtained when the adversary changes the measurement of only one remote terminal:

$$I_d = |(|I_n| + |\Delta I_n|) \angle(\theta_n + \Delta\theta_n) - |I_n| \angle\theta_n| \tag{6}$$

This equation, however, can be generalized for an arbitrary number of remote terminals. In addition to the differential current, the restraining current also changes from I_r^0 to $I_r^0 + |\Delta I_n|$. As a result, when adversaries' objective is to trip a line, they need to find $\Delta\theta_n$ and $|\Delta I_n|$, so that (3) holds. This equation is modified to (7) to demonstrate the impact of false data:

$$|(|I_n| + |\Delta I_n|) \angle(\theta_n + \Delta\theta_n) - |I_n| \angle\theta_n| > I_{op}\left(I_r^0 + |\Delta I_n|\right) \tag{7}$$

It is worth noting that to satisfy this equation, adversaries do not need any information about the relay's setting when $\Delta\theta_n$ is chosen to be around 180° and $|\Delta I_n|$ is set to a large-enough number. For TSAs, however, only θ_n varies to $\theta_n + \Delta\theta_n$, while $|\Delta I_n|$ is zero. As a result, the value of I_r remains equal to its initial value, i.e., I_r^0. Thus, during TSAs, (7) is changed to

$$|I_n| \angle(\theta_n + \Delta\theta_n) - |I_n| \angle\theta_n > I_{op}\left(I_r^0\right) \tag{8}$$

To solve the last two equations, numerical methods can be used to obtain the locus of attack vectors, i.e., the set of all $(|\Delta I_n|,\ \Delta\theta_n)$ that satisfies (7) and (8)

3 Detecting Cyber-Attacks Against LCDRs Using an ML-Based IDS

This subsection develops an ML-Based IDS for LCDRs. This IDS, which has been illustrated in Fig. 2), utilizes the local measurements of the relay as well as remote before- and after-pickup measurements to identify FDIAs and TSAs, and to distinguish them from real faults since only these events can cause an LCDR to pick up. As a result, following the implementation of this ML-Based IDS, the relay only picks up when a fault occurs in its corresponding protection zone. The developed IDS uses an ML-based classifier and M-dimensional extracted feature-vector to detect attacks regardless of the system transients. On this basis, while respecting the time frame of fault detection, the required features are extracted from local and remote data as soon as the major transient components disappear in LCDR measurements. As a result, the best models for this purpose are the ones that operate based on the temporal sequence of features. Moreover, the ML model should be independent of the probabilistic information, e.g., the density function of events, and at the same time, it should produce the required decision function directly through its training stage. Based on these requirements, Multi-Layer Perceptron (MLP) model is selected as the learning method to distinguish the attack from the internal faults for LCDRs.

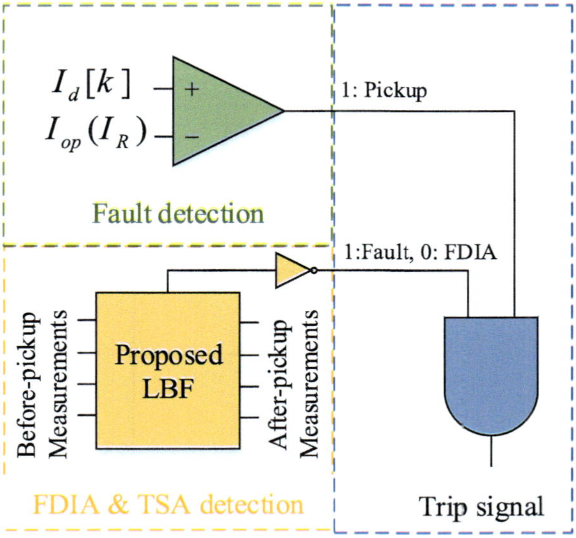

Fig. 2 Tripping logic of LCDRs after utilizing the ML-based IDS

3.1 Synthetic Data Generation for Training the MLP Model

The MLP model should calculate its optimal weights based on a training phase. Training should be performed using optimal features that enables the MLP model to accurately categorize the events into faults and attacks. To do so, a large and comprehensive set of data that includes various scenarios of faults and attacks is required. The rest of this subsection explains the under-study system, as well as the process for generating the synthetic data for feature selection and training of the MLP model.

3.1.1 Test System

The performance of the ML-Based IDS will be evaluated using the 39-bus New England system, Fig. 3. Readers are encouraged to find the details of this system in [15]. In this grid, Line 6–11, whose length is 100 km, is protected using an LCDR. The

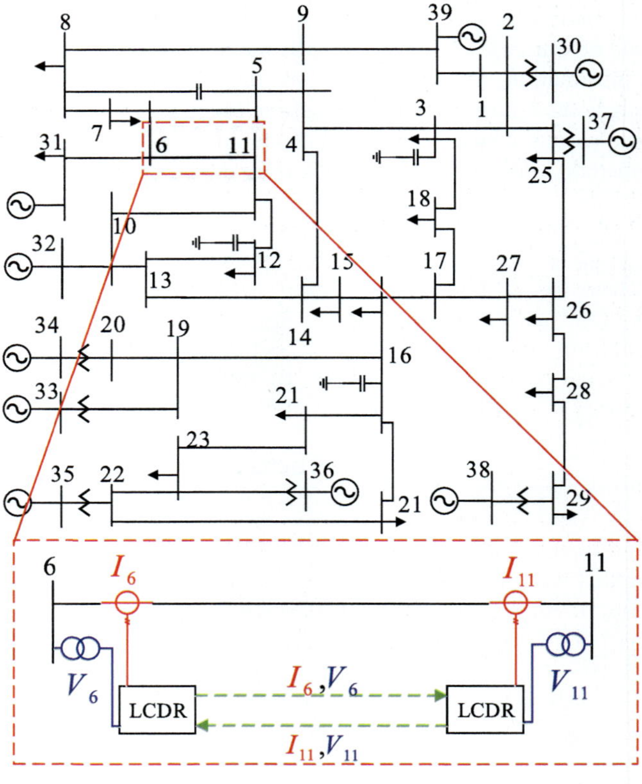

Fig. 3 New-England 10-generator 39-bus test grid and the line protected by LCDRs

focus of this study is on the LCDR located at bus 11, which is going to be upgraded using the ML-Based IDS. Both LCDRs at the terminals of Line 6–11 are configured using the default settings of commercial relays, i.e., $I_{d0} = 0.05$ kA, $I_b = 0.585$ kA, $k_1 = 0.2$, $k_2 = 0.4$ [7]. When the system operates normally, the currents equal to ($0.2824\angle 111.25°$) kA and ($0.2619\angle -78.01°$) kA, respectively, flow into line 6–11 from Buses 11 and 6. These currents result in the restraining currents of $I_r^0 = 0.544$ kA. As a result, I_{op} for the LCDR on Bus 11 is 0.158 kA.

3.1.2 Generating FDIA and TSA Cases

In order to generate data for training purposes, a set of attack vectors are required. This set can be obtained by using all ($|\Delta I_6|$, $\Delta\theta_6$) that satisfy (7) or (8) for the under-study LCDR. These vectors should be calculated and added to the measurements transmitted from Bus 6 to the LCDR of Bus 11. Figure 4 demonstrates the locus of attack vectors obtained by using (3). In this figure, the distance between each point and the origin is equal to $|\Delta I_6|$, whereas the angle of that point-which is calculated using the positive part of the real axis—indicates $\Delta\theta_6$. Figure 4 illustrates the vectors of all possible FDIAs and TSAs that can be leveraged against the measurements of Bus 6 by adversaries to shift the operating point of under-study LCDR inside the trip zone and trip Line 6–11. When the value of $\Delta\theta_6$ is in the range of $[35.11,\ 324.8]$, the relay issues the trip signal for any value of $|\Delta I_6|$. On the other hand, when $\Delta\theta_6$ is in the range of $[-35.11,\ 35.11]$, the adversary must choose a $|\Delta I_6|$ that is not within the white area. For instance, if the attacker chooses $\Delta\theta_6 = 0$, the value of $|\Delta I_6|$ should be more than 0.251 kA. Additionally, attacks against time synchronization of measurements can also be launched from the vectors illustrated in Fig. 4. Since during a TSA, $|\Delta I_6| = 0$, the time-stamps must be manipulated by the adversary so that it cause $\Delta\theta_6 \in [35.11,\ 324.8]$. Therefore, when the adversary changes the time-stamps to achieve a value for Δt_6 in the range of $[1.625,\ 15.037]$ ms, the relay will send the trip signal. Since the vectors presented in Fig. 4 account for all possible considered threats against under-study LCDR, 3000 vectors from these vectors presented in this figure are randomly selected. Then, for each attack vector, the following information is recorded:

1. the phasor of voltages following the execution of the attack.
2. the magnitude of voltages before the attack.
3. the phasor of currents after execution of the attack.
4. the magnitudes of currents before initiation of the attack.
5. the active power entering the line.
6. The reactive power entering the line.

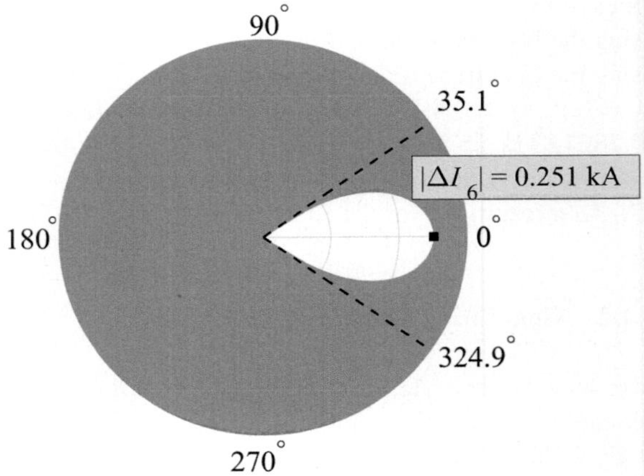

Fig. 4 The locus of attack vectors used for generating FDIA cases

3.1.3 Generating Fault Cases

In the next step, EMTDC/PSCAD is utilized to simulate fault cases for training of the model. To this aim, 3366 cases of internal faults are simulated on the under-study line, and the above-mentioned data are gathered for each fault scenario. These cases include faults with all possible types, i.e., AG, BG, CG, ABG, ACG, BCG, AB, AC, BC, ABC, ABCG, happening at different distances from Bus 11, e.g., 10, 20, 30, 40, 50, 60, 70, 80, and 90 km, and with a wide range of resistances, i.e., $\{0.01, 0.1, 1, 2, 3, 4, 5, 6, 7, 8, 9, 10, 15, 20, 25, 30, 35, 40, 45, 50, 55, 60, 65, 70, 75, 80, 85, 90, 95, 100, 150, 200, 250, 300\}\ \Omega$.

3.2 Selecting Optimal Features

Selecting an optimal set of features is of paramount importance, since in the majority of ML techniques only a subset of features positively affects the accuracy of training and the others result in overfitting or over-complicating the model. As a result, the accuracy of classification of faults and attacks depends significantly on the selected features. This chapter utilizes the Recursive Feature Elimination (RFE) technique [16] for this task. To optimize the weight vectors, the MLP model is initially trained using a baseline set of features. Then, it rates each feature according to its importance score for the MLP model's classification accuracy, taking into account the gradient of the cost function. The least-important features are removed in the next step until the classification accuracy reaches a certain limit or a pre-defined number of features are selected. To apply this feature selection method, 84 features, which are shown in Table 1, are used as an initial list. In this table, parameters V, I, P, and Q denote voltage, current, active power, and reactive power, respectively. Measurements at Bus 6 and 11 are, respectively, demonstrated by subscripts 1 and

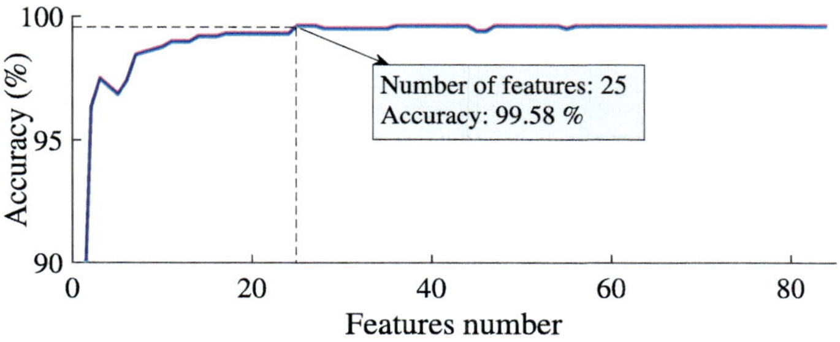

Fig. 5 Accuracy of the ML-based IDS based number of selected features

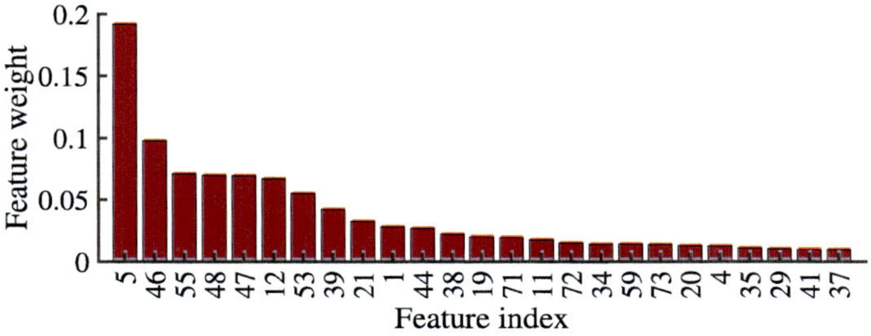

Fig. 6 Importance scores of selected features

2. Subscript dif is used for differential features, which denotes the difference in the values of certain parameters measured at both terminals of the line after the relay picks up. Additionally, subscript sup is used for superimposed features, which represents the difference in the values of certain parameters measured after and before the relay picks up. Symmetrical components and the three phases are presented with superscripts $\{+, -, 0\}$ and $\{A, B, C\}$, respectively. Moreover, $|.|$ and $(\angle.)$ indicate the magnitude and phase angle of parameters. For example, $\left|I_{dif}\right|^{-}$ represents the difference between the magnitude of the negative-sequence current at Buses 6 and 11, and $\left(\angle V^{A}_{sup,1}\right)$ indicates the angle of the superimposed voltage at phase A for Bus 11. All 84 selected features—among which 36 are differential variables and the rest are superimposed—are recorded for 4366 scenarios of faults and attacks, detailed in the previous section. The extracted features are then fed to RFE method to select the optimal set of features. As the accuracy plot of Fig. 5 shows, the maximum accuracy for classification of attack and faults, i.e., 99.58% is achieved when 25 of the features are used. These features are highlighted with grey colour in Table 1. Additionally, the importance score of each feature—i.e., the difference between the model accuracy with and without considering that feature, normalized based on the accuracy when all 25 features are considered—are illustrated in Fig. 6.

Table 1 Initial and optimal set of features considered in this study

Differential features				Superimposed features			
Extracted from symmetrical components		Extracted from Phasor values		Related to Bus 11		Related to Bus 6	
Feature #	Feature	Feature #	Feature	Feature #	Feature	Feature #	Feature
1, 2, 3	$\lvert V_{dif}^{+}\rvert, \lvert V_{dif}^{-}\rvert, \lvert V_{dif}^{0}\rvert$	19, 20, 21	$\lvert V_{dif}^{A}\rvert, \lvert V_{dif}^{B}\rvert, \lvert V_{dif}^{C}\rvert$	37, 38, 39	$\lvert V_{sup,1}^{+}\rvert, \lvert V_{sup,1}^{-}\rvert, \lvert V_{sup,1}^{0}\rvert$	61, 62, 63	$\lvert V_{sup,2}^{+}\rvert, \lvert V_{sup,2}^{-}\rvert, \lvert V_{sup,2}^{0}\rvert$
4, 5, 6	$\angle V_{dif}^{+}, \angle V_{dif}^{-}, \angle V_{dif}^{0}$	22, 23, 24	$\angle V_{dif}^{A}, \angle V_{dif}^{B}, \angle V_{dif}^{C}$	40, 41, 42	$\lvert V_{sup,1}^{A}\rvert, \lvert V_{sup,1}^{B}\rvert, \lvert V_{sup,1}^{C}\rvert$	64, 65, 66	$\lvert V_{sup,2}^{A}\rvert, \lvert V_{sup,2}^{B}\rvert, \lvert V_{sup,2}^{C}\rvert$
7, 8, 9	$\lvert I_{dif}^{+}\rvert, \lvert I_{dif}^{-}\rvert, \lvert I_{dif}^{0}\rvert$	25, 26, 27	$\lvert I_{dif}^{A}\rvert, \lvert I_{dif}^{B}\rvert, \lvert I_{dif}^{C}\rvert$	43, 44, 45	$\lvert I_{sup,1}^{+}\rvert, \lvert I_{sup,1}^{-}\rvert, \lvert I_{sup,1}^{0}\rvert$	67, 68, 69	$\lvert I_{sup,2}^{+}\rvert, \lvert I_{sup,2}^{-}\rvert, \lvert I_{sup,2}^{0}\rvert$
10, 11, 12	$\angle I_{dif}^{+}, \angle I_{dif}^{-}, \angle I_{dif}^{0}$	28, 29, 30	$\angle I_{dif}^{A}, \angle I_{dif}^{B}, \angle I_{dif}^{C}$	46, 47, 48	$\lvert I_{sup,1}^{A}\rvert, \lvert I_{sup,1}^{B}\rvert, \lvert I_{sup,1}^{C}\rvert$	70, 71, 72	$\lvert I_{sup,2}^{A}\rvert, \lvert I_{sup,2}^{B}\rvert, \lvert I_{sup,2}^{C}\rvert$
13, 14, 15	$P_{dif}^{+}, P_{dif}^{-}, P_{dif}^{0}$	31, 32, 33	$P_{dif}^{A}, P_{dif}^{B}, P_{dif}^{C}$	49, 50, 51	$P_{sup,1}^{+}, P_{sup,1}^{-}, P_{sup,1}^{0}$	73, 74, 75	$P_{sup,2}^{+}, P_{sup,2}^{-}, P_{sup,2}^{0}$
16, 17, 18	$Q_{dif}^{+}, Q_{dif}^{-}, Q_{dif}^{0}$	34, 35, 36	$Q_{dif}^{A}, Q_{dif}^{B}, Q_{dif}^{C}$	52, 53, 54	$P_{sup,1}^{A}, P_{sup,1}^{B}, P_{sup,1}^{C}$	76, 77, 78	$P_{sup,2}^{A}, P_{sup,2}^{B}, P_{sup,2}^{C}$
				55, 56, 57	$Q_{sup,1}^{+}, Q_{sup,1}^{-}, Q_{sup,1}^{0}$	79, 80, 81	$Q_{sup,2}^{+}, Q_{sup,2}^{-}, Q_{sup,2}^{0}$
				58, 59, 60	$Q_{sup,1}^{A}, Q_{sup,1}^{B}, Q_{sup,1}^{C}$	82, 83, 84	$Q_{sup,2}^{A}, Q_{sup,2}^{B}, Q_{sup,2}^{C}$

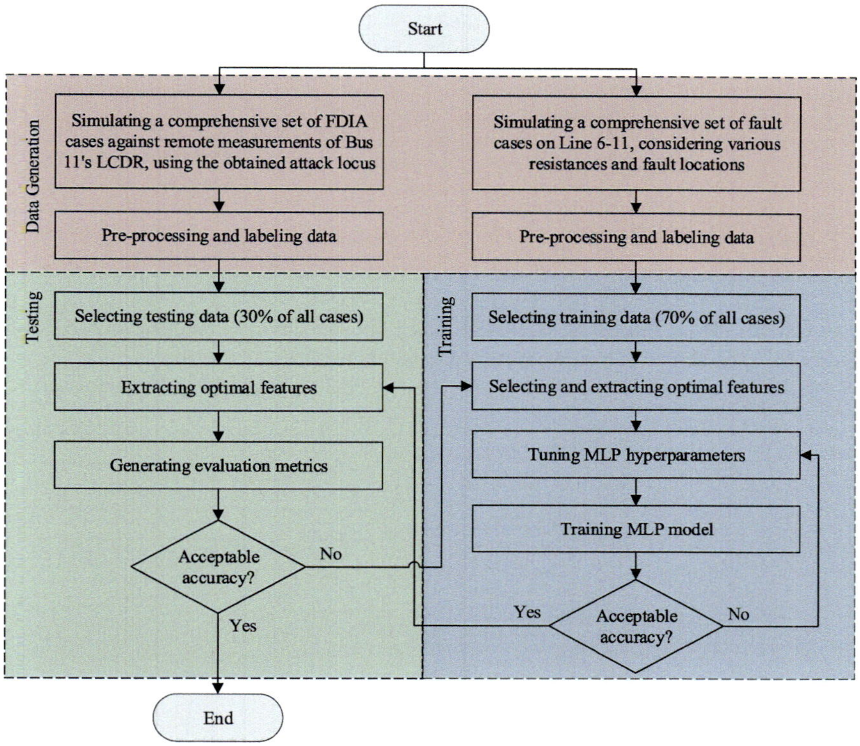

Fig. 7 Flowchart of training and testing processes

3.3 Training the MLP Model

Following the selection of the optimal features, the Keras, which is a Python-based neural-network library on top of TensorFlow [17] is used. Additionally, Adam optimizer [18] is employed to run the backpropagation technique—which calculates the parameters of the MLP model, e.g., neurons' weight and bias, such that the difference between the actual and classified labels are minimized. Using this technique, the optimal number of layers and nodes are obtained to be three layers each with 512 nodes. To train the model, 70% of the previously-discussed cases—2100 attacks and 2356 for faults—are used for training, whereas the remaining 30% are utilized for performance evaluation. The procedure of training and testing is summarized in Fig. 7.

Given that the backpropagation technique utilizes an optimization technique to calculate the hyper parameters, the improvement of the MLP method to minimize its associated objective function is shown in Fig. 8a. This figure illustrates how the MLP model's learning improves in each epoch. Moreover, the performance of

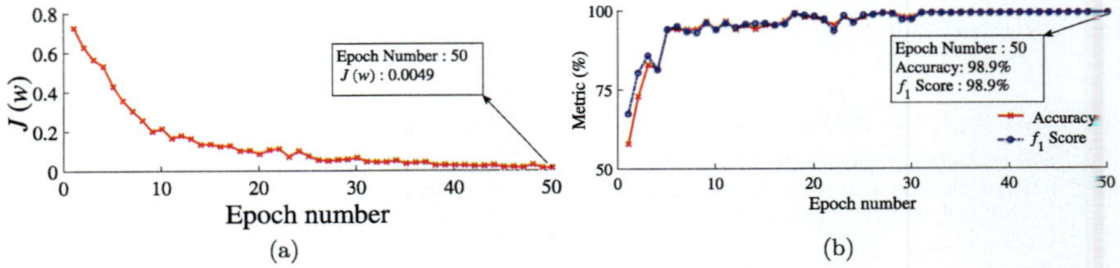

Fig. 8 Training results, **a** Error objective function, **b** Performance metrics

the developed MLP model is evaluated using six metrics, i.e., accuracy, precision, sensitivity, specificity, and f_1 Score, as defined below:

$$\text{Accuracy} = \frac{TP + TN}{TP + FP + TN + FN} \tag{9a}$$

$$\text{Precision} = \frac{TP}{TP + FP} \tag{9b}$$

$$\text{Sensitivity} = \frac{TP}{TP + FN} \tag{9c}$$

$$\text{Specificity} = \frac{TN}{TN + FP} \tag{9d}$$

$$f_1 \text{ Score} = \frac{2 \times \text{Precision} \times \text{Sensitivity}}{\text{Precision} + \text{Sensitivity}} \tag{9e}$$

The definition of parameters TP, TN, FP, and FN, which are used in above metrics, are provided in Table 2 [19]. Figure 8b shows how the f_1 Score and Accuracy metrics improve during the training phase of the MLP model: Both scores achieve 98.9% in the last epoch, demonstrating that the trained model correctly classifies faults and attacks.

Table 2 Performance measuring terms

Term	Description
True Positive (TP)	Attack cases that are correctly classified as attacks
True Negative (TN)	Fault cases that are correctly classified as faults
False Positive (FP)	Fault cases that are incorrectly classified as attacks
False Negative (FN)	Attack cases that are incorrectly classified as faults

3.4 Performance Evaluation

As discussed in the previous section, 70% of the simulated cases are used for training. This section discusses the performance of the developed MLP method for the remaining 30% of cases, which are used for testing. Table 3 demonstrates the confusion matrix of the remaining cases, where the matrix's diagonal elements represent TP and TN, and show the probability of correct classification. On the other hand, the possibilities of incorrect classification, which are obtained by using FP and FN rates, are shown by off-diagonal elements. Using the probabilities shown in Table 3, the accuracy, precision, sensitivity, specificity, and f_1 Score are 97.42%, 95.42%, 99.89%, 95.20%, and 97.35%, respectively. Such metrics confirm the performance of the developed MLP in separating faults and attacks.

Despite the high accuracy of the developed MLP method, i.e., 99.89%, the possibility of confusing faults with attacks is about 5%. In such situations, the protection system does not operate, and faults sustain in the system, so compromising the dependability of LCDRs. This misclassification occurred mostly when the fault impedance was more than 50 Ω and its location was close to the remote terminal of the relay. In fact, most of these faults, i.e., 75% of them, have occurred when the fault impedance was more than 100 Ω and the rest have happened when the resistance was between 50–100 Ω. It should be noted that although the developed IDS might compromise LCDRs' dependability in such situations, it makes them more resilient to cyber-attacks. This compromise, however, is not of high concern, since (i) backup relays—e.g., over-current and/or distance elements—are often used for LCDRs, and they can clear the misclassified faults at the expense of increasing the clearance time, and (ii) all misclassifications are related to high-impedance faults.

4 Detecting Cyber-Attacks Against LCDRs Using an Anomaly-Based IDS

This section develops a model-based IDS for LCDRs. This approach compares the estimated and locally measured voltages at an LCDR's terminal in both positive- and negative-sequence to identify attacks and distinguish them from real faults. To estimate the voltage, this method employs an Unknown Input Observer (UIO) and uses the faulty line's State-Space Model (SSM). During real faults on the line, the

Table 3 Confusion matrix for test cases

Actual cases	Classified cases	
	Faults (%)	Attacks (%)
Faults	95.2	4.8
Attacks	0.11	99.89

difference between the local voltages that is measured and the one that is estimated remain near zero for each sequence, because the SSM on which the UIO functions appropriately depicts the line. Despite this, during FDIAs, the SSM mismatch results in a substantial disparity between measured and estimated values in both sequences.

On this basis, first, the SSM of transmission lines under faults is developed in Sect. 4.1. Then, Sect. 4.2 designs a UIO based on the faulty line's SSM. Afterward, Sect. 4.3 elaborates on how the attacks are detected and differentiated from faults. Finally, Sect. 4.4 evaluates the effectiveness of the developed model-based IDS.

4.1 Transmission Lines Models for Faulty Lines

In this section, the SSMs of a faulty transmission line are developed for both positive- and negative-sequence. The acquired models are used later on to identify intrusions against LCDRs. Figure 9 illustrates the model of a faulty transmission line in positive- or negative-sequence. The fault happens at the distance of xl from terminal T_1, where l is the length of the line and $0 < x < 1$. To locate the fault and determine x, a fault location technique—similar to the one presented in [20] for LCDRs—can be used. Parameters i_1^s, i_2^s, i_{11}^s, i_{22}^s, v_1^s, and v_2^s denote the Sampled Values (SVs) of voltages and currents in positive- and negative-sequence, i.e., $s \in \{+, -\}$. These currents and voltages are computed from phase SVs, as follows

$$\Phi^+(t) = \frac{1}{3}\left[\Phi_a(t) + \Phi_b\left(t - \frac{f_s}{3f}\right) + \Phi_c\left(t - \frac{2f_s}{3f}\right)\right] \quad (10a)$$

$$\Phi^-(t) = \frac{1}{3}\left[\Phi_a(t) + \Phi_b\left(t - \frac{2f_s}{3f}\right) + \Phi_c\left(t - \frac{f_s}{3f}\right)\right] \quad (10b)$$

where Φ signifies the instantaneous values of a current or a voltage; subscripts a, b, and c denote the three phases; and f_s and f are the sampling and fundamental frequencies. Thus, $f_s/3f$ and $2f_s/3f$ are equal to 120° and 240°, respectively [21].

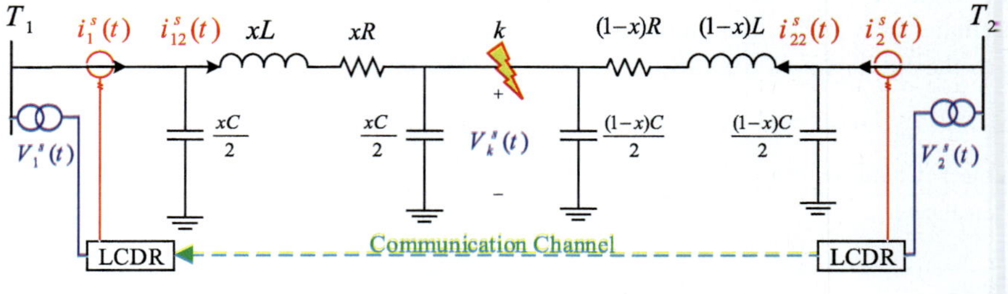

Fig. 9 The model of a line in the presence of an internal fault

To develop the SSM of the line shown in Fig. 9, fault can be implemented as a current source with the current of i_f^s for sequence s at the location of the fault [22]. Therefore, the differential equations listed below can be formulated for each sequence:

$$i_{12}^s(t) - i_1^s(t) + \frac{xC}{2}\frac{dv_1^s(t)}{dt} = 0 \tag{11a}$$

$$v_1^s(t) - xRi_{12}^s(t) - xL\frac{di_{12}^s(t)}{dt} - v_k^s(t) = 0 \tag{11b}$$

$$i_{22}^s(t) - i_2^s(t) + \frac{(1-x)C}{2}\frac{dv_2^s(t)}{dt} = 0 \tag{11c}$$

$$v_2^s(t) - (1-x)Ri_{22}^s(t) - (1-x)L\frac{di_{22}^s(t)}{dt} - v_k^s(t) = 0 \tag{11d}$$

$$i_{12}^s(t) + i_{22}^s(t) + i_f^s(t) - \frac{C}{2}\frac{dv_k^s(t)}{dt} = 0 \tag{11e}$$

In these equations, the fault current $i_f^s(t)$ is is unknown. By writing these equations in a matrix form, the SSM of the line in Fig. 9 is obtained for each sequence as follows:

$$\underbrace{\begin{bmatrix} \dot{v}_1^s(t) \\ \dot{v}_2^s(t) \\ \dot{v}_k^s(t) \\ \dot{i}_{12}^s(t) \\ \dot{i}_{22}^s(t) \end{bmatrix}}_{\dot{\mathbb{X}}^s(t)} = \underbrace{\begin{bmatrix} 0 & 0 & 0 & \frac{-2}{xC} & 0 \\ 0 & 0 & 0 & 0 & \frac{-2}{(1-x)C} \\ 0 & 0 & 0 & \frac{2}{C} & \frac{2}{C} \\ \frac{1}{xL} & 0 & \frac{-1}{xL} & \frac{-R}{L} & 0 \\ 0 & \frac{1}{(1-x)L} & \frac{-1}{(1-x)L} & 0 & \frac{-R}{L} \end{bmatrix}}_{\mathbb{A}_C} \underbrace{\begin{bmatrix} v_1^s(t) \\ v_2^s(t) \\ v_k^s(t) \\ i_{12}^s(t) \\ i_{22}^s(t) \end{bmatrix}}_{\mathbb{X}^s(t)} + \underbrace{\begin{bmatrix} \frac{2}{xC} & 0 \\ 0 & \frac{2}{(1-x)C} \\ 0 & 0 \\ 0 & 0 \\ 0 & 0 \end{bmatrix}}_{\mathbb{B}_{c,n}^T} \underbrace{\begin{bmatrix} i_1^s(t) \\ i_2^s(t) \end{bmatrix}}_{\mathbb{U}_n^s(t)} + \underbrace{\begin{bmatrix} 0 \\ 0 \\ \frac{2}{C} \\ 0 \\ 0 \end{bmatrix}}_{\mathbb{B}_{c,u}^T} i_f^s(t) \tag{12a}$$

$$\mathbb{Y}^s(t) = \underbrace{\begin{bmatrix} 1\,0\,0\,0\,0 \\ 0\,1\,0\,0\,0 \end{bmatrix}}_{\mathbb{C}} \mathbb{X}^s(t) \tag{12b}$$

where, $\mathbb{X}^s$ is the state vector for sequence s; $\mathbb{A}_c$, $\mathbb{B}_{c,n}$, and $\mathbb{B}_{c,u}$ are the state, known input, and unknown input matrices; $\mathbb{C}$ is the output matrix; $\mathbb{U}_n^s$ is the input vector that contains current measurements from terminals $T1$ and $T2$, which are both known parameters; $\mathbb{Y}^s$ is the output vector for sequence s, which contains voltage measurements.

To implement the IDS in practice, matrices A_c, $B_{c,n}$, and $B_{c,u}$ in (9) must be discretized using

$$\mathbb{B}_n = \int_{\tau=0}^{T_s} e^{\mathbb{A}_c \times \tau}\mathbb{B}_{c,n} d\tau \tag{13a}$$

$$\mathbb{B}_u = \int_{\tau=0}^{T_s} e^{\mathbb{A}_c \times \tau} \mathbb{B}_{c,u} d\tau \tag{13b}$$

$$\mathbb{A} = e^{\mathbb{A}_c \times T_s} \tag{13c}$$

where T_s represents the discretization time step [23]. As a result, the discretized model of the system presented in (9) becomes

$$\mathbb{X}^s[k+1] = \mathbb{A}\mathbb{X}^s[k] + \mathbb{B}_n \mathbb{U}_n^s[k] + \mathbb{B}_u i_f^s[k] \tag{14a}$$

$$\mathbb{Y}^s[k] = \mathbb{C}\mathbb{X}^s[k] \tag{14b}$$

where $\mathbb{X}^s[k] \in \mathbb{R}^n$ is the state vector, $\mathbb{U}_n[k] \in \mathbb{R}^m$ is the vector of known inputs, and $\mathbb{Y}[k] \in \mathbb{R}^p$ is the output vector, all at time step k. Additionally, $i_s^f[k] \in R$ is an unknown input. To address the problem of not having i_f^s available, a UIO can be used, as explained in the next section.

4.2 State Estimation Using UIOs

A UIO is an appropriate tool to estimate the states of the state-space equation presented in (9) accurately without requiring the fault current, i.e., i_f^s. UIOs estimate the states of a system by utilizing the system outputs $\mathbb{Y}^s$ and known inputs, i.e., $\mathbb{U}_n^s$, without requiring the unknown inputs [24]. As shown in [25], this great feature is achieved by introducing a delay, signified by α, to the UIO. This delay depends on system parameters, and it is explained in the remaining of this section. Thus, a UIO requires a time frame of samples with the length of $(\alpha + 1)$, i.e., from time step k to $k + \alpha$, to estimate the states of the system at time step k. The most recent sampling moment in this time frame is $k + \alpha$. The output vector during this time frame, i.e., $\mathbb{Y}^s[k]$ to $\mathbb{Y}^s[k+\alpha]$, are used to estimate $\mathbb{X}^s[k]$.

In order to develop a UIO, the system outputs provided by (9) need to be constructed in the form of a matrix for the considered time frame, i.e., from time step k to $k + \alpha$, as follows:

$$\mathcal{Y}^s[k:k+\alpha] = \mathcal{O}_\alpha \mathbb{X}^s[k] + \mathcal{J}_{n,\alpha} \mathcal{U}_n^s[k:k+\alpha] + \mathcal{J}_{u,\alpha} i_f^s[k:k+\alpha] \tag{15}$$

where,

$$\mathcal{Y}^s[k:k+\alpha] = \left[\, \mathbb{Y}^s[k]^T \; \mathbb{Y}^s[k+1]^T \; \cdots \; \mathbb{Y}^s[k+\alpha]^T \,\right]^T \tag{16a}$$

$$\mathcal{O}_\alpha = \left[\, \mathbb{C}^T \; (\mathbb{C}\mathbb{A})^T \cdots (\mathbb{C}\mathbb{A}^\alpha)^T \,\right]^T \tag{16b}$$

$$\mathcal{J}_{u,\alpha} = \begin{bmatrix} 0 & 0 & \cdots 0 \\ \mathbb{C}\mathbb{B}_u & 0 & \cdots 0 \\ \vdots & \vdots & \ddots \vdots \\ \mathbb{C}\mathbb{A}^{\alpha-1}\mathbb{B}_u & \mathbb{C}\mathbb{A}^{\alpha-2}\mathbb{B}_u & \cdots 0 \end{bmatrix} \tag{16c}$$

$$\mathcal{J}_{n,\alpha} = \begin{bmatrix} 0 & 0 & \cdots 0 \\ \mathbb{C}\mathbb{B}_n & 0 & \cdots 0 \\ \vdots & \vdots & \ddots \vdots \\ \mathbb{C}\mathbb{A}^{\alpha-1}\mathbb{B}_n & \mathbb{C}\mathbb{A}^{\alpha-2}\mathbb{B}_n & \cdots 0 \end{bmatrix} \tag{16d}$$

$$\mathcal{U}_n^s\left[k : k+\alpha\right] = \left[\, \mathbb{U}_n^s\left[k\right]^T \; \mathbb{U}_n^s\left[k+1\right]^T \; \cdots \; \mathbb{U}_n^s\left[k+\alpha\right]^T \,\right]^T \tag{16e}$$

$$i_f^s\left[k : k+\alpha\right] = \left[\, i_f^s\left[k\right] \; i_f^s\left[k+1\right] \; \cdots \; i_f^s\left[k+\alpha\right] \right]^T \tag{16f}$$

Using (15), a UIO with the following formulation can be used to estimate $\mathbb{X}^s$ at time step $k+1$ [10]:

$$\hat{\mathbb{X}}^s\left[k+1\right] = \mathbb{A}\hat{\mathbb{X}}^s\left[k\right] + \mathbb{B}_n\mathbb{U}_n^s\left[k\right] + \mathcal{L}\left(\mathcal{Y}^s\left[k : k+\alpha\right] - \mathcal{O}_\alpha\hat{\mathbb{X}}^s\left[k\right] - \mathcal{J}_{n,\alpha}\mathcal{U}_n^s\left[k : k+\alpha\right]\right) \tag{17}$$

where $\hat{\mathbb{X}}^s\left[k\right]$ is the estimation of $\mathbb{X}^s\left[k\right]$ and $\mathcal{L}$ is the UIO's gain, which can be designed using the procedure explained in [10, 26] to ensure that when $k \to \infty$, the UIO's error—i.e., $e^s\left[k+1\right] = \hat{\mathbb{X}}^s[k+1] - \mathbb{X}^s[k+1]$—tends to zero. Additionally, $\mathcal{L}$ guarantees the UIO's stability. Therefore, after designing the UIO's gain, the states of the system at each time step can be estimated using a UIO that operates based on (17). The estimated states can be utilized to detect FDIAs.

4.3 FDIA Diagnosis Using UIOs

This section presents a method for detecting FDIAs against LCDRs. First, Sect. 4.3.1 elaborates on the mismatch between the SSMs of a faulty line and attacks; then, Sect. 4.3.2 makes use of this discrepancy to recognize attacks and distinguish them from faults. Finally, Sect. 4.3.3 develops an auxiliary function to improve the accuracy of the developed IDS.

4.3.1 Line Model During FDIAs

If an FDIA targets an LCDR of a line, it can manipulate the remote measurements of the LCDR as shown in Fig. 10. In fact, this manipulation of measurements can be modelled by two inputs, i.e., h_v^s and h_i^s, which signify the difference between

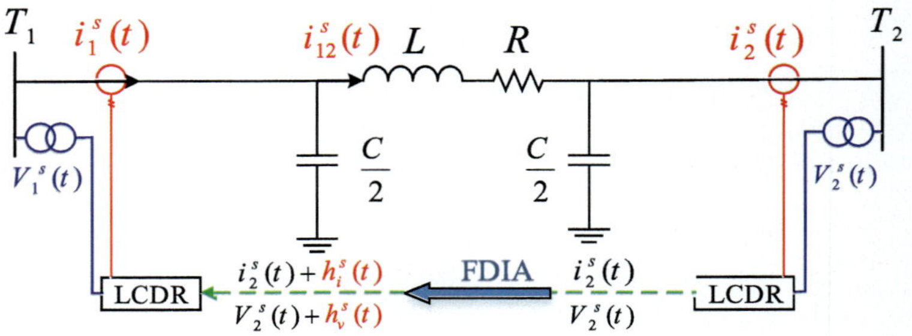

Fig. 10 The model of a line under FDIAs

the authentic and manipulated values of $v_2^s(t)$ and $i_2^s(t)$, respectively. Rewriting the differential equations of (11) in the presence of these two new parameters yields

$$\underbrace{\begin{bmatrix} \dot{v}_1^s(t) \\ \dot{i}_{12}^s(t) \\ \dot{v}_2^s(t) \end{bmatrix}}_{\dot{\overline{\mathbb{X}}}^s(t)} = \underbrace{\begin{bmatrix} 0 & \frac{-2}{C} & 0 \\ \frac{1}{L} & \frac{-R}{L} & \frac{-1}{L} \\ 0 & \frac{2}{C} & 0 \end{bmatrix}}_{\overline{\mathbb{A}}_c} \underbrace{\begin{bmatrix} v_1^s(t) \\ i_{12}^s(t) \\ v_2^s(t) \end{bmatrix}}_{\overline{\mathbb{X}}^s(t)} + \underbrace{\begin{bmatrix} \frac{2}{C} & 0 \\ 0 & 0 \\ 0 & \frac{2}{C} \end{bmatrix}}_{\overline{\mathbb{B}}_{c,n}} \underbrace{\begin{bmatrix} i_1^s(t) \\ i_2^s(t) \end{bmatrix}}_{\mathbb{U}_n^s(t)} + \underbrace{\begin{bmatrix} 0 & 0 \\ 0 & 0 \\ \frac{2}{C} & 0 \end{bmatrix}}_{\overline{\mathbb{B}}_{c,ca}} \underbrace{\begin{bmatrix} h_i^s(t) \\ h_v^s(t) \end{bmatrix}}_{\overline{\mathbb{U}}_{ca}^s(t)} \tag{18a}$$

$$\bar{\mathbb{Y}}^s(t) = \underbrace{\begin{bmatrix} 1 & 0 & 0 \\ 0 & 0 & 1 \end{bmatrix}}_{\overline{\mathbb{C}}} \overline{\mathbb{X}}^s(t) + \underbrace{\begin{bmatrix} 0 & 0 \\ 0 & 1 \end{bmatrix}}_{\overline{\mathbb{D}}_{ca}} \overline{\mathbb{U}}_{ca}^s(t) \tag{18b}$$

Thus, as seen in (18b), replacing $v_2^s(t)$ with $h_v^s(t) + v_2^s(t)$ affects $v_2^s(t)$ in $\bar{\mathbb{Y}}^s(t)$. Similarly, replacing $i_2^s(t)$ with $h_i^s(t) + i_2^s(t)$ in (18a) impacts the states. It should be noted that in (18) and in the rest of this chapter, the parameters of the SSM of the line under FDIAs are provided by an over-line to differentiate them from the parameters of the model under faults (i.e., (9)).

By discretizing (18) for each sequence, the discretized SSM of an attacked line is obtained as follows:

$$\overline{\mathbb{X}}^s[k+1] = \overline{\mathbb{A}}\,\overline{\mathbb{X}}^s[k] + \overline{\mathbb{B}}_n \mathbb{U}_n^s[k] + \overline{\mathbb{B}}_{ca} \overline{\mathbb{U}}_{ca}^s[k] \tag{19a}$$

$$\bar{\mathbb{Y}}^s[k] = \overline{\mathbb{C}}\,\overline{\mathbb{X}}^s[k] + \overline{\mathbb{D}}_{ca} \overline{\mathbb{U}}_{ca}^s[k] \tag{19b}$$

where $\overline{\mathbb{A}}$, $\overline{\mathbb{B}}_n$, and $\overline{\mathbb{B}}_{ca}$ are the discretized versions of $\overline{\mathbb{A}}_c$, $\overline{\mathbb{B}}_{c,n}$ and $\overline{\mathbb{B}}_{c,ca}$, and $\overline{\mathbb{U}}_{ca}[k]$ is the attack input. In addition to other differences, (19b) has two more components than (14), i.e., $\overline{\mathbb{B}}_{ca}\overline{\mathbb{U}}_{ca}^s[k]$ and $\overline{\mathbb{D}}_{ca}\overline{\mathbb{U}}_{ca}^s[k]$, which represent the effects of FDIAs on the line model. Thus, the SSMs of a faulty line and FDIAs are different. These discrepancies can be exploited to differentiate between faults and FDIAs, as shown in the following subsections.

4.3.2 Attack Detection Scheme

As previously mentioned, the SSMs of a faulty line differs from a line's model under FDIAs. This discrepancy can be utilized to differentiate between faults and FDIAs. On this basis, positive- and negative-sequence submodules are developed. Each submodule includes a UIO that is modelled based on the SSM of the line during a fault in its associated sequence. Given that an LCDR should only pick-up when a fault occurs, the suggested approach treats every pick-up as if it were the result of a fault, and then applies the fault localization methodology described in [20]. During a real fault, this approach locates it by utilizing local and remote measurements. However, during FDIAs, since there is no actual fault, i.e., the LCDR's pickup is the consequence of tampering with remote measurements, this method yields a nonsensical fault location. Using its corresponding UIO, each submodule calculates the states of the system at each time step and utilizes them to compute the error between the estimated and measured voltages. Because the real model of the line is the same as the model used to build UIOs during faults, the UIOs' error for both sequences, i.e., $e^s[k]$, approaches zero, as discussed in Sect. 4.2. However, during FDIAs, these two models mismatch, resulting in a substantial estimation inaccuracy, as detailed below.

To obtain the estimation error, i.e., $e^s\,[k+1] = \hat{\mathbb{X}}^s[k+1] - \mathbb{X}^s[k+1]$, $\hat{\mathbb{X}}^s[k+1]$, for each submodule during FDIAs, (17) should be used. Everything in this equation stays the same during FDIAs, with the exception of the system outputs, i.e., $\mathcal{Y}^s\,[k:k+\alpha]$, which should be determined using the system's under-attack SSM, i.e., (19).

To obtain $\mathcal{Y}^s\,[k:k+\alpha]$ during FDIAs, the outputs at time step k to $k+\alpha$ should be written in a matrix, as shown below

$$\overline{\mathcal{Y}}^s_{ca}\,[k:k+\alpha] = \overline{\mathcal{O}}_\alpha \overline{\mathbb{X}}^s\,[k] + \overline{\mathcal{J}}_{n,\alpha}\mathcal{U}^s_n\,[k:k+\alpha] + \overline{\mathcal{J}}_{ca,\alpha}\overline{\mathcal{U}}^s_{ca}\,[k:k+\alpha] \tag{20}$$

in which, $\overline{\mathcal{Y}}^s_{ca}\,[k:k+\alpha]$ signifies the system output during attacks from time-step k to $k+\alpha$, and

$$\overline{\mathcal{O}}_\alpha = \left[\overline{\mathbb{C}}^T \left(\overline{\mathbb{C}}\,\overline{\mathbb{A}}\right)^T \cdots \left(\overline{\mathbb{C}}\,\overline{\mathbb{A}}^\alpha\right)^T\right]^T \tag{21a}$$

$$\overline{\mathcal{J}}_{n,\alpha} = \begin{bmatrix} 0 & 0 & \cdots 0 \\ \overline{\mathbb{C}}\,\overline{\mathbb{B}}_n & 0 & \cdots 0 \\ \vdots & \vdots & \ddots \vdots \\ \overline{\mathbb{C}}\,\overline{\mathbb{A}}^{\alpha-1}\overline{\mathbb{B}}_n & \overline{\mathbb{C}}\,\overline{\mathbb{A}}^{\alpha-2}\overline{\mathbb{B}}_n & \cdots 0 \end{bmatrix} \tag{21b}$$

$$\overline{\mathcal{U}}_{ca}^{s}[k:k+\alpha]=\left[\mathbb{U}_{ca}^{s}[k]^{T}\ \mathbb{U}_{ca}^{s}[k+1]^{T}\ \cdots\ \mathbb{U}_{ca}^{s}[k+\alpha]^{T}\right]^{T} \tag{21c}$$

$$\overline{\mathcal{J}}_{ca,\alpha}=\begin{bmatrix} \overline{\mathbb{D}}_{ca} & 0 & 0 & \cdots & 0 \\ \overline{\mathbb{C}}\,\overline{\mathbb{B}}_{u} & \overline{\mathbb{D}}_{ca} & 0 & \cdots & 0 \\ \overline{\mathbb{C}}\,\overline{\mathbb{A}}\,\overline{\mathbb{B}}_{u} & \overline{\mathbb{C}}\,\overline{\mathbb{B}}_{u} & \overline{\mathbb{D}}_{ca} & \cdots & 0 \\ \vdots & \vdots & \vdots & \ddots & \vdots \\ \overline{\mathbb{C}}\,\overline{\mathbb{A}}^{(\alpha-1)}\overline{\mathbb{B}}_{u} & \overline{\mathbb{C}}\,\overline{\mathbb{A}}^{(\alpha-2)}\overline{\mathbb{B}}_{u} & \overline{\mathbb{C}}\,\overline{\mathbb{A}}^{(\alpha-3)}\overline{\mathbb{B}}_{u} & \cdots & \overline{\mathbb{D}}_{ca} \end{bmatrix} \tag{21d}$$

During FDIAs, $\mathcal{Y}^{s}[k:k+\alpha]$ in (17) is replaced with $\overline{\mathcal{Y}}_{ca}^{s}[k:k+\alpha]$. The estimation error for each submodule during FDIAs can be obtained using (20) and (14a), as follows

$$\begin{aligned} e^{s}[k+1] &= \overline{\mathbb{A}}e^{s}[k]+\mathcal{L}\overline{\mathcal{J}}_{ca,\alpha}\overline{\mathcal{U}}_{ca}^{s}[k:k+\alpha]+\mathcal{L}\overline{\mathcal{O}}_{\alpha}\overline{\mathbb{X}}^{s}[k]-\mathcal{L}\mathcal{O}_{\alpha}\mathbb{X}^{s}[k] \\ &\quad +\mathcal{L}\left[\overline{\mathcal{J}}_{n,\alpha}-\mathcal{J}_{n,\alpha}\right]\mathcal{U}_{n}^{s}[k:k+\alpha] \end{aligned} \tag{22}$$

As demonstrated in (22), during FDIAs, the error of UIO becomes non-zero due to the addition of the extra terms, resulting in inaccurate state estimation for each of the positive- and negative-sequence submodules. This can be used by each submodule to detect attacks using the Residual Function (RF) defined below:

$$r^{s}[k]=\frac{v_{1}^{s}[k]-\hat{v}_{1}^{s}[k]}{v_{1}^{s}[k]} \tag{23}$$

where $\hat{v}_{1}^{s}[k]$ signifies the estimated local voltage and is compared with $v_{1}^{s}[k]$, which is the sole trustworthy system state that is measured locally. As a result, $e^{s}[k]$ approaches zero in the absence of FDIAs, culminating in $r^{s}[k]\rightarrow 0$. However, $r^{s}[k]$ rises during FDIAs, since the estimate error increases. Thus, the RF of each submodule can be tracked and compared with a detection threshold in order to detect attacks. As a result, an FDIA is under way if

$$\left|r^{s}[k]\right|>tr^{s} \tag{24}$$

in which, $|.|$ signifies the absolute value operator, and tr^{s} is the threshold defined for sequence s, and is used d to detect attacks. The detection thresholds must be selected such that false-positive alarm and false-negative alarm rates during internal faults and FDIAs, respectively, are minimized.

4.3.3 Negative-Sequence Deactivation During Balanced Faults

This section introduces a function that is added to the developed model-based IDS to improve its accuracy. This function disables the negative-sequence submodule during balanced faults to avoid erroneous attack detection. This function detects

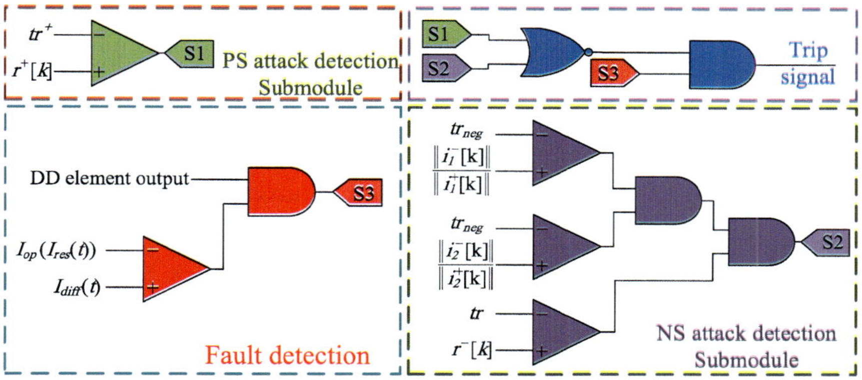

Fig. 11 Tripping logic of LCDRs after augmenting them with the developed model-based IDS

small negative-sequence currents, similar to 67NEG element, by using the following equation:

$$\frac{\left\| i_n^-[k] \right\|}{\left\| i_n^+[k] \right\|} \leq tr_{neg} \quad n \in \{1, 2\} \tag{25}$$

where tr_{neg} is the threshold, and is set to 2% [27]. If the condition of (25) is met for both terminals, the negative-sequence submodule is deactivated. Figure 11 shows how the logic of LCDRs should be modified after augmenting them with the developed model-based IDS.

4.4 *Performance Evaluation*

This section investigates the performance of the developed model-based IDS using PSCAD/EMTDC program on the New-England 10-generator 39-bus grid, shown in Fig. 3. In this system, LCDRs of Buses 11 and 6 protect Line 6–11 with 100 km length, and are set according to the default settings of [7]. The developed model-based IDS is implemented in these LCDRs, allowing remote measurements from either Bus 6 or Bus 11 to be targeted by cyber-attacks. In this section, the focus is on the LCDR that is installed on Bus 11. The attack detection thresholds, i.e., tr^+ and tr^-, are selected equal to 10% and 12.4%, respectively.

4.4.1 Evaluation Scenarios

This section provides four scenarios to assess the effectiveness of the developed IDS. Scenarios 1 and 2 present two internal faults that occur on Line 6–11 at $t = 10$ ms;

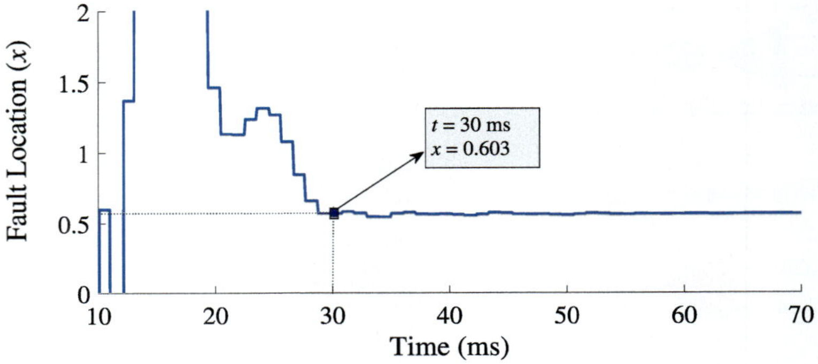

Fig. 12 The obtained fault location

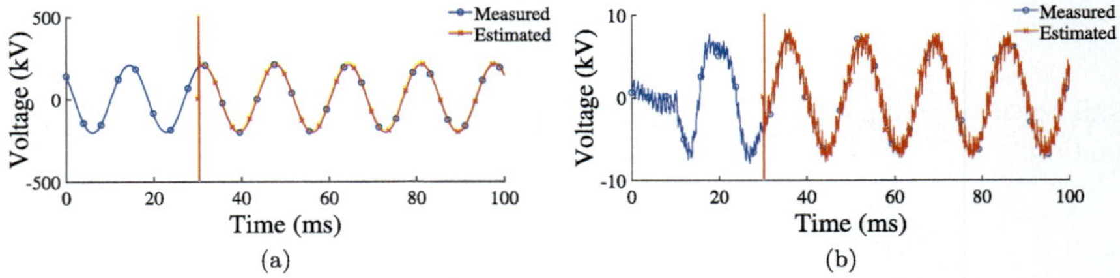

Fig. 13 Voltages of Bus 11 during Scenario 1, **a** positive-sequence and **b** negative-sequence

nevertheless, Scenarios 3 and 4 include two FDIAs that target remote measurements of the LCDR placed at Bus 11. The FDIAs start at $t = 10$ ms.

• **Scenario 1:** The performance of the developed method is evaluated in this scenario in the presence of a phase-A-to-ground fault, which occurs at $xl = 60$ km with the resistance of 100 Ω. After the occurrence of the fault, the relay sends the trip command at $t = 10$ ms, and the fault location is identified after 20 ms by utilizing the measurements at both ends of the line with more than 99% accuracy. Since the fault is asymmetrical, the values of $\left\|i_{11}^{-}[k]\right\| / \left\|i_{11}^{+}[k]\right\|$ and $\left\|i_{6}^{-}[k]\right\| / \left\|i_{6}^{+}[k]\right\|$ are, respectively, 1.94% and 2.54%. This shows that both positive- and negative-sequence submodules should function. Following calculation of x, the UIOs of positive- and negative-sequence submodules calculate $v_{11}^{+}[k]$ and $v_{11}^{-}[k]$, and compute the RFs. The estimated and measured voltages of the two submodules are shown in Fig. 13. It can be observed that $r^s[k]$ is initially small, but ultimately reach 0.30% and 3.31%, respectively, for positive and negative sequences in about 2 ms. As it can be seen from Fig. 14, $r^+[k] < tr^+$ and $r^-[k] < tr^-$ for $t \geq 32$ ms, respectively. As a result, the relay correctly identifies the internal fault, and issues the trip signal 25 ms after the fault happens.

• **Scenario 2:** In this scenario, a phase-A-to-ground fault with the impedance of 200 Ω happens at $xl = 60$ km. The main aim here is to analyze the impact of high-impedance faults on the RF of the UIO. The results show that from the two submodules, only positive-sequence operate and negative-sequence remains inac-

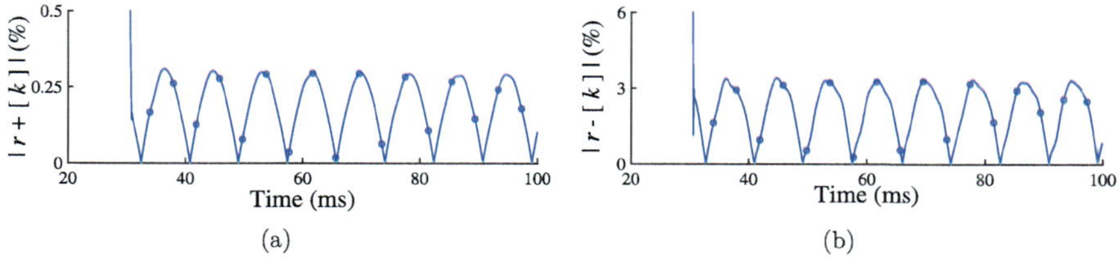

Fig. 14 RFs of Scenario 1, **a** positive-sequence and **b** negative-sequence

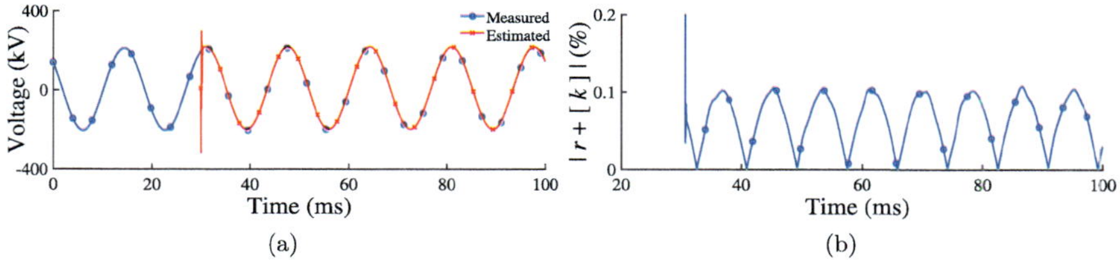

Fig. 15 Positive-sequence results for Scenario 2, **a** voltages of Bus 11 and **b** RF

tive since both $\|i_{11}^-[k]\| \,/\, \|i_{11}^+[k]\| = 1.21\%$ and $\|i_6^-[k]\| \,/\, \|i_6^+[k]\| = 1.50\%$ are less than the specified threshold (this is in contrast to previous scenario). The estimated and measured voltages of positive-sequence for Bus 11 and the RF of the positive-sequence submodule are, respectively, demonstrated in Fig. 15b. Similar to Scenario 1, $r^+[k] < tr^+$ for $t \geq 32$ ms. As a result, the IDS verifies that the relay's pickup is caused by an internal fault, and therefore the relay trips the line.

• **Scenario 3:** In this scenario, an adversary manipulates the remote current measurements, i.e., $i_6[k]$, to mislead the relay into tripping the line. During such an attack, remote measurements are multiplied by 1.96, which is the minimum number for moving the trajectory of the LCDR to the trip zone. The developed IDS receives the pickup signal from the relay and tries to calculate the fault location. However, the calculated location, i.e., x, is meaningless, since there is no actual fault in the system. Additionally, given that the remote measurements are symmetrically manipulated, $\|i_{11}^-[k]\| \,/\, \|i_{11}^+[k]\|$ and $\|i_6^-[k]\| \,/\, \|i_6^+[k]\|$ are both almost zero. Thus, in this scenario, only the positive-sequence submodule operates. The estimated and measured voltages at Bus 11 are demonstrated in Fig. 16a. It can be observed from this figure that the meaningless value of x and the under-attack remote measurements significantly increase the estimation error. Additionally, as Fig. 16b depicts, the RF of the positive-sequence submodule exceeds tr^+ and increases up to 24.45%. Thus, the attack is identified, and the relay is prevented from issuing the trip signal.

• **Scenario 4:** In this scenario, the effectiveness of the developed IDS is examined in the presence of a replay attack. In this attack, the adversary utilizes the fault data of Scenario 1 and sends it to the LCDR at $t = 10$ ms. As a result, for the local bus, $\|i_{11}^-[k]\| \,/\, \|i_{11}^+[k]\|$ is negligible. However, for the remote bus, $\|i_6^-[k]\| \,/\, \|i_6^+[k]\|$ reaches 2.54%, demonstrating that in this case the positive- and negative-sequence

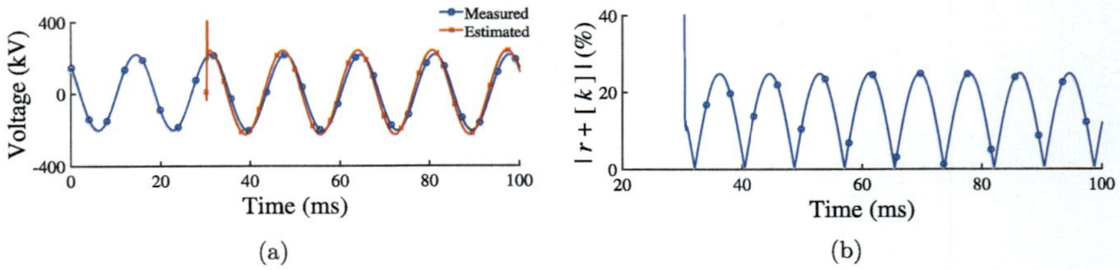

Fig. 16 Positive-sequence results for Scenario 3, **a** Voltages of bus 11 and **b** RF

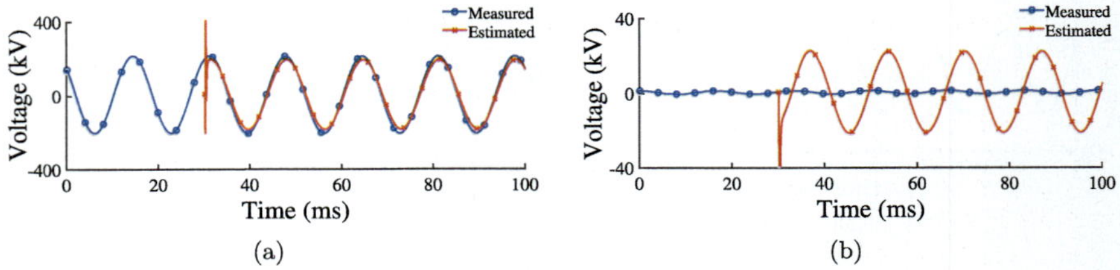

Fig. 17 Voltages of bus 11 in Scenario 4, **a** positive-sequence and **b** negative-sequence

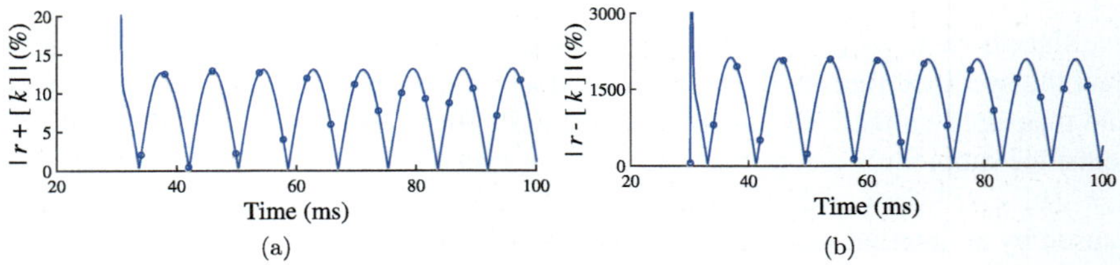

Fig. 18 The RFs associated with Scenario 4, **a** positive-sequence and **b** negative-sequence

submodules should both operate. The estimated and measured local voltages at Bus 11 as acquired by the positive- and negative-sequence submodules of the local LCDR are shown in Fig. 17. There are substantial differences between the estimated and measured voltages, especially for the negative-sequence. Moreover, as Fig. 18 demonstrates, for both submodules the errors are more than the thresholds, i.e., $r^+[k]$ and $r^-[k]$, respectively, reach 12.5 and 2090%. As a result, the developed IDS detects the FDIA and prevents the issue of the tripping command.

5 Conclusion

In the second part of this two-chapter set, the working principle and cyber-vulnerabilities of LCDRs were discussed. Additionally, two types of IDSs, i.e., a model-based IDS and an ML-based one, were explained and developed for detecting attacks against LCDRs. Simulation-based studies on the New-England 10-generator

39-bus grid showed that both methods can accurately detect attacks upon their occurrence, and can differentiate them from faults.

Acknowledgements This research is financially supported by the Public Safety Canada under grants PSIMS 22168.

References

1. S.H. Horowitz, A.G. Phadke, *Power System Relaying* (Wiley, 2014)
2. B. Kasztenny, N. Fischer, K. Fodero, A. Zvarych, Communications and data synchronization for line current differential schemes, in *proceedings of the 38th Annual Western Protective Relay Conference, Spokane* (2011), pp. 1–4
3. IEEE guide for application of digital line current differential relays using digital communication, in *IEEE Std C37.243-2015* (2015), pp. 1–72
4. B. Moussa, M. Debbabi, C. Assi, Security assessment of time synchronization mechanisms for the smart grid. IEEE Commun. Surv. Tutor. **18**(3), 1952–1973 (2016)
5. Y. Yang, H.-Q. Xu, L. Gao, Y.-B. Yuan, K. McLaughlin, S. Sezer, Multidimensional intrusion detection system for iec 61850-based scada networks. IEEE Trans. Power Delivery **32**(2), 1068–1078 (2017)
6. A. Ameli, A. Hooshyar, E.F. El-Saadany, Development of a cyber-resilient line current differential relay. IEEE Trans. Industr. Inf. **15**(1), 305–318 (2019)
7. L90 Line Differential – UR Series Instruction Manual, GE Multilin, Markham, ON, Canada, 2006, version 4.9x, https://www.gegridsolutions.com/products/manuals/l90/l90man-m2.pdf
8. SEL-387L Relay – Line Current Differential Instruction Manual, Schweitzer Engineering Laboratories (Pullman, Washington, 2016), https://selinc.com/products/387L
9. SIPROTEC 7SD61–Differential protection relay for two line ends (Nuremberg, Germany, 2013), http://w3.siemens.com/smartgrid/global/en/products-systems-solutions/protection/line-differential/pages/7sd61.aspx
10. A. Ameli, A. Hooshyar, E.F. El-Saadany, A.M. Youssef, An intrusion detection method for line current differential relays. IEEE Trans. Inf. Forensics Secur. **15**, 329–344 (2020)
11. A. Ameli, A. Kirakosyan, K.A. Saleh, E.F. El-Saadany, Vulnerabilities of line current differential relays to cyber-attacks, in *2019 IEEE Power Energy Society Innovative Smart Grid Technologies Conference (ISGT)*, 1–5 (2019)
12. A. Ameli, A. Ayad, E.F. El-Saadany, M.M.A. Salama, A. Youssef, A learning-based framework for detecting cyber-attacks against line current differential relays. IEEE Trans. Power Delivery **36**(4), 2274–2286 (2021)
13. H. Miller, J. Burger, N. Fischer, B. Kasztenny, Modern line current differential protection solutions, in *2010 63rd Annual Conference for Protective Relay Engineers* (2010), pp. 1–25
14. Z. Zhang, S. Gong, A.D. Dimitrovski, H. Li, Time synchronization attack in smart grid: impact and analysis **4**(1), 87–98 (2013)
15. M. Pai, *Energy Function Analysis for Power System Stability* (Springer Science & Business Media, 2012)
16. I. Guyon, J. Weston, S. Barnhill, V. Vapnik, Gene selection for cancer classification using support vector machines. Mach. Learn. **46**(1–3), 389–422 (2002)
17. A. Géron, *Hands-On Machine Learning with Scikit-Learn, Keras, and TensorFlow: Concepts, Tools, and Techniques to Build Intelligent Systems* (O'Reilly Media, 2019)
18. I. Bello, B. Zoph, V. Vasudevan, Q.V. Le, Neural optimizer search with reinforcement learning, in *Proceedings of the 34th International Conference on Machine Learning*, vol. 70 (JMLR. org, 2017), pp. 459–468
19. S. Ahmed, Y. Lee, S. Hyun, I. Koo, Unsupervised machine learning-based detection of covert data integrity assault in smart grid networks utilizing isolation forest **14**(10), 2765–2777 (2019)

20. J. Izykowski, E. Rosolowski, M.M. Saha, M. Fulczyk, P. Balcerek, A fault-location method for application with current differential relays of three-terminal lines **22**(4), 2099–2107 (2007)
21. R.P. Medeiros, F.B. Costa, K.M. Silva, Power transformer differential protection using the boundary discrete wavelet transform **31**(5), 2083–2095 (2016)
22. A.S. Dobakhshari, A.M. Ranjbar, A novel method for fault location of transmission lines by wide-area voltage measurements considering measurement errors **6**(2), 874–884 (2015)
23. K. Ogata, *Discrete-Time Control Systems, Chap. 5* (Prentice-Hall International, 1995)
24. L. Silverman, Inversion of multivariable linear systems **14**(3), 270–276 (1969)
25. A. Saberi, A.A. Stoorvogel, P. Sannuti, Exact, almost and optimal input decoupled (delayed) observers **73**(7), 552–581 (2000)
26. S. Sundaram, C.N. Hadjicostis, Delayed observers for linear systems with unknown inputs **52**(2), 334–339 (2007)
27. A. Hooshyar, R. Iravani, A new directional element for microgrid protection **9**(6), 6862–6876 (2018)

Semi-supervised Deep Learning-Driven Anomaly Detection Schemes for Cyber-Attack Detection in Smart Grids

Abdelkader Dairi, Fouzi Harrou, Benamar Bouyeddou, Sidi-Mohammed Senouci, and Ying Sun

Abstract Modern power systems are continuously exposed to malicious cyber-attacks. Analyzing industrial control system (ICS) traffic data plays a central role in detecting and defending against cyber-attacks. Detection approaches based on system modeling require effectively modeling the complex behavior of the critical infrastructures, which remains a challenge, especially for large-scale systems. Alternatively, data-driven approaches which rely on data collected from the inspected system have become appealing due to the availability of big data that supports machine learning methods to achieve outstanding performance. This chapter presents an enhanced cyber-attack detection strategy using unlabeled data for ICS traffic monitoring and detecting suspicious data transmissions. Importantly, we designed two semi-supervised hybrid deep learning-based anomaly detection methods for intrusion detection in ICS traffic of smart grid. The first approach is a Gated recurrent

A. Dairi
Computer Science Department, University of Science and Technology of Oran-Mohamed Boudiaf (USTO-MB), El Mnaouar, BP 1505, Bir El Djir 31000, Algeria
e-mail: abdelkader.dairi@univ-usto.dz

F. Harrou (✉) · Y. Sun
CEMSE Division, King Abdullah University of Science and Technology (KAUST), Thuwal 23955-6900, Saudi Arabia
e-mail: fouzi.harrou@kaust.edu.sa

Y. Sun
e-mail: ying.sun@kaust.edu.sa

B. Bouyeddou
STIC Lab, Department of Telecommunications, Abou Bekr Belkaid University, Tlemcen, Algeria
e-mail: benamar.bouyeddou@univ-saida.dz

LESM Lab., Department of Electronics, Faculty of Technology, University of Saida-Dr Moulay Tahar, Saida, Algeria

S.-M. Senouci
DRIVE Laboratory, University of Burgundy, Nevers, France
e-mail: Sidi-Mohammed.Senouci@u-bourgogne.fr

H. Haes Alhelou et al. (eds.), *Power Systems Cybersecurity*, Power Systems,
https://doi.org/10.1007/978-3-031-20360-2_11

unit (GRU)-based stacked autoencoder (AE-GRU), and the second is constructed using a generative adversarial network (GAN) model with a recurrent neural network (RNN) for both generator and discriminator that we called GAN-RNN. The employment of GRU and RNN in AE and GAN models is expected to improve the ability of these models to learn the temporal dependencies of multivariate data. These models are used for feature extraction and anomaly detection methods (Isolation forest, Local outlier factor, One-Class SVM, and Elliptical Envelope) for cyber-attack in power systems. These approaches only employ normal events data for training without labeled attack types, making them more attractive for detecting cyber-attack in practice. The detection performance of these approaches is demonstrated on IEC 60870-5-104 (aka IEC 104) control communication that is often utilized for substation control in smart grids. Results showed that GAN-GRU and AE-GRU-based LOF methods achieved enhanced detection with an averaged F1-score of 0.98, among others.

Keywords Cyber-attack detection · Protocol IEC 104 · Deep learning · Semi-supervised methods · Anomaly detection

1 Introduction

Smart grids play a central role in the efficient management and control of the produced energy. They are designed to transform traditional electricity grids from totally centralized and isolated systems to fully connected and decentralized systems that rely on distributed generation, transmission, distribution, and monitoring processes [1]. Substantially, modern power grids come up with many enhancements, including self-healing, offering more services to consumers (e.g., consumers with a multitude of services), improving power quality and reliability, integrating different sources of energy (Fig. 1), and enabling robust and automatic control and supervision procedures [2, 3]. Power grids have become more dependent on common information technology computing and networking infrastructure for conducting all aspects of operation and maintenance, which significantly increased their vulnerability to anomalies and cyber-attacks. Since intra-SCADA communications are based principally on DNP3 (Distributed network protocol DNP3), International Electro-technical commission IEC60870-5, Profibus IEC 61850 [1], the backbone network involves different wired and wireless technologies and protocols including Zigbee, WiFi, IP-based networks and cellular networks (e.g., 4G) [4]. However, with the evolving of power grids into a cyber-physical systems, their vulnerabilities to cyber-attacks increased more than before [5]. Generally speaking, cyber-attacks in power systems could result in the loss of availability and may have a real severe impact on physical processes (e.g., human life, environment, damage of equipment, etc.), loss of productivity, and revenue.

Power grids are generally placed in remote sites, and they are remotely monitored and controlled via SCADA (Supervisory Control and Data Acquisition) Systems

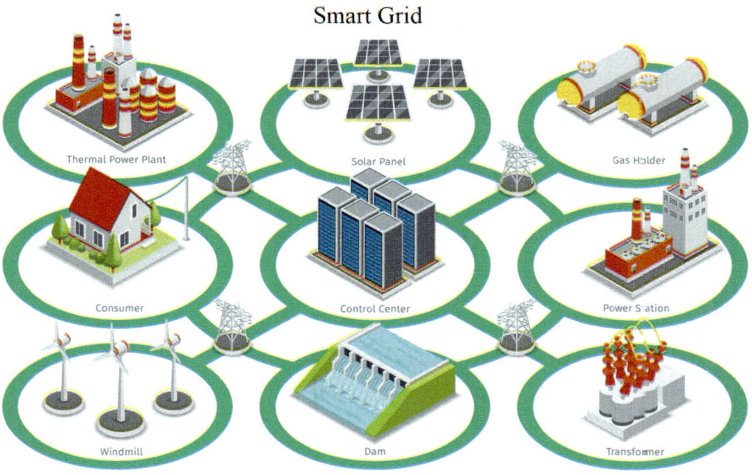

Fig. 1 Smart grid

[6, 7]. The central role of SCADA systems is gathering and analyzing data, communicating, and controlling the operation of systems in real-time [8, 9]. Essentially, SCADA systems comprise several key components, including the Human Machine Interface (HMI), the Master Terminal Unit (MTU), and Remote Terminal Units (RTU) [9, 10]. Specifically, HMI enables operators to monitor the inspected process's state and adjust its control settings. The role of the MTUs, the heart of the SCADA system, is to store and process information gathered from field devices and communicate control signals. RTU receives commands from the MTU to control the local process, acquire data from field devices, and continuously transmit it to the MTU. Each RTU is usually connected to numerous sensors and actuators managing a local process or field equipments. The communication between these components can be wired or wireless via the internet. The common protocols (ModBus, Profibus, DNP3) used in the communication between these components present many vulnerabilities regarding the authentication mechanisms between the MTU and the other components, the integrity of the transmitted packets, and the anti-replay mechanisms. Modern power systems are continuously exposed to malicious cyber-attacks and anomalies, which are challenging and hard to identify [11–13]. If not detected accurately and timely, cyber-attacks and anomalies in industrial systems could decrease plants' productivity, efficiency, and safety, cause severe economic losses, and damage the attacked system. For example, disruption of electric power operations could significantly impact national security and the economy. Therefore, designing an accurate and sensitive intrusion and anomaly detection method is undoubtedly necessary to ensure the productivity and safety of power systems against cyber-attacks and anomalies.

Numerous intentional cyberattacks targeting power systems have occurred in the last two decades [5, 14–16]. For example, in December 2015, a cyber-attack on the Ukraine power grid caused circuit breakers at 30 substations to trip, cutting power to around 225,000 customers [17]. A denial of service (DoS) targeted the telephone system and communication network, making the call center unavailable to customers. In addition, the malware implanted on the human-machine interface (HMI) was employed for deleting software on the system, which prevented the operator from characterizing the extent of the power outage and hampered repair actions. SCADA equipment was rendered inoperable, and power restoration had to be completed manually. It could have resulted in severe damage to the power grid. These attacks highlight the lack of a security-driven technique for building up and maintaining power systems. Hackers exploit vulnerabilities in these critical systems to penetrate and gain access to SCADA networks that monitor physical processes, collect, manipulate and destroy critical data exchanged between facilities and operators, and implant malicious malware to disrupt the normal operating conditions of systems [18]. Moreover, attacks can be implemented on different layers (networks, mac, and physical) by exploiting the hosted protocols (e.g., DNP3, Modbus, Transmission control protocol TCP, User datagram protocol UDP, Internet control message protocol ICMP, and Hypertext transfer protocol HTTP) [1].

Power grids are exposed to a variety of cyber-attacks, which can be grouped into three groups according to the attacker's targets [2, 19]. Attacks within the first group attempt to violate confidentiality and potentially privacy by infiltrating the system and gaining access to devices and stored/exchanged data [19]. Hence, they can collect, manipulate, change and exploit information about the grid's operation and consumers (personal information, consumption profile, bills). To this end, various techniques are elaborated, including traffic sniffing, eavesdropping, man-in-the-middle-based attacks, and IP (Internet protocol)/ARP (Address resolution protocol) spoofing [2]. The second group of attacks consists of data integrity attacks (DIA) [20]. In DIA attacks, malicious attackers try to falsify the sensor measurements by compromising device settings and commands and injecting false data [21, 22]. The third group relies on the availability, and their main objective is to make targeted devices and data inaccessible temporarily or permanently and at least delay their responses. They can manifest in different forms, like link failures, bandwidth exhaustion, flooding, and malformed data structure [23]. They are commonly referred to as DOS (Denial of service) attacks and DDOS (distributed DOS) if initiated using multiple sources. In practice, several techniques have been used to create DOS situations, including Jamming, buffer overflow, TCP-based floorings, UDP-based amplification, and malwares [24, 25].

Smart grid networks play a central role by ensuring the distribution and transmission of electric supply to consumers. Connecting smart grids with the internet generated much space for various types of anomaly injection and cyber-physical attacks. In addition, numerous industrial control protocols, such as Modbus, Goose, or IEC 104, are now used to operate within a smart grid substation and interconnect the control system with power equipment. However, this makes a smart grid network a potential target for outside hackers. Protecting smart grid networks is indispensable to ensuring security and energy production, and it has recently become even more critical than ever before [17]. This is a challenging task because smart grid networks are continuously exposed to malicious attacks coming from the outside and inside the network.

1.1 Contribution

This chapter presents a semi-supervised deep learning-based approach for detecting suspicious communications in ICS/SCADA networks. Unlike supervised methods, semi-supervised cyber-attack detection methods need only the data of normal events to train the detection model, making them more attractive for detecting cyber-attacks in smart grid networks since it is not always easy to get accurately labeled data. Of course, the major contributions of this chapter are summarized bellow.

- This chapter introduces two effective deep learning-driven cyber-attack detection schemes for ICS traffic monitoring and detecting suspicious data transmissions in a smart grid network. Importantly, the proposed approaches combine the advantages of deep learning models and semi-supervised anomaly detection methods for intrusion detection in ICS traffic of smart grid. It is worth noting that these semi-supervised anomaly detection methods need only normal instances of ICS traffic (i.e., unlabeled data), which make them appropriate for detecting unknown attacks. We used the proposed deep learning to learn important features in ICS communication traffic data and the sensitivity of semi-supervised anomaly detection methods for cyber-attack detection. The first approach is a Gated recurrent unit (GRU)-based stacked autoencoder (AE-GRU), and the second is constructed using a generative adversarial network (GAN) model with a recurrent neural network (RNN) for both generator and discriminator that we called GAN-RNN. The employment of GRU and RNN in AE and GAN models is expected to improve the ability of these models to learn the temporal dependencies of multivariate data. These models are used for features extraction and anomaly detection methods (Isolation forest, Local outlier factor, One-Class SVM, and Elliptical Envelope) for cyber-attack in power systems. As we know, this is the first work applying semi-supervised deep learning-driven anomaly detection algorithms to detect attacks in ICS flow data of smart grids. At first, the considered deep learning-based models are built based on training data (normal ICS traffic) and then used to detect cyber-attacks that can be launched from inside or outside the network.
- Furthermore, we compared the detection performance of the proposed AE-GRU-based and GAN-RNN-based anomaly detection schemes to that of the standalone anomaly detection methods (i.e., Isolation forest (iF), 1SVM, Elliptic Envelope (EE), and Local outlier factor(LOF)). We assessed the effectiveness of these approaches on IEC 60870-5-104 (aka IEC 104) control communication that is often utilized for substation control in smart grids. Four statistical indices are employed to compare the discrimination accuracy of the considered methods: accu-

racy, precision, F1-score, and the Area Under the Curve (AUC). Results revealed the promising performance of the proposed approaches in detecting different types of attacks.

The rest of the paper is structured as follows. Section 2 highlights literature reviews on the related works, and Sect. 3 introduces the proposed deep learning-based malicious attack detection methods. Section 4 assesses the proposed method on IEC 60870-5-104 (aka IEC 104) control communication that is commonly utilized for the substation control in smart grids. Finally, Sect. 5 concludes this study and sheds light on potential future research lines.

2 Related Work

Protecting smart grid networks from malicious attacks gained much consideration in the last two decades [26–30]. For instance, in [31], Matoušek et al. investigated the utility of ICS flow monitoring in dealing with internal and external attacks against smart grids. First, the authors proposed a new ICS flow using extended IPFIX flows with application layer headers. After that, the constructed flows are used with statistical and deterministic probabilistic automata (DPA) techniques for anomaly-based attack detection. The proposed techniques were validated using IEC 60870-5-104 standard when considering rogue devices, anomalous traffics, unauthorized data downloads, and scanning attacks. Results had shown promising outcomes that depend on some tuning parameters of DPA. In [32], Jarmakiewicz et al. developed a cyber-security analysis system for a domestic power grid environment. The system correlates all information gathered from control subsystems (e.g., SCADA, IDS, and firewalls) to detect attacks and enables consistently necessary information for identifying eventual threats. Tests demonstrated that the system could detect attacks and anomalous behaviors in the grid. The general performances of such systems mainly rely on the deployed technologies and their detection capabilities. In [33], Hong et al. proposed two anomaly-based systems to tackle attacks on substations in power grids. The first system is host-based, which aims to determine the type and number of attacks affecting the substations according to several anomaly indicators (e.g., intrusions on the user interface or Intelligent Electronic Devices (IEDs), file system modifications, and IED settings alteration). The system assumed that all these indicators were available from other security mechanisms. The second system is a network-based system implemented to detect multicast message attacks, namely Generic Object Oriented Substation Event (GOOSE) and Sampled Measured Value (SMV). After being filtered, the traffic characteristics in terms of the number of messages per observation time, sequence and state number, timestamp, and data integrity are used to reveal anomalies and, eventually, attacks exploiting both messages. Besides the definition of the decision rules about these parameters that can be a challenging task in practice, they are commonly compromised by attackers.

The authors in [34] introduced pattern and flow-driven anomaly detection for communication pattern anomaly detection. Specifically, the pattern-based approach is employed to detect anomalies in communication patterns among hosts, and the flow-based approach for monitoring traffic patterns for individual flows. Results based on normal traffic reveal the promising performance of these detection schemes by reaching lower false alarms than general enterprise systems. In addition, results indicate the capacity of these schemes in detecting some common attacks launched on the MODBUS servers, including DoS attacks, scanning, system degradation, modified data, and DoS attack. In [35], Yans et al. designed an intrusion detection system by combining signature-based and model-based techniques for SCADA systems that support IEC 60870-5-104 standard. Specifically, the signature-based technique is employed to detect known attacks, and the detection signatures are defined accordingly. The model-based detector is applied to predict unknown attacks. To identify attack types, cause of transmission, and length fields are used to build a protocol-based model, and TCP connection requests, server's port numbers, and legitimate users are used to build traffic-based models. However, this IDS can detect a limited number of attacks. In [36], Lin et al. focused on the intra-SCADA exchanges and how they can be utilized to uncover related anomalies. Inter-arrival times are generated and clustered in categorical classes to construct traffic models. Then the Probabilistic suffix tree (PST) is applied to find out latent models. In [37], Kleinmann et al. introduced an attack detection scheme based on statechart. The procedure starts by converting traffic data to signal and discerns peaks and the corresponding cyclic model. Then, individual DFA (Deterministic finite automata) are created per cyclic model. Finally, the resulting DFA is used to construct the system statechart, which will be involved in attack detection. In [38], an innovative approach using a private blockchain system to detect anomalies in a smart grid network and a novel Linear Support Vector Machine Anomaly Detection (LSVMAD) approach in a fog computing (FC) environment is proposed. The LSVMAD approach achieved a detection accuracy of 89% in the FC environment and 78% in the cloud.

3 Methodology

This section presents the materials needed to design the proposed semi-supervised detection schemes.

3.1 Deep Recurrent Autoencoder

Traditional autoencoders were designed especially for dimensionality reduction applied successfully to images, where the training approach is unsupervised without labeling data [39–41]. The key concept of the AE is replicating its input at its output. An AE comprises an encoder and a decoder, which can have multiple layers (Fig. 2).

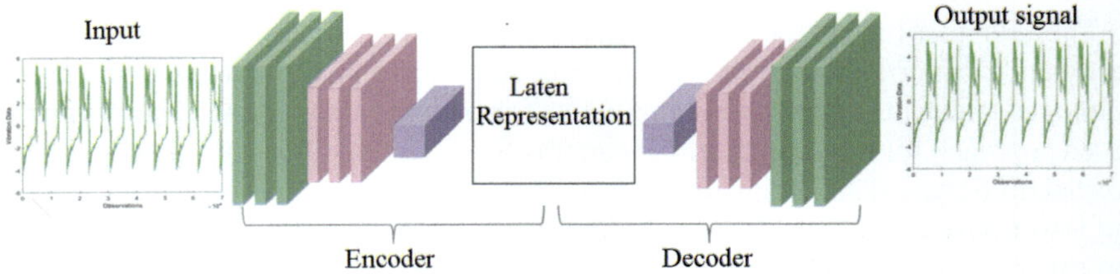

Fig. 2 Basic illustration of an AE model

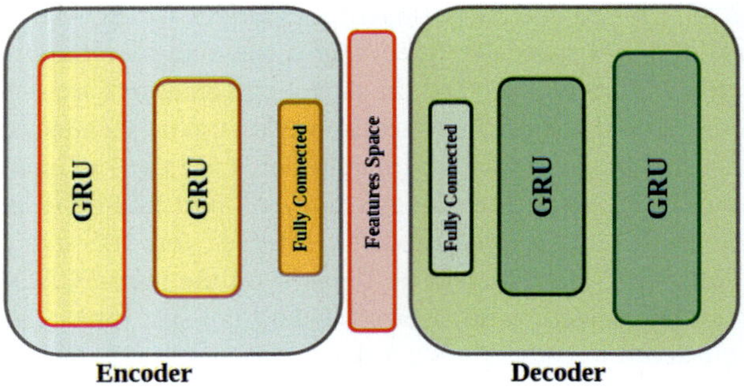

Fig. 3 The AE-GRU structure

It is trained in an unsupervised manner without using labeled data. The training stage aims to optimize a cost function that quantifies the deviation between the input X and its reconstruction at the output $\hat{X}$. For more details on AEs, see [40, 42].

The existing recurrent model learning, such as Recurrent Neural Network (RNN) [43], Long short-term memory (LSTM) [43, 44], Gated recurrent units (GRUs) [45] are built-in supervised way. They are dedicated to capturing the time dependencies in a time series (or data sequence), with a memory cell and gates mechanism that help to remember historical data features. In this first proposed approach, we merge the desirable features of the AE and RNN-based models in one hybrid architecture to improve the encoding of time-series data in an unsupervised manner. To this end, the minimization of the reconstruction error measured using Binary Cross-Entropy (BCE) is expressed as follow:

$$\mathcal{L}_{AE-GRU} = -(X \log(\hat{X}) + (1 - X) \log(1 - \hat{X})). \quad (1)$$

The loss function represented by the BCE considers the input as binary or probability data distribution. The autoencoders are composed of two parts encoder and decoder; we incorporate two stacked Gated Recurrent Unit (GRU) into the encoder and the decoder (Fig. 3). The main objective of integrating GRU is to serve as a feature extractor for time series (normal traffic) by training to reconstruct efficiently the normal data input in an unsupervised manner.

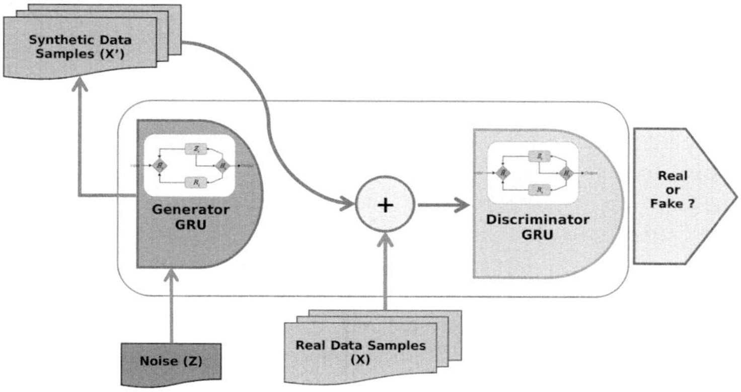

Fig. 4 The GAN-GRU structure

3.2 *The Hybrid GAN-GRU Deep Learning Method*

The second proposed approach is based on GAN model, which has recently emerged as an effective and efficient deep learning model for data generation and learning data representations from unlabeled data [46–48]. GAN has been successfully applied in various areas, such as image data generation and learning and time-series prediction [49]. Conventionally, GANs contain two neural networks called a generator $\mathcal{G}$ and a discriminator $\mathcal{D}$, which are placed in an adversarial way and make GANs very flexible. They are trained in an unsupervised way, making them very attractive as labeling is an expensive task. Moreover, the GAN's generator and discriminator could be trained via only backpropagation. GAN adopts a clever procedure in training: the generator model is trained to continually generate fake data, while the discriminator model seeks to identify between true and fake (generated) data (Fig. 4). In the discriminative model, $\mathcal{D}$, learning is based on two sources of data: the training dataset (true data) and noisy data (simulated) produced by the generative model $\mathcal{G}$. After the training process, the discriminator will be able to distinguish between true and simulated data.

GAN training optimize a cost function $\mathcal{V}(\mathcal{G}, \mathcal{D})$ as:

$$\min_{G} \max_{D} \mathcal{V}(\mathcal{G}, \mathcal{D}) = E\ p_{data(\mathbf{x})} log\mathcal{D}(\mathbf{x}) + E\ p_{g(\mathbf{x})} log(1 - \mathcal{D}(\mathcal{G}(\mathbf{z}))) \quad (2)$$

where $p_z(z)$ describe a prior on input noise variables and $p_{data}(x)$ represents the probability distribution of the true data. The distribution of samples is denoted by $p_g(x)$ produced by the generator. During the GAN training, the generator is encouraged to generate a distribution $p_g(x)$ similar to $p_{data}(x)$ of the real data. Indeed the generator helps the discriminator classify new data points (true, generated). After completing the training, the $p_g(x)$ becomes similar to $p_{data}(x)$ of the real data, which is the distribution of the historical data learned during the training by generator and discriminator.

In the GAN model, the generator and discriminator models can be any kind of neural networks, such as recurrent networks (e.g., RNN, LSTM, or GRU), making the GAN model very flexible. For instance, in the GAN-RNN, the generative and discriminative models are recurrent neural networks hence the name GAN-RNN. Indeed, the GANs are not designed for time-series modeling; however, their ability for the data distribution approximation makes them able to predict the next values of a given data sequence and thus can be investigated for improving feature extraction. The central idea of the second approach is to adopt GRUs and arrange them in an adversarial way in one architecture denoted by Generative is G-GRU, and the discriminative is D-GRU (Fig. 4). The training procedure of the proposed GAN-GRU approach remains similar to the traditional GAN model.

The proposed deep learning models will be used to model ICS traffic flow without attacks. Stable and periodic features characterize ICS communications, which is not the case with the common Internet traffic [50]. After that, the trained models can be used in anomaly detection methods to detect potential attacks in new arrival data.

3.3 Semi-supervised Anomaly Detection Methods

Four anomaly detection methods have been adopted in our study, namely: One-Class SVM (1SVM), Local outlier factor (LOF), Isolation forest (iF), and Elliptic Envelope (EE). Those methods are constructed using a semi-supervised training procedure, meaning that only normal data (traffic without attack) are used in training, while the testing could include both normal and abnormal data without labeling (in the presence of attacks). The detection stage aims to distinguish between normal and abnormal traffic communications on a smart grid.

The 1SVM algorithm is one of the most popular anomaly detection techniques, known for its insensitivity to noise measurements and outliers in training. Crucially, the 1SVM is based on two essential concepts, maximizing the margin and mapping the data to a high dimensional feature space induced by a kernel function [51]. It should be noted that 1SVM is a semi-supervised binary classifier [39, 52]. More specifically, the 1SVM is constructed using unlabeled training data that contains inliers samples (anomaly-free data). In the training stage, the 1SVM process consists of estimating a boundary area, which contains most of the training data. This is carried out by determining a hyperplane with the largest distance to the nearest training data [53]. After that, the designed 1SVM is used to identify anomalies (outliers) by checking if a new test data falls within this boundary or not. Of course, testing data points are declared normal (inlier) if they are within the previously defined boundary; otherwise, they are identified as an anomaly (outliers). The 1SVM procedure assures finding a hyperplane that produces a good data separation by using kernel tricks. The 1SVM detector has been successfully applied in various applications, including fall detection [54], COVID-19 detection from blood tests [55], and fault detection in photovoltaic systems [56].

The Isolation Forest approach was primarily designed by Lui in 2008 [57] and improved later in 2011 [58] to deal with anomaly detection problems where only normal observations are available. Importantly, it is an unsupervised anomaly detection approach since it is designed without the need for labeled data. The essence of the approach is founded on the principle of the Decision Tree algorithm, and it identifies anomalies by isolating outliers from the data [58]. The iF is based on the well-known Random Forest, which consists of a set (ensemble) of decision trees constructed during the training phase [59]. Isolation Forest can be considered an ensemble learning approach to deal with classification and regression problems.

In this study, we considered two other commonly used anomaly detection schemes, namely Elliptical Envelope (EE) [60] , and Local Outlier Factor (LOF) [61]. In the LOF detector, an anomaly score is computed for each observation by measuring the local divergence of the density of a given sample compared to its neighbors. In this study, the number of neighbors used in LOF is 35. In the EE detector, which aims to fit an ellipse around the data using minimum covariance determinant (MCD), the proportion of points to be included in the support of the raw MCD estimate is 0.9.

In this chapter, we will compare the performance of the AE-GRU and GAN-GRU-based 1SVM, LOF, i-Forest, and EE methods for cyber-attack detection in a smart grid. Next, we will discuss how these two models (i.e., AE-GRU and GAN-GRU) can be used for cyber-attack detection.

3.4 The Proposed Deep-Learning-Driven Anomaly Detection Framework

This study presents semi-supervised hybrid deep learning methods to detect cyber attacks in smart grids, particularly monitoring smart grid communication networks. The proposed approach constitutes a framework for online monitoring of the communications (traffic) within an industrial network based on IEC-60780-5-104 (aka IEC 104), often used for smart grids' substation control [62]. Importantly, the IEC-104 protocol, widely adopted in Europe, represents an international standard of data transmission between a power SCADA center and outstations via TCP/IP [63]. This protocol enables connecting MTUs and RTUs using a standard TCP/IP network. Specifically, the IEC-104 protocol ensures data transmission in two directions: from an RTU to the MTU and vice-versa. More details about IEC-104 protocol can be found in [36, 62].

Generally speaking, network communication consists of timed packets sent. Some of them wait for confirmation and have a cause of transmission. Normal communication can be seen as data sequences and time-series data. Furthermore, temporal data dependencies have to be modeled to learn the data distribution of normal traffic. In this chapter, the designed deep learning models will be used to capture the evolution in normal data. The AE-GRU and GAN-GRU models integrate GRU models in their structure, which are powerful deep learning models designed with a gating mechanism and memory cell, allowing them to learn long-range dependencies. As discussed above, for instance, in the AE-GRU model, the autoencoder performs two essential tasks I) features extraction and II) dimensionality reduction. Moreover, deep AE constructs a new compact representation that incorporates pertinent features that we call features space built during the training procedure. We combined the robustness of the deep autoencoders and the effectiveness of the recurrent neural network in order to design a deep recurrent autoencoder model, able to learn learning lengthy-time period dependencies. The proposed framework is a data-driven approach based on learning from data. The proposed hybrid model learns normal traffic, where the communication responds to the IEC-104 protocol constraints (request, response, acknowledgment).

Figure 5 illustrates the flowchart of the proposed cyber-attack detection procedure. The experiment is divided into the training phase and the detection phase. In more detail, we remove missing values and standardize numeric features in the data preprocessing step. In the training step, we train the neural network and anomaly detection procedures using attack-free data. Finally, the trained model is verified using the testing data.

In the training stage, the data is preprocessed by normalization and then arranging data into a sequence with a given length. The next step consists of learning to encode the data sequence into a compact representation that contains pertinent features. Furthermore, the training procedure aims to minimize the reconstruction error of the encoded data sequence; this step is repeated until the convergence of the trained model and until the stabilization of the reconstruction error. In fact, the result of this step is a feature space of normal traffic communications with no attack, which is used to feed the anomaly detection considered methods in this study. Indeed, the anomaly detection considered methods are trained using a compact version of the normal data and tested later with data containing attacks and normal communication to evaluate the detection performance.

Here, we evaluate two types of autoencoders, the AE-GRU and the GAN-GRU. Recall here that the training of the AE-GRU is accomplished by minimizing the reconstruction error of the training dataset using the cross-entropy loss function. In the GAN-GRU, the discriminator is trained to figure out how to distinguish between real data (training dataset) from fake generated by the generator using noisy data. In other words, the discriminator learns to recognize the generator's flaws.

The anomaly detection approaches used are trained based on the resulting feature space containing the encoding of normal data. Furthermore, when a data sequence belongs to a given attack category, its numerical signature is encoded with model parameters learned using normal data, it will be certainly different from normal encoding, which can be sensed by the detectors.

In this study, five statistical scores are employed to quantify the performance of the studied methods computed using a 2×2 confusion matrix: Accuracy, Precision, Recall, F1-score, and Area under curve (AUC) [64, 65]. For a binary detection problem, the number of true positives (TP), false positives (FP), false negatives (FN), and true negatives (TN) are used to compute the evaluation metrics.

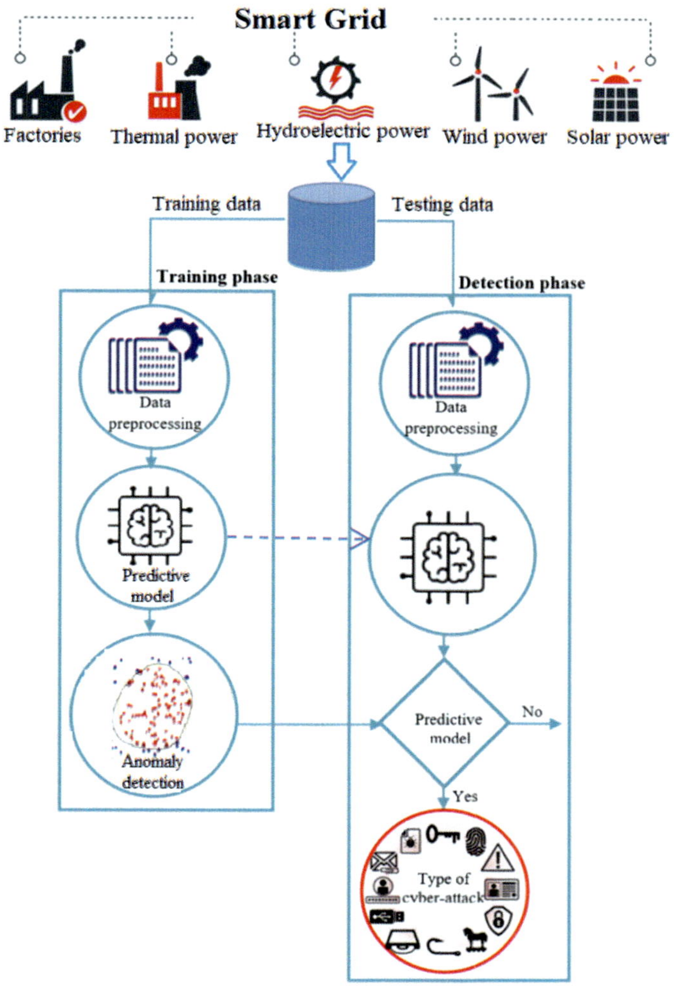

Fig. 5 Flowchart of the detection framework

$$\text{Accuracy} = \frac{\text{TP} + \text{TN}}{\text{TP} + \text{FP} + \text{TN} + \text{FN}}. \tag{3}$$

$$\text{Recall} = \frac{\text{TP}}{\text{TP} + \text{FN}}. \tag{4}$$

$$\text{Precision} = \frac{\text{TP}}{\text{TP} + \text{FP}}. \tag{5}$$

$$\text{F1}-\text{score}=2\frac{\text{Precision.Recall}}{\text{Precision}+\text{Recall}}=\frac{2\text{TP}}{2\text{TP}+\text{FP}+\text{FN}}. \quad (6)$$

4 Results and Discussion

We conducted several experiments to evaluate the performance of the considered anomaly detection models used to design the intrusion detection framework. Indeed, we assess the intrusion detection system performance with several attacks, namely: one man-in-the-middle (MITM) attack, DOS, Connection loss, injection, spoofing, scanning, switching, replaying RTU, blocking HMI, replaying HMI, value changing, masquerading and Rouge devices. The present study adopted two deep recurrent auto-encoders combined with four anomaly detection methods to design a data-driven intrusion detection framework. In order to validate the effectiveness of the proposed approach, we used three datasets of smart grid ICS containing normal and various attacks on IEC 104 communications.

4.1 Data Description

This part is devoted to assessing the efficiency of the proposed deep learning-driven approaches in detecting cyber-attacks in the ICS traffic flow of smart grids. The study is accomplished through actual data from a publicly available database provided in [66]. Three different datasets comprise normal traffic data, and different types of attacks are considered in this study. The first dataset, called BUT-IEC104-I, was generated by from Brno University of technology [31, 67] and contains traffic records from a real smart grid that supports the industrial network standards IEC 608705-104. The second data, called VRT-IEC 104, represent various attacks on IEC 104 communication created using IEC virtual testbed developed at Brno University of Technology. The dataset was captured at the HMI side in the topology. The third data, called GICS-MMS, comprises a set of cyber security attacks on MMS communication that were created manually on the real-life communication provided by the G-ICS labs, University of Grenoble Alps, France (http://lig-g-ics.imag.fr). More details about the used dataset can be found in [66].

Next, the developed AE-GRU-based andGAN-GRU-based 1SVM, iF, LOF, and EE methods will be assessed using three different datasets.

Case (1): Cyber-Attack Detection in BUT-IEC104-I Dataset The dataset was created by Matousek et al. [31, 67] from Brno University of technology. It provides traffic records from an actual smart grid that supports the industrial network standards IEC 608705-104, commonly known as IEC 104 and IEC 61850. Traffic monitoring has been carried out using the IPFIX traffic monitor. The captures consist of IPFIX flow added to application protocols headers. The resulting data includes several traffic features: IP addresses, ports, object id, and other derived characteristics (e.g., start and end times and quantity of exchanged data). Besides the normal traffic, this dataset contains the following scenarios with attacks.

- **DOS attack against an IEC 104 control station:** DOS attack has intended to crash a control station and collapse the grid accordingly. The attacker gets access using a spoofed IP address and floods the victim with 1049 messages in 30 min.
- **Switching attack:** A malware-based attack consists of switching on/off the targeted devices. Within this attack, 72 packets were sent in an interval time of 10 min.
- **Injection commands attacks:** in this case, the attacker manipulates a connected device, then changes its settings or inserts false commands to generate unusual actions and eventually creates different sorts of anomalies. Two scenarios of 5 and 15 min in which 83 and 221 packets were injected, respectively.
- **Connection loss attacks:** In this case, the attacker tries to disconnect targeted devices and break down the attached communications. Two scenarios were implemented. In the first one, connection loss during 10 min resulted in 146 missing packets. In the second scenario, connection loss for one hour, causing losses of packets.
- **Rogue devices attack:** this attack takes place when unauthorized devices are attributed to the communication network, allowing them to share random messages with legitimate devices, which can also cause in random responses and unpredicted actions, which are generally destructive. The dataset includes an attack of 30 min, generating 417 messages.

At first, the models were trained based on training data. The training data contains normal IEC 104 communication without attacks; it contains 58930 packets recorded in 2 days and 19:55 hours of traffic. At each time point, the extracted from the IEC 104 packet contains 16 features of traffic variables, including source IP address, destination IP address, source port, destination port, and destination port. The values of the tuned parameters of the trained models are listed in Table 1. All the hyperparameters are computed during the models training by the minimization of the cross-entropy of the reconstructed error.

Detection results of the proposed AE-GRU and GAN-GRU-based intrusion detection methods when applied to the but-iec104-i dataset are listed in Tables 2 and 3, respectively. The adopted anomaly detection approaches are trained using the features space of normal data, and testing data contain both normal traffic with different attacks. For the DOS attack, Table 2 indicates that the AE-GRU-based EE method recorded the best performance, while the AE-GRU-based 1SVM comes in the second place with an F1-score of 0.9968, followed by the AE-GRU-iForest with an F1-score of 0.9052. The AE-GRU-LOF scores the best detection score for the Switching attack with an F1-score of 0.942, followed by the AE-GRU-iForest with an F1-score of 0.9404. Here, the AE-GRU-EE detection performance reaches an F1-score of 0.9281. Results in Table 2 show that the injection attack has been detected by all adopted anomaly detection methods with an F1-score around 0.98.

The AE-GRU-EE method dominates the other models for detecting connection loss attacks with an F1-score of 0.9643. AE-GRU-based LOF and iForest follow

Table 1 Hyper-parameters used for the experiment

Model	Parameter	Value
	Input features	9
	Activation function	ReLu
	Loss function	Cross Entropy
	Optimizer	Rmsprop
	Epoch	300
	Batch size	250
	Timesteps	12
AE-GRU & GAN-RNN		
Encoder	Layers	3
	Layer 1	GRU (units = 128)
	Layer 2	GRU (units = 16)
	Layer 3	Dense (units = features)
Decoder	Layers	
	Layer 1	Dense (units = Features)
	Layer 2	GRU (units = 16)
	Layer 3	GRU (units = 128)
iF	Contamination	0.01
	Estimators	150
LOF	Contamination	0.01
	Novelty	TRUE
EE	Contamination	0.01
	Support_fraction	0.995
1SVM	Kernel	RBF
	ν	0.0015
	γ	0.25

it by reaching, respectively, F1-score values of 0.9493 and 0.9493. Moreover, the AE-GRU-1SVM scheme provides the best detection for Rogue Devices attacks with an F1-score of 0.9968, while the AE-GRU-based iForest and LOF schemes provide comparable performance with an F1-score of 0.9903. Of course, we noticed that the AE-GRU-1SVM scheme is providing poor results for both switching and connection loss attacks even after tuning its hyper-parameters ν and γ, which are very sensitive and impact the detection, in contrast to the other methods where the most important parameter is the contamination that indicates the percentage of abnormal data (outliers) used during the training.

From Table 3, there is not a single model that is uniformly superior to others. We observe that GAN-GRU-based iForest and LOF schemes reach satisfactory detection performance for the five considered attacks. For instance, the GAN-GRU-LOF scheme obtained the best performance for DOS attack detection with 0.9936, followed by the GAN-GRU-EE scheme with an F1-score of 0.987 and *One-Class SVM*

Table 2 AE-GRU-based anomaly detection schemes using but-iec104-i dataset

Attack	Method	Accuracy	Precision	Recall	F1-score
DOS	iForest	0.8875	1	0.8269	0.9052
DOS	LOF	0.65	0.65	1	0.7879
DOS	1SVM	0.9958	1	0.9936	0.9968
DOS	EE	1	1	1	1
Switch	iForest	0.8903	0.9856	0.8991	0.9404
Switch	LOF	0.8945	1	0.8904	0.942
Switch	1SVM	0.1181	1	0.0833	0.1538
Switch	EE	0.8692	0.9852	0.8772	0.9281
Injection	iForest	0.9619	1	0.9616	0.9804
Injection	LOF	0.9619	1	0.9616	0.9804
Injection	1SVM	0.9771	1	0.977	0.9884
Injection	EE	0.9619	1	0.9616	0.9804
ConnLoss	iForest	0.8017	0.9789	0.8158	0.8899
ConnLoss	LOF	0.9052	1	0.9035	0.9493
ConnLoss	1SVM	0.3103	1	0.2982	0.4594
ConnLoss	EE	0.931	0.9818	0.9474	0.9643
Rogue Dev	iForest	0.9853	1	0.9808	0.9903
Rogue Dev	LOF	0.9853	1	0.9808	0.9903
Rogue Dev	1SVM	0.9951	1	0.9936	0.9968
Rogue Dev	EE	0.7647	0.7647	1	0.8667

with 0.9315. It is worth noting that combining the proposed hybrid models with anomaly detection methods provides a promising tool in detecting attacks in ICS traffic of smart grids.

For comparison purposes, we applied the standalone anomaly detection schemes, including iF, 1SVM, EE, and LOF, to detect cyber-attacks based on the but-iec104-i dataset (Table 4). Results reveal that the LOF detector dominated the 1SVM, EE, and LOF detectors. From results in Tables 2, 3 and 4, we observe that the proposed AE-GRU and GAN-GRU-based LOF schemes achieved the best results compared the other considered intrusion detection schemes.

To conclude the assessment of the proposed AE-GRU and GAN-GRU-based intrusion detection schemes and the standalone anomaly detection schemes (i.e., iF, 1SVM, LOF, and EE) when applied to the but-iec104-i dataset, Table 5 lists the averaged evaluation scores for each scheme. Results show that combining the proposed hybrid deep learning models with anomaly detection methods provides a promising tool for monitoring ICS traffic flow without using labeled data. It can be seen that the GAN-GRU-based LOF and EE schemes dominate the other models by reaching an F-score of 0.96. They are followed by the AE-GRU-based EE scheme

Table 3 GAN-GRU adopted anomaly detection schemes using but-iec104-i dataset

Attack	Method	Accuracy	Precision	Recall	F1-score
DOS	iForest	0.8	1	0.6923	0.8182
DOS	LOF	0.9917	1	0.9872	0.9936
DOS	1SVM	0.9167	1	0.8718	0.9315
DOS	EE	0.9833	1	0.9744	0.987
Switch	iForest	0.8439	0.9848	0.8509	0.913
Switch	LOF	0.9873	1	0.9868	0.9934
Switch	1SVM	0.1814	1	0.1491	0.2595
Switch	EE	0.9789	0.9869	0.9912	0.989
Injection	iForest	0.9553	0.9932	0.9616	0.9771
Injection	LOF	0.9891	1	0.989	0.9945
Injection	1SVM	0.0229	1	0.0164	0.0323
Injection	EE	0.9826	1	0.9825	0.9912
ConnLoss	iForest	0.7069	1	0.7018	0.8248
ConnLoss	LOF	0.9052	1	0.9035	0.9493
ConnLoss	1SVM	0.8194	1	0.8158	0.8986
ConnLoss	EE	0.9677	0.9804	0.9868	0.9836
Rogue Dev	iForest	0.9853	1	0.9808	0.9903
Rogue Dev	LOF	0.7549	0.7624	0.9872	0.8604
Rogue Dev	1SVM	0.25	1	0.0192	0.0377
Rogue Dev	EE	0.7549	0.7624	0.9872	0.8604

with an F-score of 0.96. We also observe that the standalone LOF scheme obtained satisfactory detection with an averaged F1-score of 0.95.

Case (2): Cyber-Attack Detection in G-ICS Dataset The second experiment evaluates the proposed cyber-attack detection schemes using the G-ICS created University of Grenoble Alps, France. The G-ICS data contains a set of cyber security attacks on MMS communication, including lost connection, injection, scanning, and interrupt attacks. In addition, this data comprises normal communication data, which is used for training the proposed models.

The data contains two lost connections, with 58 missing packets in the first and 63 missing packets in the second. The data include scanning attacks, also known as reconnaissance attacks, which are performed to obtain necessary information about network topology and components. Here, the attack was launched at 07:50:41,51 and ended at 07:53:01:51. In the G-ICS, We find also interrupt attacks, where an attacker tries to interrupt MMS services, and modification attacks, where an attacker tries modifying packets.

Tables 6 and 7 shows detection results obtained by AE-GRU and GAN-GRU-based anomaly detection schmes using G-ICS dataset. As can be seen in Table 6, for the modification attacks, the AE-GRU-1SVM achieved the best performance,

Table 4 Standalone anomaly detection methods based on using but-iec104-i dataset

Attack	Method	Accuracy	Precision	Recall	F1-score
DOS	iForest	0.5709	0.8391	0.6411	0.7269
DOS	LOF	0.8905	0.8905	1	0.9421
DOS	1SVM	0.8869	0.8901	0.9959	0.94
DOS	EE	0.8905	0.8905	1	0.9421
Switch	iForest	0.4718	0.9176	0.4927	0.6411
Switch	LOF	0.9576	0.9576	1	0.9783
Switch	1SVM	0.9556	0.9575	0.9979	0.9773
Switch	EE	0.9576	0.9576	1	0.9783
Injection	iForest	0.457	0.8631	0.4927	0.6273
Injection	LOF	0.9275	0.9275	1	0.9624
Injection	1SVM	0.8203	0.9188	0.8844	0.9013
Injection	EE	0.9275	0.9275	1	0.9624
ConnLoss	iForest	0.3472	0.6003	0.449	0.5137
ConnLoss	LOF	0.7624	0.7671	0.9917	0.8651
ConnLoss	1SVM	0.7136	0.7547	0.9292	0.8329
ConnLoss	EE	0.4672	0.6682	0.6083	0.6368
Rogue Dev	iForest	0.4298	0.7685	0.4927	0.6004
Rogue Dev	LOF	1	1	1	1
Rogue Dev	1SVM	0.769	0.855	0.8844	0.8695
Rogue Dev	EE	0.8696	0.8696	1	0.9303

Table 5 Averaged performance of AE-GRU and GAN-GRU-based methods and standalone anomaly detection methods using but-iec104-i dataset

Method	Accuracy	Precision	Recall	F1-score
AE-GRU-iForest	0.91	0.99	0.90	0.94
AE-GRU-LOF	0.90	0.94	0.95	0.94
AE-GRU-1SVM	0.68	1.00	0.67	0.72
AE-GRU-EE	0.91	0.95	0.96	0.95
GAN-GRU-iForest	0.86	1.00	0.84	0.90
GAN-GRU-LOF	0.93	0.95	0.97	0.96
GAN-GRU-1SVM	0.44	1.00	0.37	0.43
GAN-GRU-EE	0.93	0.95	0.98	0.96
iForest	0.46	0.80	0.51	0.62
LOF	0.91	0.91	1.00	0.95
1SVM	0.84	0.88	0.95	0.91
EE	0.82	0.86	0.92	0.89

Table 6 AE-GRU adopted anomaly detection approaches using GICS dataset

Attack	Method	Accuracy	Precision	Recall	F1-score
Modification	iForest	0.9538	1	0.9483	0.9735
Modification	LOF	0.8462	0.8871	0.9483	0.9167
Modification	1SVM	1	1	1	1
Modification	EE	0.9154	0.9134	1	0.9547
ConnLoss	iForest	0.9576	0.9912	0.9658	0.9783
ConnLoss	LOF	0.9322	1	0.9316	0.9646
ConnLoss	1SVM	0.6102	0.9863	0.6154	0.7579
ConnLoss	EE	0.9831	0.9915	0.9915	0.9915
scanning	iForest	0.9524	1	0.9483	0.9735
scanning	LOF	0.9524	0.9508	1	0.9748
scanning	1SVM	0.9524	0.9508	1	0.9748
scanning	EE	0.8571	0.9455	0.8966	0.9204
Injection	iForest	0.7273	1	0.7143	0.8333
Injection	LOF	0.8333	1	0.8254	0.9043
Injection	1SVM	0.9697	1	0.9683	0.9839
Injection	EE	1	1	1	1

followed by the AE-GRU-iForest with an F-score of 0.9735 and the AE-GRU-EE with an F-score of 0.9547, while the AE-GRU-LOF recorded an F1-score of 0.9167. The AE-GRU-EE recorded the best performance for the connection loss attack with an F1-score of 0.9915, followed by the AE-GRU-based iForest and LOF schemes with an F1-score of 0.9783, and 0.9646, respectively. In contrast, the AE-GRU-1SVM did not record a good performance, with an F1-score of 0.7579. However, for scan resources attack, all adopted methods have recorded a high F1-score greater than 0.97, except the AE-GRU-EE with 0.9204. For injection attack detection, AE-GRU-EE dominates the other detectors. It is followed by the AE-GRU-based 1SVM, LOF, and iForest schemes with F1-score values of 0.9839, 0.9043, and 0.8333, respectively. Similar conclusions hold true also when using the GAN-GRU-based anomaly detection schemes (Table 7). There is no single approach dominating all models for the considered attacks.

In this case study, we also evaluate the standalone anomaly detection schemes to detect cyber-attacks in the GICS dataset (Table 8). The detection results demonstrate the suitable performance of the four detectors in identifying the four types of attacks that occurred in the GICS dataset. The EE scheme dominates the other standalone anomaly detection schemes (iF, 1SVM, and LOF) in this experiment by reaching an F1-score above 0.94. The LOF scheme follows it with an F1-score above 0.93 (Table 8).

As shown above, no unique approach is uniformly superior to others in detecting different attacks. Table 9 presents the aggregated performances of each cyber-attack detection scheme. In terms of all evaluation scores computed, the GAN-GRU-EE

Table 7 GAN-RNN adopted anomaly detection approaches using GICS dataset

Attack	Method	Accuracy	Precision	Recall	F1-score
ConnLoss	iForest	0.6695	0.9875	0.6752	0.802
ConnLoss	LOF	0.8898	1	0.8889	0.9412
ConnLoss	1SVM	0.7881	1	0.7863	0.8804
ConnLoss	EE	0.9831	0.9915	0.9915	0.9915
Injection	iForest	0.8485	1	0.8413	0.9138
Injection	LOF	0.9545	0.9545	1	0.9767
Injection	1SVM	0.9545	1	0.9524	0.9756
Injection	EE	1	1	1	1
scanning	iForest	0.9206	1	0.9138	0.955
scanning	LOF	0.9048	0.9483	0.9483	0.9483
scanning	1SVM	0.9841	0.9831	1	0.9915
scanning	EE	0.9365	0.9355	1	0.9667
Modification	iForest	0.8231	1	0.8017	0.8899
Modification	LOF	0.8769	1	0.8621	0.9259
Modification	1SVM	0.9692	0.9667	1	0.9831
Modification	EE	0.9769	0.9748	1	0.9872

Table 8 Standalone anomaly detection approaches using GICS dataset

Attack	Method	Accuracy	Precision	Recall	F1-score
Modification	iForest	0.8033	0.9125	0.8618	0.8864
Modification	LOF	0.8782	0.9174	0.9487	0.9328
Modification	1SVM	0.9175	0.9163	0.9986	0.9557
Modification	EE	0.8909	0.8909	1	0.9423
ConnLoss	iForest	0.6554	0.9978	0.6538	0.79
ConnLoss	LOF	0.9421	0.9925	0.9487	0.9701
ConnLoss	1SVM	0.8178	0.9897	0.8248	0.8998
ConnLoss	EE	0.9887	0.9915	0.9972	0.9943
Scanning	iForest	0.9235	0.9225	1	0.9597
Scanning	LOF	0.9183	0.9176	1	0.957
Scanning	1SVM	0.3268	0.7961	0.3504	0.4866
Scanning	EE	0.9183	0.9176	1	0.957
Injection	iForest	0.7136	1	0.6904	0.8168
Injection	LOF	0.9249	0.9249	1	0.961
Injection	1SVM	0.9249	0.9249	1	0.961
Injection	EE	0.9249	0.9249	1	0.961

Table 9 Averaged performance of AE-GRU and GAN-GRU-based methods using G-ICS dataset

Method	Accuracy	Precision	Recall	F1-score
AE-GRU-iForest	0.90	1.00	0.89	0.94
AE-GRU-LOF	0.89	0.96	0.93	0.94
AE-GRU-1SVM	0.88	0.98	0.90	0.93
AE-GRU-EE	0.94	0.96	0.97	0.97
GAN-GRU-iForest	0.82	1.00	0.81	0.89
GAN-GRU-LOF	0.91	0.98	0.92	0.95
GAN-GRU-1SVM	0.92	0.99	0.93	0.96
GAN-GRU-EE	0.97	0.98	1.00	0.99
iForest	0.77	0.96	0.80	0.86
LOF	0.92	0.94	0.97	0.96
1SVM	0.75	0.91	0.79	0.83
EE	0.92	0.92	1.00	0.96

scheme is the best approach to detect attacks in the G-ICS dataset with high efficiency by obtaining an averaged F1-score of 0.99. It is followed by the AE-GRU-EE scheme with an averaged F1-score of 0.97. It could be due to the capability of the hybrid deep learning models (i.e., AE-GRU and GAN-GRU) to extract relevant features from ICS traffic data and the sensitivity of the anomaly detection method. From Table 9, we also observe that standalone anomaly detection schemes (i.e., LOF and EE) show satisfactory performance in finding attack events in this experiment with an averaged F1-score of 0.96. Overall, we should highlight the potential of merging hybrid deep learning models with anomaly detection methods in detecting cyber-attacks in ICS traffic flow.

These results prove that the amalgamation of the hybrid deep learning models with anomaly detection methods provided satisfactory results in identifying cyber-attacks in ICS traffic flow.

Case (3): Cyber-Attack Detection in vrt-iec104 Dataset The final experiment aims to assess the potential of the proposed technique in detecting various attacks on IEC 104 communication created using the IEC virtual testbed developed by Peter Grofcik at Brno University of Technology [66]. Since IEC 104 packets may include multiple APDUs (application protocol data units) and ASDUs (application service data units), the IPFIX probe separates these packets into so-called virtual IEC 104 packets used for further analysis. The dataset was captured at the HMI side in the topology [66]. The propose models have been trained using three days of normal communication data with no attacks (i.e., total of 381.666 virtual packets). Then, we assessed the trained models using data with different types of attacks, including report-block-HMI attacks, HMI-MITM attacks, reply-HMI attacks, value-change-

HMI attacks,and masquerating attacks. More details about the considered attacks can be find in the readme file in [66].

Table 10 reports the detection results of the AE-GRU-based anomaly detection methods when applied to the vrt-iec104 dataset. Results indicate the AE-GRU-LOF recorded the highest detection accuracy for detecting Command Act attacks, with an F1-score of 0.9831; it is followed by the AE-GRU-EE scheme with an F1-score of 0.9565. For the other attacks, we observe that the AE-GRU-based EE and LOF schemes recorded the highest F1-score greater than 0.95, while the other methods did perform well.

Table 11 lists detection results of the GAN-GRU-based anomaly detection schemes using the vrt-iec104 dataset. We observe that GAN-GRU-based iForest and LOF schemes recorded the highest F1-score of 9831 compared to GAN-GRU-based 1SVM and EE schemes that achieved an under 0.78 which is not satisfactory. For the Changes value and replaying HMI attacks, the GAN-GRU-LOF was the only method that satisfactory detected this kind of attack with an F1-score of 0.9831. The GAN-GRU-LOF reached the best performance with an F1-score of 0.9831 regarding the attack replaying RTU; after that comes the GAN-GRU-1SVM with an F1-score of 0.806. In contrast, the other methods could not detect this kind of attack properly. The same performance detection was recorded for the attacks: masquerading, change of HMI, and change of RTU, where the GAN-GRU-LOF scheme is the only able to detect the attacks with an F1-score of 0.9831.

From results in Tables 10 and 11, the main finding is that AE-GRU and GAN-GRU-based LOF schemes have maintained a high detection performance for all attacks. Moreover, the remaining methods could not keep a high detection accept for the block of HMI (Table 12).

Table 13 reports the detection performance of the standalone anomaly detection schemes when applied to the vrt-iec104 dataset. It can be seen that the standalone anomaly detection methods are not able to correctly detect cyber-attacks in the vrt-iec104 dataset (Table 13). They provide unsuited detection performance, making them inappropriate to be adopted in practice for this application.

Table 14 displays the aggregated performances of each approach based on the three considered datasets. We can see that the AE-GRU and GAN-GRU-based LOF schemes achieved the best detection performance of all attacks, with an F1-score of 0.98. They dominate the other investigated deep learning-based anomaly detection methods and the standalone anomaly detection methods.

5 Conclusion

Accurately detecting attacks in ICS communication networks is undoubtedly necessary for designing modern smart grids and ensuring safe and reliable operation. This chapter introduced data-driven anomaly detection approaches for intrusion detection in a smart grid environment by monitoring ICS traffic communication. At first, we designed two-hybrid deep learning models, AE-GRU and GAN-GRU, by integrating

Table 10 AE-GRU adopted anomaly detection approaches using vrt-iec104 dataset

Attack	Method	Accuracy	Precision	Recall	F1-score
command Act	iForest	0.7396	0.7188	0.9583	0.8214
command Act	LOF	0.9792	1	0.9667	0.9831
command Act	1SVM	0.5729	0.6092	0.8833	0.7211
command Act	EE	0.9479	1	0.9167	0.9565
Modification	iForest	0.1282	0.1239	0.975	0.2199
Modification	LOF	0.9958	1	0.9667	0.9831
Modification	1SVM	0.0987	0.1006	0.775	0.1781
Modification	EE	0.9958	1	0.9667	0.9831
block-HMI	iForest	0.4579	0.124	0.95	0.2194
block-HMI	LOF	0.9973	1	0.9667	0.9831
block-HMI	1SVM	0.0608	0.0614	0.75	0.1135
block-HMI	EE	0.9973	1	0.9667	0.9831
replaying HMI	iForest	0.2989	0.0933	0.9417	0.1698
replaying HMI	LOF	0.9975	1	0.9667	0.9831
replaying HMI	1SVM	0.0647	0.0638	0.825	0.1184
replaying HMI	EE	0.9937	1	0.9167	0.9565
replaying RTU	iForest	0.3438	0.2452	0.95	0.3898
replaying RTU	LOF	0.9926	1	0.9667	0.9831
replaying RTU	1SVM	0.3768	0.2459	0.8833	0.3847
replaying RTU	EE	0.9798	1	0.9083	0.9519
change of HMI	iForest	0.1571	0.1064	0.9583	0.1915
change of HMI	LOF	0.9965	1	0.9667	0.9831
change of HMI	1SVM	0.2335	0.0928	0.725	0.1645
change of HMI	EE	0.9913	1	0.9167	0.9565
change of RTU	iForest	0.2478	0.0997	0.95	0.1805
change of RTU	LOF	0.9971	1	0.9667	0.9831
change of RTU	1SVM	0.077	0.0778	0.8833	0.143
change of RTU	EE	0.9964	1	0.9583	0.9787
masquerating	iForest	0.3289	0.0497	0.9417	0.0944
masquerating	LOF	0.9985	1	0.9583	0.9787
masquerating	1SVM	0.2067	0.0282	0.6083	0.0539
masquerating	EE	0.9966	1	0.9083	0.9519

AE and GRU's desirable characteristics with the GRU model's capacity to capture the time-dependence in ICS traffic data. Then, these two models have been combined with semi-supervised anomaly detection techniques to design semi-supervised deep recurrent-based anomaly detection schemes for ICS communication monitoring in smart grids. The AE-GRU and GAN-GRU models aim to automatically learn and capture the relevant features from ICS traffic flow, and the anomaly detection schemes

Table 11 GAN-RNN adopted anomaly detection approaches using vrt-iec104 dataset

Attack	Method	Accuracy	Precision	Recall	F1-score
command Act	iForest	0.9792	1	0.9667	0.9831
command Act	LOF	0.9792	1	0.9667	0.9831
command Act	1SVM	0.7188	0.7845	0.7583	0.7712
command Act	EE	0.5885	0.6108	0.9417	0.741
changes Value	iForest	0.3792	0.1612	0.9333	0.2749
changes Value	LOF	0.9958	1	0.9667	0.9831
changes Value	1SVM	0.6029	0.1943	0.6833	0.3026
changes Value	EE	0.1786	0.1247	0.9167	0.2195
block-HMI	iForest	0.996	1	0.95	0.9744
block-HMI	LOF	0.9973	1	0.9667	0.9831
block-HMI	1SVM	0.9646	1	0.5583	0.7166
block-HMI	EE	0.9967	1	0.9583	0.9787
replaying HMI	iForest	0.5615	0.1418	0.9417	0.2465
replaying HMI	LOF	0.9975	1	0.9667	0.9831
replaying HMI	1SVM	0.8598	0.2735	0.5083	0.3556
replaying HMI	EE	0.1523	0.0766	0.9167	0.1414
replaying RTU	iForest	0.3989	0.2653	0.975	0.4171
replaying RTU	LOF	0.9926	1	0.9667	0.9831
replaying RTU	1SVM	0.9283	1	0.675	0.806
replaying RTU	EE	0.204	0.2075	0.925	0.339
change of HMI	iForest	0.2101	0.1143	0.975	0.2046
change of HMI	LOF	0.9965	1	0.9667	0.9831
change of HMI	1SVM	0.9158	0.5804	0.6917	0.6312
change of HMI	EE	0.3168	0.129	0.9667	0.2276
change of RTU	iForest	0.96	1	0.5417	0.7027
change of RTU	LOF	0.9971	1	0.9667	0.9831
change of RTU	1SVM	0.915	0.5101	0.6333	0.5651
change of RTU	EE	0.2515	0.1016	0.9667	0.1839
masquerating	iForest	0.9817	1	0.5083	0.674
masquerating	LOF	0.9988	1	0.9667	0.9831
masquerating	1SVM	0.9202	0.1835	0.3333	0.2367
masquerating	EE	0.1064	0.0387	0.9667	0.0744

Table 12 Averaged performance of AE-GRU and GAN-GRU-based methods using vrt-iec104 dataset

Method	Accuracy	Precision	Recall	F1-score
AE-GRU-iForest	0.34	0.20	0.95	0.29
AE-GRU-LOF	0.99	1.00	0.97	0.98
AE-GRU-1SVM	0.21	0.16	0.79	0.23
AE-GRU-EE	0.99	1.00	0.93	0.96
GAN-GRU-iForest	0.68	0.59	0.85	0.56
GAN-GRU-LOF	0.99	1.00	0.97	0.98
GAN-GRU-1SVM	0.85	0.57	0.61	0.55
GAN-GRU-EE	0.35	0.29	0.94	0.36
iForest	0.37	0.17	0.75	0.26
LOF	0.18	0.17	0.98	0.26
1SVM	0.21	0.17	0.94	0.26
EE	0.19	0.17	0.98	0.26

(i.e., iForest, LOF, 1SVM, and EE) are applied to extracted features to detect cyber-attacks based. Three publically available datasets are used to assess the performance of the proposed approaches. Four statistical scores have been utilized to judge the detection accuracy of the studied methods, including accuracy, precision, recall, and F1-score. Results revealed that the proposed AE-GRU-based EE and LOF methods offer superior detection performance of different cyber-attack types and dominate the other investigated methods.

Despite the improved detection performance of the proposed semi-supervised deep learning-based techniques for detecting anomalies in ICS traffic on the IEC 104 communication, future works will be aimed to extend further the range of their application to monitor ICS communication of industrial systems. Further, we plan to improve the robustness of the hybrid deep learning models (AE-GRU and GAN-GRU) to noisy measurements by developing a wavelet-based hybrid deep learning detector. To this end, we will use wavelet decomposition to capture multivariate information in the time and frequency domains and then employ a hybrid deep learning model to extract relevant features that will be fed to the anomaly detection methods for cyber-attack detection. It is expected that by applying wavelet-based multi-scale denoising to the received signals, noise effects will be reduced, thus improving cyber-attack detection. Another interesting direction for future work is the design of unsupervised cyber-attack detection strategy by integrating unsupervised deep learning methods as features extractors, such as deep variational auto-encoder [68], with the sensitivity of statistical monitoring charts, such as Generalized Likelihood Ratio Test [69].

Table 13 Standalone anomaly detection approaches using vrt-iec104 dataset

Attack	Method	Accuracy	Precision	Recall	F1-score
Command Act	iForest	0.6307	0.6344	0.98	0.7702
Command Act	LOF	0.6358	0.6374	0.982	0.773
Command Act	1SVM	0.616	0.6314	0.942	0.756
Command Act	EE	0.6316	0.6316	1	0.7742
Modification	iForest	0.133	0.1288	0.9787	0.2276
Modification	LOF	0.133	0.1292	0.982	0.2284
Modification	1SVM	0.1739	0.1307	0.942	0.2296
Modification	EE	0.1306	0.1306	1	0.231
Block-HMI	iForest	0.4786	0.0814	0.5127	0.1405
Block-HMI	LOF	0.0849	0.0826	0.99	0.1525
Block-HMI	1SVM	0.0834	0.0831	0.9987	0.1534
Block-HMI	EE	0.0831	0.0831	1	0.1534
Replaying HMI	iForest	0.2225	0.2231	0.9807	0.3635
Replaying HMI	LOF	0.2227	0.2234	0.982	0.364
Replaying HMI	1SVM	0.2589	0.2266	0.942	0.3653
Replaying HMI	EE	0.2264	0.2264	1	0.3692
Replaying RTU	iForest	0.0822	0.0777	0.9793	0.144
Replaying RTU	LOF	0.0865	0.0782	0.982	0.1449
Replaying RTU	1SVM	0.0978	0.0764	0.942	0.1413
Replaying RTU	EE	0.0788	0.0788	1	0.1461
Change of HMI	iForest	0.4777	0.1054	0.5127	0.1749
Change of HMI	LOF	0.1097	0.1066	0.982	0.1923
Change of HMI	1SVM	0.1539	0.108	0.942	0.1938
Change of HMI	EE	0.1079	0.1079	1	0.1948
Change of RTU	iForest	0.4777	0.0884	0.5127	0.1508
Change of RTU	LOF	0.0932	0.0894	0.982	0.1639
Change of RTU	1SVM	0.1385	0.0905	0.942	0.1651
Change of RTU	EE	0.0905	0.0905	1	0.166
Masquerating	iForest	0.477	0.0378	0.5127	0.0704
Masquerating	LOF	0.0413	0.0383	0.99	0.0737
Masquerating	1SVM	0.1467	0.0386	0.8833	0.074
Masquerating	EE	0.1506	0.0381	0.866	0.073

Table 14 Averaged performance of AE-GRU and GAN-GRU-based methods using the three datasets

Method	Accuracy	Precision	Recall	F1-score
AE-GRU-iForest	0.34	0.20	0.95	0.29
AE-GRU-LOF	0.99	1.00	0.97	0.98
AE-GRU-1SVM	0.21	0.16	0.79	0.23
AE-GRU-EE	0.99	1.00	0.93	0.96
GAN-GRU-iForest	0.68	0.59	0.85	0.56
GAN-GRU-LOF	0.99	1.00	0.97	0.98
GAN-GRU-1SVM	0.85	0.57	0.61	0.55
GAN-GRU-EE	0.35	0.29	0.94	0.36
iForest	0.53	0.64	0.69	0.58
LOF	0.67	0.67	0.99	0.72
1SVM	0.60	0.65	0.90	0.67
EE	0.64	0.65	0.97	0.70

Acknowledgements This publication is based upon work supported by King Abdullah University of Science and Technology (KAUST), Office of Sponsored Research (OSR) under Award No: OSR-2019-CRG7-3800.

References

1. W. Wang, Z. Lu, Cyber security in the smart grid: survey and challenges. Comput. Netw. **57**(5), 1344–1371 (2013)
2. M.Z. Gunduz, R. Das, Cyber-security on smart grid: threats and potential solutions. Comput. Netw. **169**, 107094 (2020)
3. R. Leszczyna, A review of standards with cybersecurity requirements for smart grid. Comput. Secur. **77**, 262–276 (2018)
4. R.K. Pandey, M. Misra, Cyber security threats-smart grid infrastructure, in *National Power Systems Conference (NPSC)*, vol. 2016 (IEEE, 2016), pp. 1–6
5. A. Teymouri, A. Mehrizi-Sani, C.-C. Liu, Cyber security risk assessment of solar pv units with reactive power capability, in *IECON 2018-44th Annual Conference of the IEEE Industrial Electronics Society* (IEEE, 2018), pp. 2872–2877
6. W. Wang, F. Harrou, B. Bouyeddou, S.-M. Senouci, Y. Sun, A stacked deep learning approach to cyber-attacks detection in industrial systems: application to power system and gas pipeline systems. Clust. Comput. **25**(1), 561–578 (2022)
7. M. Stănculescu, S. Deleanu, P.C. Andrei, H. Andrei, A case study of an industrial power plant under cyberattack: simulation and analysis. Energies **14**(9), 2568 (2021)
8. A.A.Z. Khan, G. Serpen, Intrusion detection and identification system design and performance evaluation for industrial scada networks (2020), arXiv preprint arXiv:2012.09707
9. J.R. Vacca, *Cyber Security and IT Infrastructure Protection*. Syngress (2013)
10. M. Touhiduzzaman, S.N.G. Gourisetti, C. Eppinger, A. Somani, A review of cybersecurity risk and consequences for critical infrastructure. 2019 Resilience Week (RWS) **1**, 7–13 (2019)

11. J. Jiang, X. Zhao, S. Wallace, E. Cotilla-Sanchez, R. Bass, Mining pmu data streams to improve electric power system resilience, in *Proceedings of the Fourth IEEE/ACM International Conference on Big Data Computing, Applications and Technologies* (2017), pp. 95–102
12. C. Konstantinou, M. Sazos, M. Maniatakos, Attacking the smart grid using public information, in *17th Latin-American Test Symposium (LATS)*, vol. 2016. (IEEE, 2016), pp. 105–110
13. S. Basumallik, R. Ma, S. Eftekharnejad, Packet-data anomaly detection in pmu-based state estimator using convolutional neural network. Int. J. Electrical Power Energy Syst. **107**, 690–702 (2019)
14. W. Wang, F. Harrou, B. Bouyeddou, S.-M. Senouci, Y. Sun, Cyber-attacks detection in industrial systems using artificial intelligence-driven methods. Int. J. Critic. Infrastruct. Protect. 100542 (2022)
15. A. Walker, J. Desai, D. Saleem, T. Gunda, *Cybersecurity in Photovoltaic Plant Operations, National Renewable Energy Lab* (NREL), Golden, CO (United States), Technical Report, 2021)
16. J. Ye, A. Giani, A. Elasser, S.K. Mazumder, C. Farnell, H.A. Mantooth, T. Kim, J. Liu, B. Chen, G.-S. Seo et al., *A review of cyber-physical security for photovoltaic systems* (IEEE J. Emerg. Select, Topics Power Electron, 2021)
17. C.-C. Sun, A. Hahn, C.-C. Liu, Cyber security of a power grid: state-of-the-art. Int. J. Electr. Power Energy Syst. **99**, 45–56 (2018)
18. Z. El Mrabet, N. Kaabouch, H. El Ghazi, H. El Ghazi, Cyber-security in smart grid: survey and challenges. Comput. Electr. Eng. **67**, 469–482 (2018)
19. F. Nejabatkhah, Y.W. Li, H. Liang, and R. Reza Ahrabi, Cyber-security of smart microgrids: a survey. Energies **14**(1), 27 (2020)
20. Y. Zhang, L. Wang, Z. Liu, W. Wei, A cyber-insurance scheme for water distribution systems considering malicious cyberattacks. IEEE Trans. Inf. Forens. Secur. **16**, 1855–1867 (2020)
21. A. Giani, E. Bitar, M. Garcia, M. McQueen, P. Khargonekar, K. Poolla, Smart grid data integrity attacks: characterizations and countermeasures π, in *2011 IEEE International Conference on Smart Grid Communications (SmartGridComm)* (IEEE, 2011), pp. 232–237
22. D. An, Q. Yang, W. Liu, Y. Zhang, Defending against data integrity attacks in smart grid: A deep reinforcement learning-based approach. IEEE Access **7**, 110 835–110 845 (2019)
23. P. Srikantha, D. Kundur, Denial of service attacks and mitigation for stability in cyber-enabled power grid, in *IEEE Power and Energy Society Innovative Smart Grid Technologies Conference (ISGT)*, vol. 2015 (IEEE, 2015), pp. 1–5
24. M.Z. Gunduz, R. Das, Analysis of cyber-attacks on smart grid applications, in *2018 International Conference on Artificial Intelligence and Data Processing (IDAP)* (IEEE, 2018), pp. 1–5
25. A. Huseinovic, S. Mrdovic, K. Bicakci, S. Uludag, A taxonomy of the emerging denial-of-service attacks in the smart grid and countermeasures, in *26th Telecommunications Forum (TELFOR)*, vol. 2018 (IEEE, 2018), pp. 1–4
26. Z. Lu, X. Lu, W. Wang, C. Wang, Review and evaluation of security threats on the communication networks in the smart grid, in *2010-Milcom, Military Communications Conference*, vol. 2010 (IEEE, 2010), pp. 1830–1835
27. S.A. Yadav, S.R. Kumar, S. Sharma, A. Singh, A review of possibilities and solutions of cyber attacks in smart grids, in *2016 International Conference on Innovation and Challenges in Cyber Security (ICICCS-INBUSH)* (IEEE, 2016), pp. 60–63
28. M. Caselli, E. Zambon, F. Kargl, Sequence-aware intrusion detection in industrial control systems, in *Proceedings of the 1st ACM Workshop on Cyber-Physical System Security* (2015), pp. 13–24
29. H. Yoo, T. Shon, Novel approach for detecting network anomalies for substation automation based on iec 61850. Multimedia Tools Appl. **74**(1), 303–318 (2015)
30. E. Maynard, K. McLaughlin, B. Haberler, Towards understanding man-in-the-middle attacks on iec 60870-5-104 scada networks, in *2nd International Symposium for ICS and SCADA Cyber Security Research 2014 (ICS-CSR 2014) 2* (2014), pp. 30–42
31. P. Matoušek, O. Ryšavý, M. Grégr, V. Havlena, Flow based monitoring of ics communication in the smart grid. J. Inf. Secur. Appl. **54**, 102535 (2020)

32. J. Jarmakiewicz, K. Parobczak, K. Maślanka, Cybersecurity protection for power grid control infrastructures. Int. J. Crit. Infrastruct. Prot. **18**, 20–33 (2017)
33. J. Hong, C.-C. Liu, M. Govindarasu, Integrated anomaly detection for cyber security of the substations. IEEE Trans. Smart Grid **5**(4), 1643–1653 (2014)
34. A. Valdes, S. Cheung, "Communication pattern anomaly detection in process control systems, in *2009 IEEE Conference on Technologies for Homeland Security* (IEEE, 2009), pp. 22–29
35. Y. Yang, K. McLaughlin, T. Littler, S. Sezer, B. Pranggono, H. Wang, Intrusion detection system for iec 60870-5-104 based scada networks, in *IEEE Power and Energy Society General Meeting*, vol. 2013 (IEEE, 2013), 1–5
36. C.-Y. Lin, S. Nadjm-Tehrani, Understanding iec-60870-5-104 traffic patterns in scada networks, in *Proceedings of the 4th ACM Workshop on Cyber-Physical System Security* (2018), pp. 51–60
37. A. Kleinmann, A. Wool, Automatic construction of statechart-based anomaly detection models for multi-threaded scada via spectral analysis, in *Proceedings of the 2nd ACM Workshop on Cyber-Physical Systems Security and Privacy* (2016), pp. 1–12
38. S. Shukla, S. Thakur, J.G. Breslin, Anomaly detection in smart grid network using fc-based blockchain model and linear svm, in *International Conference on Machine Learning, Optimization, and Data Science* (Springer, 2021), pp. 157–171
39. F. Harrou, Y. Sun, A.S. Hering, M. Madakyaru, A. Dairi, Unsupervised deep learning-based process monitoring methods, in *Statistical Process Monitoring Using Advanced Data-Driven and Deep Learning Approaches* (Elsevier, 2021), pp. 193–223
40. A. Dairi, F. Harrou, Y. Sun, M. Senouci, Obstacle detection for intelligent transportation systems using deep stacked autoencoder and k-nearest neighbor scheme. IEEE Sens. J. **18**(12), 5122–5132 (2018)
41. A. Dairi, F. Harrou, M. Senouci, Y. Sun, Unsupervised obstacle detection in driving environments using deep-learning-based stereovision. Robot. Auton. Syst. **100**, 287–301 (2018)
42. D. Charte, F. Charte, S. García, M.J. del Jesus, F. Herrera, A practical tutorial on autoencoders for nonlinear feature fusion: taxonomy, models, software and guidelines. Inf. Fus. **44**, 78–96 (2018)
43. S. Hochreiter, J. Schmidhuber, Long short-term memory. Neural Comput. **9**(8), 1735–1780 (1997)
44. F. Harrou, F. Kadri, Y. Sun, Forecasting of photovoltaic solar power production using lstm approach, in *Advanced Statistical Modeling, Forecasting, and Fault Detection in Renewable Energy Systems* (2020), p. 3
45. A. Zeroual, F. Harrou, A. Dairi, Y. Sun, Deep learning methods for forecasting covid-19 time-series data: a comparative study. Chaos, Solitons Fractals **140**, 110121 (2020)
46. A. Creswell, T. White, V. Dumoulin, K. Arulkumaran, B. Sengupta, A.A. Bharath, Generative adversarial networks: an overview. IEEE Signal Process. Mag. **35**(1), 53–65 (2018)
47. L. Zhu, Y. Chen, P. Ghamisi, J.A. Benediktsson, Generative adversarial networks for hyperspectral image classification. IEEE Trans. Geosci. Remote Sens. **56**(9), 5046–5063 (2018)
48. I. Goodfellow, J. Pouget-Abadie, M. Mirza, B. Xu, D. Warde-Farley, S. Ozair, A. Courville, Y. Bengio, Generative adversarial networks. Commun. ACM **63**(11), 139–144 (2020)
49. F. Kadri, A. Dairi, F. Harrou, Y. Sun, Towards accurate prediction of patient length of stay at emergency department: a gan-driven deep learning framework. J. Ambient Intell. Human. Comput. 1–15 (2022)
50. R.R.R. Barbosa, R. Sadre, A. Pras, Towards periodicity based anomaly detection in scada networks, in *Proceedings of 2012 IEEE 17th International Conference on Emerging Technologies and Factory Automation (ETFA 2012)* (IEEE, 2012), pp. 1–4
51. H.J. Shin, D.-H. Eom, S.-S. Kim, One-class support vector machines-an application in machine fault detection and classification. Comput. Ind. Eng. **48**(2), 395–408 (2005)
52. F. Harrou, N. Zerrouki, A. Dairi, Y. Sun, A. Houacine, Automatic human fall detection using multiple tri-axial accelerometers, in *2021 International Conference on Innovation and Intelligence for Informatics, Computing, and Technologies (3ICT)* (2021), pp. 74–78
53. B. Schölkopf, J.C. Platt, J. Shawe-Taylor, A.J. Smola, R.C. Williamson, Estimating the support of a high-dimensional distribution. Neural Comput. **13**(7), 1443–1471 (2001)

54. F. Harrou, N. Zerrouki, A. Dairi, Y. Sun, A. Houacine, Automatic human fall detection using multiple tri-axial accelerometers, in *2021 International Conference on Innovation and Intelligence for Informatics, Computing, and Technologies (3ICT)* (IEEE, 2021), pp. 74–78
55. A. Dairi, F. Harrou, Y. Sun, Deep generative learning-based 1-svm detectors for unsupervised covid-19 infection detection using blood tests. IEEE Trans. Instrum. Meas. **71**, 1–11 (2021)
56. F. Harrou, A. Dairi, B. Taghezouit, Y. Sun, An unsupervised monitoring procedure for detecting anomalies in photovoltaic systems using a one-class support vector machine. Sol. Energy **179**, 48–58 (2019)
57. F.T. Liu, K.M. Ting, Z.-H. Zhou, Isolation forest, in *Eighth IEEE International Conference on Data Mining*, vol. 2008 (IEEE, 2008), pp. 413–422
58. F.T. Liu, K.M. Ting, Z.-H. Zhou, Isolation-based anomaly detection. ACM Trans. Knowl. Discov. Data (TKDD) **6**(1), 1–39 (2012)
59. L. Breiman, Random forests. Mach. Learn. **45**(1), 5–32 (2001)
60. P.J. Rousseeuw, K.V. Driessen, A fast algorithm for the minimum covariance determinant estimator. Technometrics **41**(3), 212–223 (1999)
61. M.M. Breunig, H.-P. Kriegel, R.T. Ng, J. Sander, Lof: identifying density-based local outliers," in *Proceedings of the 2000 ACM SIGMOD International Conference on Management of Data* (2000), pp. 93–104
62. C.-Y. Lin, S. Nadjm-Tehrani, A comparative analysis of emulated and real iec-104 spontaneous traffic in power system networks, in *International Workshop on Cyber-Physical Security for Critical Infrastructures Protection* (Springer, 2020), pp. 207–223
63. G. Clarke, D. Reynders, E. Wright, Practical modern SCADA protocols: DNP3, 60870.5 and related systems. Newnes (2004)
64. F. Harrou, B. Khaldi, Y. Sun, F. Cherif, An efficient statistical strategy to monitor a robot swarm. IEEE Sens. J. **20**(4), 2214–2223 (2019)
65. D.M. Powers, Evaluation: from precision, recall and f-measure to roc, informedness, markedness and correlation (2020), arXiv preprint arXiv:2010.16061
66. P. Matoušek, O. Ryšavý, P. Grofčík, ICS dataset for smart grid anomaly detection (2022). https://dx.doi.org/10.21227/1trw-n685
67. P. Matoušek, V. Havlena, L. Holík, Efficient modelling of ics communication for anomaly detection using probabilistic automata, in *IFIP/IEEE International Symposium on Integrated Network Management (IM)*, vol. 2021 (IEEE, 2021), pp. 81–89
68. A. Dairi, F. Harrou, Y. Sun, S. Khadraoui, Short-term forecasting of photovoltaic solar power production using variational auto-encoder driven deep learning approach. Appl. Sci. **10**(23), 8400 (2020)
69. F. Harrou, Y. Sun, A.S. Hering, M. Madakyaru et al., *Statistical Process Monitoring Using Advanced Data-Driven and Deep Learning Approaches: Theory and Practical Applications* (Elsevier, 2020)

Vertical Approach Anomaly Detection Using Local Outlier Factor

Nils Jakob Johannesen, Mohan Lal Kolhe, and Morten Goodwin

Abstract Detection of anomalies based on smart meter data is crucial to identify potential risks and unusual events at an early stage. In addition anomaly detection can be used as a tool to detect unwanted outliers, caused by operational failures and technical faults, for the pre-processing of data for machine learning, to detect concept drift as well as enhancing cyber-security in smart electrical grid operations. It is known that anomalies are defined through their contextual appearance. Hence, anomalies are divided into point, conceptual and contextual anomalies. In this work the contextual anomaly detection is examined, through a novel type of load forecasting known as vertical approach. This chapter explores the use of anomaly detection in the relevant learning systems for machine learning in smart electrical grid operation and management through data from New South Wales region in Australia. The presented vertical time approach uses seasonal data for training and inference, as opposed to continuous time approach that utilizes all data in a continuum from the start of the dataset until the time used for inference. It is observed that Local Outlier Factor identifies different local outliers given different vertical approaches. In addition, the local outlier factor score vary vertically. An anomaly is defined as a deviation from an established normal pattern. Spotting an anomaly depends on the ability to defy what is normal. Anomaly detection systems aim at finding these

N. J. Johannesen (✉) · M. L. Kolhe
Department of Engineering and Science, University of Agder, PO Box 422, 4604 Kristiansand, Norway
e-mail: nils.j.johannesen@usn.no
URL: http://www.usn.no

M. L. Kolhe
e-mail: mohan.l.kolhe@uia.no
URL: http://www.uia.no

N. J. Johannesen
Department of Electrical Engineering, IT and Cybernetics University of Southest-Norway, PO Box 4, 3199 Borre, Norway

M. Goodwin
Department of Information and Communication Technology, University of Agder, PO Box 422, 4604 Kristiansand, Norway
e-mail: morten.goodwin@uia.no

H. Haes Alhelou et al. (eds.), *Power Systems Cybersecurity*, Power Systems,
https://doi.org/10.1007/978-3-031-20360-2_12

anomalies. Anomaly detection systems are in high demand, despite the fact that there is no clear validation approach. These systems rely on deep domain expertise.

Keywords Cybersecurity · Anomaly detection · Smart grid · Local outlier factor

1 Introduction

Machine learning (ML) can provide electrical load demand forecasting, giving information about future loads, which provides essential input to other applications such as Demand Response, Topology Optimization and Anomaly Detection, facilitating the integration of intermittent clean energy sources.

An anomaly is defined as a deviation from an established normal pattern. Spotting an anomaly depends on the ability to defy what is normal. Anomaly detection systems aim at finding these anomalies. Anomaly detection systems are in high demand despite the fact that there is no clear validation approach. These systems rely on deep domain expertise.

In the safe operation and management of the smart grid there is a need for efficient and reliable detection of anomalies. The data used in grid operation is of such an amount, that it is not possible to do so manually or by visual inspection alone, and there is a need for efficient, automated and accurate anomaly detection methods [1].

The available advanced information and communicating platform and computational capability renders smart grid prone to attacks with extreme social, financial and physical effects. The smart grid concept enables the utilization of smart appliances in homes and electric vehicles for providing support for frequency regulation and voltage regulation. Cyber threats could affect the ancillary services that are being delivered from the aggregators, which might lead to stability and security issues resulting in brownout or massive blackouts [2].

To coordinate and manage the increase of renewable energy sources, such as wind, sun and hydro, they can be operated using gried-tied voltage source converters (VSCs) [3]. VSCs regulate voltage and frequency locally. The VSCs enable the operation of intelligent microgrids (MGs), and vulnerable for attack. In the distributed power network the attack can disrupt the frequency regulation, voltage stability and the power flow management [4].

The implementation of two way communication by the use of sensors and intelligent agents such as advanced metering infrastructure (AMI) as well as load aggregation, make these attractive objects for cyber attacks. Sensors can be penetrated using a Trojan Horse, to manipulate the adversary inside the control platform, and change reference inputs in both outer and secondary control for VSCs. The attacker can here change acquisition gains, that create bias in the measurements report.

The enhanced agent topology of a smart electrical grid, reveals the mentioned vulnerabilities. The valuable question is; what is the price of the smartness of the smart electrical grid, in terms of the security of the supply?

Anomaly detection is an important first step in supervised machine learning processing to search out erroneous data, where the data was recorded due to an error or disrupted by other causes. Examples of such erroneous data could be the product of a sensor fault, downtime due to maintenance or when time series data is recorded in the changes from daylight saving time. This first step in data cleaning process has been known to enhance any forecasting algorithm [5, 6].

From the field of Time Series Analysis and different correlation studies, it is known that seasonality affects the electrical load consumption. Time series analysis includes this knowledge in parametric methods such as Seasonal AutoRegressive Moving Average [7, 8]. The authors of this book chapter has also included drift due to seasonal variations, in the method vertical axis approach [9, 10].

In the Sect. 2 of this book chapter the different learning systems in anomaly detection are explained, as well as the main categories of anomalies. Section 3 covers the literature review. Section 4 gives the mathematical foundation for LOF. Section 5 presents the methodology used, and finally in Sect. 6 are given the results and conclusion.

2 Learning Systems in Anomaly Detection

Machine learning algorithms are divided into unsupervised, supervised and semi-supervised learning. Unsupervised learning is when the algorithm is learning without knowing the target for its learning. The algorithm is trying to make sense amongst the features of the data, as in discovering natural groups within the data by clustering techniques. In supervised learning the algorithm know the target of prediction during the training stage of the algorithm. When training the algorithm features are divided into dependent and independent features. The independent features involves the basis of the decision making, the input vector, and the dependent variable is the target vector. In supervised learning the patters between input and output is learned in the training phase and based on this result the models parameters are identified. Semi-supervised learning method is a hybrid between the mentioned supervised and unsupervised learning, and it involves elements from both supervised and unsupervised learning- Semi-supervised learning uses partially labeled and unlabeled instances to detect anomalies. One instance is in autoencoders where only the data that depicts normal behavior are labeled and used to train the algorithm. Based on the assumption of what is normal it will flag anomalous datapoints when exposed to data that differs from normal data [11].

To learn patterns across timescales, sliding window technique is introduced. Sliding window method is used when doing time series forecasting, and it introduces a shift in the input and output variable relation. This can be done through dynamic or static modeling. For this work we will explain the static modeling of sliding window method.

In the Sliding window method the training and test set input and target vectors timestamp is shifted using a sliding technique that allows the number of timesteps shifted, to equal the predictive window. In practical terms it means that the algorithm reads a features input for a certain timeframe and then matches it with the selected predictive window. In the training phase of the modeling, the algorithm will then learn the patterns that suggest the following behaviour ahead in time. In sliding window adaptation the 'Feature Data' columns is the input (time of day, dry-bulb temperature, humidity, previous loads, etc.) and the target vector is 'Load Data', and the 'Load Data' is shifted 48 timesteps back (half-hourly values), and this is how the algorithm is fed the information when doing day ahead forecasting.

The sliding window method explained here is used in supervised learning for time series. Alternative versions of sliding window method have been adopted to deal with concept drift. In machine learning concept drift occurs when the underlying distribution of the data changes over time changes over time, making the model unfit to predict for future events [12]. An adaptive sliding window method, that calculates an adoptive window-size on the fly is has been proposed to deal with concept drift [13]. Another adaptation raises the complexity by using multi sliding window detecting growth length over several windows detecting the drift length by adjoining several windows finding the optimal window length useful for online learning [14].

In Recurrent Neural Networks like Long Short-term Memory networks (LSTM) different gates are used to remember and forget time-occurrence over different time windows. To deal with concept drift in LSTMs it has been proposed a novel forgetting mechanism for anomaly detection [15]. It has also been proposed to narrow down the scope, by critical lines detecting distribution change. The first step in the proposed method is to reduced dimensionality through an orthogonal transformation of the data reducing the feature space to its principal components [16]. After reducing the feature space the distributions are compared by two-sample Kolmogorov-Smirnov Test (KS test) [17].

Anomalies depend on the structure, distribution and type of data, as well as carnality of relationship of the data [18]. Basically there are 3 main types of anomalies: Point, Contextual and Collective anomalies. Point anomaly is an instant, that can be regarded anomalous, amongst other instances in the data. They are extreme, and caused instantly. In electric load demand point anomalies when other sources of energy (such as district heating) are curtailed [19]. Collective Anomaly is an instance that is defined an anomaly based on the context. Meaning that, in some context (e.g based on the temperature of a season), a particular behaviour is normal, but put in a different context (a different season) it qualifies as abnormal behaviour. Collective anomalies are data instances that deviate significantly from the entire dataset, but individual data amongst the collective instances may not qualify as an outlier.

3 Literature Review

Anomaly detection is done on any time series data. Robust statistical methods have been known to determine deviation from normal electrical load patterns. Simple statistical methods, such as Box-plots have been used in the pre-processing of electrical consumption load profiles to determine daily usage patterns [19]. Box-plot have also been used to identify anomalies on regression forecast errors in order to improve the prediction [20].

Efforts to find data-mining framework to typical electricity load patterns (TELP). TELP has been proven successful for anomaly detection in building electricity consumption data. The first step in TELP is to cluster data in temporal segments, such as daily electricity load profiles (DELP) using density based spatial clustering application with noise (DBSCAN). The framework is aimed to identify typical electricity load patterns and gain knowledge hidden in the patterns and to potentially be used in an early fault detection of anomalous electricity load profiles [21]. Also to detect anomalies of electricity consumption in office buildings an improved kNN is proposed, ikNN, to automatically classify consumption footprints as normal or abnormal [22].

Other techniques for identifying patterns in electricity consumption, are determining distinctive clusters within seasons, and the obtained clusters are used to create seasonal curves for the four seasons, with each having two optimal clusters representing the load demand [23]. An autoregressive integrated moving average with exogenous inputs (ARIMAX) model is used to extract weather dependency to find the residuals, then through hypothesis testing the extremities, maximum and minimums are found [24]. This procedure was reproduced, with linear regression finding the residuals and a Bayesian maximum likelihood classifier to identify anomalies [1]. Dynamic Bayesian Networks and Restricted Boltzman Machine has been proposed for anomaly detection in large-scale smart grids. Simulated on the IEEE 39, 118, and 2848 bus systems the results were verified [25]. Real-Time Mechanism for detecting false data injection attacks analyzed the change of correlation between two phasor measurement units parameters using Pearson correlation coefficient on IEEE 118 and 300-bus systems [26].

Machine learning techniques have been highlighted for their ability to differentiate between cyber-attacks and natural disturbances. By a simulating a variety of scenarios the ability for One R, Random Forest, Naive Bayes and J-Ripper to recognize attacks was investigated: Short Circuit faults; location is represented by the percentage range, Line maintenance; identified through remote relay trip command, Remote tripping command injection; the attacker operates the relay remotely that causes a breaker to open, Relay setting change; the attacker misconfigures the relay settings to cause maloperation of relays, FDIA; attacker manipulates measurements sensors. The simulated scenarios was grouped into classes; natural events, attack events, and no events [27].

4 Local Outlier Factor for Anomaly Detection

Local Outlier Factor (LOF) is a density based unsupervised anomaly detection algorithm introduced in 2000 [28]. LOF compares the local density of a point to the local

Fig. 1 An example dataset

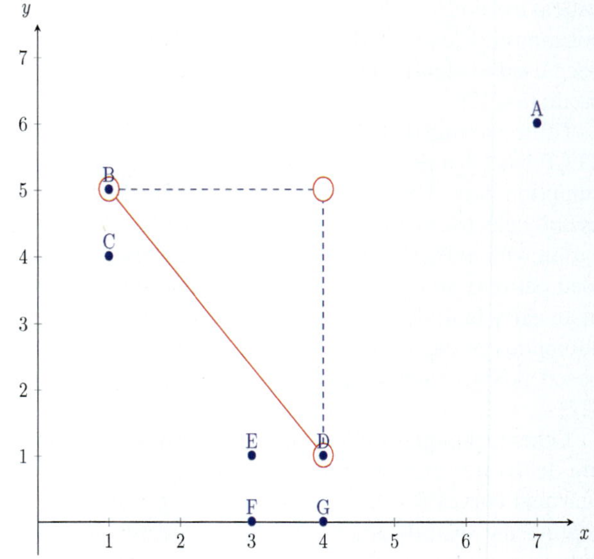

density of k of its neighbors. By comparing the local density of a point to the local density of its neighbors one can identify point that have substantially lower density than its neighbors. These points are considered to be outliers.

The first step in LOF is to compute all distances. As shown for in (Eq. 1) for an m-dimensional Euclidean space:

$$d(x_i, x_j) = \sqrt{\sum_{i,j=1}^{n} (x_i - x_j)^2} \tag{1}$$

In an example dataset consisting of n = 7 points, as shown in Fig. 1, and Table 1.

The distance x_i to x_j between B and D is:

$$BD^2 = (B_x - D_x)^2 + (B_y - D_y)^2$$
$$BD^2 = (1 - 4)^2 + (5 - 1)^2$$

$$BD^2 = 25$$

$$BD = \sqrt{25} = 5$$

The distances k to all point in the neighborhood of x, $N_{k(x_i)}$. All distances are computed in order to define the nearest neighbors distances N_k:

- AB: 6.08 AC: 6.32 AD: 5.83 AE: 6.4 AF: 7.21 AG: 6.71
- BA: 6.08 BC: 1.0 BD: 5.0 BE 4.47 BF: 5.39 BG: 5.83

Table 1 An example dataset

x	y	Letter
7	6	A
1	5	B
1	4	C
4	1	D
3	1	E
3	0	F
4	0	G

Table 2 All nearest distances in ascending order

A	B	C	D	E	F	G	k
5.83	1	1	1	1	1	1	1
6.08	4.47	3.61	1	1	1	1	2
6.32	5	4.24	1.41	1.41	1.41	1.41	3
6.4	5.39	4.47	4.24	3.61	4.47	5	4
6.71	5.83	5	4.47	4.47	5.39	5.83	5
7.21	6.08	6.32	5	6.4	7.21	6.71	6

- CA: 6:32 CB: 1.0 CD: 4.24 CE 3.61 CF: 4.47 CG: 5.0
- DA: 5.83 DB: 5.0 DC: 4.24 DE 1.0 DF: 1.41 DG: 1.0
- EA: 6.4 EB: 4.47 CC: 3.61 ED: 1.0 EF: 1.0 EG: 1.41
- FA: 7.21 FB: 5.39 FC: 4.47 FD: 1.41 FE 1.0 FG: 1.0
- GA: 6.71 GB: 5.83 GC: 5.0 GD: 1.0 GE 1.41 GF: 1.0

After computing all the distances the values are sorted from nearest in row 1, see Table 2, in ascending order to the 6th nearest neighbor in row 6. From the N_k-matrix in Table 2 reveal some intuition for the k-parameterisation. The value of k is also controlling the smoothing strength of LOF [29]. It is recommended choosing a minimum k and a maximum k, and for each point, taking the maximum LOF value over each k in that range. Several guidelines for choosing the bounds are given. The lower bound should be restricted by the minimum points to be found in a cluster, and upper bounded by how many outliers are likely to be found nearby the cluster, although this information is not always available [30]. For the minimum value, the LOF values fluctuate the points in a uniform distribution for k<10, with points in a uniform distribution sometimes showing up as outliers. When trying to quantify a bias-variance trade off, for a suitable model with a k-value that at the same time is accurate and able to generalise for future instances, some problems needs to be dealt with for the unsupervised learning due to lack of labels [31].

Once the optimal k-value is found the LOF uses the reachability distance (RD) to compute the LOF of a point x:

Table 3 Energy mix in New South Wales, Australia

Power plant	Number	Installed power (MW)
Hydro	24	4794
Wind	14	1250
Solar	9	228
Coal	8	11730
Biogas	11	56
Natural gas	19	3766

$$\text{Reachability Distance}(x_i, x_j) = max(k - distance(x_j), dist(x_i, x_j)) \quad (2)$$

If x_i is further away from x_j, than x_j's kth nearest neighbor, then x_i is the RD, if the opposite, that x_j's kth nearest neighbor the actual distance between x_i and x_j then $dist(x_i, x_j)$ is used as RD, this is known as the smoothing factor.

When all RD's are computed the next step is to find the Local Reachability Density (LRD). This is found by taking the inverse of the average RD:

$$LRD(x_i) = \left(\sum_{x_j \epsilon N(x_i)} \left\{ \frac{RD(x_i, x_j}{|N(x_i)|} \right\} \right)^{-1} \quad (3)$$

If the $LRD_{(x_i)}$ is high then the point x_i is in a dense neighborhood, and opposite, when $LRD_{(x_i)}$ is low, then x_i is in a sparse neighborhood. The LRD's of the points in $N_k x_i$, multiplied by the LRD of x_j is used to compute the LOF of point x_i:

$$LOF(x_i) = \frac{\sum_{x_j} \epsilon N(x_i) LRD(x_j)}{|N(x_i)|} * \frac{1}{LRD(x_i)} \quad (4)$$

If the average value of the LRD's in $N_k x_i$ is large and the LRD of x_i is small, then x_i is an outlier.

5 Methodology

In this work data from New South Wales, Sydney region, is considered. New South Wales, Sydney region electrical load profile data set [32] includes meteorological parameters (e.g. DryBulb and WetBulb Temperature, Humidity, Electricity price and time of use) [33]. Data is gathered from 2006–2011. The overall energy mix in New South Wales consists mainly of Coal, Natural Gas, Hydro and other renewable energy sources as shown in Table 3.

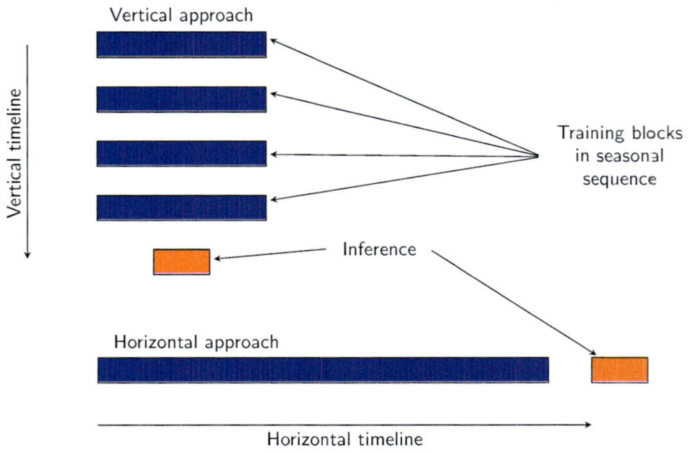

Fig. 2 Vertical and horizontal approach

The methodology in this work is based on a comprehensive correlation analysis on the impact of external parameters on electrical load demand [7–10]. It is observed from these analyses that that meteorological temperatures are highly correlating to the electrical load demand. The vertical time approach uses seasonal data for training and inference. The horizontal approach uses continuous datasets, i.e., it utilizes all data in a continuum from the start of the dataset until the time period used for inference. The illustration of horizontal and vertical approaches is presented in Fig. 2.

Vertical approach can be performed with minimum amount of data compared to continuous approach. Also, the vertical time approach predictive results are compared with prediction based on continuous time series data. In vertical approach, the training set, $D = \{x_i\}_{i=1}^{N}$, is partitioned into subsets by each season of the year, and then are merged together only containing seasonally information about the load pattern. In a dataset containing time observation for five years (e.g., 2016–2020), time is separately selected season-wise, and then merged to contain only the specific season for training, $D = \{x_{spring_i}\}_{i=2016}^{2019}$.

In this study, seasons are divided by months, where Season 1 is December, January, February, and Season 4 is September, October, November. The LOF, is compared for the Horizontal and Vertical Approach, as seen in the flowchart presented in Fig. 3. First the LOF is computed on all 4 seasons, finding the optimal value of k, to be compared seasonwise, independently on each season, and finally comparing the percentage of detected anomalies.

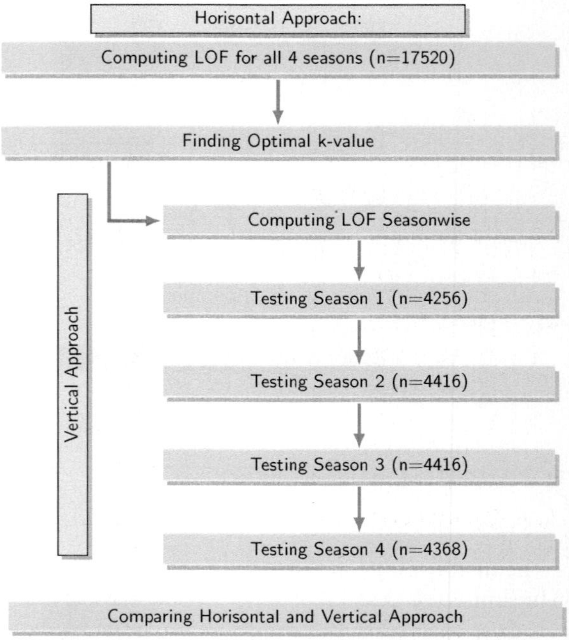

Fig. 3 Flowchart of the proposed model

6 Results, Discussion and Conclusion

The first results from step one in the flowchart for the proposed model, in Fig. 3, are illustrated in Fig. 4: The results show all the outliers from the horizontal approach, the red rings encircling these LOFs in, Fig. 4, are the radius used to describe their individual LOFs, from the threshold value of LOF <–1.5 to the maximum LOF = −2.93. This maximum LOF is recorded on the 15 th of September 2010 at 03:30 hours, and recorded a system load of 6882.9 MW. From these results the upper and lower bounds are chosen.

The data is computed from LOF using k-nearest neighbor of 6, 8 and 10, and showing 3 different LOF's (LOF <–1.5, LOF <–2 and LOF <–2.5), shown in Table 4. The results show that as the search space widens the fraction of outliers detected decreases. In the first row in Table 4 The fractions continues to decrease as for the lower level of LOF. For k = 10, the fraction of LOF is haled when regarding the –1.5 threshold compared to the –2.5 threshold.

Observing the results from the LOF's for the different k-values; when comparing for horizontal and vertical approach most of the results show that vertical approach detects a bigger fraction of anomalies. This is valid for 25 out of 32 results a fraction of 78 %. For LOF <–1.5 and k = 6, k = 8, k = 10, for LOF <–2 and k = 6, K = 8, k = 10 and finally for LOF <–2.5 and k = 6, vertical approach detects more anomalies than horizontal approach.

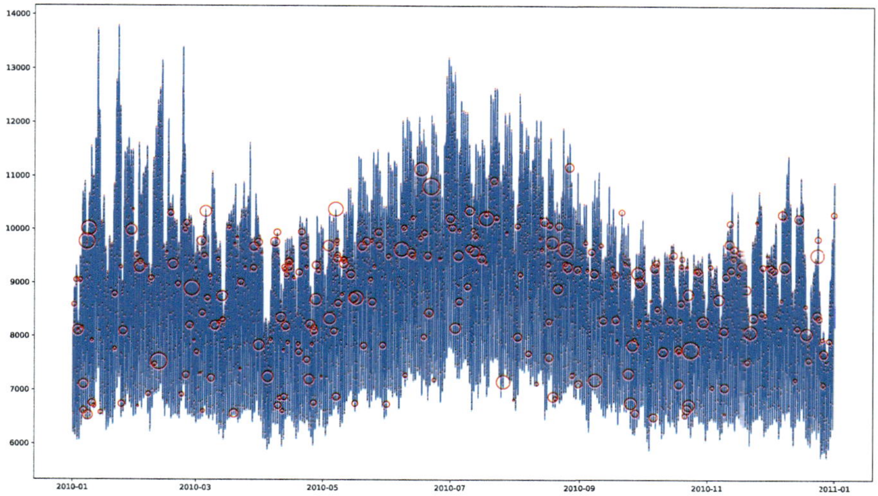

Fig. 4 Outliers detected varying radius of the red encirclement show their individual LOFs ranging from, LOF<–1.5 to the maximum LOF = –2.93

Observing the results from horizontal approach to vertical approach on Season One, in 7 out of 9 test results the fraction of detected anomalies are higher using vertical approach. In Season Two this fraction is reduced to 4 out of 8 (since the k = 8 for LOF<–2 has equal values). In Season Three vertical approach has a higher fraction on 100% of the test results, compared to horizontal approach. In Season Four in 6 out of 9 test results the fraction of detected anomalies are higher using vertical approach. These results verify observations from analysing the seasonal impact on the load electrical load demand.

Season One has a relatively higher electrical load demand as compared to Season Three. These two seasons stand out in the detection of outlier and the explanation for this is found in the correlation to external parameters. In Season Three (June-July-August) dry bulb temperatures recorded are so low that the demand for heating increases the electrical load demand. This might be an explanation to the many outliers found in Season Three. Season One has a lower load demand, but the visual inspection shows higher occasional peaks in the demand. By visual inspection, Fig 3, non of these peaks are detected by horizontal approach.

This work reflects the previous extensive correlational analysis as well as algorithmic development designed to option for changing seasonal behaviour due to the impact of external parameters. It is shown that the number and fraction of detected outliers are higher when using vertical approach, as this work proposes, compared to the traditional continuous method using horizontal approach. There is a need for evolving the impact of external parameters and seasonal behavioural patterns on electrical load demand to enhance the outlier detection for smart energy systems.

Table 4 Results

Season	LOF <-1.5			LOF <-2			LOF <–2.5		
	k = 6	k = 8	k = 10	k = 6	k = 8	k = 10	k = 6	k = 8	k = 10
All seasons	3.39	0.52	0.12	1.28	0.11	0.01	0.51	0.03	0.01
Season One	3.40	0.44	0.16	1.50	0.12	0.05	0.77	0.05	0
Season Two	3.54	0.59	0.09	1.26	0.11	0.02	0.46	0	0
Season Three	3.88	0.73	0.14	1.52	0.18	0.05	0.75	0.05	0.05
Season Four	3.34	0.60	0.21	1.09	1.09	0.07	0.46	0.21	0.12

References

1. H.N. Akouemo, R.J. Povinelli, Probabilistic anomaly detection in natural gas time series data. Int. J. Forecast. **32**(3), 948–956 (2016). https://www.sciencedirect.com/science/article/pii/S016920701500076X
2. K. Huang, C. Zhou, Y.-C. Tian, S. Yang, Y. Qin, Assessing the physical impact of cyberattacks on industrial cyber-physical systems. IEEE Trans. Industr. Electron. **65**(10), 8153–8162 (2018)
3. F. Blaabjerg, R. Teodorescu, M. Liserre, A.V. Timbus, Overview of control and grid synchronization for distributed power generation systems. IEEE Trans. Industr. Electron. **53**(5), 1398–1409 (2006)
4. S. Sahoo, T. Dragičević, F. Blaabjerg, Cyber security in control of grid-tied power electronic converters-challenges and vulnerabilities. IEEE J. Emerg. Select. Top. Power Electron. **1** (2019)
5. I. Chang, G.C. Tiao, C. Chen, Estimation of time series parameters in the presence of outliers. Technometrics **30**(2), 193–204 (1988). http://www.jstor.org/stable/1270165
6. N.J. Johannesen, M. Kolhe, M. Goodwin, Deregulated electric energy price forecasting in nordpool market using regression techniques, in *2019 IEEE Sustainable Power and Energy Conference (iSPEC)* (2019), pp. 1932–1938
7. N.J. Johannesen, M. Kolhe, M. Goodwin, Comparison of regression tools for regional electric load forecasting, in *2018 3rd International Conference on Smart and Sustainable Technologies (SpliTech)* (IEEE, 2018), pp. 1–6
8. N.J. Johannesen, M.L. Kolhe, Application of regression tools for load prediction in distributed network for flexible analysis, in *Flexibility in Electric Power Distribution Networks* (CRC Press, 2021)
9. N.J. Johannesen, M. Kolhe, M. Goodwin, Relative evaluation of regression tools for urban area electrical energy demand forecasting. J. Cleaner Product. **218**, 555–564 (2019). https://www.sciencedirect.com/science/article/pii/S0959652619301192
10. N.J. Johannesen, M.L. Kolhe, M. Goodwin, Smart load prediction analysis for distributed power network of holiday cabins in norwegian rural area. J. Cleaner Product. **266**, 121423 (2020). https://www.sciencedirect.com/science/article/pii/S0959652620314700
11. S. Alla, S.K. Adari, *Beginning Anomaly Detection Using Python-Based Deep Learning* (Springer, 2019)
12. J.A. Gama, I. Žliobaitundefined, A. Bifet, M. Pechenizkiy, A. Bouchachia, A survey on concept drift adaptation. ACM Comput. Surv. **46**(4) (2014). https://doi.org/10.1145/2523813
13. S.Q. Du, Lei, X. Jia, detecting concept drift: an information entropy based method using an adaptive sliding window. Intell. Data Anal. **18**(3), 337–364 (2014)

14. H. Guo, H. Li, Q. Ren, W. Wang, Concept drift type identification based on multi-sliding windows. Inf. Sci. **585**, 1–23 (2022). https://www.sciencedirect.com/science/article/pii/S0020025521011439
15. R. Xu, Y. Cheng, Z. Liu, Y. Xie, Y. Yang, Improved long short-term memory based anomaly detection with concept drift adaptive method for supporting iot services. Future Gener. Comput. Syst. **112**, 228–242 (2020). https://www.sciencedirect.com/science/article/pii/S0167739X20302235
16. N.J. Johannesen, M.L. Kolhe, M. Goodwin, Comparing recurrent neural networks using principal component analysis for electrical load predictions, in *2021 6th International Conference on Smart and Sustainable Technologies (SpliTech)* (IEEE, 2021), pp. 1–6
17. M. Mohammadpourfard, Y. Weng, M. Pechenizkiy, M. Tajdinian, B. Mohammadi-Ivatloo, Ensuring cybersecurity of smart grid against data integrity attacks under concept drift. Int. J. Electr. Power Energy Syst. **119**, 105947 (2020). https://www.sciencedirect.com/science/article/pii/S0142061519331904
18. R. Foorthuis, On the nature and types of anomalies: a review of deviations in data. Int. J. Data Sci. Anal. **12**, 461–478 (2021). https://www.sciencedirect.com/science/article/pii/S1364032113007211
19. J.E. Seem, Using intelligent data analysis to detect abnormal energy consumption in buildings. Energy Build. **39**(1), 52–58 (2007). https://www.sciencedirect.com/science/article/pii/S0378778806001514
20. G.F. Martin Nascimento, F. Wurtz, P. Kuo-Peng, B. Delinchant, N. Jhoe Batistela, Outlier detection in buildings and RSQUO; power consumption data using forecast error. Energies **14**(24) (2021). https://www.mdpi.com/1996-1073/14/24/8325
21. X. Liu, Y. Ding, H. Tang, F. Xiao, A data mining-based framework for the identification of daily electricity usage patterns and anomaly detection in building electricity consumption data. Energy Build. **231**, 110601 (2021). https://www.sciencedirect.com/science/article/pii/S0378778820333879
22. Y. Himeur, A. Alsalemi, F. Bensaali, A. Amira, Smart power consumption abnormality detection in buildings using micromoments and improved k-nearest neighbors. Int. J. Intell. Syst. **36**(6), 2865–2894 (2021). https://onlinelibrary.wiley.com/doi/abs/10.1002/int.22404
23. J.D. Rhodes, W.J. Cole, C.R. Upshaw, T.F. Edgar, M.E. Webber, Clustering analysis of residential electricity demand profiles. Appl. Energy **135**, 461–471 (2014). https://www.sciencedirect.com/science/article/pii/S0306261914009349
24. H.N. Akouemo, R.J. Povinelli, Time series outlier detection and imputation, in *2014 IEEE PES General Meeting | Conference and Exposition* (2014), pp. 1–5
25. H. Karimipour, S. Geris, A. Dehghantanha, H. Leung, Intelligent anomaly detection for large-scale smart grids, in *2019 IEEE Canadian Conference of Electrical and Computer Engineering (CCECE)* (2019), pp. 1–4
26. Y. He, G.J. Mendis, J. Wei, Real-time detection of false data injection attacks in smart grid: a deep learning-based intelligent mechanism. IEEE Trans. Smart Grid **8**(5), 2505–2516 (2017)
27. M. Panthi, Anomaly detection in smart grids using machine learning techniques, in *2020 First International Conference on Power, Control and Computing Technologies (ICPC2T)* (2020), pp. 220–222
28. M.M. Breunig, H.-P. Kriegel, R.T. Ng, J. Sander, Lof: identifying density-based local outliers, in *Proceedings of the 2000 ACM SIGMOD International Conference on Management of Data*, Series. SIGMOD '00 (Association for Computing Machinery, New York, 2000), pp. 93–104. https://doi.org/10.1145/342009.335388
29. O. Alghushairy, R. Alsini, T. Soule, X. Ma, A review of local outlier factor algorithms for outlier detection in big data streams. Big Data Cognit. Comput. **5**(1) (2021). https://www.mdpi.com/2504-2289/5/1/1
30. Z. Xu, D. Kakde, A. Chaudhuri, Automatic hyperparameter tuning method for local outlier factor, with applications to anomaly detection, in *2019 IEEE International Conference on Big Data (Big Data)* (IEEE, 2019), pp. 4201–4207

31. C.C. Aggarwal, S. Sathe, Theoretical foundations and algorithms for outlier ensembles. SIGKDD Explor. Newsl. **17**(1), 24–47 (2015). https://doi.org/10.1145/2830544.2830549
32. AEMO, National electricity market data - nem (2021). https://aemo.com.au/energy-systems/electricity/national-electricity-market-nem/data-nem/aggregated-data
33. V. Dehalwar, A. Kalam, M.L. Kolhe, A. Zayegh, Electricity load forecasting for urban area using weather forecast information, in *2016 IEEE International Conference on Power and Renewable Energy (ICPRE)* (2016), pp. 355–359

A Modular Infrastructure for the Validation of Cyberattack Detection Systems

Davide Cerotti, Daniele Codetta Raiteri, Giovanna Dondossola, Lavinia Egidi, Giuliana Franceschinis, Luigi Portinale, and Roberta Terruggia

Abstract We propose a framework for the evaluation of cyberattack detection systems in which theoretical results can be tested in a realistic setup. We emulate a power control infrastructure, an attacker and a monitoring system. In this controlled environment, through a modular approach, it is possible to evaluate a variety of detection models: we inject adversarial activity, collect logs from the systems, analyze such logs and produce evidences that are later processed by artificial intelligence models that can raise alerts, and give diagnostic or predictive information. In particular, we test our framework with detection models based on Dynamic Bayesian Networks, that take into account the evolution of adversarial activities over time. The testbed allows us to effectively test the adequacy of the detection mechanisms for early warning of suspicious events; currently, it includes man-in-the-middle attacks and false data injection.

D. Cerotti · D. Codetta Raiteri · L. Egidi (✉) · G. Franceschinis · L. Portinale
Computer Science Institute - DiSIT, Università del Piemonte Orientale (UPO), Alessandria 15121, Italy
e-mail: lavinia.egidi@uniupo.it

D. Codetta Raiteri
e-mail: daniele.codetta@uniupo.it

G. Franceschinis
e-mail: giuliana.franceschinis@uniupo.it

L. Portinale
e-mail: luigi.portinale@uniupo.it

D. Cerotti (✉) · D. Codetta Raiteri · G. Franceschinis · L. Portinale
UdR UPO, Consorzio Nazionale Interuniversitario per le Telecomunicazioni (CNIT), Roma 00133, Italy
e-mail: davide.cerotti@uniupo.it

G. Dondossola · R. Terruggia
Transmission and Distribution Technologies Department, Ricerca sul Sistema Energetico (RSE), Milano 20134, Italy
e-mail: giovanna.dondossola@rse-web.it

R. Terruggia
e-mail: roberta.terruggia@rse-web.it

H. Haes Alhelou et al. (eds.), *Power Systems Cybersecurity*, Power Systems,
https://doi.org/10.1007/978-3-031-20360-2_13

Keywords Early evidence-based cyberattack detection · Power systems · Bayesian Networks · Test environment · Time-driven attack analysis

1 Introduction

Power systems are critical assets whose disruption can cause significant damage to the society investing all aspects, including human lives. In the actual context of the energy transition, the digitalisation of the power infrastructure allows the implementation of new, essential functionalities (i.e. distributed energy resource control, electric vehicle charging, demand response, etc.) to optimize the power grid management, but poses new challenges in terms of cybersecurity: the interaction of new and diverse actors by means of heterogeneous and third party networks widens the attack surface of energy infrastructures. Hence, to increase the system resilience to cyber threats, it is crucial that any adversarial activity be detected as early as possible, enabling the adoption or strengthening of defense measures. Following the European Network and Information Security (NIS) Directive 2016/1148,[1] currently under review,[2] the Member States have to ensure that essential entities, e.g. energy operators, take appropriate measures to manage the cybersecurity risks, including the adoption of methodologies to detect attacks and to actuate fast response and defense actions.

In this setting, we present a framework for the evaluation of cyberattack detection systems developed within a research cooperation between Università del Piemonte Orientale (UPO) and Ricerca sul Sistema Energetico (RSE), a company carrying out research activities in the electro-energy sector in national strategic projects of general public interest. Our research activity is aimed at the development of effective cyberattack detection tools for the energy systems, based on Artificial Intelligence methodologies [1, 2]. In a critical context such as the energy infrastructures, testing the detection tools on fully operational infrastructures is unfeasible, and therefore a significant part of our activity is the design and implementation of a framework in which theoretical results can be tested in a realistic setup.

In this paper we refer to a scenario illustrated in Fig. 1, considering Distributed Energy Resources (DERs). The increasing importance and penetration of the DERs requires to rethink the grid control logic with the acquisition of more information from the field and the control of DER functions in terms, for instance, of active and reactive power. Communications need to be implemented from and to the energy resources (see Fig. 1) in order to collect measures and send commands. This information exchange can be implemented by means of heterogeneous and third party communication channels, hence they are prone to cybersecurity issues. National and international communities are working on norms to regulate the grid connection rules, associated controls, communications and cybersecurity. For instance, in Europe the

[1] https://eur-lex.europa.eu/eli/dir/2016/1148/oj.

[2] https://www.europarl.europa.eu/RegData/etudes/BRIE/2021/689333/EPRS_BRI(2021)689333_EN.pdf.

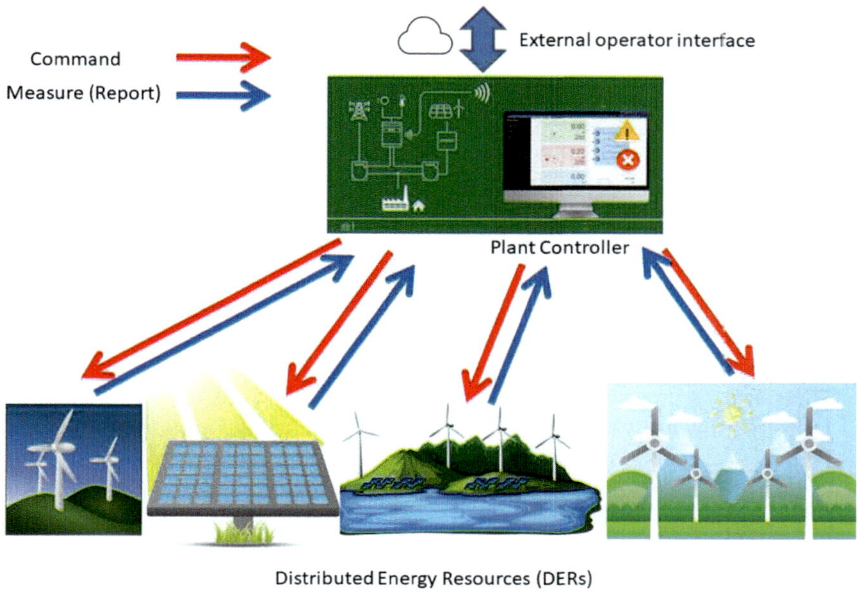

Fig. 1 Power grid with Distributed Energy Resources (DERs)

legislative EU references are, besides the NIS Directive, the future European Network Code on Cybersecurity,[3] and in Italy the Italian Law Decree on the National cybersecurity perimeter.[4]

In this article we first survey some related works (Sect. 2), then we introduce the framework (Sect. 3), explaining how attackers are emulated, giving an overview of the attacks that we implemented, and describing the way we integrated detection models. Next, the representation of the possible attacks using probabilistic graphical models (Dynamic Bayesian Networks–DBNs) is illustrated on a small but realistic example (Sect. 4). In Sect. 5 we describe some experiments in our testbed to validate our proposal. Our experiments show how a detection tool responds to different possible adversarial activities, highlight ways in which the models can be useful for monitoring the system, but also point to some aspects that require refinement. A final discussion completes the paper, presenting also future directions of research (Sect. 6).

[3] https://www.entsoe.eu/network_codes/nccs/.

[4] https://www.gazzettaufficiale.it/eli/id/2019/09/21/19G00111/sg, in Italian.

2 Related Work

The idea of a testbed for Intrusion Detection Systems (IDS) in the environment of power control systems is not new, given the risk of conducting tests on a critical system. The testbed structure proposed in [3] is quite articulated and various protocols are considered. But the attacks are injected in the system by altering traffic traces (`.pcap` files) and the goal is to test a machine learning based IDS. In contrast we propose a testbed in which an attacker is emulated, and therefore the malicious traffic produced is more realistic, we can inject entire attack processes and thus can test early detection and forecasting of adversarial actions. Moreover the modularity of our approach enables the testing of different detection systems.

In [4], a testbed for intrusion detection in an electricity generation and distribution control system, is presented. In this case, traffic is captured at the boundaries of a real Industrial Power System and comprises corporate and Operational Technology (OT) traffic. The testbed was used to verify the quality of a protection system [5]. IP traffic obtained from real control and supervisory substations was injected into the testbed domains. Hostile traffic was simulated as malformed traffic. This approach can detect traffic anomalies but is not geared at recognizing an attack process.

Another testbed for intrusion detection in the power grid is presented in [6]. In this case, the authors propose detection through a module that simulates the grid and compares measures reported with "ideal" data it computed. They simulate attacks by actually installing malware. Their focus is on single attacks and not on processes and the detection takes place when the attacker is already succeeding in destabilizing the grid.

In an older approach [7] the computer networks and the power grid are simulated. The detection tool is an anomaly detection tool, trained to recognize normal traffic in the net and to raise an alarm when the traffic is anomalous.

A much broader project is the US National SCADA Test Bed [8] to address cyber security issues of energy delivery systems. The project was started in 2003 and encompasses a number of subprojects such as quantum key distribution, risk assessment, removing certain classes of attackers, protocol compliance, development of control system standards.

The approach presented in [9] refers to an actual IDS (DETECT), rather than a testbed, however it is related with our proposal since Bayesian Networks derived from Attack Trees are used as models to recognize attacks as they evolve: events are gathered and monitored (within a given time frame), and observations are used to update the Bayesian Network (BN) parameters. The model is solved after any event-driven update and may produce an attack warning or alarm. In our case, Dynamic BNs derived from attack graphs are used in a similar way to detect emulated attacks.

For completeness, let us stress that testing IDSs is a problematic issue. Delving into this is out of the scope of this chapter, and we refer to a classical survey for an interesting analysis [10].

Other proposals study cybersecurity aspects emulating power grid components at a lower level, using hardware-in-the-loop techniques. For instance [11], focussing on DERs, is related to our work in that it focuses on attack processes and follows closely the MITRE ATT&CK project (see Sect. 3.2) but its aim is assessing the security posture of the system and it is not meant to evaluate monitoring and IDSs, that are not implemented.

The criticality of power systems, and in particular smart grids, is an incentive for much research on its cyber security aspects; see, e.g., the recent surveys [12, 13]. In particular, in our architecture we focus on DERs (see Sect. 3); current research [14–16] highlights interesting aspects of cybersecurity in this domain.

As will become clear in the following, our methodology and framework consist of coordinated components, such as the attack graph (see e.g. Fig. 5), the selection of the attacks that the synthetic attacker will carry out (see Sect. 3.1) and the model (see Fig. 9). In order to make the framework more flexible, allowing automated derivation of the three components of the attack/detection workflow, it would be useful to formalize attack scenarios by using some domain specific language. This formalization would allow us to derive in a coordinated way attack graphs of adversarial processes and, on the practical level, both the specifications to drive the emulated attacker and the corresponding detection models, thus supporting the development of more complex testbed scenarios. A good candidate domain specific language is MAL [17]. In this case, the MAL language integrates the description of assets (servers, software, network components, ...) and of possible attack techniques (mainly derived from MITRE ATT&CK) that may apply to each asset, and possible countermeasures. From the integration of such specification with the description of the assets composing the IT (or OT) infrastructure under study, attack graphs can be derived automatically and used to customize the testbed for experimenting with different scenarios.

3 The Framework

Our framework is part of a testbed that reproduces a power control infrastructure where the focus is on communication implemented by standard protocols as IEC 61850 and IEC 60870-5-104. Figure 2 represents a diagram of the main control communications involved in the testbed: the Distributed Energy Resources (DERs) send measures to a primary substation (arrow 1 in the figure) where control functions (e.g. voltage/frequency control) are executed and setpoints evaluated. These are then sent back to DERs (arrow 2) in order to be applied. The communications are implemented by IEC 61850 standard.

The primary substation aggregates the measures coming from the DERs, integrates them with other power values and sends the data to a Control Center (arrow 3). Here higher level control functionalities are performed and commands could be forwarded to the primary substation (arrow 4 in the figure). The IEC 60870-5-104 protocol is used.

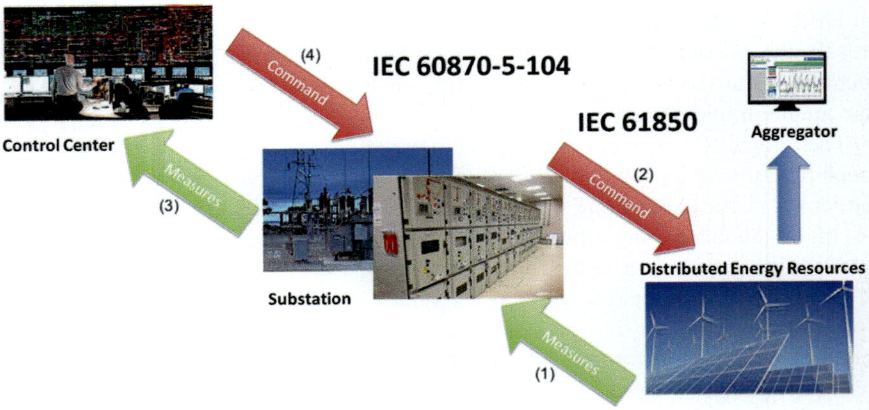

Fig. 2 Testbed communications

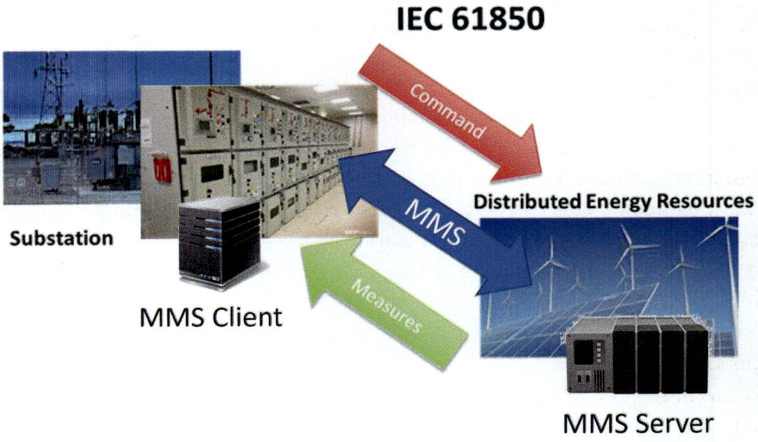

Fig. 3 MMS communication

In this chapter the focus is on DER control (see Sect. 3.2): the communications are implemented by means of the Manufacturing Message Specification (MMS) protocol as defined in IEC 61850-8-1 standard. At the substation the MMS client is active and is connected with the MMS server placed at the DER site (see Fig. 3).

In this emulated environment, we inject adversarial activity, collect logs from the systems, analyze such logs and produce evidences that are later processed by models that can raise alerts, and give diagnostic or predictive information (see Fig. 4). The whole framework enables us to effectively test the adequacy of the detection mechanisms, using the emulated attacks for validating the framework results.

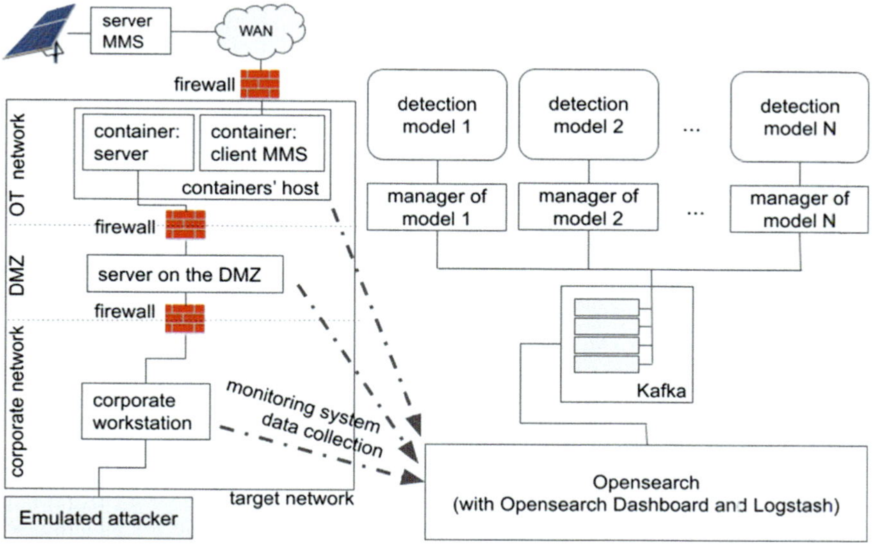

Fig. 4 The logical architecture of the framework

The adversary side is based on the tool Metasploit[5] and it can emulate different behaviours, to render more aggressive or more naive attack attempts. It is capable of taking advantage of all the choices offered by Metasploit in terms of attacks, but we have also implemented more power-specific threats. Thus the malicious traffic is created in a realistic way.

The monitoring observes activity in the target network. Log data is conveyed to an Opensearch[6] database which implements analytics, filtering the information it receives and raising alarms when specific events take place or defined thresholds are exceeded.

Such alarms are conveyed to the detection models through a Kafka[7] broker, whose aim is to ensure modularity of the framework. It enables us to feed the same data to different models in order to compare their effectiveness for the different tasks of interest, such as early detection of adversarial activity, diagnosis of the attack steps carried out so far, prediction of most likely evolution of the attack processes.

The models we have so far integrated in the framework are probabilistic graphical models (see Sect. 4).

[5] https://www.metasploit.com/.

[6] https://opensearch.org.

[7] https://kafka.apache.org.

3.1 The Attacker

In order to test the detection systems with complex attack processes, we implemented an automatic adversary (a preliminary version is described in [18]). Our adversary is fully automated in the sense that it is capable of carrying out entire attack processes without the need of human intervention.

We wanted to mimic through our attacker different adversarial profiles, to test our detection tools in various settings, ranging from uninformed adversaries to very precise ones that know very well the network they attack.

We worked at making it modular and expandable in order to enrich over time the pool of attacks that can be carried out, with the specific aim of specializing increasingly over time our tool for power systems.

We opted for an open source tool, built over well established open source software, to provide a tool that can be of use to the community at large.

We could not use directly for our aims open source tools readily available because our aim is different from the usual one: attack tools are normally used for penetration testing. On the contrary, we know well the vulnerabilities of the network we emulate. Instead, we want to carry out those attacks in order to test our monitoring and detection tools. The environment in this sense is much more controlled: we know in advance what the attacker can achieve and this is key for validating the defense mechanisms.

We built our adversary around Metasploit, that has become the toolbox of our adversary. Some of the implementation choices we made, resulted in constraints on the emulated network, but we adhered to some basic principles in order to be able to retrieve realistic data through the monitoring system. These principles can be summarized as emulating and using actual software on victim machines, including the network stack, and carrying out real attacks on these machines.

3.2 Attacks

The design of our attacks and of the detection models is framed as much as possible in the MITRE ATT&CK classification,[8] in order to maintain a common language shared in other research approaches and for adherence to real world scenarios, since the MITRE project itself is based on data from known attacks or adversarial software.

In this context, we also developed attacks specific for the power world, that implement techniques from the ICS ATT&CK matrix.[9] Moreover we implemented attack steps from the Containers ATT&CK matrix.[10]

The process chosen as testbed, considers an attacker that, having initially compromised a workstation in the corporate network, moves laterally through the DMZ

[8] https://attack.mitre.org/.

[9] https://collaborate.mitre.org/attackics/index.php/Main_Page.

[10] https://collaborate.mitre.org/matrices/enterprise/containers/.

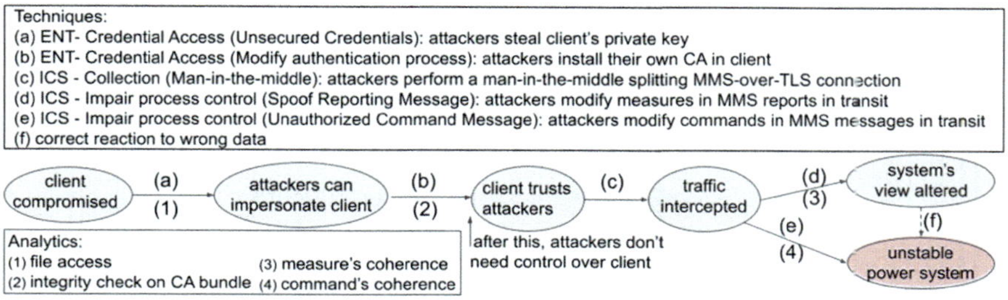

Fig. 5 Man in the middle to MMS-over-TLS and related analytics

to finally compromise a computer in the Operational Technology (OT) network, and from there proceeds to make the system unstable.

The final portion of the attack process we developed injects false data in the system: to this end, a Man in the Middle (MITM) attack to a IEC 61850/MMS over TLS connection is performed in the power control environment.

Our methodology consists in first designing attack processes that we express through Attack Graphs (AGs). Then, on one side we implement the attacks for the synthetic attacker and on the other we implement monitoring sensors. The sensors must be realistic and not specifically targeted for the attack steps we selected: for instance, we monitor the integrity of relevant configuration files or directories, but we do not search for traffic patterns exposing one single exploit. On the basis of the AG, we also design the detection models.

Figure 5 depicts an AG describing in detail the Man-in the-middle attack. Nodes denote states reached by the attacker and edges are labelled by attack steps. For each attack step we first indicate the MITRE ATT&CK tactic it aims at, in parentheses the specific technique, and then we detail the actual implementation of that technique. The tactic name is introduced by a code indicating the specific ATT&CK project we refer to; in Fig. 5 some tactics are from the Enterprise ATT&CK matrix (ENT) and some are from the ICS one. So for instance, the edge labelled (c): "*ICS–Collection (Man-in-the-middle): attackers perform a man-in-the-middle splitting MMS-over-TLS connection*" indicates an implementation of the technique Man-in-the-middle for the tactic Collection of the ICS matrix. In particular, the MITM attack is implemented splitting MMS over TLS.

The node labelled "unstable power system" is the attackers' goal that can be reached following two different attack paths: either the attackers modify reporting messages or commands. In the latter case, they can issue a wrong command that causes malfunctioning of the device driven by the MMS server. If on the other hand they alter reporting messages, the control center will have an incorrect view of the system's state. This might prompt reactions that can cause system's instability. A reaction to an altered view is not properly an attack step, and therefore it appears in the AG as a dotted arrow (labelled f), but it is an action that the attackers count on to reach their target.

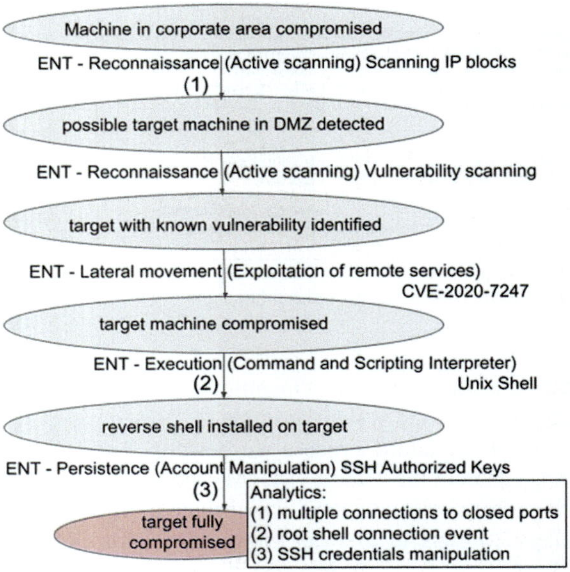

Fig. 6 Lateral movement

The numbers on some edges denote analytics to expose the attack. In Fig. 5 four analytics are proposed. The one on the leftmost edge, analytic (1), raises an alarm in case of access to the file containing the client's private key. Integrity check on the CA bundle, i.e. analytic (2), is very important because any change to that file implies modified trust. Analytics (3) and (4) are at application level: they check coherence over time of the measures received or of the commands issued.

The first portion of the whole process is a more classical lateral movement attack that proceeds scanning the network to identify possible victims and then exploits vulnerabilities of a mail server. Through these steps the attacker that has initially compromised a computer in the corporate network, most likely using standard phishing techniques, moves to the DMZ between corporate and OT networks. These attack steps are described in Fig. 6. Notice that all techniques here are from the Enterprise ATT&CK matrix.

Perhaps more interesting is the attack to the containerization software; the subprocess is detailed in Fig. 7: after compromising, with standard vulnerability exploitation, a container on a machine of the OT network, the attackers are able to escape from the container and take control of the whole host machine. Notably, they take control of all the containers running on the host, including the MMS client target of the attack in Fig. 5. In Fig. 7 most tactics are from the MITRE Container ATT&CK matrix (code CNT).

Whereas the emulated attacker is directed to pursue these specific attack paths, it is implemented to embed the attack processes that will be successful more realistically

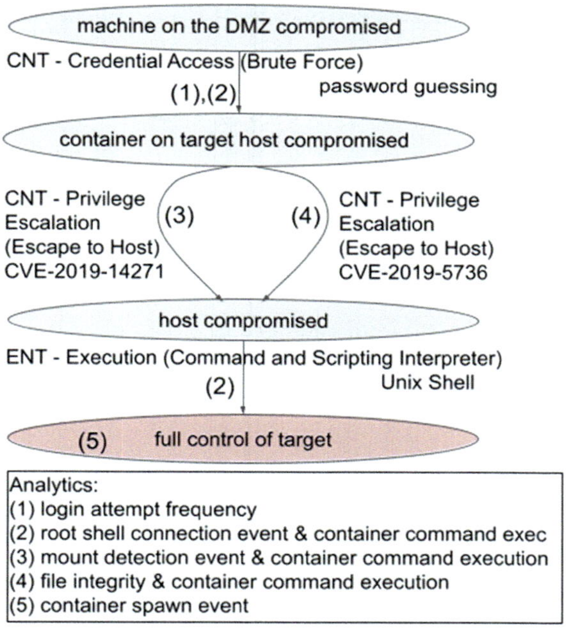

Fig. 7 Escape from a container

into a swarm of alternative attempts that are bound to fail, but create further challenges to diagnostic and predictive analyses.

3.3 *Integration of Detection Models*

The monitoring system collects data from the network and conveys them to the Opensearch database. Here a pipeline filters the data and tests them against parameter thresholds or other specific conditions, configured according to the analytics that have been planned to expose possible attacks. When evidences of suspicious activity are detected, the alerts are forwarded to the detection models through a Kafka broker, on a specific topic.

Kafka consumers read the evidences and forward them to the detection models. As anticipated, we have so far integrated only (D)BNs, and so we describe here this integration. Our software includes a Kafka consumer that reads a Kafka topic at regular intervals (that map to the instants t in the DBN), collects them in a file according to a format that is compatible with Octave.[11] Then the program starts the DBN analysis: evidences are read from the file and one of the possible inference

[11] https://www.gnu.org/software/octave.

Fig. 8 Dashboard with different views of the first experiment in Sect. 5.5 at time $t = 35$

tasks is executed (see Sect. 4.1). As the (D)BN manager receives new evidences, it starts a new analysis of the model.

The results produced by any model are redirected through Kafka back to Opensearch, where they can be visualized using OpenSearch Dashboard,[12] in a flexible way. Figure 8 is an example dashboard showing (from top left, counterclockwise) a graphical representation of the evolution of probabilities that some techniques have success, an horizontal bar graph showing a ranking of attack steps based on their probability of success, numeric results of the analysis, and a gauge showing at a glance the security posture of the whole system.

4 Probabilistic Graphical Models

The detection models we have so far integrated in the framework are Bayesian Networks (BNs) [19] and Dynamic Bayesian Networks (DBNs) [20], extending BNs by providing a discrete temporal dimension. Both BNs and DBNs belong to the family of probabilistic graphical models. In this paper we concentrate on DBNs because they are more interesting, allowing more informative analyses, especially when used to approximate continuous models, as proven by the popularity gained in the dependability field [21].

A BN is a pair $N = \langle\langle V, E\rangle, P\rangle$ where $\langle V, E\rangle$ are the nodes and the edges of a directed acyclic graph respectively, and P is a probability distribution over V. Discrete random variables $V = \{X_1, X_2, \ldots X_n\}$ are assigned to the nodes, while

[12] https://opensearch.org/docs/latest/dashboards/index/.

each edge $e \in E$ from node X to node Y represents that Y directly depends on X in a probabilistic way. This interpretation factorizes the joint probability of the variables as shown in Eq. (1). Each local distribution can be described by the Conditional Probability Table (CPT).

$$P[X_1, X_2, \ldots, X_n] = \prod_{i=1}^{n} P[X_i|Parent(X_i)] \tag{1}$$

4.1 DBN Definition and Inference

Given a BN N, a DBN is essentially a replication of N over two time slices $t - \Delta$ and t (Δ is the time discretization step), with the addition of a set of arcs representing the transition model. Let X_i^t denote the copy of variable X_i at time slice t, the transition model is defined through a distribution $P[X_i^t|X_i^{t-\Delta}, Y^{t-\Delta}, Y^t]$ where $Y^{t-\Delta}$ is any set of variables at slice $t - \Delta$ different from X_i (possibly the empty set), and Y^t is any set of variables at slice t different from X_i (possibly the empty set). An edge connecting a variable $X_i^{t-\Delta}$ in the slice $t - \Delta$ to the variable X_j^t in the slice t, is called *temporal arc* (if $i = j$ the arc connects the two instances of the same variable). The dependency of a given node on its parent nodes (possibly including its historical copy) is quantified in its CPT; an example is provided in Fig. 9.

The following inference (analysis) tasks can be performed on a DBN:

- *Prediction* can support the analyst/system in identifying, on the basis of evidences obtained from the analytics, the subset of techniques that the attacker will more likely exploit in the future, thus enabling the setup of defensive actions. It consists in computing the probability of a future state, given past evidence. In other words, the prediction task consists in computing the posterior probability at time t of a set of queried variables Q, given a stream of observations $e_{t_1}, \ldots, e_{t_k}$ from time t_1 to time t_k with $t_1 < \cdots < t_k \leq t$. Every evidence e_{t_j} consists of a (possibly different) set of instantiated variables.
- A special case of prediction, called *Filtering*, occurs when the last evidence time point and the query time are the same ($t = t_k$). Filtering can support the security analyst or system in online diagnosis and early detection.
- *Smoothing* enables *diagnosis*, allowing the determination of the probable causes of security events. The task estimates a past state, given all the evidence up to now. So, the smoothing task consists in computing the probability at time t of a set of queried variables Q, given a stream of observations $e_{t_1}, \ldots, e_{t_k}$ from time t_1 to time t_k with $t < t_1 < \ldots < t_k$.

4.2 Representing Attack Paths Timing

The DBN is a discrete time model and as such it is capable to represent and evaluate probabilities that a certain attack step is carried out by attackers provided its preconditions are true, or compute the probabilities that a certain attack process takes place. But it cannot compute **probability distributions of completion times** of the different techniques or attack paths.

This kind of information would be very useful for security analysts, supporting them in decisions on how to react when under attack. For instance, assume that the analysts know that with a very high probability, say 90%, the attackers will not gain their goal before two days. This allows them to take advantage of this delay to carefully plan the defense, activating effective protection mechanisms and further detection tools, and increasing monitoring for a better post-mortem analysis. To the contrary, the information that the time left before the attackers reach their goals is less than two hours, with a probability of, say, 80%, prompts the analysts to rush to the protection of most important resources, even with drastic measures.

The above tasks require a continuous time model, which we approximated with our discrete time models assuming that each time-step has a constant duration. Thus, we derive a discrete-time DBN in which all conditional probabilities are re-scaled with respect to such value. In this way the elapsed times between two consecutive state-transitions in the continuous model is approximated by a geometric distribution in the DBN. To obtain an accurate approximation the time-step size is computed from the set of estimated mean completion times of all the techniques, as explained in [22]. Such approach allows us to compute distribution probabilities of the *time to compromise* (TTC) of the different techniques and attack paths. In Table 1, we show the mean times to completion of each technique (and of all other event and status nodes) where one unit corresponds to 10 min.

4.3 A DBN for the MITM Attack

As an example to make the presentation more concrete, we show in Fig. 9 a DBN that models (among others) the attack processes of Fig. 5. This is only a portion of the full DBN that includes also the previous parts of the attack process, that is the lateral movement and the escape from the container, but the relevant concepts are clearly exemplified by this fragment. The central branches in the figure, starting from the low central node labelled "UnsecureCred", up to the top node, model the attack. The top node, "UnstablePS (OR)", represents the attack's goal: making the power system unstable. The branches to the right and to the left are different possible attack paths, that have not been implemented so far. We included such attacks, even though they currently cannot take place in our testbed, since we want to show how our model detects among others the attack process that is really taking place.

Table 1 Estimated completion times of each technique in Fig. 9

Techniques, events, states		Mean
Abbreviations	Full names	TTC
Masquerade	Masquerade	2
MITM	Man In The Middle	2
ModAuthProc	Modify Authentication Process	1
ModCtrlLogic	Modified Control Logic	50
ModifyProgram	Modify Program	2
SpoofRepMsg	Spoof Reporting Message	15
UnauthCommand	Unauthorized Command	40
UnsecCred	Unsecured Credentials	1
CorrReact	Correct Reaction	0
WrongLogicExec	Wrong Logic Execution	0
UnstablePS	Unstable Power System	0

Other than the top node, all nodes represent either attack steps, e.g. edges in the AG, or analytics, that in the AG are denoted as numbers on edges (cf. Figs. 5, 6 and 7). The power system is unstable (the attacker's goal occurs), if any of the three last steps below it are successful. Analytics are leaves in the DBN graph: when an attack step takes place, traces are collected by sensors and alarms are raised by the analytic. Therefore, in the DBN there is an edge from a technique to an analytic, if the occurrence of the technique potentially causes an evidence from the analytic.

Nodes labelled "CorrReact" and "WrongLogicExec", near the top, are different from the rest. They represent actions causing the attackers' plan to evolve but are not properly attack steps: "CorrReact" represents the reaction of the automatic controls to false measures injected by the attacker, corresponding to the dotted arrow in Fig. 5 (see the discussion in Sect. 3.2). "WrongLogicExecution" is similar in concept: it is the final step of an attack process where the attackers have modified the controller's logic, say with a malware, so the system will react according to the malicious logic. Also in this latter case, the event is not properly an attack step but completes the process planned by the attackers. In both cases, we consider that some protection measures have been taken, in the form of some automatic control, so these final steps don't take place necessarily but only with a certain probability.

Each attack step can occur with a certain probability, conditioned on the occurrence of its ancestors: so, for instance, the edge from the node labelled "MITM" to the one labelled "SpoofRepMsg" implies that a reporting message can be spoofed only after a MITM has been successfully established. If a node has no ancestors, we are assuming that the prerequisites for that attack step are verified. Arcs from one node to itself model the fact that if an attack step has been carried out at time t, at time $t + 1$ we acknowledge that it has already occurred. The *a priori probabilities* of nodes with no ancestors and the *conditional probabilities* of all other nodes are parameters that must be carefully set.

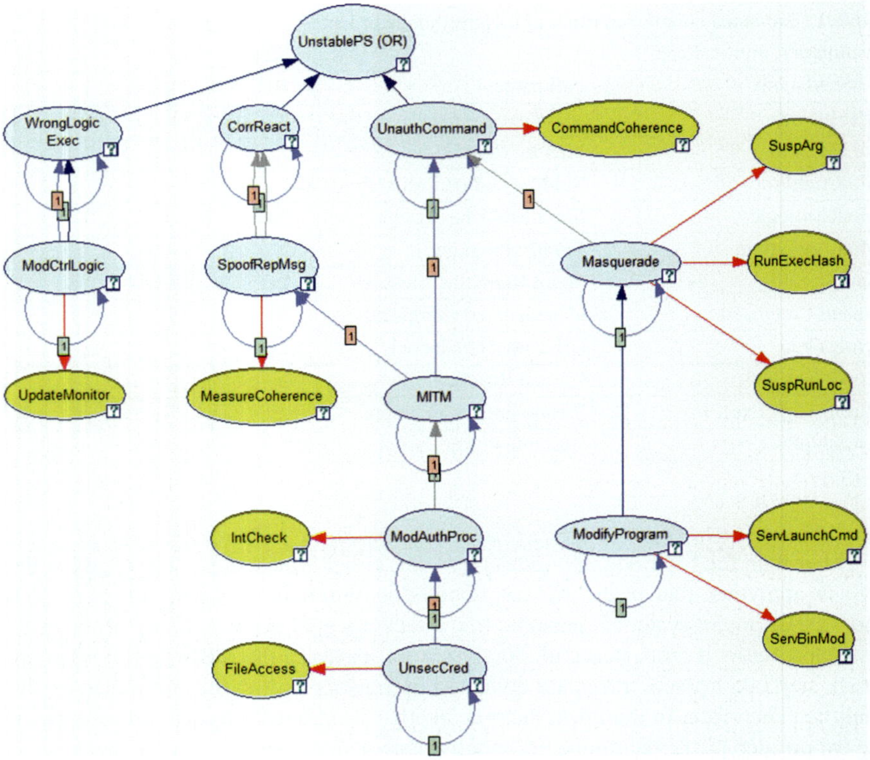

(a)

	Technique	*Analytic*	pr.
1	0	0	p
2	0	1	1-p
3	1	0	1-q
4	1	1	q

(b)

	MITM	*SpoofRep Msg* (t-1)	*SpoofRep Msg* (t)	pr.
1	0	0	0	1
2	0	0	1	0
3	0	1	0	1
4	0	1	1	0
5	1	0	0	0.93
6	1	0	1	0.07
7	1	1	0	0
8	1	1	1	1

Fig. 9 A portion of our DBN, CPTs of a generic analytic (top) and of node *SpoofRepMsg*

The table (b) at bottom-right of Fig. 9 is the CPT of node "SpoofRepMsg"; at time t, it depends on its copy at time $t - 1$ and on "MITM". Each entry of the CPT represents a specific state configuration of "MITM" and "SpoofRepMsg" $(t - 1)$, and provides the probability that "SpoofRepMsg" (t) evolves towards the state 0 or 1. In rows 1–4, "MITM" has not happened yet, so the probability that "SpoofRepMsg" occurs (state 1) is null. In the rows 5–6 "MITM" has occurred and "SpoofRepMsg" is not active yet at time $t - 1$; so "SpoofRepMsg" (t) may occur (1) or not (0) with a probability computed from the completion time in Table 1. In rows 7–8,

"SpoofRepMsg" ($t - 1$) is already active (1), so "SpoofRepMsg" (t) will maintain its state.

In the model we assume that the analytics are imperfect: a technique may go undetected (false negative), or an alarm may be raised even without the attempt to use the technique (false positive). In Fig. 9, table (a), we show the CPT of a generic analytic: $1 - p$ (resp. $1 - q$) is the probability of a false positive (resp. false negative).

5 Experiments

The experimental setup involves several Linux virtual machines running on the RSE private cloud based on Openstack. With reference to Fig. 4, the attacker, the corporate workstation, the DMZ server and the Kafka machine run Ubuntu 20.4, the MMS server and MMS client and the DBN detection model run Ubuntu 18.04. All the virtual machines have 8 GB of RAM, 4 VCPU and 80 GB of disk. All the applications run on Docker containers.

The experiments carried out involve several aspects. We have started with functional experiments showing that the integration of all components works as expected. The attacker carries out the prescribed attack processes, sensors collect data and send it to the database, the Opensearch pipeline forwards the data to the model managers through Kafka and the models are activated. In the following we report how variations in the behaviour of the attacker have led to different inferences of the model.

Recall that the attacker can carry out the attack processes in the center of Fig. 9. The "File access" analytic monitors that there are no suspicious accesses to the private key file; it is implemented using the Linux Audit Framework[13] together with the `auditd` module of Auditbeat.[14] "Integrity check" raises an alert if the CA bundle has been altered; this is implemented using the `file_integrity` module of Auditbeat. The two coherence analytics were implemented collecting MMS logs and analyzing them in Opensearch. We also implemented the "ServLaunchCmd" analytic, even if it is related to an attack process that we don't support yet, to monitor additional activity: recall that our synthetic attacker embeds the attack processes in additional, randomly selected attack attempts that are generally bound not to succeed. This analytic is implemented again using Auditbeat, and it looks for occurrences of the `Systemctl` command.

In our experiments, all the analytics that we haven't implemented yet will give no evidences. We stress that "no evidence" from an analytic is different from negative evidence. In the former case, we have not implemented the analytics and therefore the detection system has no information about possible adversarial activity in that domain. In the latter, the sensors are deployed but detect nothing suspicious. We assume a high confidence in the monitoring system and set the error probabilities to 10^{-4} ($1 - p$ and $1 - q$ in Fig. 9).

---

[13] https://linux-audit.com.

[14] https://www.elastic.co/beats/auditbeat.

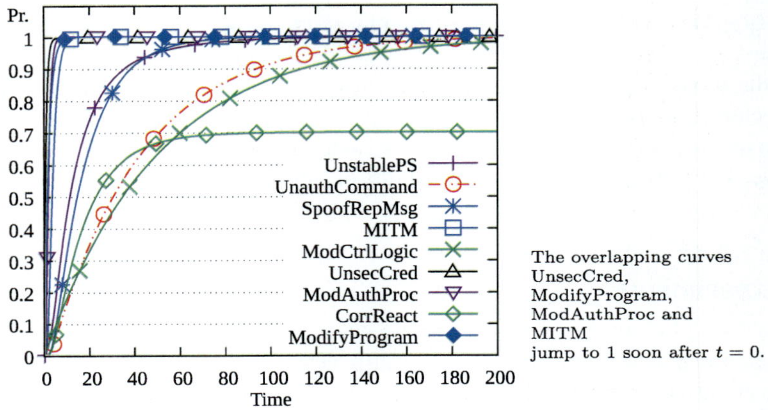

Fig. 10 Probabilities of successful technique exploitation with no evidences (the curves UnsecCred, ModifyProgram, ModAuthProc and MITM all jump to 1 soon after $t = 0$, overlapping, and therefore are difficult to recognize in the chart)

We have designed and evaluated the DBN described in this paper using the *Bayes Net Toolbox for Matlab* (BNT),[15] and, in particular, we resorted to the Boyen-Koller (BK) inference algorithm [23] for DBNs. BK can perform both exact and approximate inference by trading-off precision (accuracy) of the posterior probability estimation and computational (time and space) resources. In the present work we experimented with the version of BK able to perform exact inference on the DBN model.

5.1 No Monitoring

In the first experiment we compute the likelihood that the techniques are successfully exploited by the attacker, in the setting in which an attack is carried out, but the monitoring system is not deployed into the architecture, thus no evidences are collected. For this experiment, we turned off the implemented monitoring system. Notice that, in this case, since no monitoring has been deployed, it is irrelevant whether the synthetic attacker is at work or not, since there's no monitoring system to record its activity anyway. Results in this setting provide useful insights about the structural characteristics of the attack process according to the a priori probabilities derived from the estimated completion times. Figure 10 shows the cumulative probabilities of success of the different techniques. According to the attack process modeled in the DBN in Fig. 9, after compromising a machine in the OT network, the attacker can modify the DER control logic (labeled "ModCtrlLogic"), obtain insecurely stored credentials ("UnsecCred") or modify a program of the plant controller ("ModifyPro-

[15] https://github.com/bayesnet/bnt.

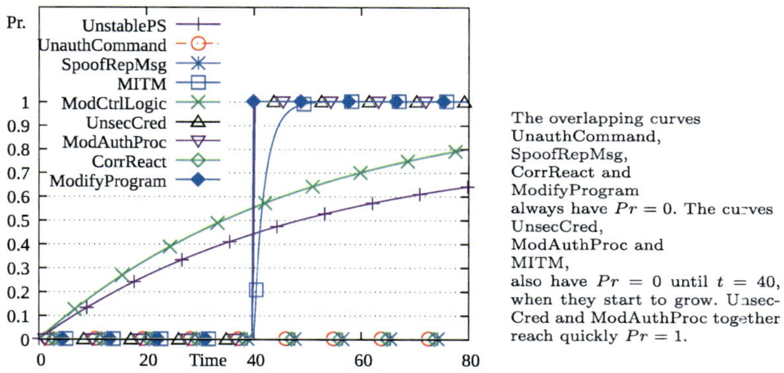

Fig. 11 Probability of successful technique exploitation when the attacker performs the steps "UnsecCred" and "ModAuthProc" just before t =40 (analytics "FileAccess" and "IntCheck" up from $t = 40$, all other implemented analytics down, the rest undefined)

gram"). The difference between the success probabilities of such techniques depends on their average completion times, thus as expected it is more likely that the attacker completes the "UnsecCred" attack step since it has the shortest average completion time, then it can be followed by "ModifyProgram' or the sequence of techniques "ModAuthProc" and "MITM". The "ModifyProgram" attack is less likely because it takes a significant higher time to be completed.

Generally, we can observe an inverse correlation between the success probability of a technique at a given time and its distance from the final goal since the chance of success of a high-level attack step depends on the success of lower steps in the attack path and it is further modulated according to its average completion time. Moreover, we can notice that in the long run all considered techniques will eventually succeed, except for the final step "CorrReact". As explained in Sect. 4.3 such step depends on the activation of a protection measure and thus it can be successful only with a given probability, fixed to 0.7 in all the experiments.

5.2 *Two Attack Steps*

Next, we turned on the monitoring system and observed how the collected evidences affected the results. We started the adversary with a delay and had it carry out the first two attack steps in rapid sequence; evidences of its activity reached the DBN at time $t = 40$. The model assumes that an attack is underway starting from $t = 0$, which models a typical safe posture of the security analyst: the attacker is never inactive, as far as we know. But, until $t = 40$, all the implemented analytics certify that there is no adversarial activity; at that point both "FileAccess" and "IntegrityCheck" raise an alert. The alerts, once raised, are never cleared in the DBN, and so the evidences are considered positive until the end of the analysis at $t = 80$ (see Sect. 6). All

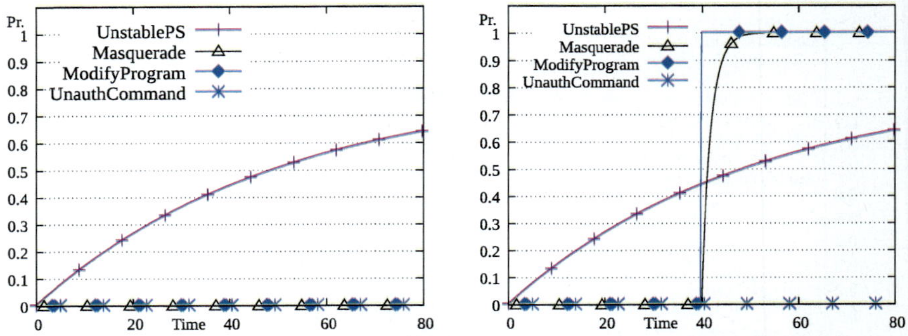

Fig. 12 Probabilities of successful technique exploitation of the techniques in the attack path starting with "ModifyProgram". At left: in the same setting of Fig. 11 (here all curves except for "UnstablePS" overlap with probability 0); at right: when the attacker also tries to stop a service triggering analytic "ServLaunchCmd" (in this case all curves except for "UnstablePS" overlap with $Pr = 0$ up to $t = 40$)

other implemented analytics are still giving no alarm. We can observe in Fig. 11 that initially, without any positive or negative evidences from the "UpdateMonitor" analytic, and negative evidence from all the others, the DBN model infers that the attacker can only attempt a "ModCtrlLogic" attack; then "WrongLogicExecution" might occur and thus there is a probability, increasing over time, that the system be compromised through attacks that cannot be detected.

When both "FileAccess" and "IntegrityCheck" start to raise alerts, the model correctly infers that the attacker is taking the central right attack path that goes from "UnsecCred" to "ModAutProc" and that the following more likely step will be the exploitation of a Man-In-The-Middle attack splitting MMS over the TLS connection. Note that, even if such step is successful, the model derives that the attacker is not able to go further along this path, since both analytics "MeasureCoherence" and "UnauthCommand" are not raising alerts. Therefore, the evidences triggered by the first attack steps carried out by our synthetic attacker have no impact on the forecasts of system instability.

5.3 Attack Processes and Unrelated Suspicious Activities

Recall that our attacker can add randomly selected attacks to the attack paths, to mimic a real attacker trying different possibilities, and create disturbances in the detection system. Moreover some unusual legitimate administration operations can be interpreted by the monitoring system as adversarial activity. In the following experiment we replicated the activity of Sect. 5.2, but the attacker also tried to stop a Tomcat server (not running) in the client MMS. This attempt triggered an alert of the "ServLaunchCmd"analytic. To highlight the differences, we plot in Fig. 12 only the probabilities of the techniques in the rightmost attack path of Fig. 9. On the left side, we show the probabilities in the setting of the experiment of Sect. 5.2. Here we see that all techniques are considered not to occur because the only analytic deployed for the first step ("ModifyProgram") is raising no alarm: all other attack steps can take place only after this one. The growing probability of an unstable system is due to the reasons discussed in Sect. 5.2.

On the right side, we show what the DBN model infers when our attacker is also trying to stop the service. This time, adversarial activity related with the "ModifyProgram" technique is detected. Thus, the detection system expects with high probability that "ModifyProgram" has been carried out, and correctly warns the security analyst. Since no monitoring for a "Masquerade" technique is deployed, the DBN can only warn that, with high probability, also that attack step will be soon carried out (as a side observation, notice that this underlines a weakness in the monitoring system). But, we have an analytic in place to detect occurrence of an unauthorized command, and this is raising no alarm. Then, even though there is suspicion that the attacker has started along this attack path, the analyst can be sure that no further steps have been taken so far and therefore no impact on the instability of the system is expected ("UnauthCommand" has zero probability to occur and the curve for "UnstablePS" is exactly as on the left-hand side). This highlights the benefit of using a detection model that focuses on attack processes rather than single attack steps. Of course the more complex the system, the more difficult it becomes avoiding false positives.

5.4 A Whole Attack Process

Here we had our attacker carry out one complete attack process: from "UnsecCred" all the way through "SpoofRepMsg". The monitoring system detects that the measures sent to the server all of a sudden are not coherent, and thus also "MeasureCoherence" signals a suspicious activity at $t = 40$. The conclusions of the model are shown in Fig. 13. In this case, we can derive that the attacker is also able to modify reporting messages. We assumed that, in this case, with probability 0.7 the automatic protections of the control system are not able to detect that something is wrong with the reported measures and to disable a reaction, thus causing the power system instability. Notice, in Fig. 13, how the probability of an unstable system increases when "MeasureCoherence" raises an alert.

This experiment highlighted one interesting aspect that requires further discussion: we had estimated (Table 1) a rather long execution time for the attack steps that injects false measures ("SpoofRepMsg"). Our attacker has altered dramatically the measures and thus the attack step was detected right away. On the one hand, this is unrealistic and says that we should parametrize the attacker more carefully. On the other hand this shows how in our model the evidences take priority over any other consideration. We discuss evidences further in Sect. 6.

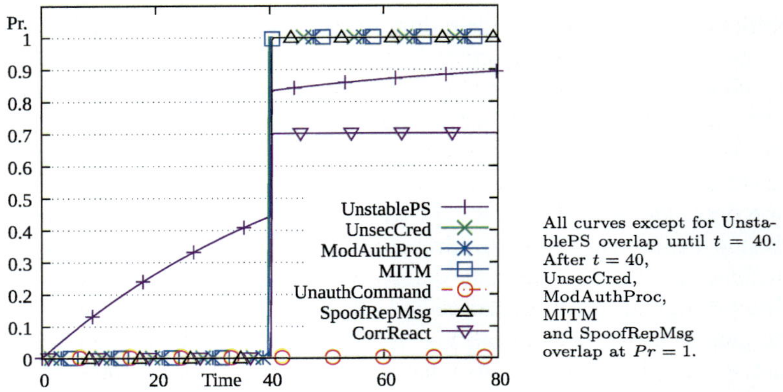

Fig. 13 Probabilities of successful technique exploitation in case of a full attack process (analytics "FileAccess", "IntCheck", "MeasureCoherence" up from $t = 40$, other implemented analytics down, the remaining analytics undefined (no evidence))

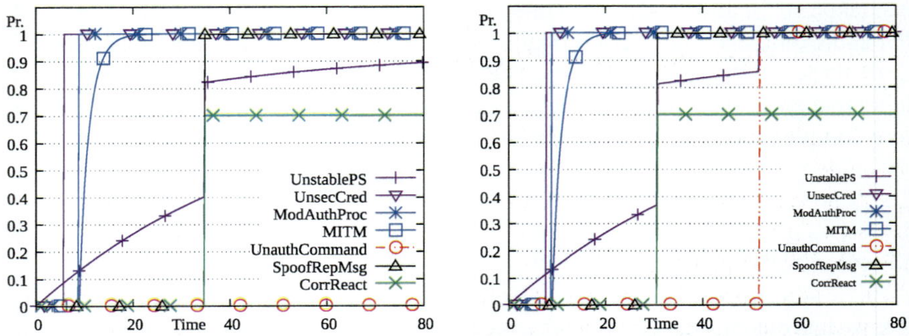

	"FileAccess"	*"IntCheck"*	*"MeasureCoherence"*	*"CommandCoherence"*
Left:	$t = 6$	$t = 9$	$t = 35$	not activated
Right:	$t = 8$	$t = 9$	$t = 31$	$t = 52$

Fig. 14 Probabilities of successful technique exploitation with a slow attack and randomized times. Analytics are activated as in the table above (other implemented analytics down, the rest undefined, i.e. give no evidence). The curve "UnsecCred" grows to $Pr = 1$ at $t = 6$ (left) and $t = 8$ (right), followed by "ModAuthProc" at $t = 9$, and by "MITM" soon after. After $t = 35$ (left) and $t = 31$ (right) all curves except "UnstablePS" and "CorrReact" overlap at $Pr = 1$. At right, after $t = 52$ all curves except "CorrReact" overlap at $Pr = 1$

5.5 *Slower Attacks*

The model can also cope with a sequence of observations spread over time. In the experiment shown in Fig. 14 (left) we had the attacker wait some random time between attack steps. So the alert from analytic "FileAccess" has been raised at $t = 6$, then at $t = 9$ "IntCheck" raised an alert, and finally "MeasureCoherence" at $t = 35$. Again all other implemented analytics remained inactive for all the time.

The DBN infers the same probabilities for the techniques we are interested in, as in Fig. 13. In Fig. 14 (left), though, since alerts are raised at different time-instants, it can be seen more clearly what the DBN infers from each alert. In particular, notice that the occurrence of a MITM starts to be expected only after it is reported that the authentication process has been modified ("ModAuthProc"). Also notice that only when the analytic for "SpoofRepMsg" becomes active, then the model infers a higher probability of reaction to the injected measures, and as a consequence also the probability of an unstable system increases abruptly.

Finally, we had the attacker carry out both the attack processes that have been implemented. Again we timed its activity randomly. We report the probabilities inferred by the DBN in Fig. 14 (right). The times at which alerts were reported have changed with respect to Fig. 14 (left) since the adversarial timings are randomized in the current experiment and the previous one. It is interesting to notice here that after the attacker injects the command (technique "UnauthCommand", detected at $t = 52$), the probability that the system will be unstable jumps to 1.

6 Conclusions and Future Directions

This chapter presents the implementation of a testbed for the evaluation of detection models in which an attacker is emulated and a monitoring system deployed. The framework is modular in that it allows testing different detection mechanisms, possibly in parallel, but also enables defining a specific profile for the attacker, and adding or removing attack steps.

The first testing phases have confirmed the interoperability of the various components and that the data flow is as expected. The models mirror the activity of the attacker. On the other hand, the experiments have exposed some shortcomings of the models: for instance, our current models don't capture the different attacker profiles, the analytics remain valid indefinitely. See below for the discussion of these issues, but bear in mind that the goal of the framework is precisely exposing possible inadequacies of the models we test and therefore these results are a validation of our testbed.

Experiments highlight how in our framework it is possible to modify the adversarial activity, including partial attack processes, unrelated suspicious activity or slower attack processes, and verify the response of the deployed detection models. As a side effect, also the effectiveness of the monitoring system is evaluated.

The achieved results will be further developed in various directions. The DBNs are so far parametrized only based on estimated completion times of the attacks. This means that there is no attempt, as yet, at modelling the attacker's profile. The profile is important when the detection model must integrate incomplete information from the monitoring system. We plan to introduce features to model different attacker profiles in the DBN, matching the various behaviours that can be configured for the synthetic attacker. In order to parametrize correctly the attacker's behaviour and the DBN that models it, it is necessary to have access to extensive data on real world attacks.

The MITRE ICS ATT&CK project is a step in that direction, but unfortunately the knowledge of the real scenario is currently limited, partly because not all attacks are detected, partly because not all of those detected are reported. Moreover, that project considers all industrial and critical infrastructures, whereas we would like to have data specific for the power infrastructure. Therefore, currently, we cannot count on using realistic data; as a consequence in the test process we also intend to determine the robustness of the models, analyzing how much the results are affected by a discrepancy between the actual adversarial behaviour and the assumptions expressed by BNs' parameters.

On the other hand, a more realistic emulation of the attacker would allow us to create realistic traffic in the network and gather traces that could in turn be used to train machine learning models, to increase our detection modules.

After these preliminary tests, our aim is to use the framework to understand the potentialities of the models we integrate in it, in particular the DBNs. In the experiments we presented here we only used filtering for the analyses, that is we concentrated on online diagnosis and early detection. Through our framework's dashboard the analysts will see graphs limited to the time instant of the observation. With DBNs we can carry out also predictive analyses for times beyond the current instant t; or diagnosis that tries to determine (past) causes of events detected by the monitoring system. Within our framework, besides testing the validity of the model results, we also intend to study an effective way to present all of this information in a way that supports the evaluation of the security posture.

Currently, when an alert is raised from an analytic in our DBN model, it remains active for the rest of the analysis. We plan on introducing an ageing mechanism for analytics, in the spirit of [9]. This would enable us to discard an alert as a false alarm if it is not followed by additional indications of suspicious activity in due time. The ageing parameter is again a delicate issue, since often the attacker chooses to act slowly precisely to avoid detection.

In conclusion, detection models offer added value with respect to the raw observation of the alerts from the monitoring system, allowing us to derive correlated information and not just the occurrence of observable events. It is key understanding how to exploit this possibility in the most effective way. Our framework is a valuable tool to help analyze and unleash these potentialities.

Let us not forget that the final aim is a valid tool for detection of adversarial activity. The implementation of a framework with the ability of collecting monitoring and log data and analysing them by means of a set of analysis modules is very important for power infrastructure management, in order to provide a tool for the early detection of attack processes integrating I/O events with more operational specific ones. The implementation in power utility control centers of a framework like the one we presented in this chapter, along with the monitoring of power values, shall give an enhanced view of the status of the infrastructure and shall enable a more appropriate control operation to maintain the stability of the power system.

Acknowledgements We thank the anonymous referees for their comments. This work is original and has been supported by a joint collaboration between RSE S.p.A. and Università del Piemonte

Orientale, partially funded by the Research Fund for the Italian Electrical System in compliance with the Decree of Minister of Economic Development April 16, 2018. During the research activity D. Cerotti was funded by a CNIT grant. D. Cerotti and L. Egidi are affiliated to the Indam National Research group GNCS, Italy.

References

1. D. Cerotti, D. Codetta, G. Dondossola, L. Egidi, G. Franceschinis, L. Portinale, R. Terruggia, Evidence-based analysis of cyber attacks to security monitored distributed energy resources. Appl. Sci. **10** (2020)
2. D. Cerotti, D. Codetta, G. Dondossola, L. Egidi, G. Franceschinis, L. Portinale, R. Terruggia, Analysis and detection of cyber attack processes targeting smart grids, in *Innovative Smart Grid Technologies Europe (ISGT)* (2019)
3. S. Lee, S. Lee, H. Yoo, S. Kwon, T. Shon, Design and implementation of cybersecurity testbed for industrial IoT systems. J. Supercomput. **74**, 4506–4520 (2018)
4. J. Jarmakiewicz, K. Maślanka, K. Parobczak, Development of cyber security testbed for critical infrastructure, in *International Conference on Military Communications and Information Systems* (IEEE, 2015), pp. 1–10
5. J. Jarmakiewicz, K. Parobczak, K. Maślanka, Cybersecurity protection for power grid control infrastructures. Int. J. Crit. Infrastruct. Prot. **18**, 20–33 (2017)
6. P. Singh, S. Garg, V. Kumar, Z. Saquib, A testbed for scada cyber security and intrusion detection, in *International Conference on Cyber Security of Smart Cities, Industrial Control System and Communications* (IEEE, 2015)
7. M. Mallouhi, Y. Al-Nashif, D. Cox, T. Chadaga, S. Hariri, A testbed for analyzing security of SCADA control systems (TASSCS), in *Innovative Smart Grid Technologies* (IEEE, 2011)
8. U.S. Department of Energy: National SCADA Test Bed. www.energy.gov/oe/technology-development/energy-delivery-systems-cybersecurity/national-scada-test-bed (2003–2022)
9. M.J. Pappaterra, F. Flammini, Bayesian networks for online cybersecurity threat detection, in *Machine Intelligence and Big Data Analytics for Cybersecurity Applications* (Springer, Cham, 2021), pp. 129–159
10. P. Mell, V. Hu, R. Lippmann, J. Haines, M. Zissman, An overview of issues in testing intrusion detection systems. National Institute of Standards and Technology ITL (2003)
11. K. Park, B. Ahn, J. Kim, D. Won, Y. Noh, J. Choi, T. Kim, An advanced persistent threat (APT)-style cyberattack testbed for distributed energy resources (DER), in *2021 IEEE Design Methodologies Conference (DMC)* (2021), pp. 1–5
12. C.-C. Sun, A. Hahn, C.-C. Liu, Cyber security of a power grid: state-of-the-art. Int. J. Electric. Power Energy Syst. **99**, 45–56 (2018)
13. R.V. Yohanandhan, R.M. Elavarasan, P. Manoharan, L. Mihet-Popa, Cyber-physical power system (CPPS): a review on modeling, simulation, and analysis with cyber security applications. IEEE Access **8**, 151019–151064 (2020)
14. E. Veith, N. Wenninghoff, S. Balduin, T. Wolgast, S. Lehnhoff, Learning to attack powergrids with DERs (2022). arXiv:2204.11352
15. S. Das, Fgan: federated generative adversarial networks for anomaly detection in network traffic (2022). arXiv:2203.11106
16. I. Zografopoulos, C. Konstantinou, N.D. Hatziargyriou, Distributed energy resources cybersecurity outlook: vulnerabilities, attacks, impacts, and mitigations (2022). arXiv:2205.11171
17. M. Ekstedt, Meta Attack Language (MAL). https://foreseeti.com/meta-attack-language/. Accessed Apr 2022
18. M. Cosentino, Threat-em: a framework for defensive system validation via adversarial behaviour emulation. Master thesis in Computer Science, Universitá del Piemonte Orientale (2022)

19. J. Pearl, *Probabilistic Reasining in Intelligent Systems: Networks of Plausible Inference* (Morgan Kaufmann, USA, 1988)
20. K.P. Murphy, Dynamic bayesian networks: representation, inference and learning. Ph.D thesis, University of California, Berkeley (2002)
21. L. Portinale, D. Codetta, *Modeling and Analysis of Dependable Systems: A Probabilistic Graphical Model Perspective* (World Sc, USA, 2015)
22. D. Cerotti, D. Codetta, G. Dondossola, L. Egidi, G. Franceschinis, L. Portinale, R. Terruggia, A modular infrastructure for the validation of cyberattack detection systems. Technical Report TR-INF-2022-05-01-UNIPMN, Computer Science Institute, UPO. https://www.di.unipmn.it/en/publications-en/technical-reports-en.html?pubid=567
23. X. Boyen, D. Koller, Tractable inference for complex stochastic processes, in *Conference on Uncertainty in Artificial Intelligence* (Morgan Kaufmann, USA, 1998), pp. 33–42

A Novel Self-learning Cybersecurity System for Smart Grids

Michalis Skoumperdis, Nikolaos Vakakis, Maria Diamantaki, Charalampos-Rafail Medentzidis, Dimitrios Karanassos, Dimosthenis Ioannidis, and Dimitrios Tzovaras

Abstract The dynamic nature of cyberattacks, as well as the incorrect predictions of Artificial Intelligence (AI)-based cybersecurity intrusion detection systems, are major impediments to the efficient protection of critical infrastructures' Information Technology (IT) and Operational Technology (OT) systems, including Electrical Power Energy Systems (EPES). This phenomenon is caused by the cyberattack detection models, which lose their effectiveness over time. The variability of cybersecurity threats makes it difficult to establish and implement a specific model, which detects all types of attacks accurately. The key to system protection is the integration of a cybersecurity framework, which simultaneously addresses new potential threats, fixes misclassified predictions and utilizes the best performing model, according to the most recent data. This work proposes a self-learning engine which is based on the SPARK data analytics framework and is integrated into a cybersecurity platform. The self-learning module provides the opportunity for annotating data to correct misclas-

M. Skoumperdis (✉) · N. Vakakis (✉) · M. Diamantaki · C.-R. Medentzidis · D. Karanassos · D. Ioannidis · D. Tzovaras
Information Technologies Institute, CERTH, Thessaloniki 57001, Greece
e-mail: skoumpmi@iti.gr

N. Vakakis
e-mail: nikovaka@iti.gr

M. Diamantaki
e-mail: mardiama@iti.gr

C.-R. Medentzidis
e-mail: charmede@iti.gr

D. Karanassos
e-mail: dkaranassos@iti.gr

D. Ioannidis
e-mail: djoannid@iti.gr

D. Tzovaras
e-mail: dimitrios.tzovaras@iti.gr

H. Haes Alhelou et al. (eds.), *Power Systems Cybersecurity*, Power Systems,
https://doi.org/10.1007/978-3-031-20360-2_14

sifications or to add intelligence regarding previously unknown attacks. Initially, the domain experts submit annotated data through a Visual Analytics (VA) & monitoring system, to start the retraining process. Three (3) machine-learning (ML) methods–Random Forest (RF), Logistic Regression (LR) and K-nearest neighbors (KNN)-as well as one (1) Deep Learning (DL) method-SDAE-are dynamically compared in terms of F_1 score and accuracy. After the completion of the retraining process, the best performing model replaces the existing one and labels the incoming data. The dynamic nature of the self-learning module implies that it gets annotations from users anytime, compares the methods during the retraining process and assigns data labelling to the most accurate model.

Keywords Cybersecurity · Smart grids · Self-learning · Human-computer interaction

1 Introduction

The traditional legacy power systems led to the increase in global electrical energy consumption. Because of the rising demand for power supply, numerous environmental, industrial and social impacts have emerged, such as the high cost of frequent blackouts and burnouts [1, 2], the inefficiency and the unreliability of power networks in energy distribution and the inability of energy suppliers both to use renewable sources and reduce their carbon footprint [2]. The transformation of the traditional electrical network to smart grid infrastructures or smart grids has solved all the aforementioned issues. The smart grid infrastructures allow factories and consumers to define the desired level of the consuming electricity using Information and Communications Technology (ICT), to achieve an optimum allocation of resources [2]. However, the utilization of ICT technology for the communication between the smart grid network's components causes potential vulnerabilities, which raises doubts about the security and the reliability of smart grid infrastructures [3]. The complexity of the smart grid infrastructures is the main reason for their vulnerability in cyberattacks [4]. Cyberattacks target either the physical devices of the power grid or the communication network [5]. In the physical space, the cyber threats affect the confidentiality, the integrity, and the availability of information transfer between the assets of the smart grid [6]. Network intrusions may lead to severe malfunctions like information leakage and the destruction of the normal modus operandi of smart grids.

Smart grid cyber attacks are divided into categories like interruption threats, interception threats, modification threats and fabrication threats [7]. The interruption threats or Denial-of-Service (DoS) attacks affect assets' availability. The dynamic DoS attacks reform their signatures according to changing traffic patterns. The reduction of DoS attacks is succeeded by using AI security algorithms such as anomaly detection methods, big pipes, and traffic filtering [1, 8, 9]. Pre-shared keys [8] and Transport Layer Security (TLS) certificates [10] protect the EPES infrastructure against interception threats, which have a serious impact on system reliability, as they may affect the financial condition of users or may place the system in bad marketing positioning. The replay attacks, which frequently affect the security of smart grid infrastructures, belong to the interception threats. This type of attack initially records the sensor data for a certain time range and then repeats the data to control the system's operation [11, 12]. Modification attacks corrupt authorized processes in EPES and can be reduced by encrypting network traffic [13] or by restricting access [8]. The False Data Injection Attacks (FDIAs), which are the most common type of modification threats, inject false data into the smart grid to gain control of the system. They can cause serious malfunctions even the shutdown of the system [11, 14]. Finally, the fabrication attacks, which create a fictitious asset on EPES, are reduced by a single data feed [1]. State-of-the-art AI algorithms protect the system from various types of cyberattacks. For example, AI algorithms model the signature of each sophisticated attack, so they not only detect the potential threats but also label the corresponding network flows. The techniques which are used to protect smart grids against cyber threats should be efficient in both existing and future, sophisticated threats [5]. In addition, the mitigation of sophisticated attacks demands the distributed and real-time monitoring of cybersecurity system components where proper knowledge can be received via visualization techniques and deep analysis [15].

Latest research on cybersecurity, proposes the utilization of deep neural networks to forecast sophisticated cyber attackers [11, 12, 14–16]. In [14] Nonlinear Auto-Regressive & Exogenous Model (NARX) neural network, which is a Proportional Integral (PI) controller-based Artificial Neural Network (ANN), accurately detects FDIAs in DC microgrids, but the effectiveness of method is not tested on other types of threats. Various AE (Autoencoder)-based methods are also designed to mitigate the implications of cyberattacks [17–19]. The efficacy of these methods is not properly ensured because either the cyberattack simulations are not performed on a real-world, industrial framework [17, 18] or the results are derived only from a predefined dataset. In [12] the proposed method comprises a deep restricted Boltzman network along with a convolutional neural network (CNN) and on the one hand it achieves high accuracy rates ranging from 98,13% to 99,51%, but on the other hand the attacks are necessarily classified to one of the following categories: Distributed Denial of Service (DDoS) attacks and replay attacks.

The improvement and the flexibility of system protection can be achieved by the integration of a cybersecurity framework, where the recognition of the new potential and more sophisticated threats in the cybersecurity module is based on the interaction of the domain experts with the EPES system [20]. More specifically, software engineers or cybersecurity experts annotate and relabel the data via VA-based tools. Manual labelling of data [21] is a time-consuming process [22] and demands long labour hours, thus becoming a suboptimal solution [23]. More advanced approaches [22–24] propose visual-based tools for proper data labelling.

To this end, this work proposes a hybrid self-learning engine, which is based on the SPARK framework [25] and could be integrated into the existing cybersecurity systems of a critical EPES infrastructure. The smart grid infrastructure is capable of capturing, parsing and streaming heterogeneous data such as industry-specific protocols and operational data. This self-learning engine not only utilizes ML or DL models for cyberattack detection, but also involves human judgement in order to compensate for possible misclassifications and add new knowledge to the system. The main contribution of this work is the reformation of the system in such a way that new threats are possible to be detected, while false predictions of existing threats are addressed. A visual-based data annotation mechanism within a VA-aided decision support system is used to feed the self-learning engine with new knowledge, in order to retrain models to become more accurate and be able to detect new threats. In addition, the tool is scalable as new categories of attacks are added perennially. For the proof of concept, initial cyberattack detection AI models are trained on a dataset which consists of network flow statistics are extracted from network traffic and comprised of six (6) different cyberattacks, related to either Modbus or MQTT application layers protocols, as well as data from normal operation for both protocols collected from a real smart home infrastructure. Three (3) of these threats concern Modbus, which is a data communications protocol for industrial electronic devices, and the rest concern MQTT which is a standard messaging protocol. The self-learning engine receives manually annotated data as input, through a VA-aided decision support system, which facilitates a domain expert to investigate data in different layers and judge whether network flows are normal or are produced as a result of potential cyber threats. The model retraining process is triggered whenever a domain expert submits new input. During the retraining process, three (3) ML methods-RF, LR and KNN-and one (1) DL method-SDAE-are dynamically compared in terms of F1 score and accuracy.

The Grid Search framework tunes the hyper-parameters of each method to find optimum parameter combinations. To optimize the computational performance induced by heavy training workloads, the training tasks are not performed serially, but are distributed among the nodes of a SPARK cluster to speed up training through parallelization. Two (2) Modbus-related attacks that are unknown to the cyberattack prediction models and an already known MQTT-related attack, were applied to produce new models, compare them and test their effectiveness in comparison with the old ones. The results from the self-learning module analysis, showed that in most cases SDAE is a more accurate model, either in detecting new types of threats or in improving the detection of already known threats.

The remainder of this paper is organized as follows. An overview of related works in self-annotating training is summarized in Sect. 2. Section 3 describes the system architecture and the cybersecurity platform. Section 4 represents the ML and the DL methods as well as the attacker use cases. Section 5 represents the cyberattack detection classification results. In Sect. 6 a deeper view into the results is provided. Finally, in Sect. 7 our conclusions are drawn.

2 Related Work

In critical EPES infrastructures, the dynamic nature of cyber threats compels the retraining of predictive models to increase system security. One of the key prerequisites for the success of such systems is to decide when and how to retrain the models. A proper solution, which has been proposed in many research works, is the dynamic data annotation by users to trigger the retraining process [20, 22, 24, 26–28]

In [22] the adaptation of the proposed method to newly collected data is half based on the Interactive Recurrent Annotation to define bounding boxes of video objects. The method receives feedback from human annotations to feed a CNN. This type of network uses a Hierarchical Correction module at each timestep. More specifically, a detector of unlabeled data reinforces the CNN model in each iteration using the model of the previous iteration. Two (2) special measures, the Similarity In Union (mSIU) and the Intersection Over Union (mIOU), are used to represent the performance of the method. In work [24], the VA module of the system represents the results of an unsupervised method and domain experts only relabel incorrectly classified data points.

The reduction of user effort in data annotation tasks is another topic discussed by many researchers. The architecture of the system which is proposed in work [20] is almost the same as our method. Instead of receiving feedback from domain experts to define anomalous data points, the method which is proposed in this work embraces a state-of-the-art anomalous detection method and the user only defines the label of data points which are considered accurately anomalous. The annotation effort is reduced and the domain experts are not confused by data points which are considered barely anomalous. In addition, the anomaly detection module of current work depicts the suspicious data without the utilization of a sampling method [20]. The authors of work [22] target not only to reduce user effort but decrease annotation time. In [24] target variables are separated in two different categories. The samples with high confidence values are classified in the first category, while the samples with low-confidence values are classified in the second category. Subsequently, domain experts annotate the data which are classified in the second category and propagate them to the system. This approach reflects a semi-automatic data annotation system. By using five (5) public datasets, they improve classification accuracy. In a similar study [29], researchers propose Human-Augmenting Labeling System (HALS), where the higher the quality of data labelling, the more annotation effort is reduced. In this study, annotation effort is reduced as the user assesses cybersecurity threats through a user-friendly dashboard.

The selection of a representative dataset is the main reason for demanding expert annotation in many research works [30–32]. The sampling methods Sampling by Uncertainty & and Density (SUD) and Sampling by Cluster (SBC) which are proposed in work [30] are based on dynamic data selection as domain experts do not select data outliers. The techniques which are proposed in work [31] assist experts to select the appropriate and most informative handwritten text excerpts for proper training. In this paper, the system combines recently annotated data by domain experts, historic data annotations and the initial datasets to train new, more robust models for cyber attack detection.

3 System Architecture

3.1 Cybersecurity Platform

As mentioned before, the self-learning module has been integrated into a cybersecurity platform. This platform supports an enriched anomaly detection system, minimizing the time needed for detecting sophisticated cyberattacks. ML and DL models are utilized to detect both cyberattacks and anomalous data points. The platform detection is based on analyzing data from different layers of the IT-OT network and applying ML/DL models to classify them in various categories. The overall architecture of the system is illustrated in Fig. 1. There are three (3) main modules, VA, Cyberattack detection and self-learning along with Retraining Engine. The VA module is responsible for displaying information to the security engineer, facilitating her/him to conduct analysis and submit input to the self-learning and Retraining Engine, which in turn encompasses the new knowledge and retains the ML/DL models to calculate the best performing and assign it to the Cyberattack detection module.

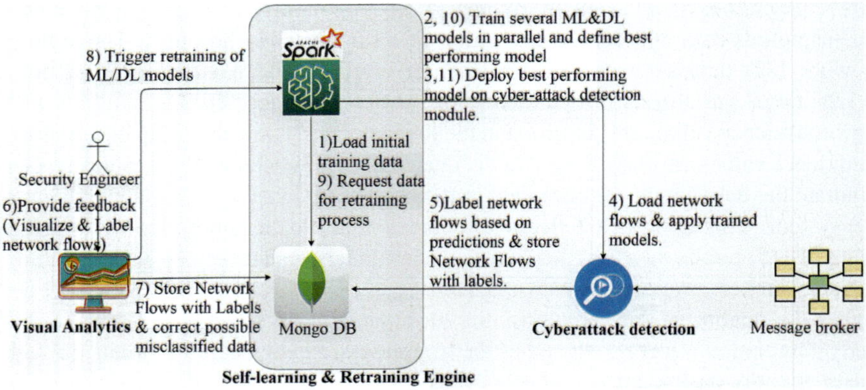

Fig. 1 Architecture of cybersecurity platform

3.2 VA Module

VA-based tools are complementary modules, offering decision support to cybersecurity engineers or domain exerts. The VA module enables users to draw conclusions regarding the network security status of the infrastructure by monitoring and visualizing data from different layers (operational data, network packet data, network flow statistics) through various visualization charts (such as linecharts, pie charts, scatter plots, graphs etc.), as well as applying analytics algorithms (such as local outlier factor, k-partite graph etc.) to investigate data in more depth. Furthermore, the VA module is backed up by its own AI engine, which applies near real-time deep packet inspection through a DL methodology, as described in [33]. This way, the user is exempted from the extremely heavy workload of having to annotate all data. Instead, s/he has to investigate only data points that are considered anomalies by the AI engine. The anomalies in the network packet data are produced as a result of high deviations between the predictions of the AI engine algorithm and the actual values. This deviation is illustrated in the VA module with a linechart as can be seen in Fig. 2. In case of detected anomalies, the VA correlates the anomalous packets with the respective network flows and highlights them in another chart as shown in Fig. 3. After investigating the data of different layers with the available visualization charts, the user is ready to judge in which category those network flows belong and label them properly. Accuracy improvement of cyberattack detection models is achieved through the interconnection between the VA-based tools, the self-learning module and the Retraining Engine. Despite the self-learning module and the Retraining Engine operating within the specific platform, they are adaptive in any system, which is relative to the protection of EPES, as long as there is the possibility to provide labels for specific data points. After the selection of suspicious network flows, the user fills out the form with the appropriate label and then submits labelled data to the Retraining Engine.

3.3 Cyberattack Detection Module

The second module of the cybersecurity platform is the Cyberattack detection module. This module is the one responsible for the deployment of the best performing model as extracted from the self-learning module & the Retraining Engine. The input of the Cyberattack detection module is unlabelled network flow data, which are received from a message broker. This module should always be updated with the best performing models according to the latest retraining process. The deployed ML/DL model makes predictions on incoming data and stores the labelled network flows in the mongoDatabase (DB) database of the self-learning and Retraining Engine used for future retraining processes.

Although the cyberattack detection module accurately detects anomalous data points, it is not ensured that the system could label such data properly. For this reason,

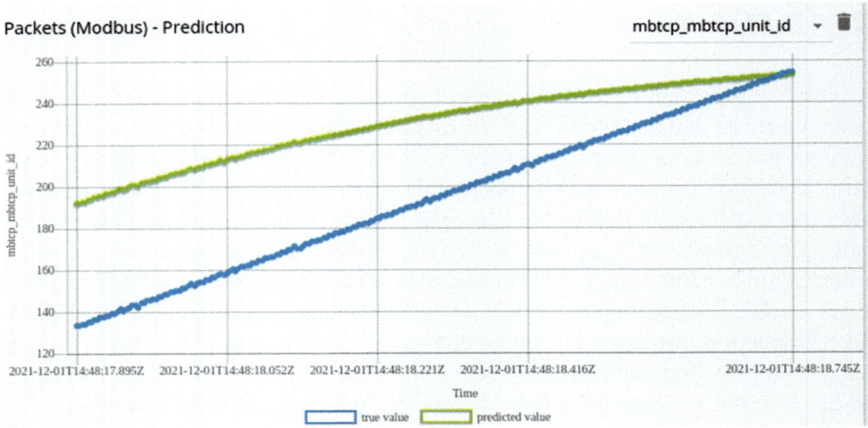

Fig. 2 Packets prediction linechart

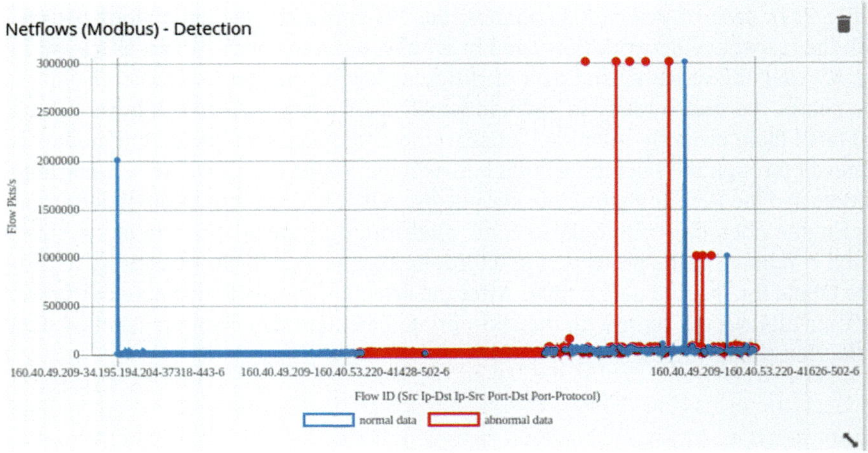

Fig. 3 Network flows annotation chart

ML models, which are assisted by human annotation, are utilized to label network flows deriving from the network communications of the interconnected devices and sensors. ML and DL models are trained on data associated with a predefined set of attacks, so it is impossible to label data from totally unknown types of attacks. Nevertheless, domain experts and security engineers have thorough experience in cybersecurity threats and specialize in annotating data properly.

The data are received from a message broker, such as Apache Kafka or Mosquitto MQTT broker. A mongoDB database is installed to retain the messages/data for further processing. The system assigns a label to incoming messages according to the most recent best performing model. Two (2) distinct collection tables named labelled

Modbus and labelled MQTT store the messages which are labelled by the best performing model for the corresponding protocol. Before the first interaction between the domain experts and the system, the messages/data are annotated according to the initial set of labels. In the time period between a new previously unknown cyberattack strikes the system and the completion of the retraining process, the data are labelled according to the set of cyberattacks which exists before the triggering of the Retraining Engine. The relabelled data are maintained in two (2) special database collections named history Modbus and history MQTT. These data are concatenated with the relabelled data of previous retrain processes. After the completion of the retraining process the system labels the incoming data based on old and recently given labels.

The self-learning module interconnects the Cyberattack detection module, the Retraining Engine and domain experts. More specifically, anomalous data points are detected and then the expert labels such data through the VA dashboard, to trigger the retraining process. After the submission of relabelled data by the user, a distinct training process is activated for each ML/DL model. Regarding the first process, SDAE model is trained, which is the most accurate model in most cases. The rest of the processes correspond to three (3) ML models, namely KNN, RF and LR. The SDAE model is trained in two different ways. In the first operation mode, the domain expert/security engineer provides a label for data corresponding to a previously unknown threat. In that case, the DL model needs to be trained from scratch, because the new data do not belong to any of the known classes. In the second operation mode, the expert/security engineer provides an already known label to correct possible false predictions from the existing model. For example, the model could raise a false positive regarding an attack and the user could label the corresponding network flows as normal, if s/he judges that the data originate from legitimate system communications. The ML models (RF, LR and KNN) are trained from scratch in any case.

To train from scratch the ML models and the SDAE model, the data on which initial models were trained, the data which were annotated in preceding retraining processes and are stored in history MQTT and history Modbus collections, the newly annotated data as well as the stored messages on labelled MQTT and labelled Modbus database collections are concatenated into a single data set. In the case where already known labels are provided, the SDAE last layer demands only the newly annotated data, along with a sample of the rest. The ML models in this case, as was mentioned before, are trained from scratch. The grid search framework tunes the hyper-parameters of models. Because each model is trained on a huge amount of data, along with a wide range of hyper-parameters, training processes demand many computing resources. To face this problem, the SPARK framework distributes each training process in a different machine. Each machine makes up a node, especially the master node saves results and models. Figure 4 depicts the flowchart of the entire self-learning process.

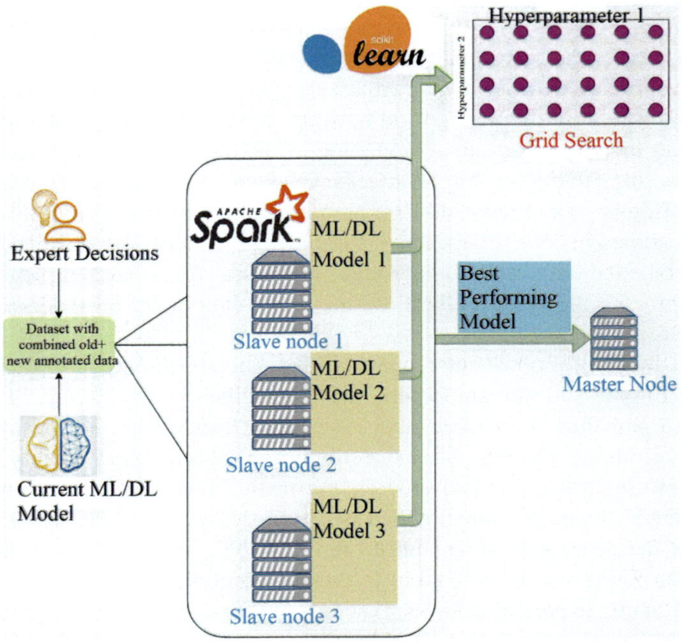

Fig. 4 Flowchart of self-learning module

4 Compared Methods and Attack Cases

4.1 SDAE

SDAE is a type of deep neural network comprising ensuing encoding layers. Ultimately, each layer is an individual Denoising Autoencoder (DAEs), which can be considered as a type of Multi-Layer Perceptron (MLP). The training process of SDAE is divided into two phases, the unsupervised layer-wise pre-training and the supervised fine-tuning.

Initially, original input data are utilized to create a higher representation, independent of they are labelled or not. Extraction of higher representation is a chain where the first trained DAE's hidden layer is the next autoencoder's input [34] and each hidden layer is trained separately. The hidden representation h is explained mathematically according to Eq. 1

$$h = e_f(\bar{W}_X + b) \tag{1}$$

where e^f is the encoding function which is defined as sigmoid, $\bar{W}^X$ is the weight matrix of layer and b is the bias term.

The transformed input y, which is the input of next encoding layer, is defined by Eq. 2.

$$y = d_f(W_X^{'} + b^{'}) \tag{2}$$

where $W_X^{'}$ is the weight matrix of the decoding layer, $b^{'}$ is a bias term and d_f is the activation function of decoder which is defined as either sigmoid or linear [35].

The SDAE architecture consists of six (6) fully connected layers and each layer includes both the encoder and decoder. Fine-tuning, which is a backpropagation phase, starts after the training of all layers and uses supervised training algorithms. The grid search method fine-tunes four (4) specific parameters (decoder_activation_function, batch_size, number_of_hidden_layers, dropout) in order for the SDAE model to perform in an optimized way. The discrete values of model parameters are: the batch_size varies between the values *two (2), four (4), eight (8), 16, 32* and *64*, the dropout between the options *0.01, 0.5* and *0.1* [36] , the number_of_hidden_layers varies between the options *25* and *50* while the decoder_activation_function varies between the options *sigmoid* and *linear* [36, 37]. The optimizer is equal to Adam, as this method was found to be effective in cases where either there are numerous parameter combinations or the input of the DL method is huge [38]. According to [34], the optimal number of layers of the SDAE method is equal to two (*2*). In this study, the early stopping method is applied during SDAE training to decide the optimal number of epochs. The model seems to be over-fitted after ten (*10*) epochs. Finally, the experiments showed that the sigmoid activation function improves the results.

4.2 KNN

The KNN classifier assigns every sample to the most common class between k samples. These samples are the most similar to the initial sample. The leaf size of a given node is its minimum number of points [39]. The grid search method fine-tunes the n_neighbors parameter between the options ten (*10*) and *20* while the leaf_size parameter is equal to *30*.

4.3 RF

The RF classification method is based on a set of decision trees. Each sample is classified according to the decision is taken by the majority of trees [40]. The n_estimators parameter reflects the number of decision trees in each RF classifier. The grid search method fine-tunes the n_estimators parameter between the options ten (*10*), *50* and *100*.

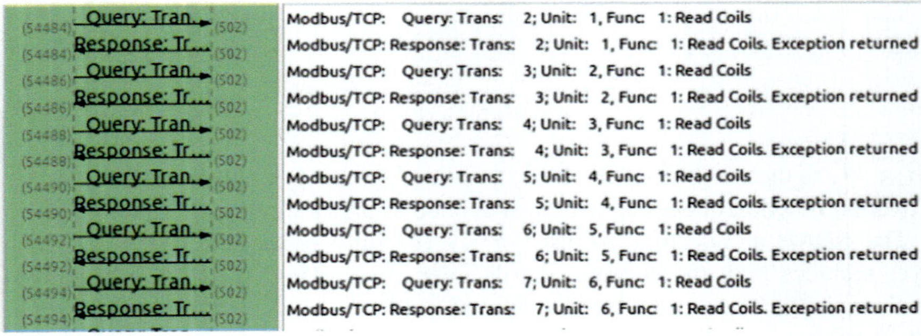

Fig. 5 Flow graph of the UID brute force attack

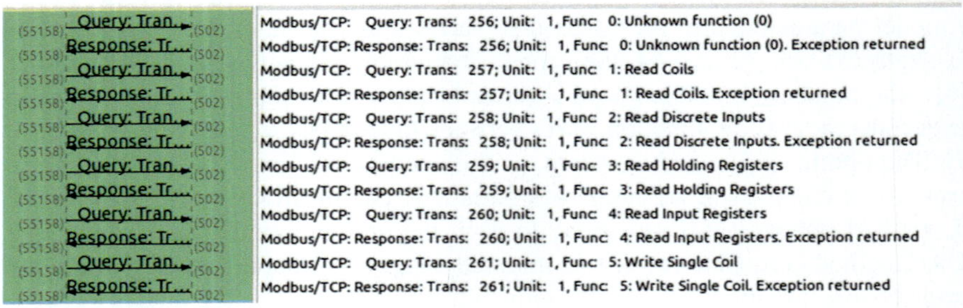

Fig. 6 Flow graph of the Enumeration Function attack

4.4 LR

The LR model is based on a logistic function which optimizes and estimates its parameters. A binary dependent variable models logistic function and an indicator variable classifies each data sample [41]. The maximum number of iterations parameter expresses the maximum number of iterative algorithmic processes. The grid search method fine-tunes the maximum_number_of_iterations parameter between the options *100* and *200*.

4.5 Attack Use Cases in Smart Grid Ecosystems

This section represents the description of attacks which are researched during the experimental procedure. The threats concern both Modbus and MQTT application layer protocols. A brief description of Modbus-related attacks is as follows:

- UID Brute Force: This reconnaissance attack enumerates which IDs are used by Modbus slave devices. Specifically, an attacker knowing the IP of the Modbus master target, tries to identify the UID of its slaves. This attack is implemented by probing every possible UID in the range 1–254 with a default function that is likely supported and see which UIDs reply to that query. As a result, consecutive packets with increasing Unit Identifier in the Modbus/Transmission Control Protocol (TCP) packet layer and function code 'Read Coils' in the Modbus packet layer appear in the Modbus network traffic, along with responses from the target with 'Read Coils' function code, and the value of the requested coil if it is used, or exception code 2 with message illegal data address, if there is a slave with the specified UID, but the requested coil is not used. If the requested UID does not correspond to a slave, exception code 11 is returned, indicating that no response was obtained from the target device. This usually means that the device is not present on the network. The attack flow in terms of packets can be seen in Fig. 5.

- Enumeration Function: This is also a reconnaissance type of attack, which aims to find out which function codes of Modbus are used. After, obtaining the UIDs of the Modbus slaves, the attacker attempts to obtain information regarding the functions that are supported by a slave she/he has identified with the previous attack, by searching through all the possible legitimate functions a Modbus slave can support. The attack consists of sequential Modbus packets from the attacker to the target, with the same Unit Identifier in the Modbus/TCP packet layer, but different function code in the Modbus packet layer for each packet. The attack flow in terms of packets can be seen in Fig. 6.
- Fuzzing Read Holding Registers: This unauthorised access attack sends a Modbus packet to read the values of specific holding registers. This is a fuzzing attack which sends read commands to a range of registers, in order to find out which ones are used and access their data. It is easy to be confused with legitimate reads of the Modbus infrastructure. The only clue that may differentiate a malicious read, is that the defined range of registers in the Modbus/TCP packet layer of the malicious request, may be wide if the attacker does not know which ones are used and include registers which are not used under normal circumstances, while in normal operation mode the range of registers corresponds to the ones that are actually used. The attack flow in terms of packets can be seen in Fig. 7.
- Write All Registers: This unauthorised access attack sends Modbus packets, in order to write values to all holding registers. This attacks belongs to the DoS attack family because it tries to write random values to all available registers and thus impede legitimate write commands and harass the Modbus normal operation. In this case, the attacker sends multiple packets with the Unit Identifier of the slave that was identified in previous attacks in the Modbus/TCP packet layer and function code 'Write Single Register', a reference number corresponding to a target register and a random value in the data field of the Modbus packet layer. If the target register reference number does not exist, the response includes Modbus exception code two (2) with message illegal data address exception. The attack flow in terms of packets can be seen in Fig. 8.
- Write All Coils: This is also a DoS attack, that floods the target system with Modbus Write Single Coil packets. In this case, the attacker sends multiple packets with the Unit Identifier of the slave that was identified in previous attacks in the Modbus/TCP packet layer and function code 'Write Single Coil', a reference

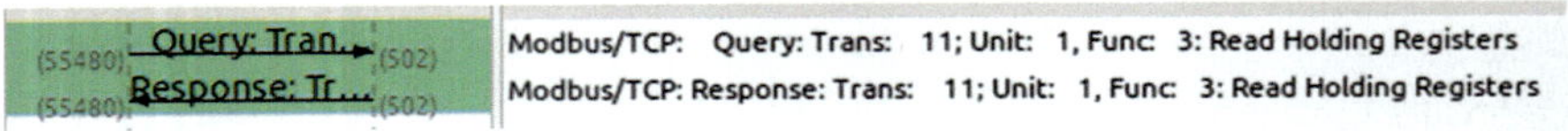

Fig. 7 Flow graph of the Fuzzing Read Holding Registers attack

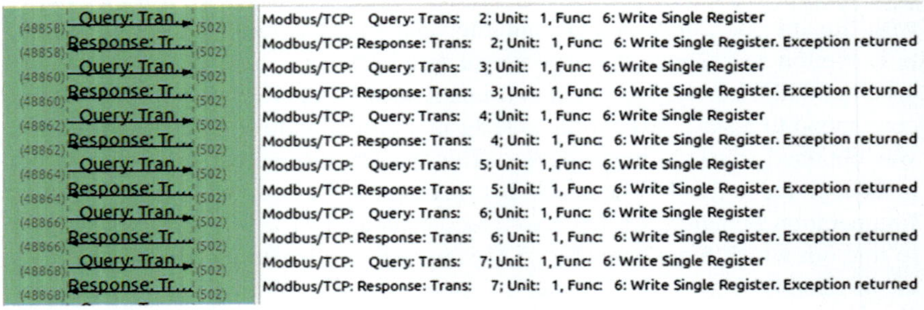

Fig. 8 Flow graph of the Write All Registers attack

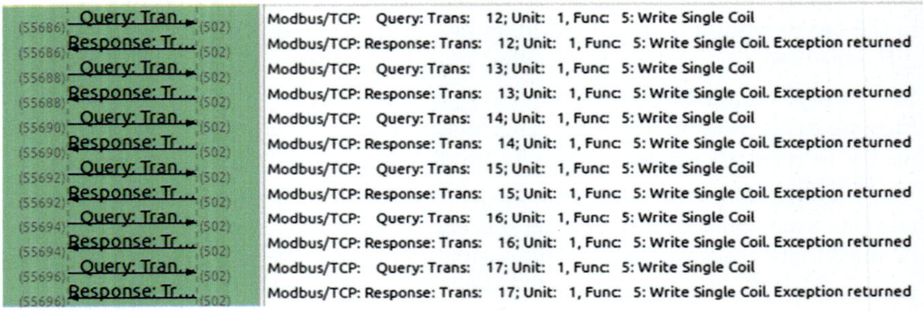

Fig. 9 Flow graph of the Write All Coils attack

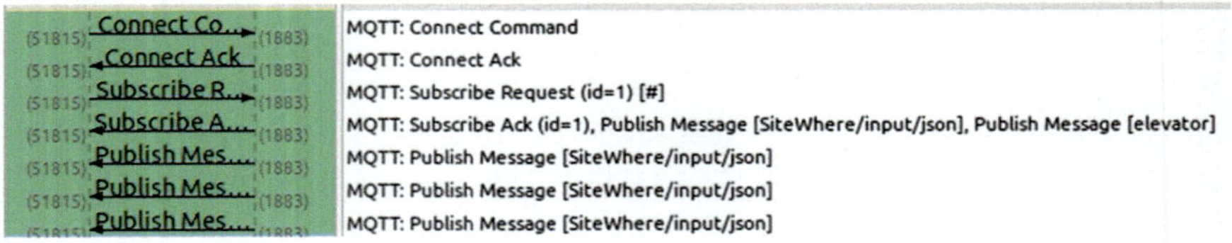

Fig. 10 Flow graph of the Unauthorized Subscribe attack

number corresponding to a target coil and 0 value in the data field of the Modbus packet layer, in order to turn it off. If the target coil reference number does not exist, the response includes Modbus exception code two (2) with message illegal data address exception. The attack flow in terms of packets can be seen in Fig. 9.

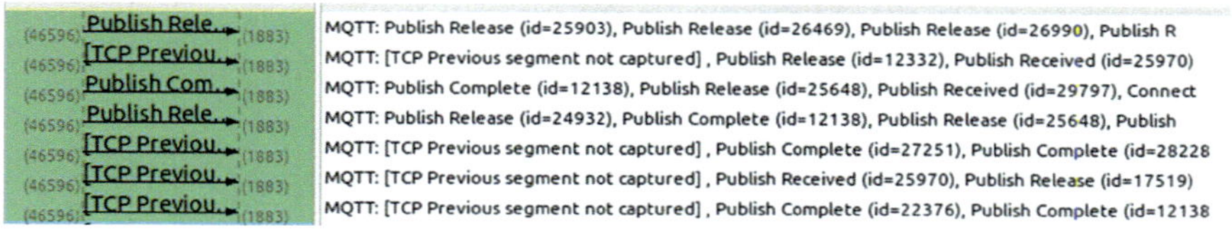

Fig. 11 Flow graph of the Large Payload attack

Additionally, a brief description of MQTT-related attacks is as follows:

- Unauthorized Subscribe: This attack leverages an MQTT broker without an authentication mechanism for clients. The attacker connects to the broker and subscribes to all topics using the wildcard '#', aiming to gain access to the messages that are being exchanged. Initially, the attacker sends a MQTT packet with message type 'Connect command' in the header flags of the MQTT packet layer without any authentication credentials set in the connect flags. After successfully being connected, the attacker sends a MQTT packet with message type 'Subscribe request' and in the header flags of the MQTT packet layer and topic flag set to '#'. After that, the attacker receives published messages from all topics of the MQTT broker with MQTT packets with message type 'Publish message' and the topic and message defined in the respective fields. The attack flow in terms of packets can be seen in Fig. 10.
- Large Payload: This DoS attack floods the system with large-size Payload Messages, trying to make the MQTT server unavailable for normal communications. The attacker, publishes several spam messages to a topic that was identified from the previous attack with no time delay. Due to the high sending speed the packets are encapsulated, thus forming messages with large payload and overloading the MQTT broker, which as a result cannot accept/forward messages from legitimate publishers. The attack flow in terms of packets can be seen in Fig. 11.
- Connection Overflows: This is again a flooding DoS attack where the attacker repeatedly sends connection packets, aiming to exhaust the MQTT server's resources. The MQTT packet with message type 'Connect' is sent by a MQTT client to initiate a MQTT session with the broker and contains client ID and other optional parameters. When broker server receives such a packet, it verifies the client identifier and other optional parameters to either allow or deny connection to the client. Malicious users can send multiple packets of that type with different client identifiers trying to exhaust server resources.The attack flow in terms of packets can be seen in Fig. 12.

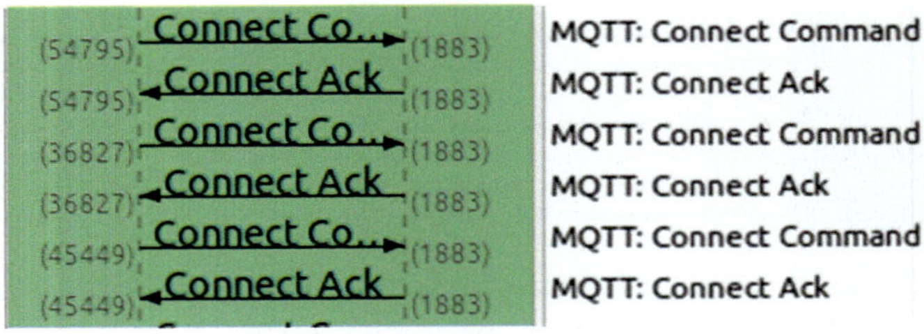

Fig. 12 Flow graph of the Connection overflow attack

5 Results

5.1 Initial Results

Initially, intrusion detection models are trained to predict specific types of cyberattacks. At this stage, the module is able to detect efficiently three (3) types of MQTT-related cyberattacks, namely Connection Overflow, Unauthorized Subscribe and Large Payload as well as three (3) types of Modbus-related cyberattacks, namely UID Brute Force, Enumeration Function and Fuzzing Read Holding Registers. Four (4) classification techniques, LR, KNN, RF, as well as SDAE are trained on network flow statistics and their results are compared to find out the best performing model. Initial data are composed of both normal and malicious network flow statistics coming from the multi-sensorial network and smart appliances of a smart home's infrastructure. The train and test data are split into an 80:20 ratio.

To achieve the simulation of a cybersecurity attack in a real device, a honeypot is struck by artificially made threats. These threats produce abnormal and malicious network flows. The anomaly detection module represents the abnormal flows through the user interface but the flows are not labelled. The domain expert realizes the nature of the cyberattack with the assistance of the visual-analytics module. Then, the expert utilizes the self-learning visual-analytics module, chooses the abnormal network flows and assigns to them the appropriate label. When the expert submits the relabelled data to the system, the Retraining Engine is started. This section represents the initial results and the results are generated after the presence of various attacks.

According to the comparative analysis, the best performance in terms of accuracy and F_1 score is achieved by the SDAE model in Modbus-related attacks and by the KNN model in MQTT-related attacks. The F_1 score is equal to the harmonic mean of precision and recall according to Eq. 3 while accuracy is the percentage of correctly classified data points out of all data points. Because data are imbalanced, the weighted average F_1 score is utilized instead of the macro F_1 score [42].

$$F_1\,score = \frac{1}{recall^{-1} + precision^{-1}} \tag{3}$$

Table 1 Results for initial training process regarding Modbus protocol

	ML Methods			DL methods
	KNN	RF	LR	SDAE
F_1 score	0.99077	0.99061	0.9857	0.99122
Accuracy	0.99582	0.99576	0.99298	0.996

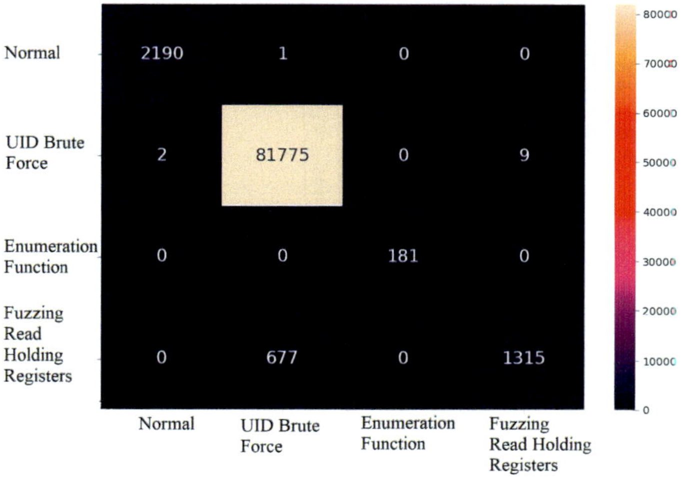

Fig. 13 Confusion matrix of SDAE which is the best performing model regarding Modbus protocol

Table 2 Results for initial training process regarding MQTT protocol

	ML Methods			DL methods
	KNN	RF	LR	SDAE
F_1 score	0.97089	0.84692	0.86037	0.90703
Accuracy	0.98579	0.9333	0.93803	0.95703

Table 1 represents the comparative results between methods in terms of F_1 score and accuracy in any label of Modbus network flow packets. Optimal parameters for the SDAE, which is the best performing model, are as follows: decoder_activation_function = *linear*, batch_size is equal to four (= *4*), n_hidden_layer = *25* and dropout = *0.05*. Figure 13 depicts confusion matrix of best performing model.

Regarding MQTT network flow packets, Table 2 represents the F_1 score and accuracy for each model. The KNN is the best performing model and its optimal parameters are n_neighbors is equal to ten (= *10*) and leaf_size = *30*. Figure 14 depicts the confusion matrix of the best performing model.

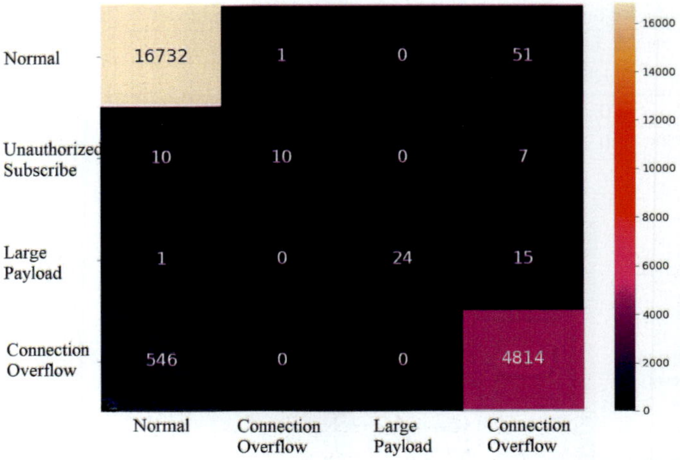

Fig. 14 Confusion matrix of KNN which is the best performing model regarding MQTT protocol

Table 3 Results in case an attack of type "Write All Registers" hits the system

	ML Methods			DL Methods
	KNN	RF	LR	SDAE
F_1 score	0.9411	0.98686	0.98684	0.94015
Accuracy	0.96046	0.99538	0.995	0.960074

5.2 *Modbus Protocol New Threats*

Subsequently, an attack of type "Write All Registers" hits the system. A domain expert annotates data points related to the attack to trigger the self-learning module. After the submission of annotated data and the completion of the retraining process, the RF model becomes the new best performing model (Table 3). It is noticeable that each model marginally differs from the rest in terms of F_1 score and accuracy. Optimal parameters for the RF model are as follows: number_of_estimators = *50*. Figure 15 represents the confusion matrix of the RF model.

In case an attack of type "Write All Coils" strikes the system, models are retrained to detect this new type of threat. Contrary to the previous case and according to Table 4, the most accurate model is the SDAE in terms of F_1 score and accuracy. Optimal parameters are as follows: decoder_activation_function = *sigmoid*, batch_size is equal to two (= *2*), n_hidden_layer = *25* and dropout = *0.01*. It is observed that the self-learning engine is dynamic and adaptive, as in each retraining circle another best performing model along with other optimized parameters can be extracted. Figure 16 represents the confusion matrix of the SDAE model.

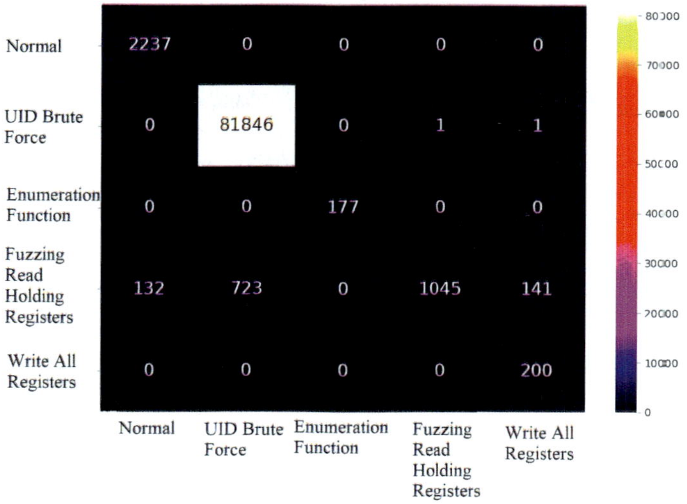

Fig. 15 Confusion matrix for RF which is the best performing model in case an attack of type "Write All Registers" hits the system

Table 4 Results in case an attack of type "Write All Coils" hits the system

	ML Methods			DL Methods
	KNN	RF	LR	SDAE
F_1 score	0.99074	0.99013	0.9903	0.99084
Accuracy	0.99663	0.99642	0.99647	0.99667

Table 5 Results for MQTT protocol in case an attack of type Connection Overflow hits the system

	ML Methods			DL Methods
	KNN	RF	LR	SDAE
F_1 score	0.9665	0.90473	0.88	0.97589
Accuracy	0.98356	0.95472	0.94348	0.98837

5.3 *Improvement of Model Effectiveness Regarding MQTT Protocol*

A major challenge faced by the proposed method is not only to detect new threats effectively but to improve the prediction accuracy of already known attacks. As it is observed by Table 2 and Fig. 14 Connection Overflow is a type of attack which has not been accurately predicted. Intuitively, the prediction accuracy will improve if a Connection Overflow attack hits the system again, and the domain expert user relabels relative data properly. Models will be retrained on more malicious data

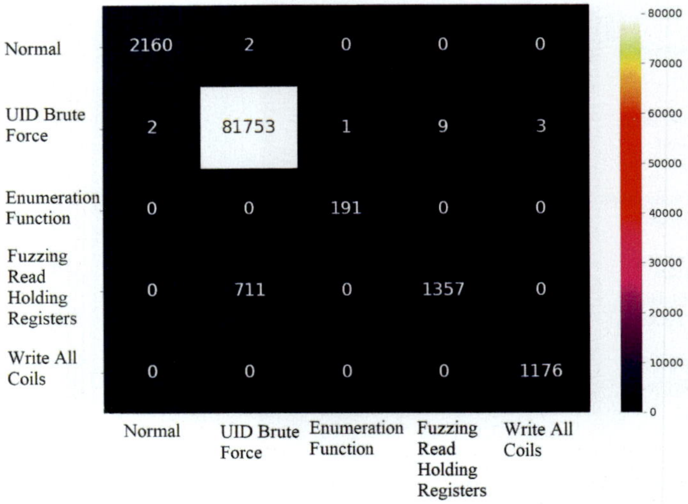

Fig. 16 Confusion matrix for SDAE which is the best performing model in case an attack of type "Write All Coils" hits the system

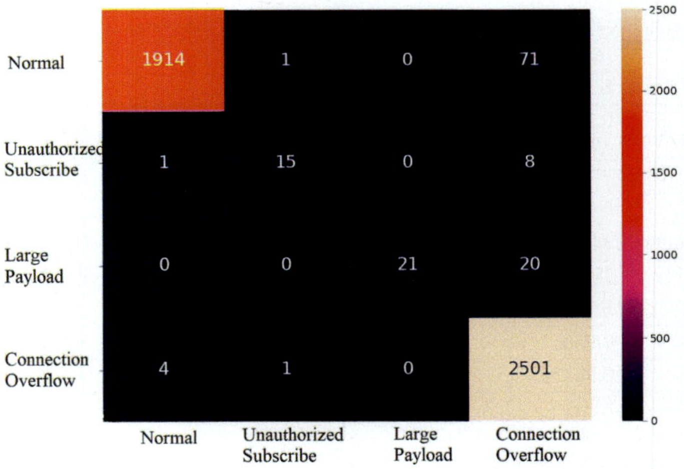

Fig. 17 Confusion matrix for SDAE which is the best performing model after an attack of type "Connection Overflow" hits the system

related to Connection Overflow, so it is expected that, after the self-learning process, models will be more efficient to detect this type of threat.

After the Connection Overflow attack hits the system, the domain expert or the security engineer labels the network flows which are produced by the presence of this threat with the appropriate label. The models of the Retraining Engine already detect this type of attack. Table 5 represents the performance of models in case the Connection Overflow threat strikes the system. The Retraining Engine dynamically replaces the KNN model with the SDAE model, which labels newly collected data (Table 2). Optimal parameters for the SDAE model are as follows: decoder_activation_function = *linear*, batch_size is equal to four (= *4*), n_hidden_layer = *25* and dropout = *0.01*. The self-learning engine improves the prediction of Connection Overflow attack as the recall measure is increased from 0.8981 after the initial training (Fig. 14) to 0,985 after the retraining process (Fig. 17). The precision measure is marginally decreased from 0.985 to 0.9619. The increase in the recall measure means that retrained best performing model makes fewer false-negative predictions. On the other hand, the decreasing of the precision measure means that the best performing model considers more frequently a normal network flow as an attack. The retraining process improves the predictability of the system corresponding to Connection Overflow threat as F1 score is increased from 0,47 to 0,487 before and after the retraining process respectively. This result is promising if it is taken into account that the results were already highly accurate.

5.4 *Performance Comparison of the System with the SPARK Framework*

In this stage, the system performance is measured before and after the integration of the SPARK framework. Table 6 represents the running time when a previously unknown attack hits the system (Write All Coils) and when the security engineer feeds a known label to the system (Connection Overflow). The presence of the SPARK framework optimizes the running time and encounters the complexity of the implementation especially in case where the user fixes misclassified predictions. The system update as well as the changing of ML/DL models' classes is done in the special case where the security engineer labels a previously unknown threat. Both the Retraining Engine and the self-learning module run on the system's backend without being noticed by the user. Consequently, the proposed method can be applied to real applications.

In our experimental setup, the SPARK cluster consists four (4) workers. Each machine/node corresponds to two (2) workers, and each worker utilizes two (2) cores. Each worker is related to the first machine/node takes up 18 GB RAM, and each worker is related to the second machine/node takes up 8 GB RAM. The master node comprises a 64-bit Intel(R) Core(TM) i5-9600K CPU and 32GB RAM. The

Table 6 Duration of the Retraining Engine with and without the SPARK framework for each case

	Duration without the SPARK framework	Duration with the SPARK framework
Write All Coils	154.63 min.	143.08 min.
Connection Overflow	7.004 min.	4.42 min.

second machine takes up 8 GB of RAM and belongs to a machine comprising an Intel(R) Core(TM) i5-8600K CPU.

6 Discussion

In smart grid infrastructures, the recognition of cyberattacks is a dynamic process. Over time, the machine and DL models which recognize cyber threats are incapable of recognizing new and more sophisticated types of attacks. The models need to be retrained with data which are generated by new cyberattacks. Moreover, the models should be trained again to become more accurate in the recognition of existing threats. In this work, a user-friendly dashboard is implemented in order to assist the domain experts to relabel suspicious data. Subsequently, the domain experts trigger the Retraining Engine by the submission of relabeled data to the system.

The main challenges which are faced in this work is the dynamic selection of best-performing ML model in cyberattack detection among KNN, RF, SDAE and LR. In addition, each model's training process includes the comparison of different combinations within a predefined set of hyperparameters. The individual models for each machine and DL method are fine-tuned by selecting the set of hyperparameters that give the best performance and finally, the overall best performing model is judged by comparing each individual model's F1 score on the test set data samples.

In the initial stage, the models are trained in datasets which include normal data along with three (3) attack cases both for MQTT and Modbus protocols. The F1 score of models is very high in any case, proving that the fine-tuning of each model is efficient. The KNN and the SDAE are the best performing models in Modbus and MQTT protocols respectively.

Two new types of Modbus attacks struck the system and the F1 score of models remained highly accurate for the detection of any threat after the retraining of models. It is impressive that the system is flexible, as in each retraining cycle the best performing model is different or has a different set of optimal hyperparameters. The newly collected data change the training process and affect the accuracy of models for each attack case.

A special research question which is generated is the degree of improvement in existing attack cases after the completion of the retraining process. An existing type of MQTT-based attack named Connection Overflow hits the platform and after the retraining of models the recall measure is increased and the precision measure is marginally decreased. The results in a future stage of research can be enriched with new attack cases to verify the effectiveness of the proposed method in numerous cases.

7 Conclusion

A key challenge in cybersecurity is the dynamic detection of new threats, as well as the maintenance of the predictive effectiveness in terms of know attacks. This study compares the performance of different ML models aiming not only to find the best combination of their hyperparameters but also to designate the most appropriate type of model according to new intelligence added to the system. The results show that differences in real-world data contribute to the incapability of a specific model to predict accurately all types of threats, no matter how state-of-the-art it is.

In this paper, a self-learning module is proposed, which assigns the labelling of new data volumes to the best performing ML or DL model after a retraining process. The retraining of models starts when human-machine interaction takes place to correct misclassified data or to label new types of attacks.

The next stage of this work and the future research challenges include a state-of-the-art drift detection method to periodically measure the effectiveness of models, thus starting the retraining process. Subsequently, domain experts will need only to label new threats.

Appendix A: List of Abbreviations

AE	Auto Encoder
AI	Artificial Intelligence
ANN	Artificial Neural Network
CNN	convolutional neural network
DB	Database
DL	Deep Learning
EPES	Electrical Power Energy Systems
DDoS	Distributed Denial of Service
DoS	Denial-of-Service
FDIAs	False Data Injection Attacks
HALS	Human-Augmenting Labeling System
ICT	Information and Communications Technology
IT	Information Technology
KNN	K-nearest neighbors
LR	Logistic Regression
mIOU	Intersection Over Union
ML	Machine Learning
MLP	Multi-Layer Perceptron
mSIU	Similarity In Union
NARX	Nonlinear Auto-Regressive Exogenous Model
OT	Operational Technology
PI	Proportional Integral
RF	Random Forest
SBC	Sampling by Cluster
SDAE	Stacked Denoising Autoencoders
SUD	Sampling by Uncertainty and Density
TCP	Transmission Control Protocol
TLS	Transport Layer Security
VA]	Visual Analytics

Acknowledgements The aforementioned work effort in this paper is conducted under the framework of both the SPEAR project, a Horizon 2020 program, funded by the European Union under the grant agreement No.787011 and the ELECTRON project, a Horizon 2020 program, funded by the European Union under the grant agreement No.101021936.

References

1. E. Bou-Harb, C. Fachkha, M. Pourzandi, M. Debbabi, C. Assi, Communication security for smart grid distribution networks. IEEE Commun. Mag. **51**(1), 42–49 (2013)
2. D. Faquir, N. Chouliaras, V. Sofia, K. Olga, L. Maglaras, Cybersecurity in smart grids, challenges and solutions. AIMS Electron. Electric. Eng. **5**(1), 24–37 (2021)
3. W. Wang, Z. Lu, Cyber security in the smart grid: Survey and challenges. Comput. Netw. **57**(5), 1344–1371 (2013)
4. D. Wei, Y. Lu, M. Jafari, P.M. Skare, K. Rohde, Protecting smart grid automation systems against cyberattacks. IEEE Trans. Smart Grid **2**(4), 782–795 (2011)

5. X. Li, X. Liang, R. Lu, X. Shen, X. Lin, H. Zhu, Securing smart grid: cyber attacks, countermeasures, and challenges. IEEE Commun. Mag. **50**(8), 38–45 (2012)
6. M.Z. Gunduz, R. Das, Analysis of cyber-attacks on smart grid applications, in *2018 International Conference on Artificial Intelligence and Data Processing (IDAP)* (IEEE, 2018), pp. 1–5
7. G. Bedi, G.K. Venayagamoorthy, R. Singh, R.R. Brooks, K.-C. Wang, Review of internet of things (iot) in electric power and energy systems. IEEE Internet Things J. **5**(2), 847–870 (2018)
8. H. Lin, S. Sambamoorthy, S. Shukla, J. Thorp, L. Mili, A study of communication and power system infrastructure interdependence on pmu-based wide area monitoring and protection, in *2012 IEEE Power and Energy Society General Meeting* (IEEE, 2012), pp. 1–7
9. Deflect. https://equalit.ie/portfolio/deflect/
10. R. Lemos, Microsoft warns of hijacked certificates, CNET Tech Ind., San Francisco, CA, USA. https://www.cnet.com/news/microsoft-warns-of-hijacked-certificates/
11. M.R. Habibi, H.R. Baghaee, T. Dragičević, F. Blaabjerg, Detection of false data injection cyber-attacks in dc microgrids based on recurrent neural networks. IEEE J. Emerg. Select. Top. Power Electron. **9**(5), 5294–5310 (2020)
12. A.A. Elsaeidy, A. Jamalipour, K.S. Munasinghe, A hybrid deep learning approach for replay and ddos attack detection in a smart city. IEEE Access **9**, 154864–154875 (2021)
13. X. Zhong, A. Ahmadi, R. Brooks, G.K. Venayagamoorthy, L. Yu, Y. Fu, Side channel analysis of multiple pmu data in electric power systems, in *2015 Clemson University Power Systems Conference (PSC)* (IEEE, 2015), pp. 1–6
14. M.R. Habibi, S. Sahoo, S. Rivera, T. Dragičević, F. Blaabjerg, Decentralized coordinated cyber-attack detection and mitigation strategy in dc microgrids based on artificial neural networks. IEEE J. Emerg. Select. Top. Power Electron. **9**(4), 4629–4638 (2021)
15. J. Zhang, L. Pan, Q.-L. Han, C. Chen, S. Wen, Y. Xiang, Deep learning based attack detection for cyber-physical system cybersecurity: a survey. IEEE/CAA J. Autom. Sinica **9**(3), 377–391 (2021)
16. I.H. Sarker, Deep cybersecurity: a comprehensive overview from neural network and deep learning perspective. SN Comput. Sci. **2**(3), 1–16 (2021)
17. M. Keshk, B. Turnbull, N. Moustafa, D. Vatsalan, K.-K.R. Choo, A privacy-preserving-framework-based blockchain and deep learning for protecting smart power networks. IEEE Trans. Ind. Inf. **16**(8), 5110–5118 (2019)
18. S. Potluri, N.F. Henry, C. Diedrich, Evaluation of hybrid deep learning techniques for ensuring security in networked control systems, in: *2017 22nd IEEE International Conference on Emerging Technologies and Factory Automation (ETFA)* (IEEE, 2017), pp. 1–8
19. J. Wang, D. Shi, Y. Li, J. Chen, H. Ding, X. Duan, Distributed framework for detecting pmu data manipulation attacks with deep autoencoders. IEEE Trans. Smart Grid **10**(4), 4401–4410 (2018)
20. S. Russo, M. Lürig, W. Hao, B. Matthews, K. Villez, Active learning for anomaly detection in environmental data. Environ. Modell. Softw. **134**, 104869 (2020)
21. H. Su, J. Deng, L. Fei-Fei, Crowdsourcing annotations for visual object detection, in *Workshops at the Twenty-Sixth AAAI Conference on Artificial Intelligence* (2012)
22. T.-N. Le, A. Sugimoto, S. Ono, H. Kawasaki, Toward interactive self-annotation for video object bounding box: recurrent self-learning and hierarchical annotation based framework, in *Proceedings of the IEEE/CVF Winter Conference on Applications of Computer Vision* (2020), pp. 3231–3240
23. F. De Sousa Ribeiro, F. Calivá, M. Swainson, K. Gudmundsson, G. Leontidis, S. Kollias, Deep bayesian self-training. Neural Comput. Appl. **32**(9), 4275–4291 (2020)
24. B.C. Benato, J.F. Gomes, A.C. Telea, A.X. Falcão, Semi-automatic data annotation guided by feature space projection. Pattern Recogn. **109**, 107612 (2021)
25. SPARK Framework. https://spark.apache.org/
26. A. Bitarafan, M. Nikdan, M. Soleymanibaghshah, 3d image segmentation with sparse annotation by self-training and internal registration. IEEE J. Biomed. Health Inf. (2020)

27. C.L. Srinidhi, S.W. Kim, F.-D. Chen, A.L. Martel, Self-supervised driven consistency training for annotation efficient histopathology image analysis (2021). arXiv:2102.03897
28. X. Wang, R. Zhang, C. Shen, T. Kong, L. Li, Dense contrastive learning for self-supervised visual pre-training, in *Proceedings of the IEEE/CVF Conference on Computer Vision and Pattern Recognition* (2021), pp. 3024–3033
29. D. van der Wal, I. Jhun, I. Laklouk, J. Nirschl, L. Richer, R. Rojansky, T. Theparee, J. Wheeler, J. Sander, F. Feng et al., Biological data annotation via a human-augmenting ai-based labeling system. NPJ Digit. Med. **4**(1), 1–7 (2021)
30. J. Zhu, H. Wang, T. Yao, B.K. Tsou, Active learning with sampling by uncertainty and density for word sense disambiguation and text classification, in *Proceedings of the 22nd International Conference on Computational Linguistics (Coling 2008)* (2008), pp. 1137–1144
31. V. Romero, J.A. Sánchez, A.H. Toselli, Active learning in handwritten text recognition using the derivational entropy, in *2018 16th International Conference on Frontiers in Handwriting Recognition (ICFHR)* (IEEE, 2018), pp. 291–296
32. S. Mussmann, P. Liang, On the relationship between data efficiency and error for uncertainty sampling, in *International Conference on Machine Learning*. PMLR (2018), pp. 3674–3682
33. M. Charalampos-Rafail, K. Thanasis, V. Vasileios, I. Dimosthenis, T. Dimitrios, S. Panagiotis, Cyber attack detection and trust management toolkit for defence-related m booktitle=
34. P. Radoglou-Grammatikis, P. Sarigiannidis, E. Iturbe, E. Rios, S. Martinez, A. Sarigiannidis, G. Eftathopoulos, Y. Spyridis, A. Sesis, N. Vakakis et al., Spear siem: a security information and event management system for the smart grid. Comput. Netw. **193**, 108008 (2021)
35. A.M. Mohsen, N.M. El-Makky, N. Ghanem, Author identification using deep learning, in *2016 15th IEEE International Conference on Machine Learning and Applications (ICMLA)* (IEEE, 2016), pp. 898–903
36. Y. Bengio, Practical recommendations for gradient-based training of deep architectures, in *Neural Networks: Tricks of the Trade* (Springer, ???, 2012), pp. 437–478
37. P. Vincent, H. Larochelle, I. Lajoie, Y. Bengio, P.-A. Manzagol, L. Bottou, Stacked denoising autoencoders: learning useful representations in a deep network with a local denoising criterion. J. Mach. Learn. Res. **11**(12) (2010)
38. D.P. Kingma, J. Ba, Adam: a method for stochastic optimization (2014). arXiv:1412.6980
39. Nearest Neighbors. https://scikit-learn.org/stable/modules/neighbors.html
40. Random Forest. https://scikit-learn.org/stable/modules/generated/sklearn.ensemble.RandomForestClassifier.html
41. Logistic Regression. https://scikit-learn.org/stable/modules/generated/sklearn.linear_model.LogisticRegression.html
42. F1 Score. https://scikit-learn.org/stable/modules/generated/sklearn.metrics.f1_score.html

Cyber-Resilience Enhancement Framework in Smart Grids

Achilleas Pasias, Thanasis Kotsiopoulos, George Lazaridis, Anastasis Drosou, Dimitrios Tzovaras, and Panagiotis Sarigiannidis

Abstract Reliability, resilience and Quality of Service are essential features of modern electric power-system operations that reflect the transition of electric energy infrastructure towards the smart grid deployment. Leveraging the Software Defined Networking technology and other cybersecurity cutting-edge technologies and algorithms, this work aims to bring a innovation in the modern Electrical Power and Energy System environment. To this end, we propose a cyber-resilience enhancement framework with the aim to modernize the traditional electrical grid and provide solutions in the domains of voltage and frequency restoration, cybersecurity and network Quality of Service. Based on the results, the framework is able to detect accurately cyberattacks and perform network path re-allocation by maximizing the Quality of Service in a more accurate way than other state of art algorithms.

Keywords Smart grids · QoS · EPES · Resilience · Cybersecurity · SDN

A. Pasias (✉) · T. Kotsiopoulos · G. Lazaridis · A. Drosou · D. Tzovaras
Information Technologies Institute, Centre of Research & Technology Hellas, 6th km Charilaou-Thermi Road, 57001 Thessaloniki, Greece
e-mail: pasiasach@iti.gr

T. Kotsiopoulos
e-mail: kotsiopoulos@iti.gr

G. Lazaridis
e-mail: glazaridis@iti.gr

A. Drosou
e-mail: drosou@iti.gr

D. Tzovaras
e-mail: Dimitrios.Tzovaras@iti.gr

T. Kotsiopoulos · P. Sarigiannidis
Department of Electrical and Computer Engineering, University of Western Macedonia, Karamanli & Ligeris Street, 50100 Kozani, Greece
e-mail: psarigiannidis@uowm.gr

H. Haes Alhelou et al. (eds.), *Power Systems Cybersecurity*, Power Systems,
https://doi.org/10.1007/978-3-031-20360-2_15

1 Nomenclature

Table 1 provides the abbreviations regarding the terminology used in this chapter.

2 Introduction

The demand for energy has increased due to the urbanization, the rising living standards and technological innovation. Apart from these factors, the Covid-19 pandemic together with the Ukrainian invasion have also aggravated the already tense situation for energy demand [1, 2]. All these aforementioned factors have increased power use to levels that, if ignored, may become extremely difficult to manage. This can be a concerning scenario not just for the provision of renewable energy but also for the global preservation of the entire ecosystem. Plain electrical grids and centrally supervised systems for electrical energy distribution have been mainly used over the years to serve the mankind for electricity needs. The term "power grid" can be also used to describe these types of systems [3]. It is worth mentioning that electrical grids throughout the world have had a similar structure, dynamics and principles since the invention of electricity, despite the technological advancements.

Traditional electrical grids are exclusively associated with a few fundamental operations, such as the electrical power generation, the transmission and the distribution [4]. Electrical power grids, in the form they appear nowadays, have several disadvantages. They appear to be unstable, with significant transmission losses, low power quality, a significant risk of brownouts and blackouts, insufficient electricity supply and barriers to the integration of distributed energy sources (DERs). Moreover, traditional electrical grids lack monitoring and overall control in real-time and only allow one-way communications. On the other hand, SG enable a two-way interaction between power providers and consumers, allowing both electricity and information to be shared in a two-way communication mode. The grid is becoming more efficient, trustworthy, stable, more secure and more eco-friendly due to the evolution of the communication networks, the introduction of more automation systems, the use of computers and more complicated control systems [5]. Renewable technologies, including wind and solar energy generation and plug-in electrical vehicles, may also be included into the national electrical system through the SG. Finally, power and network providers, using SG implementations, will have the chance to anticipate and respond to emergency issues remotely, giving the opportunity to their local repair teams to resolve critical issues, such as power outages, more quickly. Table 2 depicts identified differences between the traditional electrical grid and the smart grid.

Electric power grids are being modernized all over the world in order to integrate contemporary and cutting-edge technologies, which promise to solve a variety of issues that the traditional infrastructure is now encountering. Although these technologies modernize the EPES ecosystem, they also create additional challenges and weaknesses that malicious actors take advantage of in order to sabotage the power

Table 1 Table of Notations

AI = Artificial Intelligence
CNN = Convolutional Neural Networks
DER = Distributed Energy Resources
EDAE = Electrical Data Analysis Engine
EPES = Electrical Power Energy Systems
HDD = Hybrid Distributed and Decentralized
IDS = Intrusion Detection System
IMS = Intrusion Mitigation System
MAS = Multi Agent System
MILP = Mixed Integer Linear Programming
PMU = Phasor Measurement Unit
QoS = Quality of Service
SDN = Software Defined Networking
SG = Smart Grid
SW = Set of communication network forwarding devices (switches)
BS = Set of buses in the power transmission network
SD = Set of sensing devices in the network
GW = Set of gateways in the network
L = Set of active communication paths between hosts
$DL_c\ (h_i)$ = Delay constraint for host i
$AB_c\ (h_i)$ = Available bandwidth constraint for host i
$PL_c\ (h_i)$ = Packet loss constraint for host i
$JTR_c\ (h_i)$ = Jitter constraint for host i
$SR\ (h_i)$ = Security constraint for host i
L_{bs_i,bs_j} = Set of transmission lines between buses
k_{GW} = Device capacity of each Gateway
p_{h_i,h_j} = A full communication path between two hosts
bs_i = Bus number i
h_i = Host i
s_i = Switch i
a_i = A network asset could be switch or host
$< a_i, a_j >$ = Link between two assets of the network
$sr_{<s_i,s_j>}$ = Security risk level between two linked switches
sr_{s_i} = Security risk level of switch
sr_{h_i} = Security risk level of host
$PL{:}P_{h_i,h_j} \mapsto R^+$ = Function that maps a full communication link to its packet loss
$B{:}P_{h_i,h_j} \mapsto R^+$ = Function that maps a full communication link to its bandwidth
$AB{:}P_{h_i,h_j} \mapsto R^+$ = Function that maps a full communication link to its available bandwidth
$JTR{:}P_{h_i,h_j} \mapsto R^+$ = Function that maps a full communication link to its jitter
$DL{:}P_{h_i,h_j} \mapsto R^+$ = Function that maps a full communication link to its delay
$SR{:}P_{h_i,h_j} \mapsto N$ = Function that maps a full communication link to its security risk level
$threshold$ = Used to avoid using paths that over-collateralize some of the KPIs
$x(a, b)$ = Decision Variable of the MILP model

Table 2 Differences between traditional and smart grids

Traditional electrical grid	Smart grid
Small number of IoT sensors	IoT sensor throughout
Manual monitoring	Self-monitoring
Failures and blackouts	Adaptive and Islanding
Small number of customer choices	Large number of customer choices
Limited control	Effective control
Manual restoration	Self-healing
Electromechanical	Digital
Centralized generation	Distributed generation

supply. There are several examples of cyberattacks towards energy systems, with the first one being reported in 2015 and again in 2016, when hackers penetrated Ukraine's power grid, disrupting the control systems and creating major outages in Kiev and the other cities of west Ukraine [6]. Furthermore, another two malicious incidents were reported from the many that took place over the past years. More specifically, a cyberattack against the Colonial Pipeline [7] became one of the most widely published news in recent years. The largest fuel pipeline in the United States was targeted by a ransomware cyberattack, which led to a payment of 4.4 million dollars ransom to a ransomware group. Another example is the use of the Ryuk ransomware [8], the tool for a cyberattack on Volue ASA, a Norwegian energy corporation. According to the company, no ransom was delivered and the company dispatched instantly a cybersecurity task team to limit any damage [9]. This cybersecurity incident proves that hackers were exclusively interested in the encryption of files, databases, and applications. Due to these cybersecurity incidents in the energy domain, it is imperative that smart grid security, reliability and resilience have to be increased.

Security, reliability and resilience are three essential features of modern electric power-system operations that reflect the transition of electric energy infrastructure towards the smart grid deployment [10]. Modern EPES are progressively integrating cutting-edge technologies in order to improve the afforementioned features along with the communication and control performance, which unlocks and introduces the area of cybersecurity in their operations. Technologies such as AI, SDN and processes that restore the voltage and the frequency and execute Black Start optimal sequence re-connection of DERs and loads, are upcoming features of modern SGs [11, 12]. Furthermore, due to numerous factors, the appliance of the aforementioned technologies, the resilience and end-to-end QoS are becoming increasingly important in the delivery of electrical services to customers.

The end-to-end QoS implementation in both conventional and modern network topologies is affected by security, reliability and resilience. Specialists from both academic and business communities have presented various solutions in order to overcome existing networking's QoS restrictions, but instead many of them have either performed poorly or have not been adopted. As a solution to these networking restrictions, the SDN approach has started being adopted. The many advantages offered by the SDN solution have aroused the interest of the research community in order to enhance the QoS functionality in several domains [13]. Many researchers have introduced applications which can provide alternative network pathways between critical applications on a regular basis, ensuring that QoS and security requirements are assessed. Moreover, these applications are capable of isolating a network component, the moment it gets discovered as compromised. To this end, we propose a cyber-resilience enhancement framework in smart grids, denoted hereafter as FRANS, with the aim to enhance the cybersecurity, the resilience, the efficiency and the QoS provisioning of modern network applications in SGs.

The rest of the paper is organized as follows: Section 3 provides a literature review regarding the application of the SDN technology in SGs and methodologies for boosting the cybersecurity, reliability and resilience. Section 4 outlines the proposed FRANS architecture and thoroughly analyzes each component. Furthermore, Sect. 5 presents the evaluation setup and the results of the proposed architecture. Section 6 provides a discussion regarding the key contributions of the FRANS system and some proposed future steps. Finally Sect. 7 concludes the whole work, comments on the results and gives some future work in order to optimize the presented architecture.

3 Literature Review

The SDN technology has emerged rapidly the past few years and it has been applied successfully in smart grid communications in order to improve the system's cyber-security resilience and reliability. SDN is a novel networking technology, which differentiates itself from other conventional communication networks by separating the control plane from the data plane. The architecture of conventional networks is composed of routers and switches, which are considered as intermediary devices. Each of these aforementioned devices utilizes its own control plane, which allows them to make independent choices about the packets' forwarding policies. On the contrary, SDN incorporates the meaning of the SDN controller, which is a virtualized control plane reallocated to a centralized location [14].

The main benefits of connecting the SDN technology with SGs and generally EPES involve the ability to overview the entire network through the SDN controller, the low operational cost as well as the low equipment cost, the high level of security, the use of AI-based QoS provisioning algorithms capable of controlling and shaping the traffic data, improved scalability, flexibility and efficiency of the network [15–17]. However, there are still open issues that have to be addressed, to fully exploit the potential of SDN in EPES. The main challenges of SDN affect mainly the fields of data security, controller design, interaction of controller-application and QoS management [18, 19]. Concretely, in SDN occur difficulties in developing applications which increase network security as well as securing the SDN infrastructure. The OpenFlow standards, do not specify the certificate format to assure data integrity.

As a result, SDN security necessitates two-factor authentication and an encryption method for packet recovery in the event of failure, else a detection mechanism is required to detect and mitigate possible cyberattacks. In extend, only few protocols have been defined for the communication between applications and the controller. Effort on this is given through several projects which can be found in [18].

The end-to-end QoS management is a major consideration for Service Oriented Architecture based SDN. The authors in [19] claim that challenges on achieving high QoS management in SoA SDNs are the negotiation of the network as a service to fulfill expectations, the selection of a capable network provider who can deliver on the agreed requirements and the insurance that the network is providing, meets the QoS requirements.

Authors in [20] present a thorough analysis of current concerns, open questions and future directions while providing an outline of the methods currently used to improve smart grids' cyber resilience. These methods are organized into categories based on the cyber resilience techniques listed in NIST SP 800-160 and are connected to the various states of the typical power system resilience curve. The end result is a taxonomy, which could help researchers learn more about smart grid's cyber resiliency. In [21], authors suggest a hybrid, distributed, and decentralized (HDD) SDN architecture in order to achieve resiliency in smart grids. It offers a redundant controller architecture for fault-tolerance and fail-over operation, along with simultaneous execution of several anomaly detection methods. Furthermore, in order to help in the evolving of the next generation of wide-area monitoring, protection and control in the smart grid environments, authors in [22] propose an innovative framework for establishing a cutting-edge intrusion detection system (IDS) accompanied with an intrusion mitigation system (IMS) and an alert management system in order to create a self-healing and attack-resistant PMU network. Another work in this research area proposed a revolutionary architecture that fully utilizes SDN technology to boost the smart grid's resilience. Three primary layers make up the proposed architecture known as SDN-microSENSE architecture, namely the risk assessment layer, the intrusion detection and correlation layer and finally the self-healing layer [23].

Besides the aforementioned work, many other scientists have studied towards a cybersecure, resilient and reliable smart grid. More specifically, authors in [24] have examined a smart grid environment based on SDN, where the SDN controller, working alongside with the network switches, is able to learn about link failures within the network and then dynamically adapt to the changing network circumstances. Their suggested algorithm has the ability to recognize the used attack strategy and only choose secure communication channels. Nevertheless, in order to achieve resilient and secure operations in the event of numerous cyber threats, authors in [25] suggest an SDN-enabled microgrid infrastructure. Through this infrastructure, they look into several security applications, such as specification-based intrusion detection, real-time and uncertainty-aware communication network verification and self-healing communication network management. Last but not least, authors in [26] examine an SDN framework and they define a security score model for an IEC 61850-based

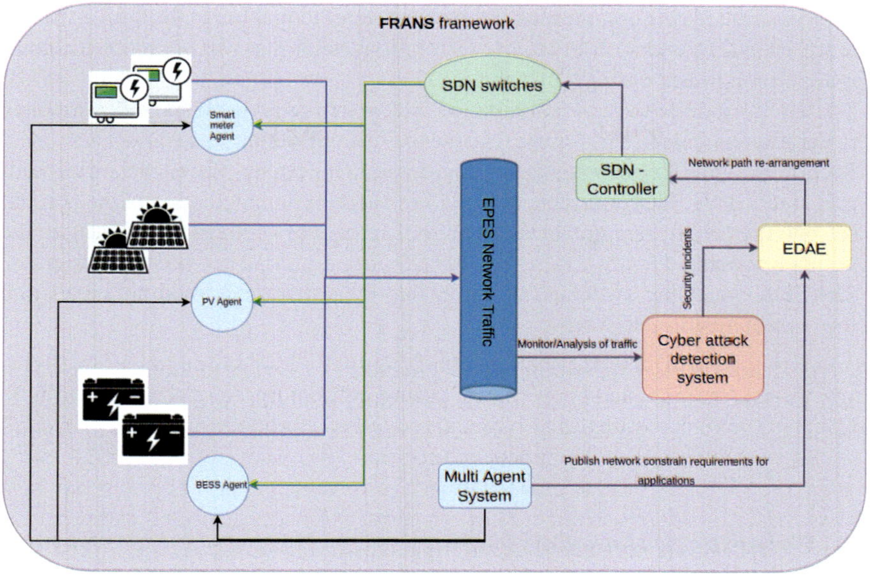

Fig. 1 FRANS internal components and communication connections

substation communication network. The security risk score model is integrated into the SDN architecture and SDN concepts are used to establish cyber resilience.

In a nutshell, to enhance the security of the SDN enabled SGs and to alleviate existing open challenges of the SDN technology, we propose the FRANS framework. FRANS aims to accurately detect cyberthreats by enabling a novel methodology which takes advantage of deep neural networks. Another target of the framework is the cyberattack mitigation and the maximization of the network's QoS Last but not least, FRANS aims to apply processes that restore the voltage and the frequency and execute Black Start optimal sequence re-connection of DERs and loads.

4 FRANS Architecture

This section is devoted to present the architecture of the FRANS framework. Figure 1 depicts the modules participating in FRANS all along with the interconnections between them.

Concretely, the framework consists of 4 modules:

- The cyberattack detection engine. This engine is responsible for analyzing the network traffic and providing security reports to EDAE.
- The Electrical Data analysis Engine. EDAE obtains the network and security requirements (constraints) for the applications of the SDN network and periodi-

cally re-arranges the communication paths in order to maximize the QoS. Furthermore, EDAE provides observability restoration techniques and is capable to apply mitigation policies during a cyberattack.

- The Multi Agent System, is a distributed system that provides voltage & frequency restoration and black start optimal sequence restoration processes in a smart grid. Specifically, MAS is responsible for coordinating the agents placed in the electrical grid assets (PVs, energy storage systems, etc.) and for the calculation of the optimal set-points in order to maintain the voltage & frequency of the electrical grid to the nominal values.
- The SDN controller-switches, which enable the software defined networking in the communication networks.

Analytical information regarding the functionalities of each module is presented in the sections below apart from EDAE and MAS modules, which are extensively presented in [11, 12] respectively and here a brief description is going to be given.

4.1 Cyberattack Detection Engine

The cyberattack detection engine consists of the network traffic sensing unit and the AI analytics module. The sensing unit captures the network traffic from the EPES network and determines whether this traffic corresponds to normal or abnormal behaviour. The development of the sensor is based on the ScaPy Python library for capturing the network packets and the Cicflowmeter tool to extract the network flow statistics. The anomaly detection process is conducted over the extracted netflow statistics. The cyberattack detection engine consists of a CNN. Figure 2 illustrates the architecture of the model, which is built using PyTorch [27]. The optimizer of the models is the Stochastic Gradient Descent with learning rate equal to 0.001.

The input to the model is based on a novel methodology, which transforms the network flows to images. Characteristically, given paired data

$$(x_1, y_1), ..., (x_n, v_n) \forall x_i, y_j \in A \tag{1}$$

where A is the set of network flow data, we compute the pairwise Pearson correlation coefficient

$$r_{xy} = \frac{\sum_{i=1}^{n}(x_i - \overline{x})(y_i - \overline{y})}{\sqrt{\sum_{i=1}^{n}(x_i - \overline{x})^2}\sqrt{\sum_{i=1}^{n}(y_i - \overline{y})^2}} \tag{2}$$

where: n is the sample size and $\overline{x}$, $\overline{y}$ is the mean of each sample. Afterwards, we form the correlation matrix B of the coefficients. Then, we create the heatmap representation of the correlation matrix, with dimensions of 720×360. Figure 3 depicts the heatmap of a netflow categorized as normal.

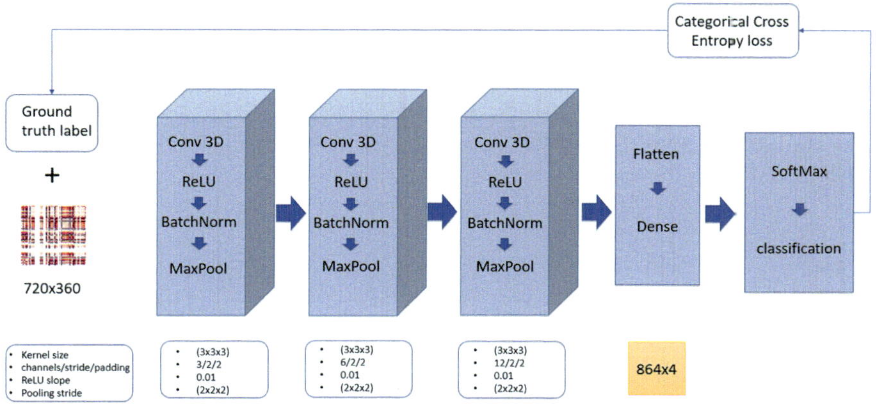

Fig. 2 The architecture of the cyberattack detection engine's model

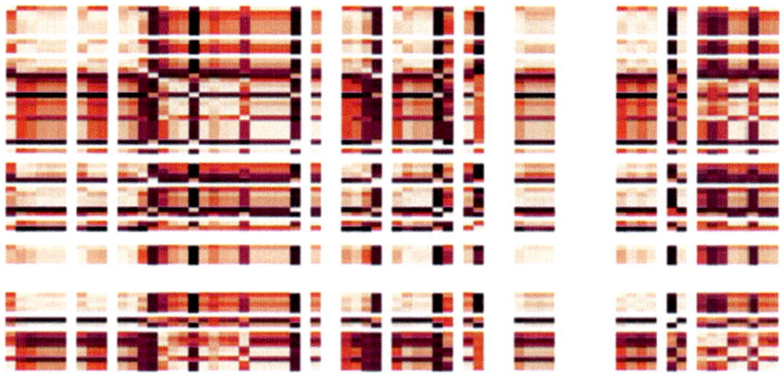

Fig. 3 Indicative input of a heatmap corresponding to "normal" condition

4.2 EDAE

EDAE's key functionality is to provide alternate network paths between essential applications on a continual basis to ensure that QoS and security requirements are fulfilled. Secondly, EDAE is responsible for isolating a network device that has been compromised. Thirdly, when a gateway (data sink) is unreachable due to a malfunction or a cyberattack, the goal is to ideally reconnect the leftover gateways with the sensing units (data source) in order to restore data acquisition in the EPES ecosystem i.e. to restore and maximize the redundant observability of the electrical grid. EDAE is comprised of two algorithms in general: A genetic algorithm for retracing communication paths, and a MILP methodology [11] to match the leftover gateways with the sensing devices.

4.2.1 Mathematical Formulation of Routing Optimization Problem

This section is devoted to present the mathematical formulation of the objective functions and the constraints that were used in order to find the optimal communication paths between the applications. See nomenclature for symbol definition.
(i) **Delay** objective function

$$min: \; J_{DL}(p) = max\left(threshold, \frac{DL\left(p_{h_i,h_j}\right)}{DL_c(h_i)}\right) \tag{3}$$

(ii) **Jitter** objective function

$$min: \; J_{JTR}(p) = max\left(threshold, \frac{JTR\left(p_{h_i,h_j}\right)}{JTR_c(h_i)}\right) \tag{4}$$

(iii) **Packet loss** objective function

$$min: \; J_{PL}(p) = \max\left(threshold, \frac{PL\left(p_{h_i,h_j}\right)}{PL_c(h_i)}\right) \tag{5}$$

(iv) **Available bandwidth** objective function

$$min: J_{AB}(p) = -\frac{AB\left(p_{h_i,h_j}\right)}{max\left(ab\langle h_i, h_j\rangle\right)} \tag{6}$$

(v) **Security** objective function

$$min: \; J_{security}(p) = \frac{SR\left(p_{h_i,h_j}\right)}{SR_c(h_i)} \tag{7}$$

$$\textbf{s.t} = \begin{cases} DL\left(p_{h_i,h_j}\right) \le min\left(DL_c(h_i),\; DL_c\left(h_j\right)\right) \\ AB\left(p_{h_i,h_j}\right) \ge max\left(AB_c(h_i),\; AB_c\left(h_j\right)\right) \\ PL\left(p_{h_i,h_j}\right) \le min\left(PL_c(h_i),\; PL_c\left(h_j\right)\right) \\ JTR\left(p_{h_i,h_j}\right) \le min\left(JTR_c(h_i),\; JTR_c\left(h_j\right)\right) \\ SR\left(p_{h_i,h_j}\right) \le min\left(SR_c(h_i),\; SR_c\left(h_j\right)\right) \\ \quad \forall p_{h_i,h_j} \in \left(L + p - p'\right) \end{cases}$$

4.2.2 Chromosome Encoding and Mutation Strategy

One of the most important aspects when someone is utilizing genetic algorithms is the appropriate selection of the chromosome mapping, which maps the solution into a chromosome encoding. The higher the length of the chromosome the bigger the volume of the search (solution) space it becomes, which can lead to the problem of search space explosion. To increase the scalability of the solver, i.e. the genetic algorithm, of the optimization problem that was presented above, the authors propose a relativistic chromosome encoding that reduces significantly the search space with respect to the standard encoding. In the following paragraphs a description of the standard and relativistic chromosome encoding is given. See Fig. 4 for a visual representation between the Standard and Relativistic chromosome representations.

Naive Chromosome Encoding: Let SW be the set of the SDN switches and p_{h_i,h_j} a valid communication path. The p_{h_i,h_j} communication path can be represented by an ordered set s representing the chain of hops that a packet sent from host h_i will follow to reach h_j. Assuming that the only known information about the network topology is the number of the switches and a validator val which validates whether a solution is valid i.e no loop is formed and the source host h_i can reach the destination host h_j, then the encoding that can be used is the naive encoding which has biggest search space, since no knowledge is incorporated into the encoding that could reduce the search space. Let V be the search space of the genetic algorithm and $\mathbf{v}$ a vector, i.e a chromosome. The length of the vector should be equal to the number of the SDN switches in order to make sure that the longest possible solution exists in the search space. Furthermore, the value of each dimension (gene) of the vector (chromosome) $\mathbf{v}$ can take values a such that $a \in SW$. Therefore, the total number of combinations given this representation is n^n where $n = Size(SW)$.

Standard Chromosome Encoding: If the knowledge that a switch sw_i can be present in a solution only once is incorporated into the chromosome encoding, then the search space is reduced to the factorial of $n = n!$. Hereafter this representation, the one that includes the information that a switch can be present in a solution only once, is defined in this work as the standard chromosome encoding.

Relativistic Chromosome Encoding: The relativistic chromosome encoding incorporates information directly from the structure of the network topology and specifically the knowledge of the neighbors of each switch sw_i N_i (ordered set), i.e. the switches $sw_j \in SW \backslash sw_i$ where sw_i is connected with. The structural knowledge of the graph that was mentioned above can constrain the values that a specific gene can take based on the value(s) of the proceeding and/or next gene respectively. The term relativistic in the name is derived from the fact that a gene g_j can take an integer value $a \in [0, Size(N_{i-1})]$ which is actually the index of the switch sw_j in the ordered set N_{i-1}, i.e. the meaning of the value of a gene g_j is related with the previous switch.

Mutation Strategy:The incorporation of the structural knowledge into the relativistic representation makes necessary to adapt the the mutation strategies to the nature of the chromosome encoding. Accordingly, the authors in this work decided to define the following four mutation strategies:
Let P_i be the pool of the available switches that a genome g_i can take values from and U_i be the list of the switches that have been already used until the g_i gene (including g_i).

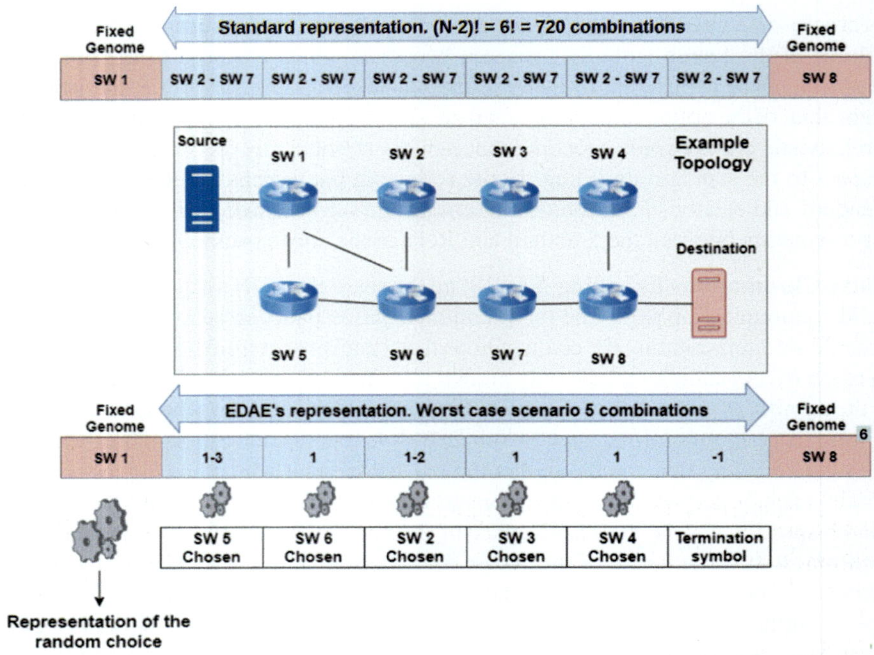

Fig. 4 Example of genome representation for a simple topology consisting of two hosts and 8 switches.Top standard representation, bottom options for relativistic representation for source strategy

(i) **Source** Strategy

In this case only the information of the source switch (g_{i-1}) is taken into account and the pool P_i for a gene g_i is defined as $P_i = N_{i-1} - U_i$.

(ii) **Destination** Strategy

In this case only the information of the destination switch (g_{i+1}) is taken into account and the pool P_i for a gene g_i is defined as $P_i = N_{i+1} - U_{i+1}$.

(iii) **Bidirectional** Strategy

In this case the information from both the source and destination switches is taken into account (g_{i-1}, g_{i+1}) and the pool P_i for a gene g_i is defined as $P_i = N_{i-1} \cap N_{i+1} - U_{i+1}$.

(iv) **Standard** Strategy

In this case no structural information is taken into account and the pool P_i for a gene g_i is defined as $P_i = SW - U_i$.

The selection of the mutation strategy is done by sampling from a discrete probability distribution $P(X)$ where $X \in [0, 3]$.

4.2.3 Mathematical Formulation of MILP Based Matchmaking Algorithm

The present paragraph presents the mathematical formulation of the proposed MILP algorithm which focuses on the matchmaking between gateways and sensing units. Let $A = (SD_1, SD_2, \ldots, SD_N)$, a set of N available Sensing Devices. Let $B = (GW_1, GW_2, \ldots, GW_M)$, a set of M available Gateways. The decision variables are the sets of SD-GW pairs:

$$\{a, b\} \forall a \in A, \forall b \in B \tag{8}$$

with binary values:

$$\{a, b\} = \begin{cases} 1, & \text{if } SD_a \text{ sends data to } GW_b \\ 0, & \text{if } SD_a \text{ does not send data to } GW_b \end{cases}$$

The objective of the problem is to select the SD-GW pairs which minimize the average communication network latency. In this case, the objective function is described as:

$$min \Sigma_{a \in A, b \in B}\, latency\,(a, b) * \{a, b\} \tag{9}$$

With respect to the following constraints:
(i) A gateway $GW_b \in B$ can receive data from at most k_b sensing devices.

$$\Sigma_{a \in A}\{a, b\} <= k_b ... \forall b \in B \tag{10}$$

(ii) Each sensing device can send data to at most one gateway:

$$\Sigma_{b \in B}\{a, b\} <= 1 ... \forall a \in A \tag{11}$$

4.3 MAS

MAS is a fault-tolerant multi-agent system that does not suffer from single point of failure (SPOF) and offers decentralized low voltage black start and voltage & frequency restoration processes, scaled to the size and complexity of the microgrid. Both main and secondary controls are performed by the tool. The operation mode of the island's DERs is defined by the primary control. When discharging, energy storage systems can switch to master operation mode (voltage source), and when charging, they can switch to slave operation mode (current source) [28]. More information about the functionalities and evaluation results are given in the following paper [12]. It should be noted that the MAS tool plays an additional role in the FRANS framework, which is to provide the network constrains to EDAE in order to calculate the optimal network paths.

4.3.1 Mathematical Formulation of Voltage and Frequency Restoration

This section is devoted to present the mathematical formulation of the functions that perform the primary and secondary control with respect to the voltage and frequency restoration.

$$f = f_{nom} - \frac{f_{nom} - f_{\min}}{P_{dis}\left(SoC\left(t\right), C_{rate}\right)} P \tag{12}$$

where the active discharge power is given by:

$$P_{dis}\left(SoC\left(t\right), C_{rate}\right) = \frac{SoC\left(t\right) - SoC_{\min}}{C_{rate}^{-1}} E_{nom} \tag{13}$$

SoC(t) is the real time value of the battery's state of charge, SoC_{min} is the minimum allowed state of charge, f_{nom} is the nominal frequency of the island, f_{min} is the minimum allowed frequency of the island, and E_{nom} is the nominal battery capacity. The voltage is calculated by:

$$V = V_{nom} - \frac{V_{\max} - V_{\min}}{2Q_{\max}} Q \tag{14}$$

Where the maximum reactive power is given by:

$$Q_{\max} = S_{nom}\sqrt{1 - PF_{\min}^{2}} \tag{15}$$

V_{nom} is the island's nominal voltage, V_{max} and V_{min} are the maximum and minimum allowed voltage, S_{nom} is the nominal apparent power, and PF_{min} is the minimum power factor.

For batteries operating in Slave mode, frequency and voltage are used as independent variables, in order to calculate the active and reactive power:

$$P = \begin{cases} \frac{P_{ch}(SoC(t))}{f_{\min} - f_{nom}} \left(f - f_{nom}\right), f \in \left[f_{\min}, f_{nom}\right] \\ \frac{P_{ch}(SoC(t))}{f_{\max} - f_{nom}} \left(f - f_{nom}\right), f \in \left[f_{nom}, f_{\max}\right] \end{cases} \tag{16}$$

$$Q = \frac{2Q_{\max}}{V_{\min} - V_{\max}} \left(V - V_{nom}\right) \tag{17}$$

where Q_{max} is given by Eq. (4) and the active charge power is calculated as follows:

$$P_{ch}\left(SoC\left(t\right)\right) = \frac{P_{ch,max}}{SoC_{\max} - SoC_{\min}} \left(SoC_{\max} - SoC\left(t\right)\right) \tag{18}$$

Regarding the PVs, they are controlled using reversed droop control. Active and reactive power are calculated as follows:

$$P = \begin{cases} P_{MPP}, \, \text{f} \in [f_{\min},\ f_{nom}] \\ P_{MPP} - \frac{P_{MPP}}{f_{\max} - f_{nom}}, \, \text{f} \in (f_{nom},\ f_{\max}] \end{cases} \tag{19}$$

$$Q = \frac{2Q_{\max}}{V_{\min} - V_{\max}} (V - V_{nom}), \, \text{V} \in [V_{\min}, V_{\max}] \tag{20}$$

where

$$Q_{\max} = P_{MPP} \sqrt{1 - PF_{\min}^2} \tag{21}$$

where *MPP* stand for the Maximum Power Point of the active power.

5 System Evaluation

This section presents the evaluation of the FRANS framework in terms of cyberattack detection and optimal network path re-allocation. The evaluation is conducted on a simulated testbed similar to the one presented in [12], using the Containernet framework [29]. Figure 5 illustrates the SDN topology. Concretely, the topology consists of 7 emulated devices: 3 PV inverters, 1 load,2 Energy Storage Systems and one Building Management System.

5.1 Evaluation Details

The objective of this paragraph is to present the details of the traffic simulation in the network, the simulation of the status of each switch, the hyper-parameters that were used for the algorithms and the process that was followed to ensure that the results are consisted.

Traffic Engineering: The network traffic between the hosts of the network are mainly ModBus/TCP packets created using the server-client stack communication offered by PyModbus [30] python library and the actual traffic between the MAS agents on the simulated microgrid. Furthermore, to stress more the available bandwidth of the network, TCP/IP traffic was also generated between the hosts of the network.

Switch QoS status: The QoS status of each switch was the same and optimal prior to any traffic generation, therefore, diversity between the status of the switch links had to be induced. The NetEm package [31] from Linux foundation was used in order to insert delay, jitter and packet loss over the network. Furthermore, to make the QoS status of the switches variable over time and to have the opportunity to evaluate

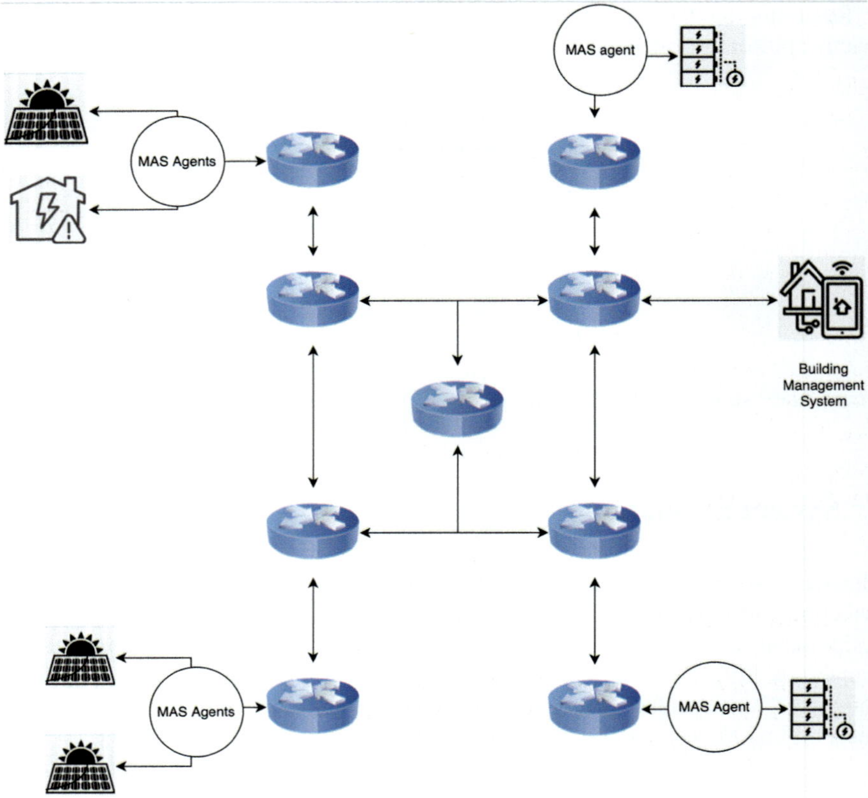

Fig. 5 The Topology used for the system evaluation

the system in multiple configurations, various profiles were created for each QoS metric that were sampled every 5 minutes for each switch link. Specifically, there were 9 different traffic generation profiles for the bandwidth metric (combination of 3 profiles for the frequency of POST action and 3 of the volume of the data), 3 different profiles were created for delay and jitter and 4 profiles for packet loss. In Table 3 the different profiles and the distribution of probabilities are shown for each QoS metrics.

Algorithm Hyperparameters: The algorithms that can be parameterized are the genetic algorithm and the detection engine model. The genetic algorithm runs for 20 generations with an initial population of 10. Furthermore, the discrete distribution probability that was used for the mutation policy can be found in Table 3. Finally, the model of the detection engine was trained with a learning rate equal to 0.001, batch size 16 with the Stochastic Gradient Descent (SGD) optimizer.

Table 3 Distribution of probabilities

Metric	Categories	Distribution
Delay (ms)	[0, 1.5, 3]	[0.5, 0.33, 0.17]
Jitter (ms)	[0.5, 1, 1.5, 2]	Uniform
Packet loss (%)	[0.15, 0.25, 0.35, 0.45]	[0.4,0.3, 0.15, 0.15]
Traffic-freq (ms)	[0.1, 0.3, 0.7]	Uniform
Traffic-load (MBps)	[4, 8, 16]	Uniform
Security (Categorical)	[1, 2, 3, 4, 5]	[0.4, 0.2, 0.15, 0.15, 0.1]
Mutation Strategy	[0, 1, 2, 3]	[0.3, 0.3, 0.3, 0.1]

Table 4 Evaluation results regarding the cyberattack detection module

Model	Precision (%)	Recall (%)	F-score (%)
Random forest	98.26	96.80	97.50
Our engine	*98.84*	*97.22*	*97.95*
Dense DNN Tanh	74.76	66.81	70.00

Experiment Execution and data Collection: The whole system was running for 2 hours in order to ensure that an adequate number of QoS profiles will be selected, which will allow for a fair comparison between the algorithms. Finally, it should be noted that in order to measure the QoS metrics, network statistics, for each switch link, were collected every 2 s via the Ryu SDN framework API [32].

5.2 Evaluation of Cyberattack Detection Engine

The evaluation of the cyberattack detection engine is conducted using the netflows gathered by the emulated PVs and the Energy Storage Systems of the aforementioned topology. The netflows correspond to ModBus TCP/IP communication protocol, which is implemented using the PyModBus Python Library [30]. For the execution of the attacks, the widely known SMOD penetration tool was selected and the attacks used were the following: the Unit Identifier (UID) Brute Force attack, the Function Enumeration attack and the Write All Registers DoS attack. The created multiclass dataset contains in total 422912 netflows. The 317184 are held as a training dataset while the rest 105728 are held as a testing dataset. Our engine is evaluated against the Dense Deep Neural Network (DNN) Tanh presented in [33] and the well known Random Forest algorithm. Table 4 presents the performance analysis in terms of the macro classification scores of Precision, Recall and F-score. Taking into account the results, the proposed model is capable of detecting in a more accurate manner the cyberattacks than the Dense DNN Tanh model and the Random Forest model.

5.3 Evaluation of Network Path Re-allocation

This section presents the comparative results of the re-routing algorithm of the FRANS's internal component named as EDAE. The algorithm is evaluated against the approach presented in [34], denoted hereafter as "NSGA". Both approaches obtain the network requirements from the MAS tool. To evaluate the performance of the aforementioned genetic algorithms, three KPIs were defined. Each KPI can be applied to all metrics. The available metrics are the delay, jitter, packet loss, bandwidth and security and the KPIS are the Success Ratio (SR), the Deviation Ratio (DR) and the Over-Allocation Ratio (OAR). Let AP be the set of all the applications, M the set of all metrics and $f_i(ap)$, where $i \in M$ and $ap \in A$, a validator that returns 1 if the i metric is matched for the specific application ap else 0:

$$f_i(ap) = \begin{cases} 1, & \text{if metric } i \text{ is satisfied for application } ap \\ 0, & \text{if metric } i \text{ is not satisfied for application } ap \end{cases}$$

Success Ratio–SR: The success ratio measures the percentage of the applications for which the requirements for a specific metric was matched during the experiment and is defined as:

$$SR_i = \frac{\Sigma_{t \in T} \Sigma_{ap \in AP} f_i(ap)}{Size(T) * Size(AP)} \tag{22}$$

where T is the set of all the timesteps for which network statistics through Ryu API where available.

Deviation Ratio–DR: The deviation ratio measures the mean value of the difference between the required value for a specific metric and the observed one across all the applications. The deviation ratio takes into account only the cases where the algorithms failed to find a network path to satisfy a specific metric for a specific application. DR is defined as:

$$DR_i = \frac{\Sigma_{f \in F} \frac{C_{i,a_p - O_i}, ap, t}{C_{i,ap}}}{Size(F)} \tag{23}$$

where F is the set of all tuples $f = (i, ap, t)$ for which the specific timestep t the metric i did not satisfied for the specific ap. Finally, $C_{i,ap}$ expresses the requirement of an application ap for a specific metric i and $O_{iap,t}$ the actual value of the metric i for ap the t timestep.

Over-Allocation Ratio–OAR: The Over-Allocation ratio expresses the same thing as DR with the difference that is now applied on the set $S = T \times AP \times M - F$, i.e the cases for which the algorithm managed to satisfy the requirements.

$$OAR_i = \frac{\Sigma_{S \in S} \frac{O_i, ap, t - C_i, a_p}{C_{iap,t}}}{Size(S)} \tag{24}$$

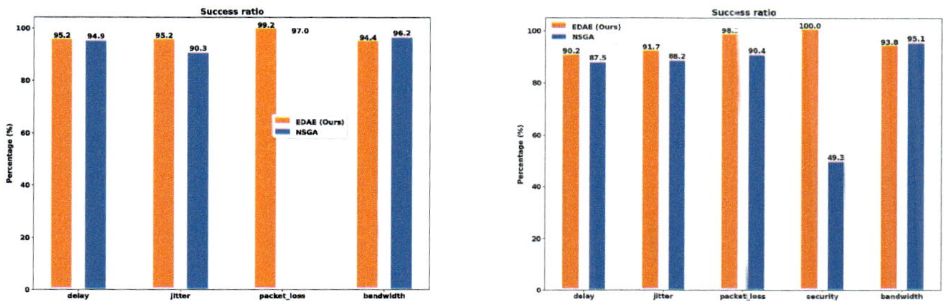

(a) (b)

Fig. 6 Success ratios **a** on non-attack scenario **b** on attack scenario

Figure 6 presents the comparative results for the SR KPI across all metrics M for both the algorithms in two different scenarios. observed that FRANS's EDAE module outperforms the NSGA for all the metrics except for the bandwidth in both scenarios. NSGA has a tendency towards communication paths that favor the bandwidth metric and outperforms EDAE by 1.8% and 1.3% in each scenario respectively. On the other hand EDAE, especially in the attack scenario, consistently outperforms the NSGA in the rest of the metrics, e.g by 50% and 7.7% for the security and packet loss metrics respectively in the attack scenario. Furthermore, despite the fact that EDAE has to take into consideration an additional metric, the security metric, which makes the problem harder to solve, the results show the superiority of EDAE.

Figure 7 depicts the results of the DR KPI for both the algorithms. The results express the deviation (error), by means of percentage, from the target metric for the solutions where the algorithms did not manage to satisfy the specific requirement of

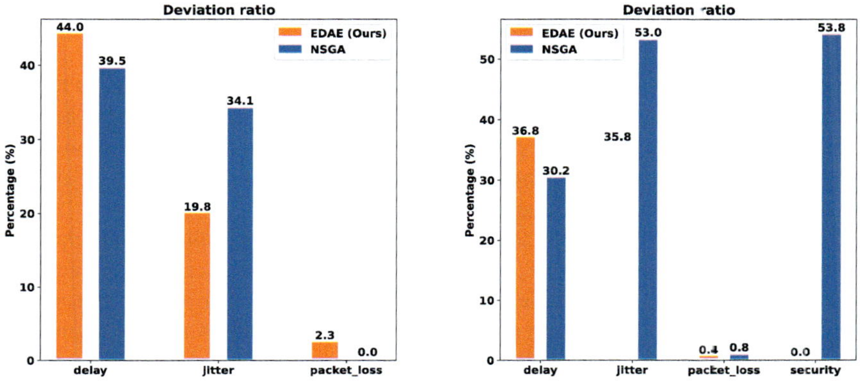

(a) (b)

Fig. 7 Deviation ratios **a** on non-attack scenario **b** on attack scenario

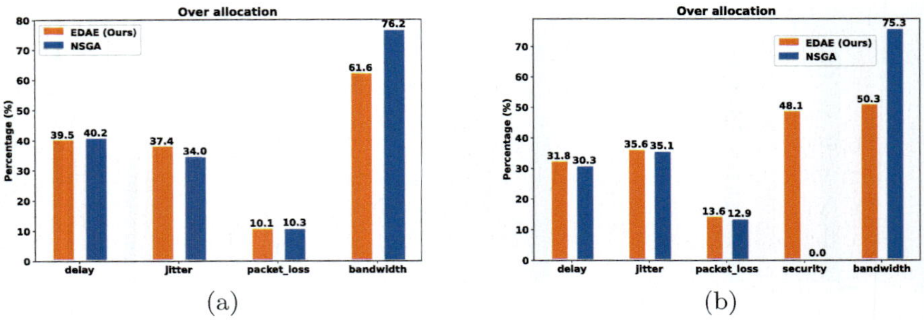

Fig. 8 Over allocation ratios **a** on non-attack scenario **b** on attack scenario

an application. In this case there is not a clear winner, depending on the scenario and the metric the best performing algorithm is different. An exception can be spotted for the jitter metric where the NSGA algorithm has consistently better results for DR KPI.

Figure 8 demonstrates the results for the OAR KPI which expresses the same value as the DR KPI but in a different set, the set for which the algorithms manages to satisfy the specific requirement of an application. The OAR KPI could be seen as a resource management KPI that expresses how optimally the resources of the network are matched to the needs of the applications. It is observed that EDAE has a bias towards security metric while NSGA towards bandwidth metric which could explain it's better performance by means of SR KPI on this metric against EDAE. Finally, for the rest of the metrics the differences are relatively small and vary on the scenario.

6 Discussion

The modernization of the EPES infrastructures is a complex process that has to take into consideration a lot of factors that are the subject of multiple scientific fields. This work focuses on two major factors. The first one is related with the integration of the DERs into the main grid, while the second one concerns the quality and security of the communication network of the EPES infrastructure.

To this end, FRANS's MAS tool proposes a distributed Voltage & Frequency restoration algorithm that exploits the capabilities of the DERs in order to increase the quality of the electrical grid. Furthermore, MAS tool consists of an additional scalable distributed black start restoration algorithm based on consensus theory augmented with AI. Moreover, the black start restoration operates on the low/medium voltage grids which can speed up the restoration process to the end user since it can be done in parallel with the traditional black start restoration process which operates on the high voltage grid.

With respect to the communication network of an EPES infrastructure, FRANS framework proposes two modules, named after EDAE and FRANS detection engine. The objective of the aforementioned modules is to detect cyberthreats, to mitigate cyberattacks, to optimize the QoS of the underlying applications and to restore the observability of the EPES, by exploiting the capabilities of the SDN technology.

The key contributions of FRANS is the effectiveness to restore the voltage & frequency to it's nominal values and the impressive scalability of the black start restoration consensus algorithm as shown in [12]. Furthermore, FRAN's EDAE module presents a novel chromosome representation that reduces the search space of the genetic algorithm, takes into consideration all the QoS aspects including the security one and finally it offers an observability restoration algorithm that takes also into account the latency. The performance of EDAE can be seen in Figs. 6, 7, 8 and in [12]. Finally, from Table 4 it is clear that the FRANS cyberattack detection engine performs better against the Random Forest and Dense DNN Tanh model.

Finally, the authors will share their thoughts which they believe that are of great scientific interest. The first thought regards the consensus algorithm of the FRANS's MAS tool. Specifically, it would be interesting to use the same consensus algorithm for both v&f and black start restoration functionalities and also to examine more secure consensus and resilient options that would be also scalable as the one presented. Additionally, the authors believe that an interesting way that EDAE's path re-allocation algorithm can be implemented is with the use of deep learning techniques. Finally, in order to make the FRANS cyberattack detection engine more resilient to future unknown attacks it would be interesting to examine zero/few shot learning techniques.

7 Conclusion

The rapid and continuous progression of electrical grids requires solutions that will be able to efficiently address the various challenges. To this end, the FRANS framework aims to bring together different technologies and modules in order to modernize the traditional electrical grid, empower it with further capabilities and improve any challenges spotted in the EPES ecosystem in the past and transform it to a smart grid. The EDAE tool, the MAS tool, the cyberattack detection system and the SDN technology, they all together form the FRANS framework, which proves through the evaluation tests conducted, to perform very well compared to similar methods. Specifically, the key contributions of FRANS with respect to the electrical part of a modern EPES are the distributed and fault tolerant voltage & frequency restoration techniques for smart grids with DERs and the scalable AI based consensus black start restoration technique for low/medium voltage grids. Furthermore, FRANS's main contributions with respect to the network communication of an EPES are the effective cyberattack detection system, the multiobjective QoS optimization genetic algorithm alongside it's low search space chromosome encoding and the exploitation of the capabilities of the SDN technology. Regarding our next steps, we aim to develop a dashboard in order to provide live monitoring features to the proposed system.

Acknowledgements The aforementioned work effort in this paper is conducted under the framework of the SDN-microSENSE project, a Horizon 2020 program, funded by the European Union under the grant agreement No. 833955. The opinions expressed in this paper are those of the authors and do not necessarily reflect the views of the European Commission.

References

1. P. Jiang, Y.V. Fan, J.J. Klemeš, Impacts of covid-19 on energy demand and consumption: challenges, lessons and emerging opportunities. Appl. Energy **285**, 116441 (2021). https://www.sciencedirect.com/science/article/pii/S030626192100009X
2. P.K. Ozili, Global economic consequence of Russian invasion of Ukraine, in Available at SSRN (2022)
3. S.W. Blume, *Electric Power System Basics for the Nonelectrical Professional* (Wiley, 2016)
4. M.F. Islam, A.M. Oo, S.H. Chowdhury, The traditional power generation and transmission system: some fundamentals to overcome challenges, in *Smart Grids* (Springer, 2013), pp. 1–21
5. N. Dorsch, F. Kurtz, H. Georg, C. Hägerling, C. Wietfeld, Software-defined networking for smart grid communications: applications, challenges and advantages, in *IEEE International Conference on Smart Grid Communications (SmartGridComm)*. IEEE, vol. 2014 (2014), pp. 422–427
6. J.E. Sullivan, D. Kamensky, How cyber-attacks in Ukraine show the vulnerability of the U.S. power grid. Electric. J. **30**(3), 30–35 (2017). https://www.sciencedirect.com/science/article/pii/S1040619017300507
7. R. Dudley, D. Golden, The colonial pipeline ransomware hackers had a secret weapon: self-promoting cybersecurity firms (2021)
8. N.A. Hassan, *Ransomware Families* (Apress, Berkeley, CA, 2019), pp. 47–68. https://doi.org/10.1007/978-1-4842-4255-1_3
9. Energy tech firm hit in ransomware attack. https://www.wsj.com/articles/energy-tech-firm-hit-in-ransomware-attack-11620764034. Accessed 18 Apr 2022
10. O. Saidani Neffati, S. Sengan, K.D. Thangavelu, S. Dilip Kumar, R. Setiawan, M. Elangovan, D. Mani, P. Velayutham, Migrating from traditional grid to smart grid in smart cities promoted in developing country. Sustain. Energy Technol. Assess. **45**, 101125 (2021). https://www.sciencedirect.com/science/article/pii/S2213138821001351
11. A. Pasias, T. Kotsiopoulos, G. Lazaridis, A. Drosou, D. Tzovaras, P. Sarigiannidis, Enabling cyber-attack mitigation techniques in a software defined network, in *IEEE International Conference on Cyber Security and Resilience (CSR)* vol. 2021 (2021), pp. 497–502
12. A. Pasias, A. Schoinas, A. Drosou, D. Tzovaras, A scalable multi-agent system for black start restoration in low voltage microgrids, in *IEEE International Conference on Cyber Security and Resilience (CSR)* vol. 2021 (2021), pp. 479–484
13. G. Lazaridis, K. Papachristou, A. Drosou, D. Ioannidis, P. Chatzimisios, D. Tzovaras, On the potential of sdn enabled network deployment in tactical environments, in *Artificial Intelligence Applications and Innovations. AIAI, IFIP WG 12.5 International Workshops*, ed. by I. Maglogiannis, J. Macintyre, L. Iliadis, vol. 2021. (Springer International Publishing, Cham, 2021), pp. 252–263
14. M.R. Belgaum, S. Musa, M.M. Alam, M.M. Su'ud, A systematic review of load balancing techniques in software-defined networking. IEEE Access **8**, 98,612–98 636 (2020)

15. M.H. Rehmani, A. Davy, B. Jennings, C. Assi, Software defined networks-based smart grid communication: a comprehensive survey. IEEE Commun. Surveys Tutor. **21**(3), 2637–2670 (2019)
16. D.A. Chekired, L. Khoukhi, H.T. Mouftah, Decentralized cloud-sdn architecture in smart grid: a dynamic pricing model. IEEE Trans. Ind. Inf. **14**(3), 1220–1231 (2017)
17. D. Kaur, G.S. Aujla, N. Kumar, A.Y. Zomaya, C. Perera, R. Ranjan, Tensor-based big data management scheme for dimensionality reduction problem in smart grid systems: sdn perspective. IEEE Trans. Knowl. Data Eng. **30**(10), 1985–1998 (2018)
18. K. Nisar, E.R. Jimson, M.H.A. Hijazi, I. Welch, R. Hassan, A.H.M. Aman, A.H. Sodhro, S. Pirbhulal, S. Khan, A survey on the architecture, application, and security of software defined networking: challenges and open issues. Internet Things **12**, 100289 (2020). https://www.sciencedirect.com/science/article/pii/S2542660520301219
19. S. Khan, F.K. Hussain, O.K. Hussain, Guaranteeing end-to-end qos provisioning in soa based sdn architecture: a survey and open issues, in *Future Generation Computer Systems*, vol. 119 (2021), pp. 176–187. https://www.sciencedirect.com/science/article/pii/S0167739X21000637
20. A. Syrmakesis, C. Alcaraz, N. Hatziargyriou, Classifying resilience approaches for protecting smart grids against cyber threats. Int. J. Inf. Secur. **05**, 1–22 (2022)
21. A. Starke, J. McNair, R. Trevizan, A. Bretas, J. Peeples, A. Zare, Toward resilient smart grid communications using distributed sdn with ml-based anomaly detection, in *Wired/Wireless Internet Communications*, ed. by K.R. Chowdhury, M. Di Felice, I. Matta, B. Sheng. (Springer International Publishing, Cham, 2018), pp. 83–94
22. V.K. Singh, E. Vaughan, J. Rivera, Sharp-net: Platform for self-healing and attack resilient pmu networks, in *IEEE Power & Energy Society Innovative Smart Grid Technologies Conference (ISGT)*, vol. 2020 (2020), pp. 1–5
23. P.R. Grammatikis, P. Sarigiannidis, C. Dalamagkas, Y. Spyridis, T. Lagkas, G. Efstathopoulos, A. Sesis, I.L. Pavon, R.T. Burgos, R. Diaz, A. Sarigiannidis, D. Papamartzivanos, S.A. Menesidou, G. Ledakis, A. Pasias, T. Kotsiopoulos, A. Drosou, O. Mavropoulos, A.C. Subirachs, P.P. Sola, J.L. Domínguez-García, M. Escalante, M.M. Alberto, B. Caracuel, F. Ramos, V. Gkioulos, S. Katsikas, H.C. Bolstad, D.-E. Archer, N. Paunovic, R. Gallart, T. Rokkas, A. Arce, Sdn-based resilient smart grid: the sdn-microsense architecture. Digital **1**(4), 173–187 (2021). https://www.mdpi.com/2673-6470/1/4/13
24. M.H. Rehmani, F. Akhtar, A. Davy, B. Jennings, Achieving resilience in sdn-based smart grid: a multi-armed bandit approach, in *2018 4th IEEE Conference on Network Softwarization and Workshops (NetSoft)* (2018), pp. 366–371
25. D. Jin, Z. Li, C. Hannon, C. Chen, J. Wang, M. Shahidehpour, C.W. Lee, Toward a cyber resilient and secure microgrid using software-defined networking. IEEE Trans. Smart Grid **8**(5), 2494–2504 (2017)
26. H. Maziku, S. Shetty, Software defined networking enabled resilience for iec 61850-based substation communication systems, in *2017 International Conference on Computing, Networking and Communications (ICNC)* (2017), pp. 690–694
27. A. Paszke, S. Gross, F. Massa, A. Lerer, J. Bradbury, G. Chanan, T. Killeen, Z. Lin, N. Gimelshein, L. Antiga, A. Desmaison, A. Kopf, E. Yang, Z. DeVito, M. Raison, A. Tejani, S. Chilamkurthy, B. Steiner, L. Fang, J. Bai, S. Chintala, Pytorch: An imperative style, high-performance deep learning library, in *Advances in Neural Information Processing Systems 32*, ed. by H. Wallach, H. Larochelle, A. Beygelzimer, F. d'Alché-Buc, E. Fox, R. Garnett (Curran Associates, Inc., 2019), pp. 8024–8035. http://papers.neurips.cc/paper/9015-pytorch-an-imperative-style-high-performance-deep-learning-library.pdf
28. A.D. Bintoudi, L. Zyglakis, A.C. Tsolakis, D. Ioannidis, L. Hadjidemetriou, L. Zacharia, N. Al-Mutlaq, M. Al-Hashem, S. Al-Agtash, E. Kyriakides et al., Hybrid multi-agent-based adaptive control scheme for ac microgrids with increased fault-tolerance needs. IET Renew. Power Gener. **14**(1), 13–26 (2020)
29. Containernetluse docker containers as hosts in mininet emulations. https://containernet.github.io. Accessed 06 May 2022

30. Pymodbus-a python modbus stack-pymodbus 2.5.0 documentation. https://pymodbus.readthedocs.io/en/latest/readme.html. Accessed 03 May 2022
31. S. Hemminger, Network emulation with netem, in *Linux Conf Au* (05, 2005)
32. Ryu sdn framework. https://ryu-sdn.org/
33. P. Radoglou Grammatikis, P. Sarigiannidis, G. Efstathopoulos, E. Panaousis, Aries: a novel multivariate intrusion detection system for smart grid. Sensors **20**(18) (2020). https://www.mdpi.com/1424-8220/20/18/5305
34. D. Li, X. Wang, Y. Jin, H. Liu, Research on QoS routing method based on NSGAII in SDN. J. Phys. Conf. Ser. **1656**, 012027 (2020)

SOAR4DER: Security Orchestration, Automation, and Response for Distributed Energy Resources

Jay Johnson, C. Birk Jones, Adrian Chavez, and Shamina Hossain-McKenzie

Abstract Monitoring data and control functionality presented by interoperable photovoltaic (PV) inverters and other Distributed Energy Resources (DER) can be used to improve site maintenance, prognostics, and grid operations. Unfortunately, DER communications present attack vectors which could lead to power systems impacts. Since adversary capabilities continually improve, avoiding catastrophic consequences requires intelligent intrusion detection and remediation systems that consider both physical and cyber features. New Security Orchestration, Automation, and Response (SOAR) technologies are equipping cyber-defenders with new capabilities to autonomously respond to network and host-based system alerts, threat hunting results, and cyber intelligence data streams. In this Chapter, we present a novel SOAR approach for DER systems, called SOAR4DER, that ingests data from multiple Intrusion Detection Systems (IDSs) to quickly block attacks and revert DER systems to known good states. Our implementation used a collection of IDS technologies on a Bump-in-the-Wire (BITW) device which incorporated physical and cyber data to detect abnormal and potential malicious behaviors. Multiple SOAR playbooks then used the IDS data streams to automatically defend the system. Laboratory testing of the SOAR4DER system showed detection and response times under 30 s for all adversary reconnaissance operations, denial-of-service attacks, malicious Modbus commands, brute force logins, and machine-in-the-middle attacks.

J. Johnson (✉) · C. B. Jones · A. Chavez · S. Hossain-McKenzie
Sandia National Laboratories, P.O. Box 5800 MS1033, Albuquerque, NM 87185-1033, USA
e-mail: jjohns2@sandia.gov

C. B. Jones
e-mail: cbirkjones@gmail.com

A. Chavez
e-mail: adrchav@sandia.gov

S. Hossain-McKenzie
e-mail: shossai@sandia.gov

H. Haes Alhelou et al. (eds.), *Power Systems Cybersecurity*, Power Systems,
https://doi.org/10.1007/978-3-031-20360-2_16

1 Introduction

Distributed Energy Resources (DER) connected to the internet or private networks require cybersecurity detection and mitigation capabilities to maintain safe electric power system (EPS) operations. The risk of adverse effects on the electric grid is evident from past studies wherein the attacker modified DER control settings [1]. Unfortunately, preventing unauthorized access to internet-connected DER devices cannot be guaranteed because adversaries continuously find new methods for gaining access to networks and equipment. After hundreds of organizations have been compromised and intrusion dwell times are still averaging more than three weeks [2], the information security community is transitioning from network perimeters to zero-trust principles [3] and "Assume Breach" cybersecurity management plans [4]. Automated intrusion detection and mitigation systems are necessary to avoid long-term consequences of DER misoperation [5].

Layered detection and response security technologies make up a portion of a comprehensive, defense-in-depth security approach to detecting and recovering from attacks. While this technology has been deployed extensively on information technology (IT) systems, it has not historically been integrated with critical infrastructure or other operational technology (OT) networks which monitor and control power, water, waste, and transportation systems. However, over the last few years, many DER owners, operators, service providers, aggregators, and utilities are standing up Security Operation Centers (SOCs) for their OT systems. SOCs are physical locations for security professionals to monitor operational activities using a variety of data from tools including:

- Security Information and Event Management (SIEM) software that ingests data from different security monitoring and detection tools while providing the cybersecurity team a single interface with detailed information,
- Network-based Intrusion Detection Systems (NIDSs), which use network data to alert on adversary actions,
- Network-based Intrusion Prevention Systems (NIPSs) that block malicious traffic before reaching its target,
- Host-based Intrusion Detection Systems (HIDSs), which track actions on endpoints (e.g., user logins, account creation/modification, binary execution, etc.)
- Host-based Intrusion Prevention Systems (HIPSs), which are a form of Endpoint Detection and Response (EDR) that monitor for system changes and take actions to stop suspicious activity, and
- Security Orchestration, Automation, and Response (SOAR) technologies that use automated response playbooks fed with one or more data streams to defend the network and assets based on pre-programmed rules. (Note: SOAR should not be mistaken for an Extended Detection and Response (XDR) technology which is similar, but represents a lighter weight platform, requiring no coding or playbook generation).

Our focus here is on SOAR because of its ability to quickly neutralize threats through automated responses. To be effective, however, SOAR must rely on the data provided by the other systems listed. Readiness to deploy SOAR technologies represents a significant step in the maturation of renewable and distributed energy SOCs.

Over time DER Management System (DERMS) security practitioners will layer new technologies into their integrated SOCs to expand their detection and response capabilities. SOAR tool efficacy is limited by the detection and threat data they ingest and their practical utility is limited by the response functions in their arsenal. Jumping to a highly sophisticated and automated security system is difficult in a single step however. The construction and expansion of DERMS SOC systems will progress from simple perimeter defenses to centralized, analytical data fusion tools to predictive, automated, and threat-based strategies. This transition will add new defense-in-depth capabilities to the SOC, as represented in Fig. 1, which was inspired by SOC maturity models from Accenture, Google Cloud, and Threat Connect [6–8]. Information represented in more mature SOC systems can be leveraged by SOAR systems to quickly execute precise mitigations.

As depicted in Fig. 1, initially, security practitioners will manually configure the network, apply perimeter defenses, and manually perform threat hunting and incident response. A well-designed system will be able to prevent an attacker locally with HIPS and anti-virus, but corporate alerting of these actions may be absent. Analysts will find it difficult to securely scale the network without centralized logging and monitoring. In the second, data-focused phase of SOC maturity, a single location for security analysis is established. This SOC will include a SIEM with security control logs (EDR, IDS. IPS, anti-virus, etc.), networking logs (firewalls, routers, wireless networks, servers, etc.), network diagrams, asset and software inventory lists, and other information from across the enterprise. At this stage, all the pertinent information can be accessed and visualized at a single location, but is likely unwieldy because there is an overwhelming quantity of data to parse. The next phase in a maturing SOC is focused on automating the analysis and response using SOAR tools. Threat intelligence is paired with vulnerability scans and malware analysis to provide real-time updates to potential attack vectors and exploits. Machine learning and artificial intelligence technologies are used to help sort through logs and prioritize human analysis and automated responses. Automated SOCs also include a wide range of defense functions including changing the network topology with software defined networking (SDN), redeploying containerized compute systems, and logging into networked equipment to modify firewall rules, managed switch configurations, etc. Realistically, the SOAR will only be able to manage known security problems that have preprogrammed playbooks; human interaction will be required for any new or evolving threats. However, offloading mundane security tasks to SOAR playbooks will give the SOC team more availability to address more serious concerns. Some things SOAR does today include removing malicious email attachments, scanning URLs, malware containment, deactivating or modifying services targeted by known bad actors, notifying users of critical patches, revoking credentials of accounts with suspicious behaviors, preventing data exfiltration, blocking IPs that are conducting

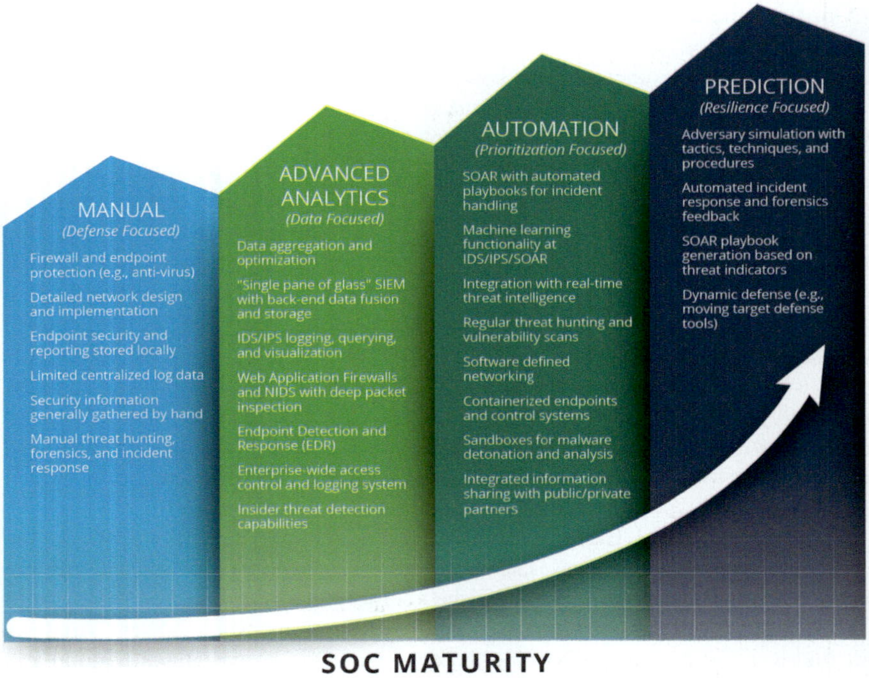

Fig. 1 Security Operations Center features with increasing maturity

DoS attacks, and scanning systems for malicious signatures (e.g., via *YARA: Yet Another Recursive Acronym* rules) [9]. In the case of detecting indicators of compromise, malware, or other adversary actions, the SOC team may share this information through previously established public or private partnerships, such as one of the Information Sharing and Analysis Centers (ISACs). In the final phase of SOC maturation, cyber-resilience is established by predicting adversary actions, the SOAR tool learns based on updated threat information and SOC operator actions, and new dynamic defense techniques are incorporated. These functionalities have not been fully realized in modern SOC environments but may be in the future.

In 2021, Idaho National Laboratory and Sandia National Laboratories conducted an industry survey of cyber-hardening technologies for the renewable energy space. The two labs queried hundreds of cybersecurity software vendors and service providers to see what IDS, SOAR, and other tools were available on the market. Those responding reported more than a dozen SOAR products applicable to the solar energy market [10–12]. Furthermore, the SANS Institute hosted a *SOAR Solutions Forum 2022* webinar in March 2022. On the webinar, Palo Alto Networks, Corelight, Greynoise, Swimlane, Splunk, and others highlighted the importance of and new advances in SOAR products [13]. The industry speakers emphasized the workload of the SOC analysts and the need for automation. Corelight stated an average SOC receives 10,000 alerts/day with 2-10 analysts reviewing the events [14].

The primary benefit of SOAR platforms is that they streamline case management, automation, information sharing, and threat intelligence management for incident response processes. Most products allow users to create Python or other scripts to automate tasks (e.g., [15, 16]), which are then layered into playbooks with features for threat feeds, incident alerts, etc. in order to provide unified situational awareness for all stakeholders (SOC analysts, IT admins, and Chief Security Officer). This move toward automated cybersecurity defenses is supported by a growing body of research and associated technologies [17]. In addition to SOAR products on the market, there are now technologies that use deep reinforcement learning to automatically update access control policies [18], automated threat sharing techniques [19], and automated penetration testing frameworks [20].

However, automated cybersecurity technologies are not without their limitations. Automated SOAR playbooks are excellent at responding to security incidents/disturbances when predictable events occur, i.e., for which indicators of compromise or signatures are known and response actions have been preprogrammed. However, for unpredictable disturbances and/or zero-day exploits with no defined signatures, playbook responses will not suffice. In those situations, human analysis will be required. If most alerts are processed automatically, SOC personnel can focus their attention on higher priority cases.

It is also important that behavior-based analyses are supported within the SOAR platform to ensure anomalies are detected when signatures are not available and to provide situational awareness into data patterns. Furthermore, for DER and general grid applications, it is increasingly important to consider both cyber and physical data. DER systems are becoming increasingly cyber-physical with the use of smart technologies, automated grid-support functions, new communication capabilities, and use of third-party software. Therefore, when a cyberattack occurs, it does not suffice to only monitor cyber data using IDSs; a hybrid approach is required that leverages both cyber and physical data to detect a broader range of attacks (e.g., insider threat), prevent physical system consequences, and understand impact propagation–as is demonstrated in this research.

This work presents a SOAR solution that uses alerts from signature- and behavior-based IDSs deployed on bump-in-the-wire (BITW) devices in front of DER equipment, as shown in Fig. 2. Past work by the same team implemented and tested a BITW approach for PV cybersecurity [21, 22], but did not implement mitigation strategies using SOAR. Although, the BITW devices could be placed at different locations in the network depending on implementation constraint and objectives. For utility-scale DER and commercial installations, there are likely to be multiple DER devices on site that are individually monitored. For the residential DER, the BITW system would be either on the homeowner Local Area Network (LAN)–as shown in Fig. 2–on an isolated virtual LAN or connected via a cellular or RF mesh backhaul to the DERMS. The BITW technology could be installed elsewhere in the network, in the DER housing, or directly on the DER operating system. The benefit

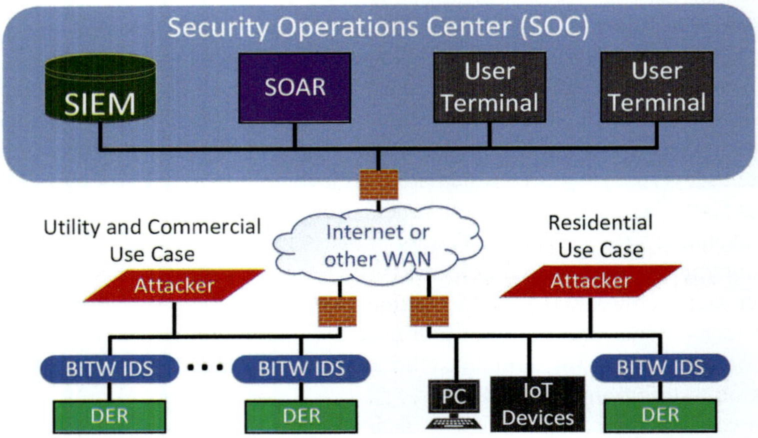

Fig. 2 Conceptual representation of DER networks integrated with IDS and SOAR technologies

of placing them upstream is reduced cost and complexity. The benefit of placing them at each DER is increasing adversary detection likelihood.

The BITW devices include multiple IDS technologies. The signature-based IDS is designed to identify packet manipulation, Address Resolution Protocol (ARP) spoofing, and other known attack types [23]. The behavior-based analysis identifies adversary reconnaissance, denial-of-service attempts, and other previously unknown actions using one or more Machine Learning (ML) algorithms to detect both known and unknown patterns through inference [24, 25]. When an attack is detected, the IDS logs pertinent information and reports it to the SIEM database. Then, the SOAR implements any mitigation measures to stop the attack and return services back to an operational state.

In this work, we demonstrate an IDS hybrid approach and SOAR playbooks to effectively detect and mitigate adversary actions with data from physical and virtualized DER equipment at Sandia National Laboratories. The primary contributions of this work are (a) establishing a reference IDS, SOC, and SOAR architecture for DER systems, (b) demonstrating the efficacy of SOAR technologies in DER environments, and (c) calculating the detection and response times for several cyberattack types using inexpensive, single-board sensor systems. The remainder of the research is structured such that Sect. 2 describes the laboratory environment for the SOAR assessments, Sect. 3 briefly discusses the findings, and Sect. 4 concludes the research.

2 Experimental Testbed and Methodology

The Distributed Energy Technologies Laboratory (DETL) at Sandia National Laboratories was configured with the SOAR, IDS, and a DER Simulator, as shown in Fig. 3. The DER simulator runs AC and DC power simulations on a Raspberry Pi 4 with

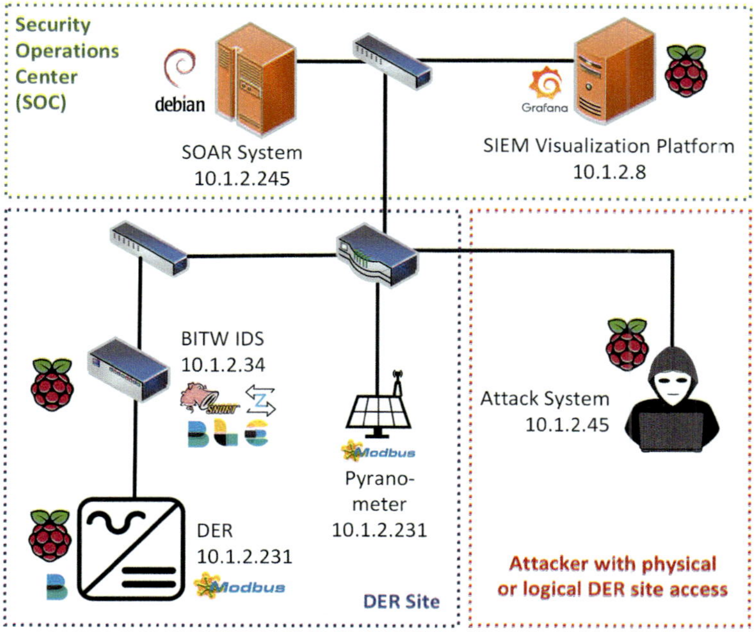

Fig. 3 DETL network used for the SOAR evaluation

prerecorded 1-second solar irradiance, grid voltage, and grid frequency profiles [26]. The DER simulator also includes multiple IEEE 1547-compliant interoperability interfaces [27, 28]; there is a SunSpec Modbus server and IEEE 2030.5 client which report production data and allow the DERMS to update the control settings in the device. For the SOAR experiments, the DER simulator was configured to represent a 100 kW utility-scale PV inverter. Prior work on IDS systems for solar inverters [29] was conducted on residential and commercial-scale PV inverters in DETL, but in this case the DER simulator was used to incorporate DER syslog monitoring and easily change the irradiance profiles for testing. The SOAR playbooks presented here could be implemented with BITW IDS systems connected to real PV inverters without any loss in functionality.

There was also a simulated pyranometer (solar irradiance meter) at the DER site that collected plane-of-array solar irradiance. These values were queried by the IDS system via Modbus to perform physical analysis of the DER operations. For convenience, these values were hosted on the DER simulator Modbus server.

The BITW IDS was hosted on a Raspberry Pi 4 with 4 GB of RAM, a 16 GB SD card, and additional flash memory to hold the Elasticsearch database. This system ran several programs, namely Zeek, Snort, Filebeat, and Logstash, to capture and analyze traffic to the DER simulator, shown in the startup and operations in Fig. 4. To begin the IDS system, Elasticsearch was initiated with the indices for storing log and alert

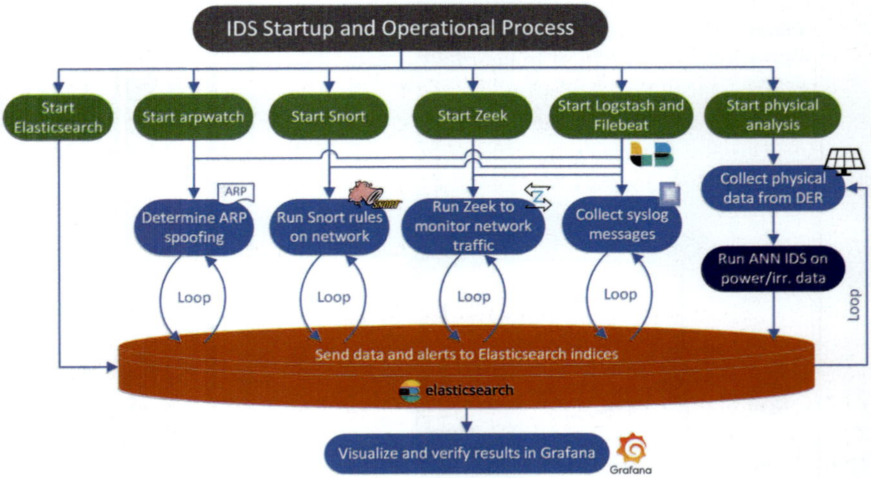

Fig. 4 Intrusion Detection System implementation

data. Zeek was used to detect signature- and behavioral-based anomalies; it collected and interpreted network traffic to create compact, high-fidelity transaction logs, file content, and other customized output [30]. For ICS applications, Zeek has a Modbus library to track Modbus sessions. When a connection was made using the Modbus protocol on port 502, sessions were automatically tracked in a file named modbus.log. Snort is an open-source packet sniffer, packet logger, and signature-based intrusion prevention system (IPS) that alarms on unique signatures of known malware or traffic that may indicate an attack [31]. A series of rules were used to define malicious network activity signatures. ICS-specific rules called QuickDraw [32], compiled by Digital Bond, were included to alarm on Modbus events such as attempted denial of service (DoS) attacks, malformed Modbus packets, and Modbus packets that have incorrect source and destination internet protocol (IP) addresses. Filebeat can monitor, collect, and forward logs specified by a user as they are written to disk in real-time [33]. In the event that Filebeat is interrupted, it will keep track of the most recent logs it already reviewed and when restored will only forward new logs that have not been processed. Additionally, Filebeat provides users with the ability to tag logs so they can be sorted and grouped with user-specified strings. Filebeat communicates alerts to Logstash for further processing and parsing. Logstash collects logs from a variety of sources, parses that data according to schemas defined by the user, and seamlessly forwards the transformed data to a database specified by the user [34]. Filebeat and Logstash work together to monitor logs specified by the user to efficiently store them in a centralized system for analysis. Filebeat and Logstash operate as network services and can easily be deployed across endpoints on a network that monitor both host and network logs. Elasticsearch [35] is a database that has the ability to interoperate with a number of log collectors (such as Filebeat and Logstash) and provide data to visually display those alerts. Visual front-end systems including Grafana [36] and Kibana [37] are often used to display network health and events detected in real time for situational awareness. For this project, Grafana was used as the SIEM visualization software, but it ran on another Raspberry Pi computer at 10.1.2.8.

The attack system was the Raspberry Pi Operating System, Raspbian, with several tools to launch representative attacks on the DER device:

- The network mapper, nmap [38], was used to collect reconnaissance information of the DER.
- hping3 [39] was used for its ability to generate TCP, UDP, ICMP and RAW-IP packet floods for DoS simulations.
- Custom Python code using PyModbus [40] and PySunSpec2 [41] packages was used to communicate Modbus to the DER.
- Hydra [42] was used for brute force username and password attacks against FTP, SSH, Telnet, and other protocols.
- Ettercap [43] was used because it contained several features for machine-in-the-middle attacks.

This system included an attack harness with a collection of bash scripts to launch one or more automated attacks. The attack harness was originally developed by Idaho National Laboratory and Sandia National Laboratories within the More Situational Awareness for Industrial Control Systems (MOSAICS) program [44]. The attack harness also included the capability to write to the Elasticsearch database on the IDS when the attack started and stopped in order to collect detection and response data.

Within the SOC, a SIEM visualization tool was created using Grafana to monitor physical data of the DER device and inspect alerts, logs, and other IDS data. For the purposes of quantifying the response times for the SOAR demonstration, the attack system and the SOAR system logged to Elasticsearch the ISO 8601 [45] timestamp, rounded to the nearest second, from the attack computer when the attack started and stopped, malicious activity detected by the SOAR, and when the SOAR system completed the response actions. An example of the physical DER data and the attack results is shown in Fig. 6. In this case, the DER simulator has been configured to operate in constant power factor mode, as shown by the straight-line on the active power (P)-reactive power (Q) plane in the SIEM. It's worth noting that while our test environment was a flat network, there would be no operational difference–aside from slightly longer communication times–for SOC systems that were connected to the DER site via SSH, VPN, or other means over the public internet, cellular connection, or radio mesh network.

The SOAR system was built in a Debian virtual machine hosted on a Windows 10 computer. The SOAR system consisted of Python code that executed in a continuous 1 second loop over each of the playbooks. Each playbook queried Elasticsearch for data of interest from the last 30 s to determine if there were any new alerts or data indicating a compromise of the DER system. If there was triggering data in Elasticsearch, the response playbook for that attack would be executed. The SOAR loop time and the sliding window of Elasticsearch queries could be adjusted as desired by the SOC team, but these settings were found to provide reasonable response characteristics for the demonstration. The SOAR playbooks demonstrated in this project are numbered in Fig. 5, and include:

1. Detecting reconnaissance using Zeek by analyzing all inbound connections to the DER.
2. Detecting denial-of-service attacks using Zeek logs and Snort rules monitoring traffic volume of inbound connections to the DER.
3. Detecting insider threats or malicious DER reconfigurations by monitoring DER power data.
4. Detecting fuzzing or other Modbus attacks by monitoring Quickdraw Snort rules.
5. Detecting secure shell (SSH) brute force login attempts by monitoring DER syslog data.
6. Detecting Address Resolution Protocol (ARP) spoofing by monitoring the ethernet/IP address pairings with arpwatch.

There are dozens of additional playbooks that could have been implemented. For example, the gray SOAR playbook in Fig. 5 would compare inbound connections to a pre-approved whitelist, but this was not implemented because it would make on-site maintenance more difficult. (Preventing these connections would be better implemented with a BITW IDS or DER host-based firewall rule too.) However, allowing all subnet IPs to connect to the DER does provide good information to the SOC team about adversary tactics, techniques, and procedures as well as the IP address(es) used by the adversary to perpetrate the attack. Ultimately, there are multiple ways to secure DER systems and the approach will vary from team to team based on the team size, SIEM/IDS/SOAR software maturity, and the desire of the security team for detailed visibility into network operations. The following subsections will detail the six SOAR4DER playbook implementations and associated test results.

2.1 SOAR Playbook 1: Reconnaissance

There are many ways in which adversary reconnaissance could be detected. In our case, we created a simple rule to inspect the recent Zeek logs to find any connections to closed port 23, which typically hosts Telnet services. Any valid DER users or tools will not be connecting to DER port 23, and it can be assumed these devices are surveilling the network for targets. More sophisticated rules could easily be created if false positives became a problem or the adversary was believed to scan different port lists. For example, training and applying behavior based techniques to analyze traffic patterns that are capable of detecting port scans is one alternative approach. Another approach could leverage Software Defined Networking (SDN) to whitelist all valid communications (source, destination, physical/logical port) and drop all other traffic. In our demonstration, any attempted connection to port 23 is flagged as reconnaissance and the source IP is blocked at the PIDMS using aebtable [46]

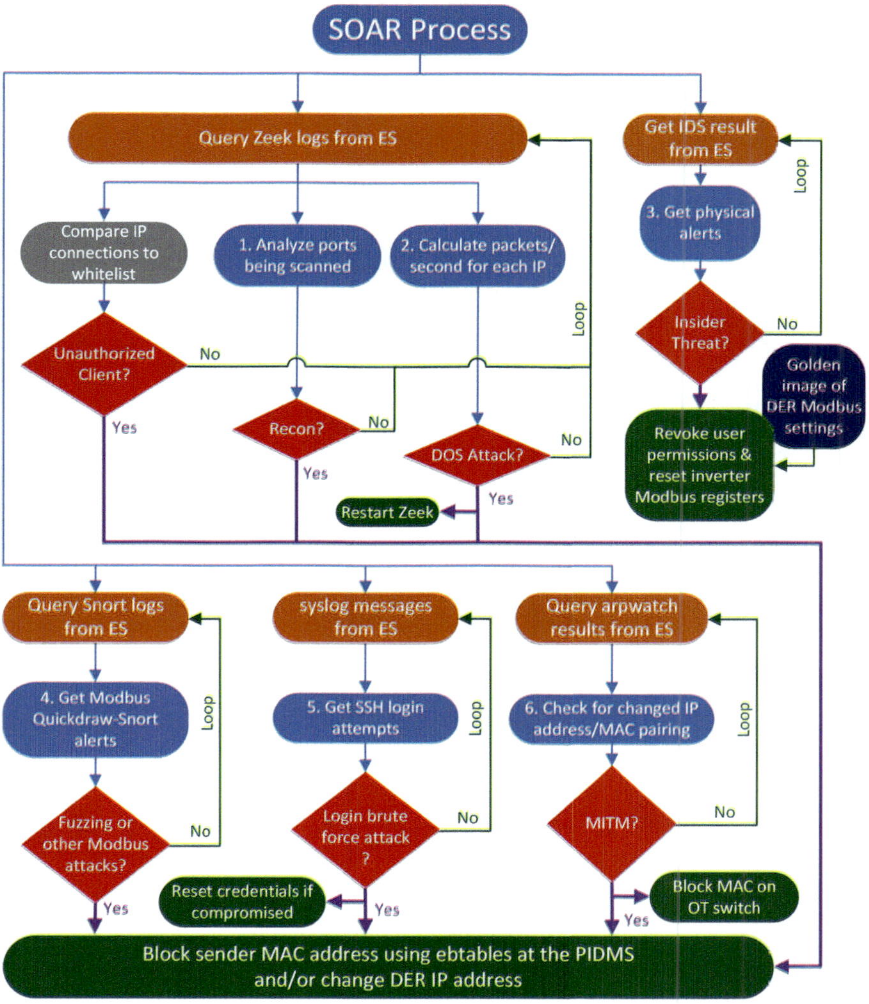

Fig. 5 SOAR implementation

ruleebtables is a Linux-based tool for filtering Ethernet frames going through network bridges. An examplebtables rule to drop traffic originating from a MAC address has the following formatebtables -A FORWARD -s < source MAC address > -j DROP. Theebtables command can be wrapped in a script where the source MAC address is provided as an argument to automate blocking a malicious endpoint. Figure 6 shows the detection logs indicating the Reconnassaince attack.

Fig. 6 Grafana SOC dashboard with DER power data and SOAR detection and response timestamps

For this experiment, the attack system issued nmap -v -T5 -Pn 10.1.2.231 -p1-100,502 command, which verbosely and quickly scanned ports 1 through 100 and 502 without host discovery. This sent significant traffic to the DER device to detect which ports were open. In fact, this triggered the DoS response initially, before the DoS rule was changed to strictly look for Modbus traffic. When port 23 was scanned, Zeek logged this connection from the 10.1.2.45 attacker IP and stored that in /var/log/bro/current/conn.log, which was pulled by Filebeat and forwarded to Logstash for further processing and parsing and then saved as an entry in Elasticsearch. Meanwhile, the SOAR system was continuously looping through all the playbooks looking for triggering patterns in the Elasticsearch database. Once traffic to DER port 23 was detected in Elasticsearch, the reconnaissance playbook was triggered, which completed the following sequence:

1. The timestamp, source IP and port number, and destination IP and port number were recorded and reported to the user.
2. The SOAR system wrote an Elasticsearch detection log entry. The timestamp was generated by SSHing into the attacking system and getting the timestamp, i.e.date–iso-8601=seconds.
3. The IDS system is logged into using SSH and a new ebtable rule was created to block the attacker's IP located in the id_orig_host field.

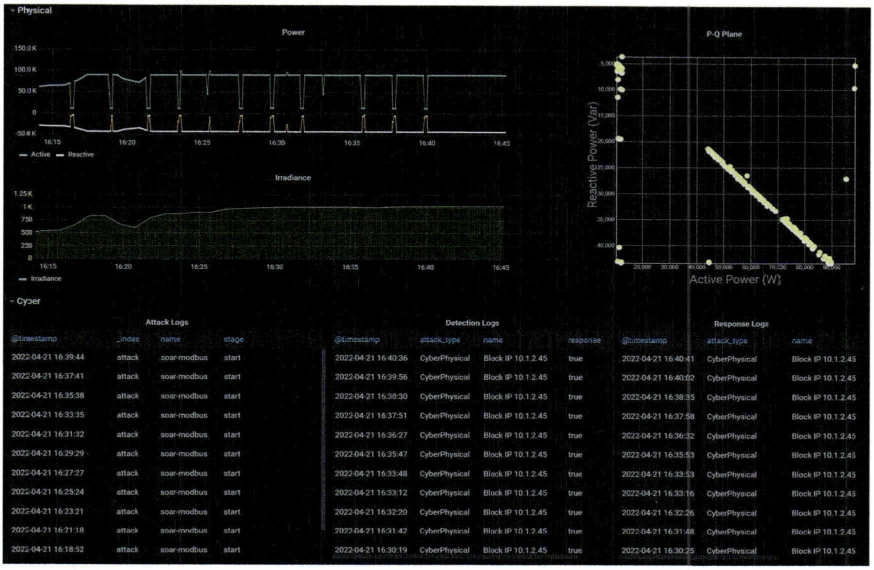

Fig. 7 Grafana SOC dashboard with DER power data and SOAR detections and responses. DER changes in output power are shown in the top plot

4. The SOAR system wrote to Elasticseach noting the response was completed using the attack system timestamp.
5. The SOAR slept for 30 s (the duration of collected Elasticsearch data) to prevent triggering a subsequent response for the same scan.
6. The IDS firewall rules were flushed in order to continue further experimentation.

In an actual SOAR implementation, the firewall rules would either be permanent or be removed after a predefined period of time (e.g., 24 hrs). The latter option would keep the environment clean of firewall rules that could prevent service or maintenance teams from working locally with the equipment. After conducting multiple, automated nmap scans on the DER device, the Grafana SIEM dashboard displayed physical DER data along with the attack, detection, and response data, as shown in Fig. 7. Using the attacker and SOAR timestamps, the detection and response times were calculated for 10 reconnaissance attacks. The results in Table 1 show that the attacks took 2–3 s, and detection was 5–9 s, and the attacker IP was blocked 8–12 s after reconnaissance probes started.

Table 1 Reconnaissance detection and response times

Test	Attack start (s)	Attack end (s)	SOAR detection (s)	SOAR response complete (s)
1	0	2	6	9 (3 s from detect)
2	0	2	6	9 (3 s from detect)
3	0	3	9	12 (3 s from detect)
4	0	2	9	12 (3 s from detect)
5	0	3	7	10 (3 s from detect)
6	0	3	5	8 (3 s from detect)
7	0	2	8	11 (3 s from detect)
8	0	3	9	11 (2 s from detect)
9	0	2	9	12 (3 s from detect)
10	0	2	7	11 (4 s from detect)

2.2 *SOAR Playbook 2: Denial-of-Service*

Denial-of-Service (DoS) and Distributed Denial-of-Service (DDoS) attacks are relatively common because they are simple to execute and botnets can be hired for this purpose [47]. Few have targeted energy infrastructure, but on March 5, 2019, a renewable energy company, sPower, lost visibility to solar and wind installations for hours as a result of a DoS attack on an unpatched Cisco firewall [48]. To create a similar attack, the attack machine issued the following attack,

```
(sleep 5; killall hping3)& hping3 -d 120 -S -p 502 –flood -V 10.1.2.231
```

which ran a daemonized `killall` command to terminate the `hping3` flood attack after 5 s. The command flooded TCP port 502 on the DER with packets containing 120 bytes of data each and the SYN flag set.

This attack is straightforward to detect because the traffic drastically increases. Based on the Elasticsearch logs, there is normally around 25-35 connections to Modbus port 502 in Zeek queries for the last 30 s. These connections are primarily made by the IDS to perform the physical data analysis. DERMS systems will also regularly query the DER to get power updates, but this was not included in the test environment, and, even if it was, the DERMS would not have a polling rate faster than 1 Hz. We determined the DoS attack rate using `tcpdump` was 36932 packets/sec for the attack machine. The Zeek system detected over 400 connections to port 502 before it consistently crashed. We analyzed the timestamps in the traffic to see if there were more than a user-selectable threshold. Since Zeek was crashing, we set the threshold conservatively at 10 packets/sec. If Zeek was operating on a more powerful computer, more of the DoS attack would be logged and the threshold could be increased significantly. The amount of RAM available on our system was 4 GB with a 1.5 GHz processor which were fully consumed during the DoS experiment.

Table 2 DoS attack detection and response times

Test	Attack start (s)	Attack end (s)	SOAR detection (s)	SOAR response complete (s)
1	0	6	15	18 (3 s from detect)
2	0	6	19	21 (2 s from detect)
3	0	6	17	19 (2 s from detect)
4	0	6	18	20 (2 s from detect)
5	0	6	20	24 (4 s from detect)
6	0	7	25	28 (3 s from detect)
7	0	9	17	20 (3 s from detect)
8	0	9	17	20 (3 s from detect)
9	0	8	21	24 (3 s from detect)
10	0	6	20	23 (3 s from detect)

Zeek recommends the fastest processor and memory that can be afforded but does not specify minimum amounts. Future research to determine the thresholds is necessary based on varying RAM and CPU combinations. Once the threshold was reached, the packet information was scanned for the source IP with the most packets and this device was blocked using an ebtable rule, as described in the reconnaissance playbook. Originally, we did not filter on DER port, but the nmap reconnaissance was tripping the DoS detection, so the Elasticsearch query was updated to only include connections to port 502.

The results for 10 DoS attacks are shown in Table 2. The attacks took 5 s to execute but the attack harness operations–including writing Elasticsearch data twice–took between 6–9 s. DoS detection took 15–25 s and the response took 18–28 s. Interestingly, our DoS detection and response times were similar to those when Cloudflare responded to a massive HTTPS DDoS attack in 2022. Cloudflare's autonomous detection and software-defined response occurred in approximately 15 s of a DDoS attack using approximately 6,000 bots in 112 countries [49]. In comparison to the reconnaissance playbook, the longer detection times are from the additional time needed to query and send a bigger Zeek Elasticsearch dataset and then parse this data in the Python SOAR playbook. The response was the same (i.e., blocking the attacker IP with ebtable rule), so the response times port detection were similar to the reconnaissance playbook. As with most of the playbooks, if reducing detection and response times were critical to the SOC team, a potentially faster approach would be to perform all the detection analytics and response actions on the IDS system locally. This eliminates the networking and SSH response latencies. After the response, the SIEM could pull in the detection and response information to maintain situational awareness of the DER system. Finally, it would be good to add a watchdog process to the IDS to verify Zeek and other security services are running so the SOAR is not responsible for restarting IDS services.

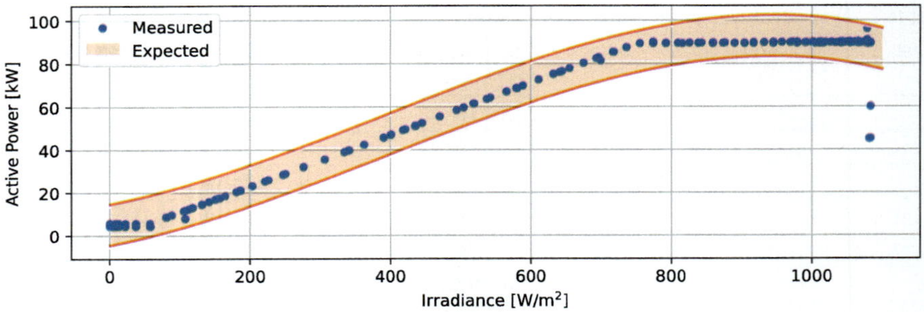

Fig. 8 Measured active power versus solar irradiance with expected power output range represented by data-driven polynomial function. If data is outside the band, it is classified as abnormal and likely the result of misconfigured DER settings or a damaged PV system

2.3 SOAR Playbook 3: Insider Threat

It is possible DER configuration changes could be made accidentally or maliciously by an operator or insider at the utility or 3rd-party. In those cases, there are valid Modbus, IEEE 2030.5, IEEE 1815, or proprietary protocol writes to the DER from a DERMS. In these cases, it is difficult to detect that adversary action from cyber data alone. But, one way to catch insider threats is to analyze the physical data for anomalies. In our testing, the PV site is assumed to be equipped with a plane-of-array pyranometer (irradiance sensor) that can be queried and used to validate the PV system's power output. To collect this data, a Python script was executed on the IDS that pulled the power and irradiance data via Modbus from the DER and pyranometer.

For the physical analysis, a custom function learned normal DER behavior, defined the expected power output of the PV array. The training data and expected output function are shown in Fig. 8 where solar irradiance, G, in W/m^2, is the x-axis and the active power, P, is the y-axis. The blue dots in Fig. 8 are measured values collected during a period of normal operations. During this period, the measured values followed a linear trend between about 50 W/m^2 and 750 W/m^2. After 750 W/m^2 the PV array reached a maximum output and never exceeded 90 kW. The knees in the curve represent (a) the startup power requirement at low irradiance, and (b) DER operations with a constant 0.90 power factor when the device reaches its current limit at higher irradiance levels. The figure also shows two outliers at high irradiance, which were likely due to errors in the DER AC simulation and were not used in the expected outcome function.

The expected outcome function, shown in Fig. 8 with the orange lines and shading, was a polynomial function fit to the measured data. This active power output was expected to resemble the function described in Eq. 1:

$$P = -0.00014G^3 + 0.16G^2 + 64G + 4031 \tag{1}$$

Table 3 Insider attack detection and response times

Test	Attack Start (s)	Attack End (s)	SOAR Detection (s)	SOAR Response Complete (s)
1	0	3	12	18 (6 s from detect)
2	0	3	12	16 (4 s from detect)
3	0	3	13	17 (4 s from detect)
4	0	3	11	15 (4 s from detect)
5	0	3	10	15 (5 s from detect)
6	0	3	13	18 (5 s from detect)
7	0	3	10	15 (5 s from detect)
8	0	3	13	18 (5 s from detect)
9	0	3	12	17 (5 s from detect)
10	0	3	5	10 (5 s from detect)

The range surrounding this function was set to be plus or minus 10% of the maximum observed active power.

This custom analysis was performed every 5 s to classify performance as either normal or abnormal. If abnormal for more than two subsequent calculations, the analysis sent a message to the detection tool log database in Elasticsearch, which would trigger the SOAR playbook to address the insider threat.

The SOAR system reverted DER control settings to a known-good "golden image" control state that was kept securely at the SOC. This golden image could be a set of default operating register values or the values from a previous, verified DER update. In the case of a more sophisticated DER management scheme, there may be many golden images for the DER devices. Securing the golden image(s) is paramount or the response will be ineffective. After reverting the DER control settings, the SOAR or SOC personnel would inspect internal log data to find the user who issued the DER changes and revoke their DERMS credentials until their intentions (accidental vs malicious) can be determined.

To test the efficacy of our tool, the test harness was programmed to set the SunSpec WMaxLimPctEna Register to 1 andWMaxLimPct to 20% of nameplate capacity. This was repeated 10 times every 2 min. The SOAR system continuously monitored the Elasticsearch database for indications in the last 30 s that there was an internal attack that changed the DER operations, but did not trip other systems. In this case, as shown in Fig. 7, the SOAR detection was triggered multiple times for a given attack because there was not a sufficient delay in the SOAR playbook after detection. This is harmless because it simply resets the DER control settings twice. Using the first detection and response, the attacks were completed in 3 s, detection was between 5–13 s, and response was completed in 10–18 s, as shown in Table 3.

Notably, one limitation of this approach is that these attacks will only be detected if DER production is high enough that the active power setpoint change produces a discrepancy in irradiance and power production. If an adversary changes this value in the early morning, late afternoon, or a very cloudy day, their actions may not be discovered for many hours, at which point it may be challenging to attribute the DER modification to a specific employee. Another approach, outlined in previous work [1], would be to compare settings to allowable ranges or the golden image. This inspection could be performed with deep packet inspection during Modbus (or other unencrypted protocol) setting changes or queried regularly from the DER device.

2.4 SOAR Playbook 4: Modbus Malicious Writes or Fuzzing

Control engineers and DERMS platforms know the correct Modbus holding registers to write for certain DER grid-support operations. Since these registers will not change for a given DER firmware version once a DERMS is correctly configured, there should be no out-of-bounds register writes unless an adversary is attacking or fuzzing the system. Fuzzing is an approach of sending random data to a device to cause it to malfunction. In our case, we created an attack using PyModbus which attempted to write to DER Modbus holding register 50,000, which did not exist on the server.

This type of attack could be detected multiple ways including reading Modbus exception codes in the network traffic, DER HIDS alerts [50], or, in our case, though Snort Modbus rules from Digital Bond QuickDraw [32]. QuickDraw is a set of ICS-specific Snort rules that generate alerts for multiple ICS protocols including Modbus and DNP3. The rules monitor traffic for attempted DoS attacks, malformed ICS packets, and ICS packets that have incorrect source and destination IP addresses. When the holding register write was attempted, the *Unauthorized Write Request* rule generated a Snort alert of the form:

```
04/21-10:04:09.270639 [**] [1:1111007:1] SCADA_IDS: Modbus TCP
- Unauthorized Write Request to a PLC [**] [Classification:
Potentially Bad Traffic] [Priority: 1] {TCP} 10.1.2.45:37833
-> 10.1.2.231:502
```

which was then captured by Filebeat, sent to Logstash to transform the data to match the database schema, and logged to an Elasticsearch index with Snort identified as the source of the entry. Once the SOAR playbook observed an *Unauthorized Write Request* Snort log in Elasticsearch, it parsed the message to determine the source IP and blocked this IP at the IDS using anebtable rule. Based on the results in Table 4, the attack harness completed the attack in 3–4 s, the detection was completed in 14–20 s, and response was completed in 18-24. There are many additional Snort rules that could be adopted in the IDS system based on the protocols being used to communicate with the DER devices. For instance, Snort includes a DNP3 (IEEE 1815) Preprocessor [51] for which rules can be created [32, 52].

Table 4 Modbus malicious writes or fuzzing attack detection and response times

Test	Attack start (s)	Attack end (s)	SOAR detection (s)	SOAR response complete (s)
1	0	4	17	21 (4 s from detect)
2	0	4	18	22 (4 s from detect)
3	0	4	17	21 (4 s from detect)
4	0	4	17	22 (5 s from detect)
5	0	4	20	24 (4 s from detect)
6	0	3	15	19 (4 s from detect)
7	0	4	16	21 (5 s from detect)
8	0	4	16	20 (4 s from detect)
9	0	4	14	19 (5 s from detect)
10	0	4	14	18 (4 s from detect)

2.5 *SOAR Playbook 5: Brute Force Attacks*

For DER equipment or other compute systems at the DER site, it is possible to brute force HTTP, SSH, Telnet, remote desktop, or other logins using simple brute force, dictionary, password spraying, or credential spraying attacks [53]. In many production DER devices and gateways, these services have been disabled, but in commercial and utility installations, there may be a jump host used for remote access to the site. In the jump host case, an additional IDS could be used to monitor connections to the OT network. DER sites where private keys or multi-factor authentication is absent and password hygiene is poor, brute force attacks could be highly effective.

To evaluate the IDS and SOAR system to brute force attacks, the first 100 passwords from theRockYou wordlist [54] was used to brute force the SSH password for userpi user using the commandhydra -l pi -P rockyou_100 -v -f ssh://10.1.2.231. Hydra then created 16 parallel SSH connections to the DER simulator attempting to log in with these common passwords. In this case, Filebeat and Logstash were added to the DER simulator to track failed authentication attempts in/var/log/auth.log and log them to Elasticsearch on the IDS. (Although, these attacks could also be detected on the BITW IDS using Zeek and filtering on port 22, if adding logging on the DER was not an option.) The elastic data entries were of the form,

Apr 25 13:09:11 raspberrypi sshd[30430]: pam_unix(sshd:auth): authentication failure; 'logname= uid=0 euid=0 tty=ssh ruser= rhost=10.1.2.45 user=pi

and if there were 5 or more failed logins detected from a remote host within the last 30 s, the brute force playbook was triggered. Once triggered, the rhost data was stripped from the Elasticsearch message and used to block the attacking system using the same technique from the other playbooks. In this case, the attack takes a significant time (73–112 s) and the detection and response is relatively quick at 7–13 and 9–16 s, as shown in Table5. This means that the attacking IP is blocked

Table 5 Brute force SSH attack detection and response times

Test	Attack start (s)	Attack end (s)	SOAR detection (s)	SOAR response complete (s)
1	0	73	9	12 (3 s from detect)
2	0	77	10	13 (3 s from detect)
3	0	77	9	12 (3 s from detect)
4	0	107	7	9 (2 s from detect)
5	0	81	12	15 (3 s from detect)
6	0	112	12	15 (3 s from detect)
7	0	79	13	15 (2 s from detect)
8	0	78	10	12 (2 s from detect)
9	0	100	8	11 (3 s from detect)
10	0	82	13	16 (3 s from detect)

before the Hydra attack can attempt all the passwords in the wordlist, which is the primary reason the attack takes so long: the attacker IP is blocked mid-attack at the IDS, and the SSH connections begin to fail. This causes Hydra to try multiple SSH login attempts before reporting connection errors. At this point, the DER SSH logs should be analyzed to see if there have been any successful logins from the attacking IP, and if so, that password should be changed enterprise-wide.

2.6 *SOAR Playbook 6: Machine-in-the-Middle*

Machine-in-the-Middle (MITM) attacks occur when an adversary reads or modifies data between two endpoints. One MITM method involves using Address Resolution Protocol (ARP) spoofing or ARP poisoning, wherein the attacker continuously broadcasts ARP messages that link the attacker's MAC address to the IP address of another host. If this is successful, the network switch will send traffic to the attacker as opposed to the legitimate destination. Once the attacker gets the data, they may read those messages or modify them before sending them on to the legitimate endpoint destination. For our demonstration, all DER and IDS traffic was redirected to the attacker. Ettercap was used for 15 s using the following MITM ARP spoofing command:

```
(sleep 15; killall ettercap)& sudo ettercap -T -iface eth0 -M
arp:remote /10.1.2.231// /10.1.2.34//
```

For those 15 s, all traffic to and from the IDS and DER will be routed through the attacker. There was no Ettercap filter applied to modify the data, but the attacker could view the data-in-flight. The IDS system was running`arpalert` to detect when the ARP table was modified. If it was changed, there would be an alert similar to,

Table 6 MITM attack detection and response times

Test	Attack start (s)	Attack end (s)	SOAR detection (s)	SOAR response complete (s)
1	0	19	6	8 (2 s from detect)
2	0	19	6	8 (2 s from detect)
3	0	19	8	9 (1 s from detect)
4	0	18	10	12 (2 s from detect)
5	0	19	6	8 (2 s from detect)
6	0	19	5	7 (2 s from detect)
7	0	19	7	9 (2 s from detect)
8	0	19	6	8 (2 s from detect)
9	0	19	7	9 (2 s from detect)
10	0	19	10	12 (2 s from detect)

```
Apr 22 11:40:10 raspberrypi arpalert[27236]: seq=451,
mac=dc:a6:32:bf:82:b0, ip=10.1.2.34, reference=,
type=mac_change, dev=br0, vendor="(null)"
```

which was detected by the MITM SOAR playbook. Once triggered, the SOAR system would blacklist the attacker IP at the IDS and on any networking equipment (e.g., switches, firewalls, etc.). For this experiment, the attacker was firewalled on the IDS only. The attacks consistently took 15 s, but the attack harness required 18–19 s to complete, as shown in Table 6. The detection time was 5–10 s and the response was 7–12 s. Notably, modern switches include similar features to prevent ARP spoofing. As an example, Cisco has a feature called ARP anti-spoofing [55]. This attack is also only successful if the attacker is on the local OT network. A remote attacker would first need to gain access to the DER site (possibly via an SSH or VPN connection) and then perform the spoofing attack.

3 Discussion

The SOAR approach successfully defended the DER device from reconnaissance operations and five different cyberattacks. By automating the detection and response mechanisms with IDS tools and SOAR playbooks, the SOC responded more quickly than a human could. The minimum, maximum, and average detection and response times for the cyber events are shown in Table 7. The worst-case time required to identify and trigger the SOAR playbook was 25 s for a DoS cyberattack. Even then, the response was completed in less than 30 s. From a DER operational standpoint, it is unlikely this disruption in availability would impact the power system or other operations, since the default IEEE 2030.5 monitoring post rate is 5 min and default polling rate for DERControls is 10 min per the Common Smart Inverter Profile

Table 7 Summary of cyberattack detection and response times

Attack	SOAR detection time (s)			SOAR response time (s)		
	Minimum	Maximum	Average	Minimum	Maximum	Average
Reconn.	5	9	7.5	8	12	10.5
DoS	15	25	18.9	18	28	21.7
Insider	5	13	10.1	10	18	15.9
Fuzzing	14	20	16.4	18	24	16.4
Brute Force	7	13	10.3	9	16	13.0
MITM	5	10	7.1	7	12	9.0

(CSIP) [56]). Automating the response also eliminated the risk of human error, such as accidentally blocking the wrong IP address(es). Therefore, the SOAR performance is encouraging and it is recommended that DER owners, operators, and monitoring organizations begin exploring this technology to improve the cyber-resilience of distributed energy assets.

In terms of implementation, the IDS could be deployed directly on the DER system or be installed on a DER gateway–if protocol translation was needed, e.g., converting between IEEE 2030.5 and SunSpec Modbus. This would eliminate the hardware expense and added complexity of integrating another device or devices at DER sites. However, the additional processing and storage requirements would need to be carefully considered in offloading the IDS operations to the DER or gateway. For example, adding the ability to wipe Elasticsearch logs after 30-90 days would help with the storage issues.

This work did not explore the robustness of the detection and response algorithms to different attacks. The detection mechanisms and playbooks were carefully crafted for the attacks on the DER system, but SOC teams and systems would not have this a priori knowledge in operational environments. Future research is necessary to investigate playbook resilience to a diverse set of cyberattack types, parameters, and sources. Similarly, the prevalence of false positives in normal, long-term operations was outside of the scope of the current work. Multi-day studies with DER owner or operator operations (e.g., monitoring production, changing settings, updating firmware) is needed to characterize the risk of falsely flagging normal activity as malicious.

Cybersecurity is an arms race. Adversaries are continuously developing new attack tools that can disable or otherwise impact OT systems or sideline IDS or SOAR tools. The recently-discovered ICS malware called PIPEDREAM includes multiple zero-day vulnerabilities affecting Programmable Logic Controllers [57]. If similar malware was designed for DER systems, it is unlikely any IDS tools would flag the attack and no preprogrammed SOAR playbook would know the appropriate response even if the attack was detected. For this reason, there is no replacing human threat hunters within these organizations. However, if known threats can be automatically disarmed, quarantined, or otherwise mitigated, that frees up the SOC team to focus on unknown threats to their equipment and networks. New malware is not only targeting ICS targets but security tools as well. A new vulnerability was recently discovered in the Snort Modbus preprocessor that "DoSs" the tool by putting it in an infinite while loop [58]. Thus, as the defensive sophistication and automation improves, any new tools and features must be secured as well or they present new attack vectors. As another example, the SolarWinds Orion Platform–an IT administration platform designed to simplify network monitoring and management–was compromised with a supply-chain attack that exposed hundreds of organizations to malicious backdoor access [59]. Therefore, any DER security products need to be carefully designed and implemented to reduce the risk of compromise.

4 Conclusion

This work demonstrated an implementation of a DER SOC with SIEM and SOAR systems. Six different adversarial cybersecurity operations were successfully identified and mitigated with several integrated monitoring, detection, and response tools located on the DER device, BITW IDS, and SOC. NIDS and HIDS data was stored in an Elasticsearch database hosted on the IDS, which was queried by the SOAR to trigger several automated playbooks that could mitigate the attack. While the specific implementations would need to be tailored to the operational environment, this lightweight approach appears to be effective to increase detection, response, and recovery time for cyberattacks. All attacks were detected and mitigated in under 30 s. Future work should explore a wider range of attacks and identify the challenges of incorporating these tools in an operational DERMS environment.

Acknowledgements Sandia National Laboratories is a multi-mission laboratory managed and operated by National Technology and Engineering Solutions of Sandia, LLC., a wholly owned subsidiary of Honeywell International, Inc., for the U.S. Department of Energy's National Nuclear Security Administration under contract DE-NA-0003525.

References

1. Johnson, J., Quiroz, J.E., Concepcion, R.J., Wilches-Bernal, F., Reno, M.J.: Power system effects and mitigation recommendations for der cyberattacks. IET Cyper-Phys. Syst. Theory Appl. **4**, 240–249 (2019)
2. FireEye Mandiant, "M-trends 2021," special report, FireEye Mandiant Services, 601 McCarthy Blvd. Milpitas, CA 95035 (2021)
3. Rose, S., Borchert, O., Mitchell, S., Connelly, S.: NIST SP 800-207: Zero trust architecture. Special publication, National Institute of Standards and Technology, 100 Bureau Drive (Mail Stop 8920) Gaithersburg, MD 20899-8920 (Aug 2020)
4. R. Pompon, *Assume Breach* (Apress, Berkeley, CA, 2016), pp. 13–21
5. J. Johnson, Roadmap for photovoltaic cyber security, Technical report SAND2017-13262, Sandia National Laboratories, Albuquerque, New Mexico 87185 and Livermore, California 94550 (Dec 2019)

6. A. Ghosh, To level up your soc game, take one logical step at a time (13 Jan 2020)
7. A. Chuvakin, A simple soar adoption maturity model (14 Jun 2022)
8. T. Connect, Security operations maturity model (10 Jun 2022)
9. Rapid7, Security orchestration and automation (soar) playbook. Rapid7 (15 Apr 2022)
10. C. Rieger, J. Gentle, A. Bochman, J. Miller, Distributed renewables cyber resilience. POWER Mag. (18 Apr 2022)
11. J. Gentle, Cybersecurity reference architecture for renewable energy: survey of vendors and application space. INL (18 Apr 2022)
12. S. Hossain-McKenzie, J. Johnson, Cyber-physical solutions for defending pv systems: proactive intrusion detection and mitigation system sensor, in *SunSpec/Sandia DER Cybersecurity Workgroup Webinar Series* (Aug 2021)
13. SOAR Solutions Forum 2022. SANS Institute
14. A. Kirk, Soar solutions: best practices & benefits of automation," in *SOAR Solutions Forum 2022* (Mar 2022)
15. Splunk Community Playbooks. GitHub
16. Cortex XSOAR Platform-Content Repository. GitHub
17. P. Fonash, P. Schneck, Cybersecurity: from months to milliseconds. Computer **48**(1), 42–50 (2015)
18. G. Fragkos, J. Johnson, E.E. Tsiropoulou, Dynamic role-based access control policy for smart grid applications: an offline deep reinforcement learning approach. IEEE Trans. Human-Mach. Syst. 1–13 (2022)
19. M.F. Haque, R. Krishnan, Toward automated cyber defense with secure sharing of structured cyber threat intelligence. Inf. Syst. Front. **23**(4), 883–896 (2021)
20. S.Y. Enoch, Z. Huang, C.Y. Moon, D. Lee, M.K. Ahn, D.S. Kim, HARMer: cyber-attacks automation and evaluation. IEEE Access **8**, 129397–129414 (2020)
21. C.B. Jones, A.R. Chavez, R. Darbali-Zamora, S. Hossain-McKenzie, Implementation of intrusion detection methods for distributed photovoltaic inverters at the grid-edge, in *2020 IEEE Power Energy Society Innovative Smart Grid Technologies Conference (ISGT)* (2020), pp. 1–5
22. C.B. Jones, A. Chavez, S. Hossain-McKenzie, N. Jacobs, A. Summers, B. Wright, Unsupervised online anomaly detection to identify cyber-attacks on internet connected photovoltaic system inverters, in *2021 IEEE Power and Energy Conference at Illinois (PECI)* (2021), pp. 1–7
23. A. Chavez, C. Lai, N. Jacobs, S. Hossain-McKenzie, C.B. Jones, J. Johnson, A. Summers, Hybrid intrusion detection system design for distributed energy resource systems, in *2019 IEEE CyberPELS (CyberPELS)* (2019), pp. 1–6
24. J. Jow, Y. Xiao, W. Han, A survey of intrusion detection systems in smart grid. Int. J. Sen. Netw. **23**, 170–186 (2017)
25. C. Lai, A. Chavez, C. Jones, N. Jacobs, S. Hossain-McKenzie, J. Johnson, A. Summers, Review of intrusion detection methods and tools for distributed energy resources, Technical report SAND2021-1737, Sandia National Laboratories, Albuquerque, New Mexico 87185 and Livermore, California 94550 (Feb 2021)
26. J. Johnson, L. Jencka, T. Ortiz, C. Jones, A. Chavez, B. Wright, A. Summers, Design considerations for distributed energy resource honeypots and canaries. Sandia Technical Report, no. SAND2021-11609
27. IEEE Standard for Interconnection and Interoperability of Distributed Energy Resources with Associated Electric Power Systems Interfaces. IEEE Std 1547-2018 (Revision of IEEE Std 1547-2003) (2018), pp. 1–138
28. J. Johnson, B. Fox, K. Kaur, J. Anandan, Evaluation of interoperable distributed energy resources to ieee 1547.1 using sunspec modbus, ieee 1815, and ieee 2030.5. IEEE Access **9**, 142129–142146 (2021)
29. S. Hossain-McKenzie, A. Chavez, N. Jacobs, C.B. Jones, A. Summers, B. Wright, Proactive intrusion detection and mitigation system: Case study on packet replay attacks in distributed energy resource systems, in *2021 IEEE Power and Energy Conference at Illinois (PECI)* (2021), pp. 1–6
30. Vern Paxon, Zeek: An Open Source Network Monitoring Tool. Online

31. Martin Roesch, Snort. Online
32. Digital Bond, Quickdraw Snort. GitHub
33. Steffen Siering, Filebeat: Lightweight shipper for Logs. Online
34. Steffen Siering, Logstash: Centralize, transform & stash your data. Online
35. Shay Banon, Elasticsearch: Search. Observe. Protect. Online
36. Grafana Labs, Grafana
37. Elasticsearch B.V., Kibana (2022)
38. G. Lyon, Nmap: the Network Mapper. nmap.org
39. Salvatore Sanfilippo, hping3
40. RiptideIO, PyModbus. GitHub
41. SunSpec Alliance, PySunSpec2. GitHub
42. van Hauser, Hydra
43. A. Ornaghi, M. Valleri, Ettercap
44. R. Scalco, B. Waugaman, More situational awareness for industrial control systems (MOSAICS), in *Integrated Cyber* (Mar 2018)
45. I.S. Organization, Iso 8601-1:2019 date and time-representations for information interchange-part 1: basic rules, *ISO* (2019)
46. Bart De Schuvmer, ebtables: firewall tool for Linux bridges. Online
47. R. Joven, E. Ananin, DDoS-for-Hire Service Powered by Bushido Botnet. https://www.fortinet.com/blog/threat-research/ddos-for-hire-service-powered-by-bushido-botnet- (Oct 2018)
48. S. Lyngaas, Utah renewables company was hit by rare cyberattack in March. https://www.cyberscoop.com/spower-power-grid-cyberattack-foia/ (Oct 2019)
49. O. Yoachimik, J. Desgats, Cloudflare blocks 15m rps https ddos attack. Cloudflare, (27 Apr 2022)
50. G. Ravikumar, A. Singh, J.R. Babu, A. Moataz, M. Govindarasu, D-ids for cyber-physical der modbus system-architecture, modeling, testbed-based evaluation, in *2020 Resilience Week (RWS)* (2020), pp. 153–159
51. Snort FAQ. (Online)
52. H. Li, G. Liu, W. Jiang, Y. Dai, Designing snort rules to detect abnormal dnp3 network data, in *2015 International Conference on Control, Automation and Information Sciences (ICCAIS)* (2015), pp. 343–348
53. Crowdstrike, Brute Force Attacks (Mar 2021). (Online)
54. Wordlists. (Online)
55. Cisco, Security configuration, cisco catalyst pon series switches. Cisco (10 Dec 2021). Online
56. Common Smart Inverter Profile: IEEE 2030.5 Implementation Guide for Smart Inverters, Version 2.1 (2018)
57. Dragos, Pipedream: Chernovite's emerging malware targeting industrial control systems. Dragos (21 Apr 2022). (Online)
58. Cisco, Multiple cisco products snort modbus denial of service vulnerability. Cisco (21 Apr 2022). (Online)
59. SolarWinds, Solarwinds security advisory. SolarWinds (6 Apr 2021). (Online)

A Study on Cyber-Physical System Architecture for Smart Grids and Its Cyber Vulnerability

N. Rajeswaran, M. Lakshmi Swarupa, Rekharani Maddula, Hassan Haes Alhelou, and Vajjala Kesava Vamsi Krishna

Abstract Technological advances resulted in overwhelming usage of electronic gadgets across the world which further resulted in increased load demands. The obvious challenges are blackouts, overloads and voltage sags. One possible and trusted solution for these problems is Smart grid. A power network supported by digital communication technology is termed as Smart grid. But this solution is not that simple as it seems to be. There are few challenges associated with this solution like cyber vulnerabilities and cyber-attacks. Therefore, an effort is made here to review comprehensively the research work carried out till date, encompassing different heuristic detection and estimation techniques. It is a convenient factor that issues in Smart grid like relay protection, Power flow control, grid security and reliability can effectively be modelled and efficiently be analyzed. An effort is made to identify the dependencies among Cyber Physical System controls in this work. There is an immediate need to protect the communications and computations carried out by the digital communication equipment in Smart grids from cyber-attack. An overview of current research efforts in expanding the Smart grid applications and its infrastructure security is presented here. The conclusions submitted here present the further scope of the research work.

N. Rajeswaran (✉)
Electrical and Electronics Engineering, Malla Reddy Institute of Engineering and Technology, Maisammaguda, Secunderabad, Telangana, India
e-mail: rajeswarann@gmail.com

M. L. Swarupa
Electrical and Electronics Engineering, CVR College of Engineering, Hyderabad, Telangana, India

R. Maddula
Department of Physics, Gokaraju Lailavathi Womens Engineering College, Hyderabad, Telangana, India

H. H. Alhelou
Department of Electrical and Computer System Engineering, Monash University, Melbourne, Australia

V. Kesava Vamsi Krishna
Department of Physics, Malla Reddy Engineering College, Maisammaguda, Secunderabad, Telangana, India

H. Haes Alhelou et al. (eds.), *Power Systems Cybersecurity*, Power Systems,
https://doi.org/10.1007/978-3-031-20360-2_17

Keywords Smart grid · Cyber physical system · Cyber vulnerabilities · Cyber-attacks · Cyber security

1 Introduction

Cyber Physical Systems (CPS) are heterogeneous technologies that deal with interactions between computing and communications systems and the regulation of related physical dynamics [1, 2]. CPS involves leveraging the economy through the integration of computing and communication technology with physical systems. The CPS research is behind, however, because it is still in its early stages and as a result lacks the requisite standards and system design. Another difficulty is that CPS's integration with the public internet raises serious questions about its security. This analysis analyses expected cybersecurity issues in addition to focusing on architectural modelling by classifying CPS [3, 4]. Additionally, this research has attempted to separate out a number of the problems and difficulties facing CPS [5].

Broadly speaking, the term "Smart Grid" (SG) refers to the integration of emerging IT and other technologies with large-scale power networks with the goal of ensuring effective, efficient, commercially viable, and environmentally sustainable electricity generation, transmission, distribution, and use [6, 7]. The conceptual model recommended by the National Institute of Standards and Technology of the United States puts forth seven important domains: bulk generation, transmission, distribution, customers, service provider, operations, and markets. SG refers to the change of the electric business in the USA from a producer-controlled network to one that is somewhat customer interactive [8–10].

Modern SGs are made up of modern communications systems, energy storage devices, electric vehicle charging stations, and distributed renewable energy resources [11]. The typical electric grids, in contrast, were developed prior to any of these technological developments. As a result, a thorough restructuring is necessary to convert to a Smart Grid that includes computing resources and connected devices. Energy production, delivery, consumption, and storage were all optimized with the use of computing resources. The urgent and challenging task is to create diversified Smart Grid architecture that implements secure communication between various SG infrastructure tiers [12–14].

1.1 Issues in SG's

With the development of internet-connected applications in SGs, significant advancements are being seen on a variety of levels, including better consumption management, production optimization, and enabling utilities to provide power with better control and reduced costs (Fig. 1). On the other hand, the same internet that provided the SGs with the benefits listed above has also opened up opportunities for cyberattacks that will disrupt the SGs [15]. The hazards vary from minor and imperceptible effects on SG caused by straightforward viruses to more significant attacks that could result in the animated freeze of the entire system. Nuclear facilities in particular are now vulnerable to everything from little viruses made by solitary people to massive and intricate cyberattacks coordinated by larger organizations [16].The development of networks and computing in recent years has made life for the average person much more comfortable. Information systems and physical objects are increasingly being integrated with more focus as a result of developing technologies like data science, the Internet of Things, cyber security, etc. [17–19]. The development of Cyber Physical Systems (CPS) was aided by these circumstances, which naturally caught the interest of governments, businesses, academic institutions, and universities.

Through the use of appropriate feedback loops, embedded computers and networks in CPS monitor and regulate the physical processes. As a result, both the computation and these physical processes have an impact. Computation, Communication, and Control are CPS's three primary building blocks (3C). Global academic academics began to pay attention to CPS, which also became an emerging topic in the industry [20].

CPS security is a critical issue because it affects the system's stability. As a result, the first focus of this work is on the CPS Security model, which also contains the

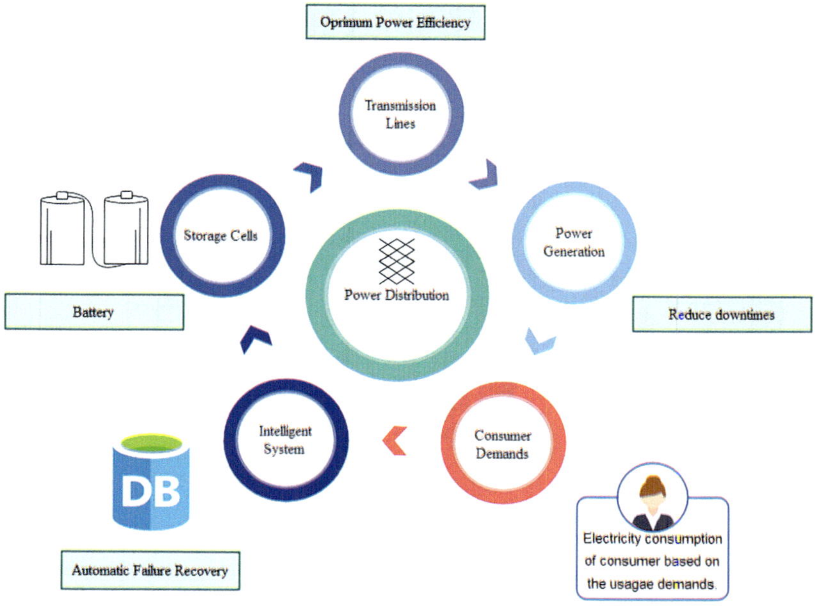

Fig. 1 Smart grid transmission network

pertinent theories, the CPS architecture, attacks and defence analysis, and application fields [21]. Security goals and the exposition of fundamental theories against the background of security theory follow. A summary of the assaults on the three layers of the CPS architecture is also provided. The attempt to produce a literature map of diverse application fields with their potential classification comes as a final step [22]. With the expanding usage of digital technologies in power grids, there is a growing need for increased networking capabilities. To proactively identify and stop the security threats, new tactics should be created.

CPS involves the design of electronic systems, Real Time Systems (RTS), and software [23]. A further essential component of CPS is the thoroughly linked sensors, computing systems, and actuators. The CPS network provides connectivity to WLANs and WWLANs, or wide-area wireless access networks. Several layers of heterogeneous CPS were proposed by Jia Shen et al. [24]. The evolution of new generation CPS with the fusion of science, technology, and engineering was made possible by the networked computing systems integrated with the physical equipment/processes. The connection of network communication and communication with the physical processes in CPS benefits the optimization of the physical processes. The diagram (Shown in Fig. 2) depicts the interplay between the human, physical, and computer elements of CPS.

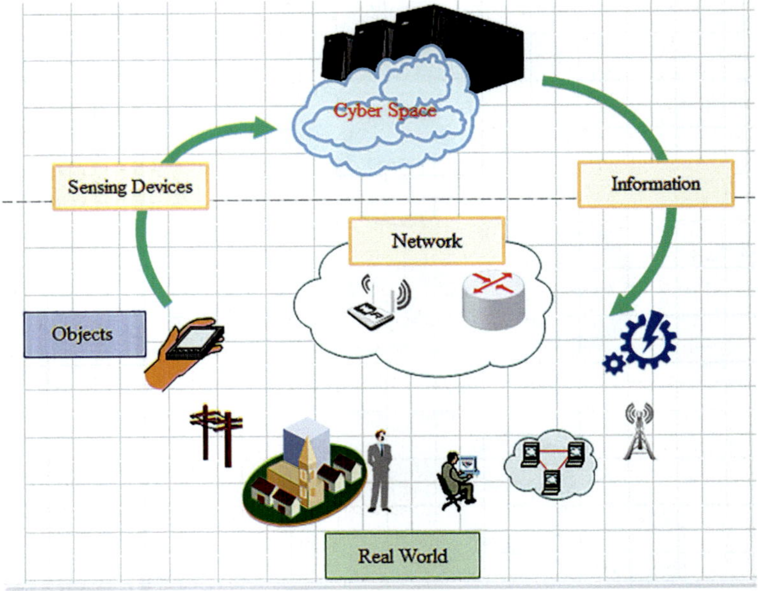

Fig. 2 Basic CPS model

2 CPS Architecture

The original CPS architecture included two layers: a physical layer and a cyber layer. While the cyber component handled the work of evaluating and processing the data from the physical part and sending the required orders, the physical part was responsible for sensing the physical environment, gathering data, and carrying out commands that it got from the cyber part. New CPS architectures that weren't hierarchical were introduced in a few papers. Three-tiered CPS architecture was attempted to be introduced [25] with a very hazy description. The three-layer CPS architecture, comprised of an execution layer, a transport layer, and a control layer, is depicted in Fig. 3 of this study. The execution layer is made up of several physical components, including sensors, actuators, RFID tags, readers, and more. The physical environment data collection and command execution are handled by this layer. Internet, LAN, and communication networks are all included in the transport layer [26]. In addition to handling data management and processing, this layer deals with real-time data transfer. The control layer serves as a link in the interaction between the cyber and physical components. The control layer gathers information from the execution layer's devices and uses that information to generate the appropriate commands. The control layer's application support layer uses middleware, entertainment platforms, cloud computing platforms, service support platforms, etc. Based on the suggested CPS architecture, this proposed design makes an effort to categories attacks for the execution, transport, and control layers [27].

Execution layer attacks target nodes such as sensors and actuators, whereas transport layer attacks target nodes such as those involved in large-scale data integration and leakage or destruction to data. Loss of user privacy, improper access control procedures, and insufficient security standards are all examples of control layer assaults. In terms of CPS application domains, researchers frequently concentrated on the SG, medical device, automotive, and aerospace industries [28]. Security is a crucial and essential key problem in the context of CPS. While administering CPS locally and on a smaller scale, it would appear that security is not a serious problem. However, if the same system implementation is expanded via the internet, several security flaws may be exposed. Borg claims that hackers are increasingly focusing their attacks on industrial machinery, particularly process control devices like Programmable Logic Controllers and local networks.

Due to a potential quality control failure, this may have an impact on the stock price of the impacted company. As a result, business executives are increasingly coming under cyber hackers' attention. When compared to the money made by cybercriminals by manipulating a company's stock price, credit card fraud is a relatively insignificant source of income. With the development of the Internet of Things, these security threats are growing and causing difficulties. When more physical infrastructures are connected to the internet, systems will become even more vulnerable.

The networks for the Internet of Things and Cyber-Physical Communication are vulnerable to the following six urgent problems:

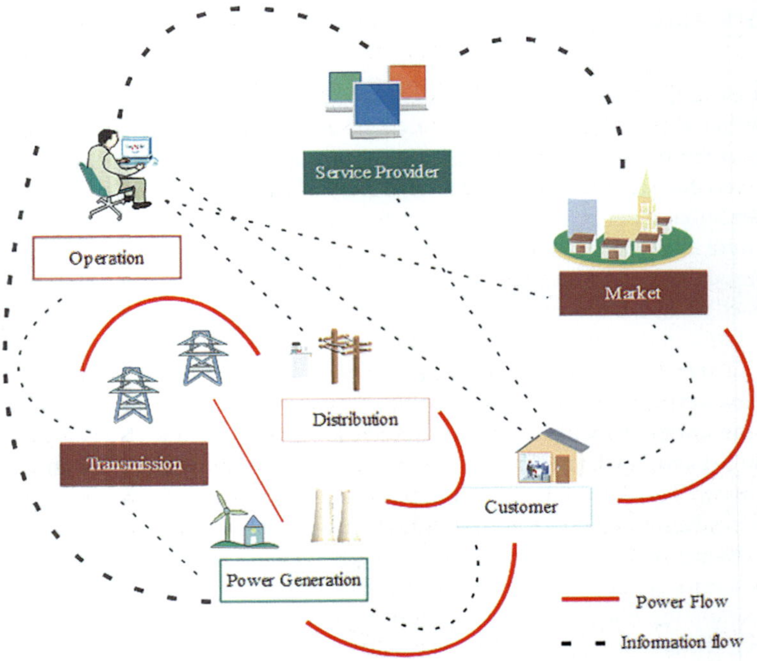

Fig. 3 Functionality of smart grid CPS

- When digital autos or infusion pumps are targeted, the effects of a security failure would be more severe and undoubtedly urgent. The lives of some people might also be destroyed as a result.
- The widespread hacking system is a full-scale cyber conflict. The worst aspect is that the hackers won't be looking to steal money or cause trouble, but rather to make IoT security vulnerable to unique threats.
- Due to unprecedented threats from hackers who could result in security vulnerabilities, some software vendors and chip makers recently offered 10-year and 7-year support terms for Internet of Things (IoT) products, whereas others either limit their support to 2 or 3 years or don't even provide any specific support contract at this time.
- Economics is another problem; in some cases, a connected product with low profit margins may need patches, updates, and security assessments; these must add costs to the product and have an impact on the profit; alternatively, if the updates are not made, they may result in security vulnerabilities.
- The frightening truth that the weak link is the vulnerability is Corman's sixth justification. When connected devices are developed using firmware, software, and hardware by many firms, the vulnerability could be caused by the telematics manufacturer failing to update the software.

- The sixth justification relates to connected devices in real-world settings, which are different from any IT system. For instance, in smart homes, there is no software expert or manager to update connected refrigerators.

The success of IoT/CPS operations in the Cyber Communication sector depends on tackling cybersecurity, which is crucial for preserving the aforementioned as well as protecting large plants, networks, and establishments.

3 CPS in Smart Grids

Smart grids can lower energy prices, assist in preventing blackouts, and thwart cyber-attacks. Smart grids are power systems that adjust to changes in demand and reconfigure as necessary to avoid overloads and other difficulties. They also provide novel, more difficult issues [29, 30]. Massive amounts of sensor data from stations must be efficiently transferred and properly evaluated in real time as power generation transitions from centralized power stations to distributed and heterogeneous systems.

3.1 Smart Grid Cyber Physical System

The SG is presented in the next part from the viewpoint of the CPS. The conventional power grid was not intended to be flexible enough to accommodate future changes like smart metering and monitoring, control, the integration of renewable energy sources, etc., whereas the SG modifies the current grid in a way that incorporates disciplines like generation, transmission, distribution, consumers, operation, and market in order to enable real-time grid monitoring. A example CPS-based SG is depicted in Fig. 3 [31]. SG-CPS plans to combine computational and physical components in order to monitor, process, and govern the Grid in real time. The SG-CPS consists of physical systems and cyber systems, with physical systems include power network infrastructures and sensors [32]. The customer domain, market domain, service provider domain, operational domain, bulk generating domain, transmission domain, and distribution domain are the seven linked domains that make up SG, according to NIST. The aforementioned domains cover sensors, actuators, intelligent electronic devices (IEDs), smart metres, distributed energy resources (DERs), and other physical and cyber components. The data exchange between the connected physical components is made possible by cyber components like software, communication networks, etc.

- Generation system
  Coal, thermal, nuclear, hydrothermal, and other conventional power-generating methods are used on a large scale. Small-scale non-conventional power-generating methods include wind, solar, geothermal, and biomass. Traditional and renewable generation sources will be integrated by SG-CPS in order to address the issue of natural resource depletion in traditional energy sources. Even the energy produced

by individual users can be integrated into the main grid, according to the roadmap for systematic integration.

- Transmission system
  Transmission lines should be used to carry the energy produced by the generation systems indicated above. Therefore, creating meaningful connection between the generation system and transmission lines for the SG-CPS becomes a crucial aspect. This is accomplished via the vital power infrastructure of the transmission system that supplies electricity to loads or microgrids.
- Distribution system
  The transmission line's high voltage is lowered to the distribution voltage at the substations. The distribution system thus joins the consumer system and the transmission system. The Distribution system is made up of the Smart Metering System, DERs, and other loads. The effectiveness and efficiency of the SG system can be increased by the usage of microgrids.
- Operation
  Since operation domain oversees the five SG-CPS essential functionalities, it is regarded as the SG's Manager. Detection, Response, Recovery, Identification, and Protection make up this group. It is one of the most crucial domains in SG-CPS and is in charge of maintaining the Power system's health.
- Service provider
  As its name suggests, this domain is in charge of offering services to users. Due to the vulnerability of this domain to cyberattacks, numerous measures need to be incorporated here.
- Market
  The issues like offering services at reasonable prices, offering security measures, etc. are taken care of in this arena. It is also in charge of creating a market that is competitive for the SG-CPS.
- Customer
  The end-user entity who uses the electricity is often the customer. The SG-CPS, however, offers the user the opportunity to produce (from renewable sources), control, and store electricity that can be included into the SG-main CPS's grid. Customers are divided into three groups: Industrial, Commercial, and Residential, depending on their need for and usage of power.

3.2 Function of SG-CPS

The SG-core CPS's functions are heavily reliant on communication networks and infrastructure. The communication network is responsible for a number of crucial tasks, such as:

- Real-time data interchange amongst the various organizations of the SG regarding grid health.
- Real-time monitoring and grid control in a dynamic environment. to deliver consistent, dependable, secure, and safe service.

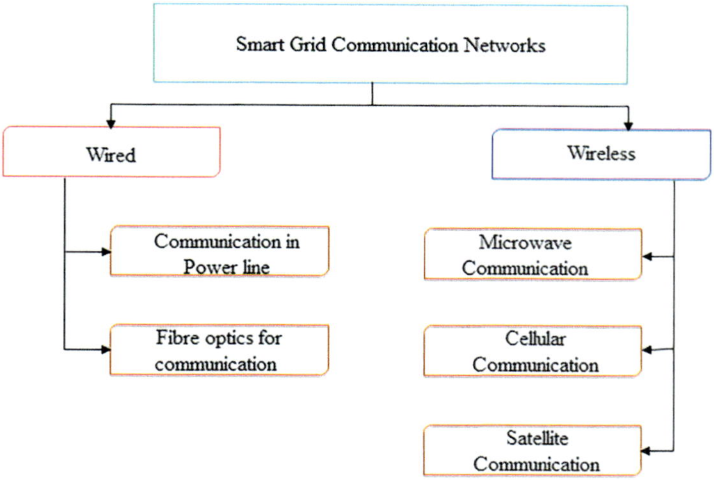

Fig. 4 SG communication network technologies

- Defense of many crucial SG components, including as the transmission, distribution, and generation infrastructures.

Figure 4 illustrates how the SG-CPS communication infrastructure needs differ from standard communication requirements. For instance, in addition to the needs of normal cellular-based commercial applications, the SG-CPS requires less delay, less packet drop, higher dependability, high resilience, better scalability, high availability, etc. [33]. The CPS communication infrastructure and its requirements are thoroughly surveyed [14]. One can infer that the communication network infrastructure in the context of SG-CPS is not extensively examined.

4 Modeling and Stability Analysis of Integrated Cps

The intricate relationships between software and hardware as well as the complex, ongoing interactions between the environment and the system in SG-CPS make it difficult to apply conventional testing methods. To lessen these difficulties, we suggest performing testing during the early stages and using executable models of the system and its surroundings. This does not imply that there aren't any challenges with the CPS model testing. Figure 5 illustrates how the complexity and variety of CPSs make it necessary to combine many modelling formalisms in order to create accurate models of each component. Additionally, the co-simulation process must be controlled, allowing it to accurately replicate the desired results, and quick, allowing thousands of simulations to be run in practical time. Mathematical modelling plays

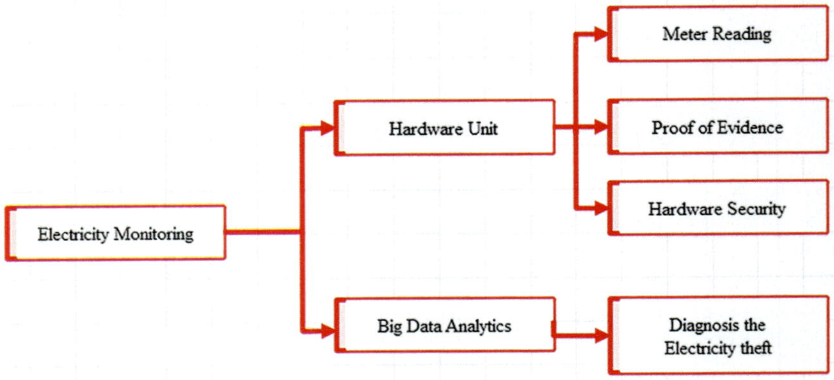

Fig. 5 A sample integrated CPS architecture

a crucial role in the design of any complex system because it generates simulation data that makes it possible to identify faults.

Mathematical modelling is crucial for the design of any complicated system, but it's especially crucial for physical systems because they can never be deterministic. Another significant worry is the system's stability. Due to the communication delays that may occur when CPS is used to manage and monitor a critical situation, packet loss during data transfer is another possibility. As a result, system stabilization would need to be given the attention it deserves in these system design concerns. The modelling and stability of CPS using the proposed Passivity Model have been studied in this research. When a storage function is present and the energy of the stored system is constrained by the.

4.1 Cyber-Attack System Modeling and Analysis

Control system networks that contain vital control systems are increasingly being connected to enterprise networks, which increases the risk of cyberattacks on those networks. In the current world, things are becoming integrated, and as the integration develops, protecting these systems becomes more crucial. Some particular examples of CPS are smart grids, pervasive healthcare systems, unmanned air vehicles, etc. The overall danger of the CPS would unquestionably be considerably higher than that of the component systems when systems are created and integrated to create it. Attacks on software have recently moved to embedded systems, and occurrences like the Stuxnet attack, which target automation systems, are quite likely since computational and physical dynamics are being connected to the internet these days.

Because the majority of CPS models employ their own proprietary protocols, there haven't been many attacks against CPS to date. However, as CPS becomes increasingly integrated with the internet, more attacks should become common place. With the link to the public internet, security would be more at risk; this would have been a much bigger worry with the new internet concept, i.e. the Internet of Things or the Internet of Everything. As more physical systems and facilities are connected to wireless sensor networks, IoT vulnerabilities are created (WSN). National infrastructure CPS systems, including national power grids, smart energy systems, sophisticated metering infrastructures, etc., are becoming more and more vulnerable as cyber security incidents have received more and more credibility as real concerns [34].

Attacks on CPS should receive considerable consideration because they involve physical systems, machinery, people, expensive establishments, and crucial infrastructures. As a result, the damage would undoubtedly be greater and might not be repairable [35]. Pervasive computing raises more security and safety concerns, although these concerns are still very important. Preschern et al. developed the one-out-of-two security model based on the study by Kai Hansen, which includes potential assaults, outcomes, and discusses attack scenarios from the attacker's perspective (Fig. 6).

Because the majority of CPS models employ their own proprietary protocols, there haven't been many attacks against CPS to date. However, as CPS becomes increasingly integrated with the internet, more attacks should become commonplace. With the link to the public internet, security would be more at risk; this would have been a much bigger worry with the new internet concept, i.e. the Internet of Things or the Internet of Everything. As more physical systems and facilities are connected to wireless sensor networks, IoT vulnerabilities are created (WSN).

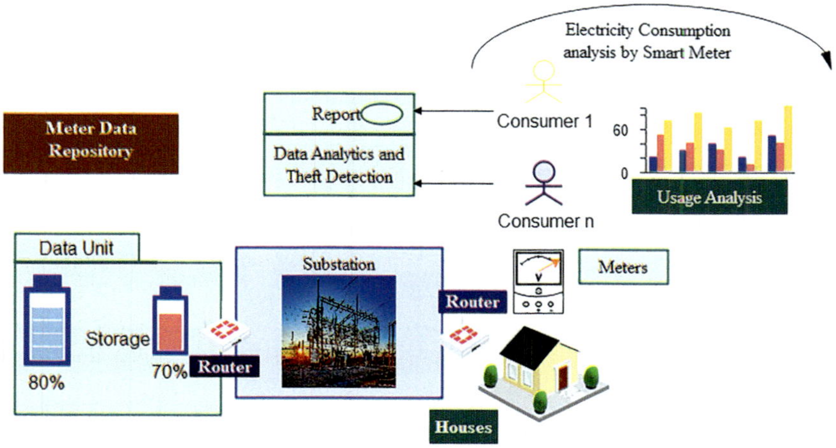

Fig. 6 Modeling of CPS based Smart Grid

As cyber security incidents have gained more and more recognition as legitimate issues, national infrastructure CPS systems, such as national power grids, smart energy systems, complex metering infrastructures, etc., are becoming more and more vulnerable. Figure 7 illustrates the necessity of interoperability, hybrid and heterogeneous modelling, operations, and simulation tools to meet system and network specifications. They ensure that the components can work together and that the large-scale network SG-CPSs can run flawlessly. Future SGs will provide dependable, affordable, open, and user-friendly options for energy and utilization. These options can be small-scale but widely dispersed sources or large-scale bases with massive distribution over great distances, supporting a variety of energy demands from electric vehicles, energy conservation, and market participation [36]. This obviously necessitates transdisciplinary and interdisciplinary work combining risk management, economic coordination, uncertainty analysis, and system security.

As of now, the operating conditions and disturbances for power network analysis have been provided based on offline experiences, so they cannot be realistically configured in accordance with the cutting-edge trends of external nonelectrical systems. This hinders the development of early warning systems as well as the adaptation of predecisions about external settings. Future analyses of power networks must take the environment and energy into account as a whole. The limitations on emissions, primary energy, and the energy market can make even a physically safe power network unstable. Energy sources, financial systems, communication networks, transportation, and products delivery networks must all work in harmony to prevent widespread blackouts. Modeling and simulation of needs a new generation framework, which will be adopted in future SG's.

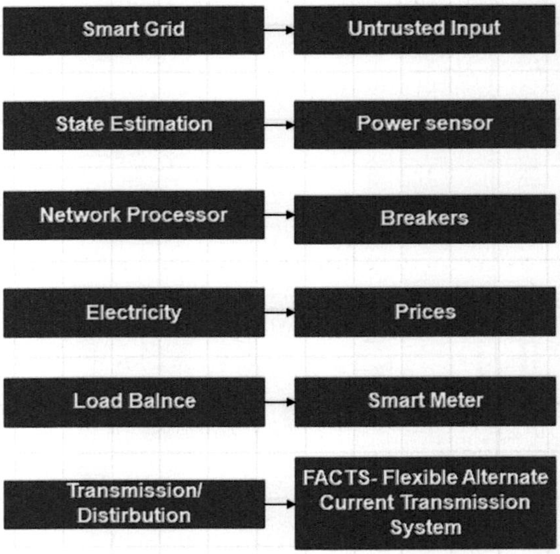

Fig. 7 Block diagram of state process

5 Conclusion

The widespread use of electronic devices as a result of technological advancements has increased load requirements. Voltage sags, overloads, and blackouts are the obvious problems. Smart grid is one viable and dependable answer to these issues. Smart grid refers to a power network enabled by digital communication technology. However, this approach is more complicated than it first appears. Future engineering systems are expected to incorporate cyber-physical systems heavily. Meanwhile, its security challenges are gaining more and more attention in both domestic and international academic study as well as in industry. We suggested a revolutionary CPS security architecture in this research and go into great detail about each of its three components. Security theories, which include security objectives and fundamental theories, are the main and most significant part. Information theory, control theory, and game theory are the fundamental ideas, while security objectives include safety, security, reliability, and resilience. The second part is the examination of assaults and countermeasures in the three-layer CPS design (control layer, transport layer, and executive layer) that we suggest. Simulation and application are the final component. Due to the complexity and scope of rising variability, CPS testing is still an arduous research area. In this article, we provide a thorough overview of the most recent CPS testing techniques and testbeds. Despite certain improvements brought by CPSs to testing theory and technology, there are still several restrictions on CPS testing's wider industrial applicability. In light of this, we identify difficulties faced by current test methodologies and testbeds and outline potential future research routes by taking a few extra requirements and limits into account. For CPS researchers to implement new CPS testing technologies or applications, the analysis and discussion in this study can offer helpful intuitions.

5.1 Future Scope

This method faces a few difficulties, such as cyberattacks and weaknesses. As a result, an effort is made here to examine the research work done so far thoroughly, taking into account various heuristic detection and estimation strategies. It is practical that problems with the smart grid may be efficiently anticipated and predicted, such as relay protection, power flow control, grid security, and dependability.

References

1. B. Luo, R. Beuran, Y. Tan, Smart Grid security: attack modeling from a CPS Perspective, in *2020 IEEE Computing, Communications and IoT Applications (ComComAp)*, (2020), pp. 1–6. https://doi.org/10.1109/ComComAp51192.2020.9398878.F

2. M. You, Q. Liu, H. Sun, New communication strategy for spectrum sharing enabled smart grid cyber-physical system. IET Cyber-Phys. Syst.: Theory Appl. **2** (2017). https://doi.org/10.1049/iet-cps.2017.0051
3. I. Horv´ath, B. Gerritsen, Cyber-physical systems: Concepts, technologies and implementation principles, in *TMCE* (2012), pp. 19–36
4. C. A. Macana, N. Quijano, E. Mojica-Nava, A survey on cyber physical energy systems and their applications on smart grids, in *Proceeding IEE PES Conference Innovative Smart Grid Technology* (2011), pp. 1–7
5. A. Sheela, S. Revathi, A. Iqbal, Cyber risks assessment for intelligent and non-intelligent attacks in power system, in *2019 2nd International Conference on Power and Embedded Drive Control (ICPEDC),* (IEEE, 2019), pp. 40–45
6. M.H. Cintuglu, O.A. Mohammed, K. Akkaya et al., A survey on smart grid cyber-physical system testbeds. IEEE Commun. Surv. Tutorials **19**(1), 446–464 (2017)
7. Facilitating whole electricity system outcomes. nationalgrideso.com/sites/eso/files/ documents/WholeElectricitySystemfinal.pdf. Accessed on 19 March 2020
8. N. Uribe-Pérez, L. Hernández, D. Vega et al., State of the art and trends review of smart metering in electricity grids. Appl. Sci. **6**(3), 68–92 (2016)
9. A. Nelson, S. Chakraborty, D. Wang, P. Singh, Q. Cui, Q, L. Yang, S. Siddharth, Cyber-physical test platform for microgrids: combining hardware, hardware-in-the-loop, and network-simulator-in-the-loop, in *Proceedings of the 2016 IEEE Power and Energy Society General Meeting (PESGM)* (Boston, 2016)
10. G. Bedi, G. Kumar, R. Singh, R.R. Brooks, K.C. Wang, Review of Internet of Things (IoT) in electric power and energy systems. IEEE Internet Things J. **5**, 847–870 (2018)
11. K.-D. Kim, P.R. Kumar, Cyber-physical systems: A perspective at the centennial, in *Proceedings of the IEEE*, vol. 100 (Special Centennial Issue) (2012), pp. 1287–1308
12. X. Yu, Y. Xue, Smart grids: a cyber-physical systems perspective. Proc. IEEE **104**, 1058–1070 (2016)
13. L. Parolini, B. Sinopoli, B. Krogh, Z. Wang, A cyber physical systems approach to data center modeling and control for energy efficiency. Proc. IEEE **100**(1), 254–268 (2012)
14. Y. Mo, T.H.J. Kim, K. Brancik et al., Cyber physical security of a smart grid infrastructure. Proc. IEEE **100**(1), 195–209 (2012)
15. z. Su, L. Xu, S. Xin, W. Li, Z. Shi, Q. Guo, A future outlook for cyber-physical power system, in *Proceedings of the 2017 IEEE Conference on Energy Internet and Energy System Integration (EI2)* (Beijing, China, 2017), pp. 1–4
16. F. Aghaee, N. Mahdian Dehkordi, N. Bayati, A. Hajizadeh, Distributed control methods and impact of communication failure in AC microgrids: a comparative review. Electronics **8**, 1265 (2019)
17. G.C. Konstantopoulos, A.T. Alexandridis, P.C. Papageorgiou, Towards the integration of modern power systems into a cyber-physical framework. Energies **13**, 2169 (2020). https://doi.org/10.3390/en13092169
18. J. Zhao, F. Wen, Y. Xue, Z. Lin, Cloud computing: Implementing an essential computing platform for future power systems, in Chinese. Autom. Electr. Power Syst. **34**(15) (2010)
19. P.P. Varaiya, F.F. Wu, J.W. Bialek, Smart operation of smart grid: Risk-limiting dispatch. Proc. IEEE **99**(1), 40–57 (2011)
20. D. Niyato, L. Xiao, P. Wang, Machine-to-machine communications for home energy management system in smart grid. IEEE Commun. Mag. **49**(4), 53–59 (2011)
21. X. Guan, B. Yang, C. Chen, W. Dai, Y. Wang, A comprehensive overview of cyber-physical systems: From perspective of feedback system. IEEE/CAA J. Autom. Sin. **3**(1) (2016)
22. B. Chen, J. Wang, M. Shahidehpour, Cyber–physical perspective on smart grid design and operation, (N. p, United Kingdom, 2018). Web. https://doi.org/10.1049/iet-cps.2017.0143.
23. Y. Wang, Research framework of technical standard system of strong and smart grid. Autom. Electr. Power Syst. **34**(22), 1–6 (2010)
24. F. Guo, L. Herrera, R. Murawski, E. Inoa, C.-L. Wang, P. Beauchamp, E. Ekici, J. Wang, Comprehensive real-time simulation of the smart grid **49**(2), 899–908 (2013)

25. E. O'Dwyer, I. Pan, S. Acha, N. Shah, Smart energy systems for sustainable smart cities: current developments, trends and future directions. Appl. Energy **237**, 581–597 (2019)
26. M. Ili´c, L. Xie, U. Khan, J. Moura, Modeling of future cyber-physical energy systems for distributed sensing and control **40**(4), 825–838 (2010)
27. K.R. Davis, C.M. Davis, S.A. Zonouz, R.B. Bobba, R. Berthier, P.W. Sauer, A Cyber-physical modeling and assessment framework for power grid infrastructures. IEEE Trans. Smart Grid **6**, 2464–2475 (2015)
28. A. Farraj, E. Hammad, D. Kundur, A cyber-physical control framework for transient stability in smart grids. IEEE Trans. Smart Grid **9**(2), 1205–1215 (2018)
29. L. Wen, K. Zhou, S. Yang, L. Li, Compression of smart meter big data: a survey. Renew. Sustain. Energy Rev. **91**, 59–69 (2018)
30. H. Li, A. Dimitrovski, J.B. Song, Z. Han, L. Qian, Communication infrastructure design in cyber physical systems with applications in smart grids: a hybrid system framework. IEEE Commun. Surv. Tutor. **16**(3), 1689–1708 (2014). [6823238]. https://doi.org/10.1109/SURV.2014.052914.00130
31. E. Lee, Cyber physical systems: design challenges, in *ISORC* (IEEE, 2008), pp. 363–369
32. Y. Susuki et al., A hybrid system approach to the analysis and design of power grid dynamic performance. Proc. IEEE (2012)
33. T. Morris et al., Engineering future cyber-physical energy systems: challenges, research needs, and roadmap, in *NAPS* (IEEE, 2009)
34. M. Wolf, D. Serpanos, Safety and security in cyber-physical systems and Internet-of-things systems. Proc. IEEE. **106**(1), 9–20 (2018). https://doi.org/10.1109/JPROC.2017.2781198
35. S. Sen, P. Pang, Architectural modeling and cybersecurity analysis of cyber-physical Systems-A Technical Review. Int. Res. J. Eng. Technol. **5** (2018)
36. C.C. Sun, C.C. Liu, J. Xie, Cyber-physical system security of a power grid: state-of-theart. Electronics **5**(3), 40 (2016)

A Study on Cybersecurity Standards for Power Systems

Sajal Sarkar

Abstract In order to have explicit cybersecurity measures and best cybersecurity practices in power systems, a number of cybersecurity standards containing generic sets of rules/guidelines for maintaining proper cybersecuirty hygiene are being designed and developed. However, appropriate study, proper understanding of scope, applicability, and implementation guidelines of these cybersecurity standards for power systems are not adequately being exercised and presented till date. In this chapter, a study is presented to understand the scope and applicability of cybersecurity standards and derive implementation guidelines in power systems. The proposed study discussed and analyzed relevant cybersecuriy standards for understanding the scope, applicability, cybersecurity process improvement for individual systems/devices deployed in present day power systems and also for a whole system setup such as substation/plant, communication setup, local/central control center.

Keywords Power systems · Cybersecurity · Standard · Intelligent Electronic Devices · Guidelines

1 Introduction

Nowadays, deployment of Information Technology (IT) along with its associated components in modern power system is very much essential to be meet functional and operational requirements for enhanced observability and visibility. However, due to inherent nature of IT there is always a possibility to make power systems vulnerable to cyber threats, which may invite operational failure and financial losses. Resultantly, all organizations face tougher pressures by stringent organizations' functional,

S. Sarkar (✉)
Information Security Department, Power Grid Corporation of India Ltd, Sector 29, Gurgaon 122001, Haryana, India
e-mail: sarkar.sajal@gmail.com

H. Haes Alhelou et al. (eds.), *Power Systems Cybersecurity*, Power Systems,
https://doi.org/10.1007/978-3-031-20360-2_18

operational and business requirements, national and international laws and regulations, expectation of interested parties in securing information, operations and assets from misuses and cyber attacks. Hence, securing information and systems has become a significant concern almost for all organizations as go to great lengths to ensure that no one can steal and leverage information and/or otherwise compromise assets and systems deployed for power systems.

Beside deployment of technical solutions and measures, it has become essential for all power sector organizations to implement and maintain a coherent set of policies, processes to manage risks of information and systems and ensure cybersecurity aspects such as availability, integrity and confidentiality of IT enabled power systems. It is also very much required to meet the cybersecurity compliance requirements for power system installation set by the concerned authority. The cybersecurity compliance requirements are generally defined issuing guidelines and regulations (e.g., [1, 2]), notifying act [3, 4], formulating new standards [5], adopting well practiced standards [6], etc. by the concerned authority of a country. In addition, selection of devices, capability of devices, designing of solution architecture, deployment of technical solutions, and implementation of measures, best practices, and processes are also needed to be done as per defined criteria laid in the relevant cybersecurity standards [2, 6–9] for ensuring cybersecurity aspects and requirements.

In order to ensure the above mentioned cybersecurity requirements and best practices in power system and its associated components and processes, a number of cybersecurity standards are being designed and developed. However, as per author's knowledge goes there is no works available, where cybersecurity standards have been studied and presented appropriately for understanding the scope and applicability, deriving the implementation guidance, and highlighting the scope of process improvement in power systems till date. Hence, it is very much required to study all relevant cybersecurity standards for modern power systems and accordingly analyze the same for listing down the scope, applicability, and implementation guidelines. Therefore, in this chapter, a study is presented to understand the scope and applicability of cybersecurity standards and derive the implementation guidelines for modern power systems. The chapter first sets a context for the study by explaining a modern power system consisted of a substation, a communication network, and a control center along with the popular 9Bus system. The proposed study is then presented the suitability of standards for individual devices and a whole setup of power system. The study also presents and discusses scope of process improvement in power systems based on cybersecurity standards.

The reminder of this chapter is organized as follows. A background of the proposed work is presented in Sect. 2. Section 3 describes existing relevant standards of cybersecurity for power system mentioning the scope and applicability of the standards. Mapping of cybersecurity parameters in forms of implementation guidelines for power system are discussed in Sect. 4. Finally, the chapter is concluded in Sect. 5.

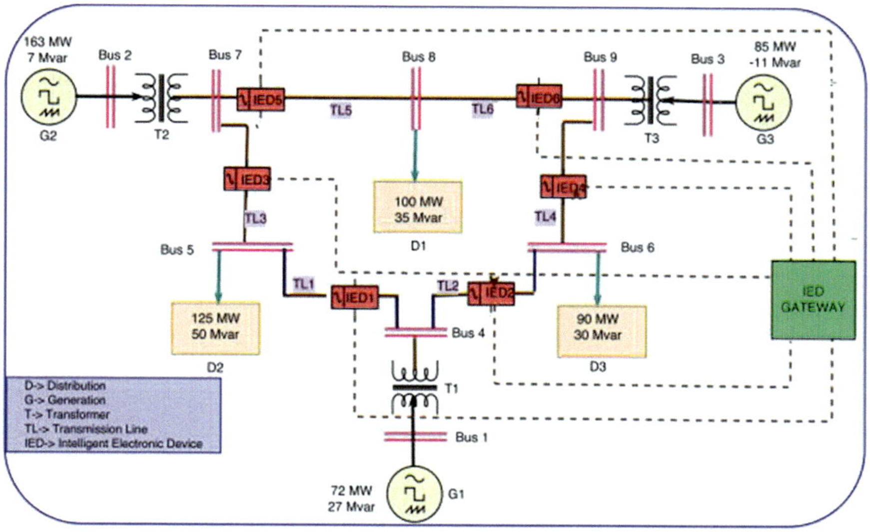

Fig. 1 WECC/IEEE 9bus system

2 Background

2.1 *Power System*

A power system is a collection of physical systems which generate, transmit, supply, and distribute power. The physical systems and its associated processes form the power grid, which supplies power to domestic consumers and industries. In the power grid, generators generate power which is carried through transmission systems to the load centers. Power is distributed by distribution to the domestic consumers and industries. The electric grid generally consists of generating station, transmission systems, and distribution systems. A typical electric grid in forms of single line diagram is shown in Fig. 1, which is also popularly known as Western Electricity Coordinating Council (WECC)/IEEE 9Bus power system. Figure 1 shows that the 9Bus power system consisted of six (TL1-TL6) transmission lines, three loads (D1-D3), three transformers (T1-T3), and three generators (G1-G3). G2 is connected with PV-bus 2 and G3 is connected with PV-bus. The slack bus 1 is connected with G1. The bus bars 8, 5 and 6 are connected with D1, D2 and D3 respectively.

2.2 Power System Automation (PSA)

PSA automatically controls the power systems and its associated processes of generation, transmission, supply and distribution using instrumentation, control devices (e.g., IEDs-Intelligent Electronic Devices), communications protocols, etc. The PSA enable us to control, monitor, and operate power systems remotely either from a local control center within the substation or from a central control center using IED's data. PSA works in a coordinated fashion based on data, supervision, and control all together. Therefore, in order to achieve these requirements a set of software (e,g., SCADA-Supervisory Control And Data Acquisition/DCS-Distributed control systems) are generally used in power system automation. SCADA/DCS is generally installed in computing systems at control center for coordination purposes. Data from IEDs is polled by the SCADA/DCS system periodically using IEC 61850, DNP3, Modbus, IEC 60870-5-101, IEC 60870-5-104, etc. and the Polled data is sent to the supervisory controller for operations and monitoring of power systems.

2.3 Network Architecture in Power System

In PSA, implementation of network architecture for generation, transmission, and distribution systems may vary and it may have different forms. However, to have commonality in variable forms of implementation, reference models are generally being used for description and to have a generic terminology and platform. As mentioned in [10, 11], Purdue Enterprise Reference Architecture (PERA) [12] is widely used reference model. Here, inline with PERA, a network architecture diagram is considered (shown in Fig. 2 [11]) for power system to study cybersecurity standards and understanding the scope and applicability of standards for it. The considered network architecture consists of substation control centers, communication network, a central control center and a 9Bus power system. Figure 2 shows that substation control centers are connected with the communication network via router-cum firewall and the communication network is connected with the central control center via a firewall. It is important to note that IED gateway connects other IEDs deployed at the location of generation, transmission, distribution systems which are shown in Fig. 1.

2.4 Cybersecurity in Power System

In the past years, cybersecurity was not a concerned for power systems rather safety in power systems was a major issue to be taken care of. However, recent power system mostly depends on IT systems in need of functional and operational requirements where cybersecurity [14, 18] needs to be considered. So, IT enabled power system

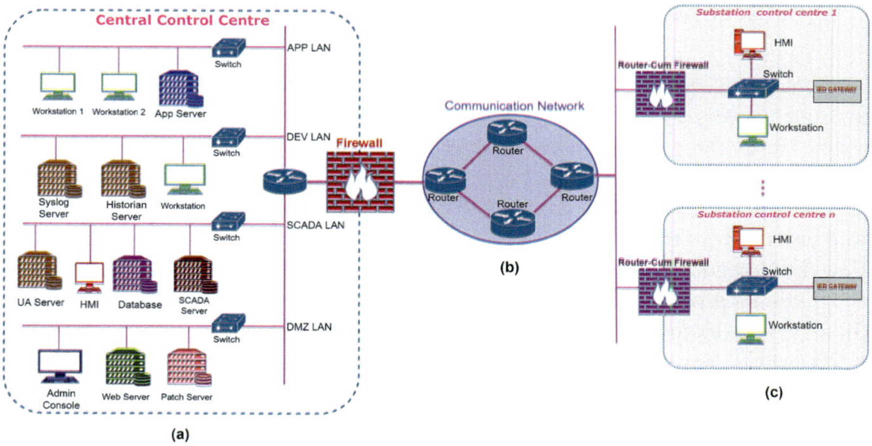

Fig. 2 Network architecture diagram of modern power system

is an interconnected system as shown in Figs. 2 and 1 and there are many cybersecurity challenges. Additionally, cybersecurity aspects are required to be in order of confidentiality, integrity, and availability in the IT network, wherein the power system network requires cybersecurity aspects in order of availability, integrity, and confidentiality. Therefore, traditional IT network's cybersecurity approaches must be redesigned for the power system network to fulfill the cybersecurity aspects. Moreover, the power system protection and control is always a top priority as it plays vital role in critical infrastructures of any country. Hence, the deployment of IT systems and associated components for controlling, operating, and managing power system not only brings the benefits, but it also make power system vulnerable to cyber threats and cyberattacks.

Hence, modern power system is heavily facing cybersecurity challenges and issues during its deployment processes and day-to-day operations. In order to meet these challenges and issues deployment of cutting edge technologies and solutions is being considered and maintained. However, in the recent years an increased trend of cyber threats and attacks is being witnessed in different sectors including power sector ([19, 20], etc.). As a result, it is now imperative to deploy multiple security solutions and follow proactive cybersecurity measures for secure power system operations and functions. It is therefore cybersecurity is not a single shot solution rather it is a continuous process to be followed for ensuring and enhancing the cybersecurity measures. Hence, cybersecurity standards are required to be considered for selection and installation of devices, deployment of cutting edge technologies and solutions, designing networks, selection and configuration of protocols, etc. in modern power system installation. As mentioned in the second paragraph of Introduction Section that it is also very much required to meet the cybersecurity compliance requirements for power system installation set by the concerned authority.

3 Cybersecurity Standards for Power System

In order to have proper understanding of cybersecurity accordance with different requirements, it is essential to find and select standards for all components such as devices, computing systems, network devices, even for whole substation, communication network, central control center, etc. A number of cybersecurity standards are being published and made available by IEEE, NIST, ISO, IEC, etc. which can be followed and used for meeting cybersecurity requirements in power systems. One can follow these standards to ensure cybersecurity for different components and deployed devices and its processes associated with day-to-day operations. However, it is very difficult to identify and select appropriate standards for implementations and meeting the compliance requirements. A such selection of appropriate standard and its implementation for power system is a time consuming and costly affair. It is also requires proper understanding and continuous engagement of skilled manpower. Hence, in this chapter a systematic study on cybersecurity standards for power system is presented and discussed to bring out proper understanding with respect to the requirements, scope and applicability, identification and selection of appropriate standards, finally deriving implementation guidelines, and meeting the compliance requirements. A consolidated list of standards is also presented in Table 1 with mentioning the scope and applicability of these standards for power system and its components.

3.1 IEEE1686

IEEE has published IEEE-1686 in 2007 and revised in 2013 [7]. The standard is not certifiable rather it provides guidance of IED capabilities and implementation features which are required to be configured in an IED deployed in PSA. Specifically, the standard covers the features and functions are to be embedded in IEDs by the manufacturer for accommodating critical infrastructure protection measures. The processes of data retrieval from an IED, accessibility and configuration of an IED, updation of firmware are also explained in the standard. It also states about safeguarding, auditing mechanisms, and alarming indications which are to be inbuilt in the IEDs. The standard has made a provision for user to define a security program using available features based on which alerts can be generated if an IED does not meet the defined requirement and accordingly the defensive technical and/or procedural measures may be taken. The secure transmission (by encryption) of data is also part of this standard. Therefore, the deployed IEDs in WESC 9bus (as shown in Fig. 1) and substation control center (Fig. 2c) are needed to be compliant with the features and functions mentioned in the standard for fulfilling the cybersecurity requirements.

Table 1 Cybersecurity standards, scope, applicability for power systems

Standards	Scope	Applicability	Responsibility	Type
IEEE1686	IEDs	Devices	Asset owner, OEM	Implementable
IEEEC37.240	Substation	Automation, protection and control system	Asset owner, system integrator	Implementable
IS16335	Substation	Automation, protection and control system	Asset owner, system integrator	Implementable
IEC62351	Substation, communication, control center	Communication protocol	Asset owner, system integrator	Implementable
ISO/IEC62443	Substation, communication, control center	Substation, network design, asset management	Asset owner, OEM, system integrator	Implementable, certifiable
ISO/IEC27001	Substation, communication, control center	Process associated with substation, communication setup, control center	Asset owner, system integrator/auditors	Implementable, certifiable
ISO/IEC27019	Substation, communication, control center	Process associated with substation, communication setup, control center	Asset owner, system integrator/auditors	Implementable, certifiable
ISO20243	Substation, communication, control center	Software, firmware, and products	Asset owner, supplier, manufacturer	Certifiable
ISO28000	Substation, communication, control center	Security assurance for supply chain issues	Asset owner, supplier, manufacturer	Certifiable
NERC-CIP	Substation, communication setup, control center	All system components and devices	Asset owner, system integrator	Implementable
NIST SP 800-82	Substation, communication setup, control center	All system components and devices	Asset owner, system integrator	Implementable
IEEEC37.118 and C37.118.1	Substation, control center	Phasor measurement unit, communication protocols	Asset owner, system integrator	Implementable

3.2 IEEE37.240

IEEE has published another standard, IEEE-37.240 in 2014 for providing guidance and technical requirements for cybersecurity in substation of generation, transmission and distribution [13]. IEEE37.240 provides details guidelines and practices to be followed for ensuring cybersecurity in modern power substation being established with automation, protection, and control systems. It describes and presents engineering practices are required to be applied for achieving high level of cybersecurity. It defines cybersecurity in forms of trust and assurance of data in use, data in motion, and data at rest. Cyber incident and its corresponding required response to be taken are also covered in the standard. As per the standard, to achieve certain cybersecurity level there is a need of a balance between economic feasibility and technical feasibility by implementing cybersecurity measures and the balance should address the existing risks in the substation. Design and implementation of cybersecurity measures should not hamper the legitimate system accessibility and operations, especially during restoration activity and times of emergency. From the above it can be noticed that the standard has covered and described the essential cybersecurity guidance and requirements which are required for implementation in a substation. So, the standard is for cybersecurity implementation in substation and it is not a certifiable standard. Hence, the systems of WESC 9Bus (Fig. 1) and substation control center (Fig. 2c) need to be compliant with the features and functions mentioned in the standard.

3.3 IS16335

The Bureau of Indian Standards (BIS) has approved a standard with name IS16335 for Power Control Systems-Security Requirements [5]. Indian power sector is being growth at an enormous pace. Various interconnection projects are on-going to strengthen the existing transmission infrastructure to build a robust national grid. The need for interconnecting power networks is in turn driving the need for interconnection of information networks. Therefore, it is imperative to utilize the capabilities provided by modern information and communication technologies. However, with increasing interconnection and complexities of power systems, there is an increase trend of cyber threats perception and cyberattacks and in the overall management of the power grid. Hence, due to deployment of modern technologies and widespread use of communication technology in the power grid, there is a need to address the cybersecurity threats. The standard has been formulated keeping in view all possible cybersecurity threats and cyberattacks may take place in operations of power system. So, the standard is applicable for all four components are shown in Figs. 1 and 2.

3.4 IEC62351

In order to have proper cybersecurity in power system communications, Technical Committee 57 of IEC has developed an implementable standard IEC62351 [14]. This standard provides detailed guidance of security for protocols such as IEC 61968 series, IEC 61970 series, IEC 60870-6 series, IEC 60870-5 series, and IEC 61850 series, which are generally used in power systems. It has defined different security objectives viz. user and data authentication, authorization of user, detection of intrusion, prevention of spoofing and playback, prevention of eavesdropping, etc. This standard has many parts which are shown in Table 2 and it has wide range of coverage of communication protocols in power systems. It is worth to mention that relevant parts, which cover security requirements for communication protocols are only shown in the following Table 2.

3.5 ISO/IEC27001

ISO/IEC27001 is a certifiable standard for information security management systems (ISMS) of ISO/IEC27000 family [6]. In this standard, a list of auditing and implementation guidance and checklist is mentioned for auditor and implementer. Implementation and certification of this standard helps organizations to build and maintain ISMS within the organizations. Other benefits of the standard are protection of information, mitigation and management of information security risks doing risk assessment and treatment, fulfilling the compliance requirements, increasing information security awareness, asset inventory, process improvement, etc.

The standard has ten short clauses such as scope of the standard, document references, reuse of the definitions and terms, organizational context, leadership and high-level support for policy, planning, risk assessment and treatment, support, making an ISMS operational, performance reviewing, and corrective action along with a list of controls and objectives. The standard has 114 controls under 35 categories and 14 groups.

Generally, most of the organizations may have implemented a number of controls which are typically addressed certain IT security aspects without consideration of protection of non-IT assets such as documents, paperwork, function of human resource, proprietary knowledge, etc. So, it is seen that without proper implementation of ISMS, these security controls may be somewhat disjointed and disorganized. In addition, business continuity planning and physical security, Human Resources practices, etc. may be managed quite independently of IT. However, there may be a little reference of ISMS in need of assigning ISMS roles and responsibilities in the organization chart. Therefore, a systematic examination of risk of information security, consideration of vulnerabilities, threats, and impacts as per ISMS helps to organize and manage information security measures and control within the organization. Hence, implementation of information security controls and carrying out risk

Table 2 Parts, purposes, and applicability of IEC 62351

Part	Scope	Protocol/Purpose	Applicability
IEC62351-3	Security profiles	TCP/IP, TLS encryption, node authentication using X.509 certificates, message authentication	Substation control center, communication network and central control center
IEC62351-4	Security profiles	MMS (e.g., IEC 61850)	9 Bus system and substation control center
IEC62351-5	Security profiles	IEC 60870-5 (e.g., IEC 60870-5-104), TLS for TCP/IP profiles and encryption for serial profiles	Communication between substation and central control center
IEC62351-6	Security for IEC 61850 profiles	VLAN for GOOSE and RFC2030 for SNTP	Substation network
IEC62351-7	Security through network and system management	SNMP based on specific management information base	All three components of Fig. 2
IEC62351-8	Role-based access control	Control of users and agents for data objects in power systems	All three components of Fig. 2
IEC62351-9	Key management	Correct and safe passwords, key encryption, life cycle of cryptographic information (enrollment, creation, distribution, installation, usage, storage and removal)	All three components of Fig. 2
IEC62351-10	Secure network architecture	Secure network design, identification of critical points in network, data encryption, user authentication, IPSec VPN, SFTP, HTTPS, etc.	Control centers, substation automation, communication network
IEC62351-11	Security for XML Files	Embedding XML content into an XML container, date of issue and access control, X.509 based XML data authentication, XML data encryption	Usage of XML files and data

assessment and treatment by defining scope to address risks is very much required by adopting a management process to ensure information security controls.

ISO/IEC27001 certification has a three-stages audit process. Stage 1 familiarizes the auditees with the auditors and vice versa. A preliminary review of key documentation such as information security policy, risk assessment and treatment plan, Statement of Applicability (SoA), etc. are also done in Stage 1. Stage 2 is for formal compliance audit, checking of scope of ISMS and implemented controls, risk assessment and treatment, record of awareness, etc. against the requirements specified in the standard. In Stage 2, the auditors also seeks documentary approved evidence to confirm that the system has been properly implemented as per required controls and measures for operations. The Stage 3 is conducted as certification audits by the Lead Auditors and on successful completion, the auditor recommend for ISMS certification as per ISO/IEC27001 which also requires annual re-assessment audits to confirm the ISMS continuation of the certificate.

So, this standard is also applicable to all four components such as 9Bus system (Fig. 1), substation control center, communication network, and central control center (Fig. 2) for process audit, risk assessment and treatment, maintaining asset inventory, implementation of information security control, etc. and ISMS certification as per the standard.

3.6 ISO/IEC27019

ISO/IEC27019 is also a certifiable security standard [15] of ISO/IEC27000 family and it is specifically published with mentioning the requirements of energy utility industry. It has mentioned additional 14 controls to be attended and maintained for the systems. It is applied for energy utility industry to control power systems and supporting processes associated distribution, transmission, storage and generation.

All power sector organizations require to implement these 14 controls mentioned in ISO/IEC27019 along with implemented ISMS certification as per ISO/IEC27001. However, this is a newly published standard and its implementation and certification is not started yet.

3.7 ISA/IEC62443

ISA/IEC62443 is a family of standards for cybersecurity of Industrial Automation and Control System (IACS) [9] and it was initially formulated by the International Society of Automation (ISA). This family of standards help to reduce the risk of IACS and safeguard the exposure of IACS networks to cyberthreats. The standards cover cybersecurity considering process and technical aspects in IACS deployed for process industries such as energy, petroleum refining, manufacturing, chemicals processing, water, and pharmaceuticals. It has fourteen sub-standards which are divided

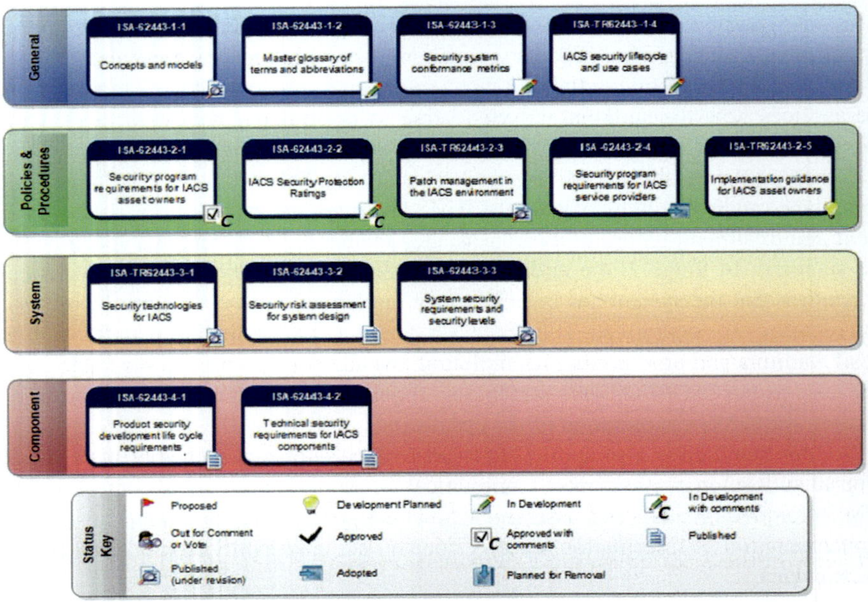

Fig. 3 Parts, Standards, and the status of various works of the ISA/IEC62443

into four parts namely, *general*, *policies and procedures*, *system*, and *component*. Figure 3 shows different parts, standards, and the status of various works of the ISA/IEC62443 family [21]. The concepts, use cases, policies, and procedures are described in the first two groups, whereas, the technical requirements for networks and system components are defined in the system and component respectively. The standards guide asset owner, system integrator, manufacturer, professional, and operator of IACS networks through the necessary best practices, controls, and requirements for a secure IACS.

The standard defines seven numbers of foundational requirements such as access control, role and authorization, integrity, confidentiality, network segmentation, monitoring and analysis, and backup and recovery. In this standard, four security levels (SL) are set for systems (IEC62443-3-3) and products (IEC62443-4-2) to indicate classes of attackers resistance. The security levels are to evaluated as per technical requirements. The security levels are:

- SL 1: Protection against accidental/unintentional misuse such as insider (well-intention, carelessness) system faults, contingencies, etc.
- SL 2: Protection against intentional misuse by simple means and motivation with limited resources and skills such as insider (disgruntled employee or contractor) or intruder (thrill seeking, hobbyist, script kiddie).

- SL 3: Protection against intentional misuse by sophisticated means and moderate motivation with moderate resources and IACS-specific knowledge such as cyber criminal, terrorists, competitors, etc.
- SL 4: Protection against intentional misuse using sophisticated means and high motivation with extensive resources and IACS-specific knowledge such as Nation-states/Groups and Government organizations.

IEC62443 also considers the following four different maturity levels (ML) as per the CMMI framework [22] for processes and technical requirements:

- ML 1-Initial: product suppliers develop product on ad-hoc basis without any documentation.
- ML 2-Managed: product supplier manages product development as per documentation and must have appropriate expertise to demonstrate the process following the written procedures.
- ML 3-Defined: product supplier follows a practiced and repeatable process of product development and documentation.
- ML 4-Improving: product suppliers follows an appropriate and effective process metrics to monitor performance of the process and has the provision of continuous improvement.

The standards also have mentioned about *Defense in Depth (DiD)*, *Zone*, *Conduit*, *redundancy* for meeting the security levels in system design. DiD is for achieving levels of security in the system. On the other hand, assets are required to be grouped in different zones with common security requirements for achieving defined security levels. The security level for a zone which can also be divided into sub-zones is determined based on the risk analysis. However, elements are grouped by conduits based on the allowable communication between two zones. Redundancy ensures the availability in case of system failures because of vulnerability exploitation or failure of security measures.

ISA/IEC62443 has certification provision for processes, systems, and products used in IACS. There are few certifying companies also available to certify product and process by testing, inspection, and certification based on ISA/IEC62443. In order to become a certifying company, there are few criteria and requirements to be fuilfilled by following a process defined in the standard.

The ISA Security Compliance Institute (ISCI) is created a conformity assessment scheme for IACSs, components and processes aligned with the IEC62443 standards. The available certification offered by ISCI under the ISASecure are shown in Table 3.

It can be concluded from the above discussion that IEC62443 covers cybersecurity aspects for process and technical in IACS. The usage of ISA/IEC 62443 in process industry is somewhat mature. However, power system organizations very recently are started to use and implement systems following the standard. It is worth to mention that ISA/IEC62443 standard is included in the cybersecurity guidelines approved CEA, Govt. of India [1]. In addition, the standard can also be mapped for considered power system's components such as 9Bus system (shown in Fig. 1), substation control center, communication network, and central control center (shown in Fig. 2).

Table 3 IEC 62443 cybersecurity certification schemes

Based on	Classification	Program name	Applicable tos
IEC62443-4-1	Device process certification	Security development process	OEM New Product development
IEC62443-2-4	System process certification	System integrator process	System integrator
IEC62443-4-1, IEC62443-4-2	Device and application certification	Security device certification	OEM product
IEC62443-2-4, IEC62443-3-3	OEM system certification	System security certification	OEM system
IEC62443-2-4, IEC62443-3-3	Integrated system certification	Integrated system certification	Integrated system
IEC62443-4-1	ISCI SDLA	Security development process	OEM new product development
IEC62443-4-1, IEC62443-4-2	ISCI EDSA	Security device certification	OEM product
IEC62443-2-4, IEC62443-3-3	ISCI SSA	System security certification	OEM system

For example products which are developed as per ISA/IEC62443 (such as IEDs, computing devices, network devices) are to be procured and the network, systems, etc. must be designed considering DiD, zones and conduits for achieving security levels. Risk assessment and analysis must also be carried as per ISA/IEC62443-3-3.

3.8 ISO/IEC20243

There is a need to develop a method for responding to the changing landscape and the increased trends of sophistication of cybersecurity attacks over the supply chain of software, firmware, and products. So, the Open Group has prepared a standard with the name of the Open Trusted Technology Provider Standard (O-TTPS) to mitigate tainted and counterfeit products maliciously and accordingly ISO/IEC has approved ISO/IEC 20243:2015 [23, 24]. The standard has formulated requirements, guidelines, and recommendations for the commercial off-the-shelf (COTS) integrity of software, firmware, and products and supply chain security. The purpose of the standard is to help software and firmware developers and providers to produce secure products and accordingly provide enough confidence in software, firmware, and products. Almost all organizations fully rely on COTS software, firmware, and products to setup their system and accordingly operate it. However, these software, firmware, and products are developed and manufactured at different sites in multiple countries. Therefore risk mitigation of counterfeit and tainted components in the software, firmware, and products is to be assured based on supply chain security and product integrity as per O-TTPS.

In order to ensure supply chain cybersecurity issues in power system, the supplier organizations of power sector may need to be certified their firmware, software and products for conformance as per the Trusted Technology Provider Accreditation Program of the standard. Generally, third party assessors assessed the products as the standard for listing the supplier organization in the accreditation register of the Open Group. The accreditation policy and assessment procedures govern the process of third party assessment. The software, firmware, and products for considered system setup shall be procured as per this standard.

3.9 ISO28000

ISO28000 is a standard [8] published by ISO for specifying requirements of security assurance in supply chain. The standard has been published in three times viz. first version in 2005, second version in 2007 and the latest version in 2022. The specified requirements in the standard for security assurance in supply chain are generic and intended to be applicable to all organizations. However, applicability of these requirements completely depends on environment and complexity of the organization. In the second version of the standard, the Plan-Do-Check-Act structure was adopted to bring the elements from the first version of standard. It was developed for organizations of varying scale who can apply the standard for handling supply chains of various degrees of complexity. Adopting the ISO28000 brings organizational and operational benefits such as management practices, integrated enterprise resilience, brand recognition, improved supply chain performance, improved risk management integration, etc. The supplier organizations of power sector need to be compliant their systems, services, products as per this standard.

3.10 NERC-CIP

The North American Electric Reliability Corporation Critical Infrastructure Protection (NERC CIP) [2] came up with a set of standards aimed at regulating, enforcing, monitoring and managing cybersecurity of Bulk Electric System (BES) in the North America. A cybersecurity framework is brought for identifying and ensuring security of critical assets which are being used for reliable and efficient supply of electricity in the country. The CIP of NERC coordinates for improving the security of power system NERC CIP includes the United States, several provinces in Canada and one state in Mexico. The standards govern critical infrastructure of all entities such as owners, operators and users of any part of the system, which impact the reliability of BES. These standards carry the force of regulations, meaning they are required by law. That's why these standards are also known as NERC CIP requirements. All entities that fall under the purview of NERC CIP must comply with these standards.

The NERC CIP standards require utility companies in North America to establish and adhere to a baseline set of cybersecurity measures. The goal is to ensure that appropriate security controls are in place to protect BES and its users and customers from all threats that may affect its timely and effective functioning. These threats may include cyber attacks, cyber vandalism or acts of cyberterrorism. Entities must perform a risk analysis of assets deployed in the system setup. They must also define a process for changing the configuration access monitoring of critical assets. In addition, NERC CIP mentioned of the use of firewalls for blocking unused and vulnerable ports and enforcement of IT controls for protecting the access of critical assets. In addition, they must deploy systems to monitor security events and implement comprehensive contingency plans to respond to cyber attacks, natural disasters and other unplanned events that may affect the functioning of BES. To ensure the delivery of consistent and effective power to all recipients, NERC and its regional bodies take compliance seriously. Therefore, NERC's compliance monitoring and enforcement program tracks, assesses and enforces the uniform compliance of covered entities via regular audits and spot checks. All North American entities must comply with NERC CIP standards. Failure to comply may result in monetary fines, sanctions or other actions. Penalties may vary from country to country since NERC is a transnational organization. The fundamental standards and sub-standards under NERC CIP specify the requirements given in Table 4 that utilities must follow to identify critical assets, create control mechanisms, enforce the logical and physical security of their systems, and recover any affected assets following a cybersecurity incident.

Although NERC-CIP is specifically formulated for US and Canada, it can be used for implementation by other countries as well. The versions and parts of NERC-CIP are free accessible. While implementing power system as per NERC-CIP by any organizations, context, compliance of regulatory requirements, economical and technical requirements, cybersecurity aspects, etc. are required to be taken into account.

3.11 NIST SP 800-82

NIST SP 800-82 provides guidance for cybersecurity in industrial control systems (ICS) including SCADA/DCS and other related systems for control functions [25]. The overview of different ICSs, typical system topologies, threat identification, vulnerabilities assessment and mitigation, recommendation of cybersecurity countermeasures are listed down for easy understanding and implementation. Potential risk for different ICS varies and its impact is also different. In order to mitigate this varying potential risk of ICS, NIST SP 800-82 provides different techniques and methods. It has also suggested to perform risk assessment of the systems and then tailor the recommended solution and guidance for meeting the cybersecurity requirements, business and operations in ICS deployed for electric, water, chemical, pharmaceutical, etc.

Table 4 NERC-CIP

Part	Topic	Scope/Purpose
CIP-001	Sabotage reporting	Identification, analysis an reporting
CIP-002	BES cyber system identification and categorization	Identifying and categorizing BES cyber systems/assets
CIP-003	Security management controls	Establishing clear accountability (delegating authority and identifying a senior manager) to protect North American BES cyber systems
CIP-004	Personnel and training	Training of staff and contractors in forms of cybersecurity awareness and training
CIP-005	Electronic Security Perimeter (ESP)	BES cyber systems protection from maloperation and instability deploying appropriate ESP
CIP-006	BES physical security	Addresses operational and physical controls for physical security
CIP-007	System security management	Defining technical, operational and procedural elements to secure all systems within ESPs
CIP-008	Incidence reporting and response planning	Preparing entities for cyber incidents and providing guidelines on how to respond to them with a cybersecurity incident response plan
CIP-009	BES recovery plans	Addressing how entities can recover from a cybersecurity incident in BES cybersecurity systems
CIP-010	Configuration management and vulnerability assessments	Outlining the security policy requirements to prevent and detect any unauthorized changes
CIP-011	Information protection	Specifying the requirements to identify information that could impact the functioning of BES
CIP-012	Cyber security communications between control centers	Ensuring data confidentiality and integrity transmitting between control centers
CIP-013	Supply chain risk management	Implementation of security control for supply chain risk mitigation in the BES operations
CIP-014	Physical security	Identification and protection of transmission stations and control centers

3.12 IEEEC37.118 and IEEEC37.118.1

IEEE has published C37.118 [26] standard in 2005 and revised the same with name C37.118.1 [27] in 2011 for Synchrophasor Measurements for Power Systems. The standard defines frequency, synchrophasors, and rate of change of frequency (ROCOF) for synchronized phasor measurements under all operating conditions in power system. It also defines the time tag and synchronization requirements for power system. In addition, the standard also defines a data communication protocol mentioning the message formats for communication in a real-time system. It has mentioned a method to evaluate and quantify these measurements and requirements for compliance considering a phasor measurement unit (PMU) with the standard under both steady-state and dynamic conditions. The standard does not specify limits to accuracy under transient conditions,measurement response time, software, hardware, or a method for computing phasor.

The main purpose of the standard is for real time interfacing of measurement equipment based on phasor and frequency with the associated systems in a substation. The synchronization and formats requirements allow to correlate phasors from various sources and compare with similar data from different measurement systems. So, the taken measurements of power system can be accurately and readily combined for system analysis and operations.

3.13 Other Cybersecurity Standards

The European Union Agency for Cybersecurity (ENISA) [16] has published documents and studies outcomes on cybersecurity for smart and critical infrastructure (e.g. power systems) along with other sectors. It is also contributes in cybersecurity policy formulation for European Union, for enhancement of trustworthiness in ICT services, products, and processes by means of certifications. In Japan, Cybersecurity Division of Ministry of Economy, Trade and Industry has formulated and finalized a cybersecurity framework for cyber physical system [17]. This framework has three parts focusing on concept, policy and method to drive the cybersecurity in cyber physical system. In Australia, the Australian Cyber Security Center [28] provides timely and tailored advice and support to uplift the cybersecurity in Critical Infrastructure, control systems, and operational technology. It helps asset owner for evaluating and identifying security vulnerabilities and help to mitigate the vulnerabilities for risk reduction. On the other hand, the US National Institute of Standards and Technology (NIST) has published National Initiative for Cybersecurity Education (NICE) Cybersecurity Workforce Framework [29] and NIST Cybersecurity Framework [30] for setting of best cybersecurity practices, standards and recommendations for increased and enhanced cybersecurity measures by individual organizations.

Table 5 Mapping of cybersecurity parameters with the relevant standards

Parameter	Relevant standards
IED selection and its cybersecurity capability	IEEE1686
Substation automaton and protection	IEEE C37.240
Selection and design of PMU	IEEE C37.118 and C37.118.1
Network architecture design	NERC-CIP, ISA/IEC26443
Information security policies	ISO/IEC27001
hline Asset management	ISO/IEC27001, ISA/IEC62443
Access control	ISO/IEC27001
Incident management	NERC-CIP
Risk management	ISO/IEC27001, ISA/IEC62443
Risk assessment	ISO/IEC27001, ISA/IEC62443
Cybersecurity assessment	ISA/IEC62443
Awareness and training	ISO/IEC27001, NERC-CIP
Information protection	NERC-CIP
Power system communication	IEC62351
Cybersecurity analysis	NIST SP 800-82
Recovery planning	NERC-CIP
Firewall policy configuration	NERC-CIP
Patch management in IACS	IEC62443-2-3
Access control	IEC62443
Secure product and component development	IEC62443-4
Cybersecurity and testing of cyber assets	IEC62443, IEC 62351
Cybersecurity audit	ISO/IEC27001, IEC62443
Business continuity Plan	ISO/IEC27001, NERC-CIP
Compliance	ISO/IEC27001
Supply chain issue	ISO28000, NERC-CIP-013
Mitigation of malicious tainted and counterfeit products	ISO/IEC20243

4 Implementation Guidelines and Certification as per Standards

In this Section a consolidated guidelines are discussed for implementing cybersecurity measures in power system and certification for it as per relevant standards. It is well understood that cybersecurity measures must be implemented and ensured by deploying up to date technologies and solutions which are required to be selected as per criteria defined in the standards. Selection of software, firmware, systems, devices is also to be done by means of requirements mentioned in the standards. In Table 5, the required parameters and activities for ensuring cybersecurity in power system are shown along with standards which are set the criteria, requirements and guidance.

For example, if a power system organization requires to design, develop, implement information/cyber security policy, it should follow ISO/IEC27001 standard for implementing security controls. The organization must also implement additional security controls specific to energy utility mentioned in ISO/IEC27019. To define the security objectives organization needs to follow ISO/IEC27001 standard, where complete process and steps are mentioned regarding security objectives and achievement as per defined objectives. Similarly, if there is a need of selection of IEDs with consideration of cybersecurity capabilities and product certification, one must follow the standards IEEE1686 and IEC62443 respectively. Organization needs to be designed and developed a security program for ensuring cybersecurity measures and best practices by following IEC62433 which elaborately covers security program creation and implementation process. The risk assessment and treatment must definitely be carried out as per ISO/IEC27001 and IEC62443. On the other hand, incident management, information protection, recovery planning must be carried out as per defined criteria mentioned in NERC-CIP. Firewall policy configuration, format of firewall report shall preferably be ensured as per requirements and guidelines given in NERC-CIP. The patch management of IACS shall be carried out as per IEC62443-2-3.

5 Summary

This chapter presents a systematic study on cybersecurity standards in power systems for proper understanding the scope and applicability and implementation guidelines for meeting cybersecurity requirements. The relevant cybersecurity standards for power systems are studied and analyzed for identification and selection of appropriate standards, bringing out the requirements, listing down the guidelines for practicing and maintaining proper cybersecuirty hygiene in power systems. The proposed study covers understanding of cybersecuriy standards along with scope and applicability of those standards for individual systems/devices such as Intelligent Electronic Device (IED), computing systems, communication and networking devices, cybersecurity devices, communication setup, local/central control center deployed in power systems. The study also presents and discusses the scope of process improvement in power systems. In future, effort will be given for deeper exploration of cybersecurity standards (e.g., IEC62443) for better understanding and applicability, and implementation guidance for power systems.

Acknowledgements The author would like to thank Power Grid Corporation of India Ltd. The author is also like to mention that idea presented and discussed in this chapter is of author's views and not anything of the management of Power Grid Corporation of India Ltd.

References

1. CEA, CEA (Cyber security in Power Sector) Guidelines. Information Technology & Cyber Security Division, Central Electricity Authority, India (2021)
2. NERC-CIP, NERC-CIP: The North American Electric Reliability Corporation Critical Infrastructure Protection (2005, 2009)
3. BSI, Germany IT Security Act, Federal Cyber Security Authority, Germany (2015)
4. ITA-2000, The Information Technology Act (India 2000)
5. IS 16335, IS 16335: 2015 Power Control Systems-Security Requirements, Bureau of Indian Standard (2015)
6. ISO/IEC27001, ISO/IEC 27001:2013 Information technology–Security techniques–Information security management systems–Requirements, in *ISO/IEC 27001:2013(en) Information technology-Security techniques-Information security management systems-Requirements, 2013*
7. IEEE, IEEE Standard for Intelligent Electronic Devices Cyber Security Capabilities, PE/PSCC-Power System Communications and Cybersecurity, IEEE Power & Energy Society (2013)
8. ISO 28000:2022, Specification for security management systems for the supply chain. International Organization for Standardization (2022)
9. ISA/IEC 62443, ISA/IEC 62443, Industrial Security Assessment and Certification
10. S . Sarkar, A. Agrawal, Y.M. Teo, E.C. Chang, VOTNET: hybrid simulation of virtual operational technology network for cybersecurity assessment, in *IEEE Winter Simulation Conference (WSC)* (IEEE, 2018), pp. 1168–1179
11. S. Sarkar, Y.M. Teo, E.C. Chang, A cybersecurity assessment framework for virtual operational technology in power system automation. Simul. Model. Pract. Theory **117**, 102453 (2022)
12. P. Didier, F. Macias, J. Harstad, R. Antholine, S.A. Johnston, S. Piyevsky, M. Schillace, G. Wilcox, D. Zaniewski, S. Zuponcic, Converged plantwide ethernet (CPWE) design and implementation guide. Cisco Systems and Rockwell Automation (2011)
13. IEEE37.240, IEEE standard cybersecurity requirements for substation automation, protection, and control systems. IEEE Power & Energy Society (2014)
14. F. Cleveland, IEC 62351 security standards for the power system information infrastructure,*IEC TC57 WG15 Security Standards* (14) (2012)
15. ISO/IEC27019, ISO/IEC27019:2017 Information technology-security techniques-information security controls for the energy utility industry (2017)
16. ENISA, The European Union Agency for Cybersecurity, ENISA (2004)
17. CPSF, The Cyber Physical Security Framework(CPSF), Version 1.0, Cybersecurity Division, Commerce and Information Policy Bureau, Ministry of Economy, Trade and Industry (April 2019)
18. E. Cole, SANS ICS Attack Surfaces, in *SANS Industrial Control System* (2017)
19. C. Bing, S. Kelly, Cyber attack shuts down top U.S. fuel pipeline network. Reuters (8 May 2021)
20. G. Liang, S.R. Weller, J. Zhao, F. Luo, Z.Y. Dong, The, Ukraine Blackout: implications for false data injection attacks. IEEE Trans. Power Syst. **32**(4), 2017 (2015)
21. ISA99, https://www.isa.org/standards-and-publications/isa-standards/isa-standards-committees/isa99. Accessed May 2022
22. CMMI, CMMI for Systems Engineering/Software Engineering, Version 1.02, CMU/SEI-2000-TR-019, ESC-TR-2000-019, Software Engineering Institute, Carnegie Mellon University, Pittsburgh, PA (2000)
23. ISO/IEC 20243-1, ISO/IEC 20243-1:2018 Information technology-Open Trusted Technology Provider TM Standard (O-TTPS)-Mitigating maliciously tainted and counterfeit products-Part 1: Requirements and recommendations, ISO/IEC (2018)
24. ISO/IEC 20243-2, ISO/IEC 20243-2:2018 Information technology-Open Trusted Technology ProviderTM Standard (O-TTPS)-Mitigating maliciously tainted and counterfeit products-Part 2: Assessment procedures for the O-TTPS, ISO/IEC (2018)

25. National Institute of Standards and Technology, Guide to Industrial Control Systems (ICS) Security (NIST Special Publication 800-82, Rev. 2) (9 Feb 2015)
26. IEEE C37.118-2005, C37.118-2005-IEEE Standard for Synchrophasor Measurements for Power Systems (IEEE 2005)
27. IEEE C37.118.1-2011, C37.118.1-2011- EEE Standard for Synchrophasor Measurements for Power Systems (IEEE, 2011)
28. ACSC, https://en.wikipedia.org/wiki/Australian_Cyber_Security_Centre. Accessed May 2022
29. W. Newhouse, S. Keith, B. Scribner, G. Witte, National Initiative for Cybersecurity Education (NICE) Cybersecurity Workforce Framework (2017)
30. Cybersecurity Framework, NIST. https://en.wikipedia.org/wiki/NIST_Cybersecurity_Framework. Accessed May 2022

Printed by Printforce, the Netherlands